DEMOSTHENES

AGAINST MEIDIAS

(Oration 21)

DEMOSTHENES

Against Meidias

(Oration 21)

EDITED WITH
INTRODUCTION, TRANSLATION, AND COMMENTARY
BY
DOUGLAS M. MACDOWELL

CLARENDON PRESS · OXFORD
1990

Oxford University Press, Walton Street, Oxford OX2 6DP
Oxford New York Toronto
Delhi Bombay Calcutta Madras Karachi
Petaling Jaya Singapore Hong Kong Tokyo
Nairobi Dar es Salaam Cape Town
Melbourne Auckland
and associated companies in
Berlin Ibadan

Oxford is a trade mark of Oxford University Press

Published in the United States
by Oxford University Press, New York

British Library Cataloguing in Publication Data
Demosthenes, 384–322 B.C.
Against Meidias: (oration 21).
1. Speeches in Greek, to ca. 500 — Greek–English parallel texts
I. Title II. MacDowell, D. M. (Douglas Maurice)
885'.–1
ISBN 0–19–814763–5

Library of Congress Cataloging in Publication Data
Demosthenes. [Against Meidias. English & Greek]
Demosthenes, against Meidias (oration 21)/edited with introduction, translation, and commentary by Douglas M. MacDowell.
Text in English and Greek.
Bibliography: Includes index.
I. MacDowell, Douglas M. (Douglas Maurice) II. Title.
III. Title: Against Meidias.
PA3951.E5M76 1989 885'.01–dc19 89–3087
ISBN 0–19–814763–5

Typeset by H Charlesworth & Co Ltd, Huddersfield, England
Printed in Great Britain by
Biddles Ltd., Guildford and Kings Lynn
Bookcraft Ltd., Bath and Midsomer Norton

PREFACE

DEMOSTHENES has been much neglected by our generation. Some good work has recently been done on the shorter speeches, but no new commentary in English on any of the major orations has appeared for many years. There are various reasons: fewer students now practise the art of Greek prose composition, for which Demosthenes is a prime model; the political history of the fourth century is less popular than that of the sixth or fifth. But it is time for a revival of interest in this author, not only for his masterly use of the Greek language and for the factual information about Athens that he provides, but also for his practical approach to perennial problems of conduct and justice.

The oration *Against Meidias*, besides telling a lively story, is particularly full of information about law and about festivals, and gives vital evidence for the concept of *hybris*. But its study also involves some curious problems about the composition and preservation of the text as we have it, which I discuss in the introduction. The translation which I have included is not meant to be read separately, but to be used as an aid to understanding the Greek text; there may be some readers of this volume whose knowledge of Greek is weak, but I do not expect it to be read by those who know no Greek at all.

The text printed here takes account of many manuscripts not previously collated. I have been able to include information about some papyrus fragments in advance of their publication in the Oxyrhynchus series, thanks to the courtesy of the Egypt Exploration Society of London and the generous assistance of Mr. P. J. Parsons and Dr. H.-C. Günther. Professor M. R. Dilts kindly gave me a typescript of his edition of the scholia for use before it appeared in print. Most of my collation of medieval manuscripts was done from photographs or microfilm, but I was able to inspect *in situ* all the manuscripts in Britain and many of those in Italy. I am grateful for the assistance given me by the librarians in all the libraries I visited, for a term's leave of absence from the University of Glasgow to make the visits, for a contribution from the British Academy to the costs of travel, and especially for the help of my colleagues in Glasgow who undertook extra teaching during my absence.

The book was completed in February 1988, and does not take account of publications which I had not seen by then.

Glasgow D.M.M.

CONTENTS

ABBREVIATIONS AND BIBLIOGRAPHY

ANCIENT TEXTS

Figures in bold type, e.g. **123**, refer to sections of *Meidias*. An added n. refers to my commentary; thus **205** n. *οὗτος* means 'see my note on *οὗτος* in section **205**'.

References to other orations in the Demosthenic corpus are in ordinary type, e.g. 18.123. The author's name is not given unless it is needed to avoid ambiguity after a reference to another author; when it is given, the use of D. does not necessarily imply that Demosthenes himself is the author of the oration concerned. *Pr.* and *Ep.* without an author's name mean the prooemia and the epistles attributed to Demosthenes.

Σ means the scholia on Demosthenes. I give first the reference to the Demosthenic passage to which the scholium is attached (unless it is the passage already under consideration) and then in brackets the number of the scholium in the edition by M. R. Dilts. Thus Σ 4.46 (102b Dilts) means the scholium on section 46 of the *First Philippic* to which Dilts assigns the number 102b. Σ (Patm.) refers to the Patmian scholia; see p. 83 n. 1.

The abbreviations used for other ancient texts will be familiar to most readers of this book. If help is needed, it may be obtained from the lists in Liddell and Scott's *Greek–English Lexicon*, revised by Jones (which I abbreviate to LSJ); but some of my abbreviations have a less Latinized form than theirs. I list here a few which might cause doubt.

AB	*Anecdota Graeca* ed. I. Bekker
Ais.	Aiskhines
Ant.	Antiphon the orator
AP	The *Athenaion Politeia* attributed to Aristotle
Lex. Cant.	*Lexicon Rhetoricum Cantabrigiense* ed. E. O. Houtsma (reprinted in *Lexica Graeca Minora* ed. K. Latte and H. Erbse)
Lex. Rhet.	*Lexeis Rhetorikai*, cited by page and line of *AB* 1
Lyk.	Lykourgos the orator
ML	*A Selection of Greek Historical Inscriptions* ed. R. Meiggs and D. M. Lewis, cited by inscription number and line
Plu. *Eth.*	Plutarch *Ethika* (*Moralia*)

Pol.	Polydeukes (Pollux) *Onomastikon*
X. *Apom.*	Xenophon *Apomnemoneumata* (*Memorabilia*)

The Greek rhetoricians, used mainly for testimonia (see p. 82), are generally cited by page and line. The editor's name and the number of the volume are included in brackets when necessary. The following editions are used, all except Walz and Ballaira in the Teubner series.

Rhetores Graeci ed. C. Walz (1832–6)
Rhetores Graeci ed. L. Spengel (1853–6)
Rhetores Graeci 1.2 ed. C. Hammer (1894)
Prolegomenon Sylloge ed. H. Rabe (1931), abbreviated as Rabe *PS*
Aristides *Libri Rhetorici* ed. G. Schmid (1926)
Hermogenes ed. H. Rabe (1913)
Ioannes Sardianus *Commentarius in Aphthonium* ed. H. Rabe (1928)
Syrianus *In Hermogenem Commentaria* ed. H. Rabe (1892–3)
Tiberius *De Figuris Demosthenicis* ed. G. Ballaira (1968), cited by chapter and line

EDITIONS OF *AGAINST MEIDIAS*

The following is a list of the printed editions of *Meidias* which I have seen. In this list (D.) indicates that the edition of *Meidias* forms part of an edition of all or many of Demosthenes' orations (not necessarily all by the same editor), T = text, V = translation, C = commentary. I refer to these editions by the editor's surname (or, if the edition is anonymous, by the place and date of publication) without a title. Where no page number is given, the reference is to the text or translation or commentary at the point under consideration.

Aldus Manutius (D.): T, Venice 1504.
Anon.: T, Louvain 1525.
Oporinus, Ioannes (D.): T, Basle 1532.
Felicianus, Io. Bernardus (D.): T, Venice 1543.
Anon. (D.): T, Basle 1547.
Paulus Manutius (D.): T, Venice 1554.
Lambinus, Dionysius (D.): T, Paris 1570.
Anon.: T, London 1586.
Wolf, Hieronymus (D.): TV, Frankfurt 1604.
Taylor, Ioannes (D.): TV, Cambridge 1748, including textual notes by Palmer, Markland, and Iurinus.
Reiske, Ioannes Iacobus (D.): TC, Leipzig 1770–5.
Spalding, G. L.: TC, Berlin 1794.

Bekker, Immanuel (D.): T, Oxford 1823; revised, Leipzig 1854.
Dobson, Gulielmus Stephanus (D.): TC, London 1827.
Meier, M. E. H.: T, Halle 1831.
Buttmann, P. (revised by A. Buttmann): TC, Berlin 1833, including comments by G. Bernhardy.
Baiter, J. G., and Sauppe, H. (D.): T, Zurich 1839–43.
Voemel, J. T. (D., revised by Dübner): TV, Paris 1849.
Dindorf, Gulielmus (D.): TC, Oxford 1846–9; revised, Leipzig 1859.
Holmes, Arthur: TC, Cambridge 1862.
Whiston, Robert (D.): TC, London 1868.
Fennell, C. A. M.: TC, Cambridge 1883.
Weil, Henri (D. *Plaidoyers politiques*): TC, Paris 1883.
Blass, F. (D., Teubner series): T, Leipzig 1888.
King, John R.: TC, Oxford 1901.
Goodwin, William Watson: TC, Cambridge 1906.
Butcher, S. H. (D., Oxford Classical Text): T, Oxford 1907.
Vince, J. H. (D., Loeb series): TV, London and Cambridge Mass. 1935.
Sykutris, Ioannes (D., Teubner series): T, Leipzig 1937.
Humbert, Jean (D. *Plaidoyers politiques*, Budé series): TV, Paris 1959.

MODERN WORKS

If elucidation of the titles of periodicals is needed, it may in most cases be obtained from the list printed in each volume of *L'Année Philologique*. A few of the abbreviations which I use for periodicals and other volumes of multiple authorship are listed here.

BSA	*Annual of the British School at Athens*
Crux	*Crux, essays presented to G. E. M. de Ste. Croix* ed. P. A. Cartledge and F. D. Harvey = *History of Political Thought* 6 (1985) issue 1/2
HCT	*A Historical Commentary on Thucydides* by A. W. Gomme, A. Andrewes, and K. J. Dover
Jb. Cl. Ph.	*Jahrbücher für Classische Philologie*
Mn. I, II, III, IV	*Mnemosyne* series I, II, III, IV
RE	*Paulys Real-Encyclopädie der Classischen Altertumswissenschaft*
RIDA	*Revue Internationale des Droits de l'Antiquité* series III
ZPE	*Zeitschrift für Papyrologie und Epigraphik*

ZSSR	*Zeitschrift der Savigny–Stiftung für Rechtsgeschichte* (Romanistische Abteilung)

The volumes entitled *Symposion* with two dates contain the proceedings of the Society for Greek and Hellenistic Legal History; e.g. *Symposion 1982* (1985) contains the proceedings of the conference held in 1982, published in 1985.

The following list of books and articles is not a complete bibliography, but gives details of some works for which I use abbreviated references. Details of other works, cited only once or twice, are given in the introduction and commentary.

Avezzù, Guido. 'Gli scoli demostenici e l'*Epitafio* di Lisia nel ms. Marc. gr. 416', *Bolletino del Comitato per la Preparazione dell'Edizione Nazionale dei Classici Greci e Latini* 27 (1979) 51–67.

Batschelet-Massini, Werner. 'Neue Versuche zum demosthenischen Prosarhythmus', *Tainia, Festschrift für Roland Hampe* (ed. H. A. Cahn and E. Simon, Mainz am Rhein 1980) 1.503–28.

Benseler, Gustavus Eduardus. *De hiatu in oratoribus atticis et historicis Graecis* (Freiburg 1841).

Blass, Friedrich. *Die attische Beredsamkeit* volume 3.1 (second edition, Leipzig 1893).

—— 'Demosthenica aus neuen Papyrus', *Jb. Cl. Ph.* 145 (1892) 29–44.

Boeckh, August. 'Von den Zeitverhältnissen der demosthenischen Rede gegen Meidias', *Abhandlungen der Königlichen Akademie der Wissenschaften in Berlin aus dem Jahren 1818–1819* (1820) Historische-philologische Klasse 60–100, reprinted in his *Gesammelte kleine Schriften* (Leipzig 1871) 153–204.

Bonner, Robert J., and Smith, Gertrude. *The Administration of Justice from Homer to Aristotle* (Chicago 1930–8).

Brown, Daniel. *Das Geschäft mit dem Staat* (Hildesheim 1974).

Brunt, P. A. 'Euboea in the time of Philip II', *CQ* 19 (1969) 245–65.

Burke, Edmund M. 'Eubulus, Olynthus, and Euboea', *TAPA* 114 (1984) 111–20.

Canfora, Luciano. *Inventario dei manoscritti greci di Demostene* (Padua 1968).

Cawkwell, G. L. 'Athenian naval power in the fourth century', *CQ* 34 (1984) 334–45.

—— 'The defence of Olynthus', *CQ* 12 (1962) 122–40.

—— 'Eubulus', *JHS* 83 (1963) 47–67.

Christ, W. 'Die Attikusausgabe des Demosthenes', *Abhandlungen der Philosophisch-philologischen Classe der Königlich Bayerischen Akademie der Wissenschaften* 16.3 (1882) 153–234.

Cobet, C. G. *Miscellanea critica* (Leiden 1876).
—— *Novae lectiones* (Leiden 1858).
Davies, J. K. *Athenian Propertied Families* (Oxford 1971).
Denniston, J. D. *The Greek Particles* (second edition, Oxford 1954).
—— *Greek Prose Style* (Oxford 1952).
Dilts, Mervin R. 'Demosthenic scholia in Codex Laurentianus 59.9', *TAPA* 104 (1974) 97–102.
—— 'Demosthenic scholia in Marcianus gr. 416 and Monacensis gr. 85', *Studia Codicologica* (ed. K. Treu, Berlin 1977) 151–8.
Dobree, P. P. *Adversaria* (Cambridge 1831).
Dover, K. J. *Greek Popular Morality* (Oxford 1974).
Drerup, Engelbert. 'Antike Demosthenesausgaben', *Philologus* Supplementband 7 (1899) 531–88.
—— 'Über die bei den attischen Rednern eingelegten Urkunden', *Jb. Cl. Ph.* Supplementband 24 (1898) 221–366.
—— 'Vorläufiger Bericht über eine Studienreise zur Erforschung der Demosthenesüberlieferung', *Sitzungsberichte der Königlich Bayerischen Akademie der Wissenschaften, Philosophisch-philologische Classe* (1902) 287–323.
Erbse, Hartmut. 'Über die Midiana des Demosthenes', *Hermes* 84 (1956) 135–52, reprinted in a revised form in his *Ausgewählte Schriften zur klassischen Philologie* (Berlin 1979) 412–31.
—— 'Überlieferungsgeschichte der griechischen klassischen und hellenistischen Literatur: Demosthenes', *Geschichte der Textüberlieferung* 1 (1961) 262–4.
Fisher, N. R. E. '*Hybris* and dishonour', *Greece & Rome* 23 (1976) 177–93, 26 (1979) 32–47.
Foucart, Paul. 'Sur l'authenticité de la loi d'Évégoros', *Rev. Phil.* 1 (1877) 168–81.
Gagarin, Michael. 'The Athenian law against *hybris*', *Arktouros: Hellenic studies presented to Bernard M. W. Knox* (ed. G. W. Bowersock, W. Burkert, and M. C. J. Putnam, Berlin and New York 1979) 229–36.
—— *Drakon and Early Athenian Homicide Law* (New Haven 1981).
Gebauer, Gustavus. *De hypotacticis et paratacticis argumenti ex contrario formis* (Zwickau 1877).
Giugnoli, Giampiero. 'A proposito della Midiana demostenica: il problema cronologico', *Atene e Roma* 20 (1975) 170–83.
Golden, Mark. 'Demosthenes and the age of majority at Athens', *Phoenix* 33 (1979) 25–38.
Goodwin, W. W. *Syntax of the Moods and Tenses of the Greek Verb* (London 1889).
Hansen, Mogens Herman. *Apagoge, Endeixis and Ephegesis against*

Kakourgoi, Atimoi and Pheugontes, Odense University Classical Studies 8 (1976).

——'Two notes on the Athenian *dikai emporikai*', *Symposion 1979* (1981) 167–75.

Harrell, H. C. *Public Arbitration in Athenian Law*, University of Missouri Studies 11.1 (1936).

Harrison, A. R. W. *The Law of Athens* (Oxford 1968–71).

Herwerden, H. van. 'Ad Demosthenis orationem in Midiam', *Mn.* II 1 (1873) 306–12.

——*Exercitationes criticae* (The Hague 1862).

——'Meletemata critica ad Oratores Atticos', *Mn.* II 3 (1875) 120–40.

Irmer, Dieter. 'Beobachtungen zur Demosthenesüberlieferung', *Philologus* 112 (1968) 43–62.

——*Zur Genealogie der jüngeren Demostheneshandschriften: Untersuchungen an den Reden 8 und 9* (Hamburg 1972).

Jaeger, Werner. *Demosthenes: the origin and growth of his policy* (Berkeley and Cambridge 1938).

Jahn, Otto. 'Variarum lectionum fasciculus, 6', *Philologus* 26 (1867) 2–4.

Jordan, Borimir. *The Athenian Navy in the Classical Period*, University of California Publications: Classical Studies 13 (1975).

Knoepfler, Denis. 'Argoura: un toponyme eubéen dans la *Midienne* de Démosthène', *BCH* 105 (1981) 289–329.

Kühner, R., and Gerth, B. *Ausführliche Grammatik der griechischen Sprache: Satzlehre* (third edition, Hanover 1898–1904).

Lewis, David M. 'Notes on Attic inscriptions', *BSA* 49 (1954) 17–50, 50 (1955) 1–36.

Lipsius, J. H. *Das attische Recht und Rechtsverfahren* (Leipzig 1905–15).

Longo, Chiara Pecorella. *"Eterie" e gruppi politici nell'Atene del IV sec. a.C.* (Florence 1971).

McCabe, Donald F. *The Prose-rhythm of Demosthenes* (New York 1981).

MacDowell, Douglas M. *Athenian Homicide Law* (Manchester 1963).

——'Athenian laws about bribery', *RIDA* 30 (1983) 57–78.

——'Athenian laws about choruses', *Symposion 1982* (1985) 65–77.

——'*Hybris* in Athens', *Greece & Rome* 23 (1976) 14–31.

——*The Law in Classical Athens* (London 1978). (Cited as *Law*.)

——'The law of Periandros about symmories', *CQ* 36 (1986) 438–49.

Meisterhans, K. *Grammatik der attischen Inschriften* (third edition revised by E. Schwyzer, Berlin 1900).

Mikalson, Jon D. *The Sacred and Civil Calendar of the Athenian Year* (Princeton 1975).

Naber, S. A. 'Observationes criticae ad Demosthenem', *Mn.* II 31 (1903) 411–25.

Ostwald, Martin. *From Popular Sovereignty to the Sovereignty of Law* (Berkeley 1986).

Pasquali, Giorgio. *Storia della tradizione e critica del testo* (second edition, Florence 1962).

Passweg, Ruth. *The Manuscript Tradition: Demosthenis In Timocratem, Oration 24* (dissertation, New York University 1975).

Pearson, Lionel. *The Art of Demosthenes* (Meisenheim am Glan 1976).

Pickard-Cambridge, Arthur W. *The Dramatic Festivals of Athens* (second edition revised by John Gould and D. M. Lewis, Oxford 1968).

Radermacher, L. Review of Sykutris' edition, *Gnomon* 16 (1940) 8–14.

Rhodes, P. J. *The Athenian Boule* (Oxford 1972).

—— *A Commentary on the Aristotelian Athenaion Politeia* (Oxford 1981).

—— 'Problems in Athenian *eisphora* and liturgies', *American Journal of Ancient History* 7 (1982) 1–19.

Richards, Herbert. 'Further notes on Demosthenes', *CR* 18 (1904) 11–17.

Ronnet, Gilberte. *Étude sur le style de Démosthène dans les discours politiques* (Paris 1951).

Ruschenbusch, Eberhard. 'Die athenischen Symmorien des 4. Jh. v. Chr.', *ZPE* 31 (1978) 275–84.

—— 'Ein Beitrag zur Leiturgie und zur Eisphora' and 'Die trierarchischen Syntelien und das Vermögen der Synteliemitglieder', *ZPE* 59 (1985) 237–49.

—— 'Demosthenes' erste freiwillige Trierarchie und die Datierung des Euböaunternehmens vom Jahre 357', *ZPE* 67 (1987) 158–9.

—— 'Drei Beiträge zur öffentlichen Diaita in Athen', *Symposion 1982* (1985) 31–40.

—— 'Symmorienprobleme', *ZPE* 69 (1987) 75–81.

—— *Ὕβρεως γραφή*, *ZSSR* 82 (1965) 302–9.

Schaefer, Arnold. *Demosthenes und seine Zeit* vols. 1–3 (second edition, Leipzig 1885–7) and vol. 3.2 (Leipzig 1858; reprinted as vol. 4, Hildesheim 1967).

Schaefer, G. H. *Apparatus criticus et exegeticus ad Demosthenem* vol. 3 (London 1826).

Sealey, Raphael. 'Dionysius of Halicarnassus and some Demosthenic dates', *REG* 68 (1955) 77–120.

Thompson, H. A., and Wycherley, R. E. *The Agora of Athens*, The Athenian Agora 14 (Princeton 1972).

Threatte, Leslie. *The Grammar of Attic Inscriptions* vol. 1 (Berlin and New York 1980).

Wankel, Hermann. *Demosthenes: Rede für Ktesiphon über den Kranz* (Heidelberg 1976).

Weil, Henri. 'Études sur Démosthène, I. La guerre d'Olynthe et la guerre d'Eubée', *Rev. Phil.* 3 (1879) 1–13.

Westermann, A. *De litis instrumentis, quae exstant in Demosthenis oratione in Midiam, commentatio* (Leipzig 1844).

INTRODUCTION

I. THE QUARREL BETWEEN DEMOSTHENES AND MEIDIAS

DEMOSTHENES' first encounter with Meidias occurred when Meidias broke into his house—not an auspicious beginning for a relationship. Demosthenes was about twenty years old, and was at home with his mother and sister, when Meidias and his brother Thrasylokhos bounded in and broke down the doors of the rooms to inspect the contents, using abusive language in the presence of the lady and her young daughter. Their right to enter the house arose from the *antidosis*, or challenge to exchange property in lieu of performing a trierarchy, in which Thrasylokhos was involved with Demosthenes; but they appear to have exercised the right in an oppressive manner (**78–9**).

Admittedly we have only Demosthenes' own account of these events. Meidias is one of those unlucky persons who are known to history from the information given us by their enemies. The oration *Against Meidias* is not an impartial report but a prosecution speech, and Demosthenes may be telling lies about Meidias if he thinks he can get the jury to believe them. Does that mean that we ought simply to discard everything that he says about Meidias? Not at all. Certainly we must keep our critical judgement on the alert, but we may still find or suspect a good deal of truth in the narrative. By the time Demosthenes composed this speech he and Meidias were both public figures, and some of the incidents in the story were notorious.[1] Meidias would have been ready in his defence speech to refute and denounce any falsehood about him; it is not likely that Demosthenes could have got away with more than a quite limited amount of fabrication, and he must have known that he would destroy his own credibility if he were seen to be flagrantly lying. We may therefore accept his account on

[1] 'This happened a long time ago, but still I expect some of you remember it; for at the time the whole city knew about the proposal for exchange and this plot and the bullying' (**80**).

the whole, while maintaining our readiness to question it at any particular point where there are grounds for doing so.

Proceeding, then, to follow his account, we find that the irruption of Meidias and Thrasylokhos into the house was connected with Demosthenes' prosecution of his guardians. The year was 364/3 B.C., and he had come of age a couple of years before.[1] His father had died when he was seven, and his property had been entrusted to guardians during his minority; but when he came of age, they handed over to him much less property than he expected, and accordingly he was taking legal action against them to recover what was due to him. Thrasylokhos was not one of the guardians, though he may have been a friend of theirs; he was just a business man—fairly prosperous, no doubt, since he was appointed that year to be a trierarch. The guardians induced him, presumably by promising him a substantial favour of some kind, to use the legal procedure of *antidosis* to challenge Demosthenes either to take on the trierarchy or to exchange properties with Thrasylokhos. They probably expected the young man to jump at the latter alternative, which would have given him some undisputed assets in place of the uncertain prospects of recovering property from his guardians. In fact, however, he resisted, and carried through the prosecutions of his guardians. The affair is discussed more fully in the commentary on **78–80**, but what we should consider here is the involvement of Meidias.

Meidias was probably about ten years older than Demosthenes.[2] He and Thrasylokhos were sons of Kephisodoros of Anagyrous, but nothing is known of their father except that he too was a trierarch at least once,[3] and he may have died before the events with which we are concerned. Demosthenes tells a story that Meidias was a changeling child (**149**), but that is one allegation that we need not believe; for he would have been unable to retain his Athenian citizenship if it had been possible to impugn his parentage seriously, and attacks on parentage seem to have been merely a traditional manner of ridiculing an opponent in Athens. We have no further information about

[1] For the chronology and other problems of detail in these incidents see the notes on **78–80**. For Demosthenes' date of birth see p. 370.

[2] For Meidias' date of birth see p. 369.

[3] If his name is rightly restored in *IG* 2² 1609.29.

events in his life before those mentioned in the oration, but no doubt he devoted some of his time to making money from the silver mines, which were probably the source of his later wealth.[1]

It was Thrasylokhos, not Meidias, who challenged Demosthenes to accept a trierarchy or exchange property, but Demosthenes declares that it was Meidias who was really responsible for the plot (**78**), and now Thrasylokhos disappears from the story and the quarrel becomes one between Demosthenes and Meidias only. The two had not even known each other before the day when Meidias burst into Demosthenes' house; so Meidias' motive for that act must have been a desire to help his brother and friends rather than hostility to Demosthenes. But the events of that day led to personal enmity between them. Demosthenes, having won his legal actions against his guardians, proceeded to prosecute Meidias for slander (*κακηγορία*), presumably for the abusive language which he had directed at his mother and sister.

The course of this prosecution is described in considerable detail in **83–101**. It came before a public arbitrator named Straton. Public arbitrators were ordinary Athenian citizens in their sixtieth year, not possessing any special knowledge or qualifications. Straton was a person with no experience of legal or public affairs (**83** *ἀπράγμων*), and was probably quite overwhelmed when caught between the bully Meidias and the clever orator Demosthenes. There were some delays, and when the day at last came on which the arbitrator was to give his verdict, Meidias failed to attend; and so Straton, doubtless under heavy pressure from Demosthenes, gave a verdict against Meidias by default (*ἔρημος*). Meidias subsequently declared that this verdict had been given improperly, and accused Straton of misconduct in the office of arbitrator. Straton was convicted and disfranchised, which indicates that Meidias had some justification for his complaint; but did that have the effect of annulling the verdict that Straton had given against him for slander? Meidias probably assumed that it did, but Demosthenes may have been technically right in saying that it did not, and that Meidias now owed him the sum of 1000

[1] For Meidias' connection with the mines see **167**, *IG* 2^2 1582.44, 82, *Hesperia* 19 (1950) 243 no. 15 line 76; cf. Davies *Families* 386–7.

drachmas awarded by Straton, which Meidias refused to pay. So the next stage was that Demosthenes raised another action against Meidias, claiming the right to seize property to the value of the sum which had been awarded to him (*δίκη ἐξούλης*). Meidias raised objections, of which we are not told any details, and it appears that this case never came to trial (**81**). Probably Demosthenes realized that he was on uncertain ground and therefore did not press the case.

The incidents described so far all occurred in the years 364 and 363, or perhaps some of them dragged on into 362, but they can hardly have lasted longer. There is then an interval of about thirteen years, during which we hear of no contact between Demosthenes and Meidias. In that time they both became prominent figures in Athens. Meidias was elected to various offices, including those of hipparch and of tamias of the sacred ship *Paralos* (**171**). Demosthenes meanwhile began his famous series of speeches in the Ekklesia. It is hard to believe that in those years the two never met, but at any rate nothing happened between them which Demosthenes later thought worth recalling.

We move on, then, to 349. In the autumn of that year an appeal for military aid reached Athens from Olynthos, the leading city in Khalkidike, which was under threat of attack from Philip. Demosthenes saw this as an opportunity for the Athenians to check the growth of Macedonian power, and in his three *Olynthiac* orations did his best to persuade them to send immediate strong support to Olynthos.[1] Their response was weaker than he wished, but they did make an alliance with Olynthos, and initially they sent 2000 mercenary peltasts, 30 triremes under the command of Khares, which were perhaps already in the north Aegean, and eight other triremes 'which they helped to man' (*ἃς συνεπλήρωσαν*);[2] this phrase probably means that those eight were the triremes donated voluntarily by individual Athenians in response to an appeal (**161**). Philip proceeded to take the cities in Khalkidike one by one, and

[1] The order and chronology of the *Olynthiacs* have been much discussed; see most recently C. Eucken *Mus. Helv.* 41 (1984) 193–208. On Athenian policy see especially Cawkwell *CQ* 12 (1962) 122–40.

[2] Dion. Hal. *Amm.* 1.9 quoting Philokhoros (*F. Gr. Hist.* 328 F49); cf. Cawkwell *CQ* 12 (1962) 131.

Olynthos asked Athens for more help; but before any more was sent, the Athenians' attention was distracted by events in Euboia. These events brought Demosthenes and Meidias into conflict once again.

The cities in Euboia had been under Athenian influence since the expulsion of the Thebans from the island some years before.[1] In 349/8 Ploutarkhos,[2] the tyrant of Eretria, faced a rebellion, and he requested the Athenians to intervene.[3] Meidias was involved in this, because he was the Athenian representative of Eretria (**200** *Πλουτάρχου προξενεῖ*); presumably he either introduced Ploutarkhos' envoy to the Ekklesia or presented the request himself. It was supported by other speakers, probably including the leading politician Euboulos.[4] Demosthenes, who would have preferred Athenian resources to be directed towards Olynthos, was the only one to speak against; 'I was nearly torn apart,' he said later (5.5). It was therefore resolved to send a force under the command of Phokion.

Demosthenes and Meidias were both among the men called up for it. Meidias, it appears, had been elected to the office of phylarch for this year, commanding his tribe's troop of cavalry.[5] Demosthenes gives a comic description of Meidias hoping that he would not draw the lot for his troop to go on the expedition at all, and finally riding off to it on a silver mule-chair, with a plentiful supply of fine cloaks and cups and flagons of wine; that is what 'we hoplites' were told (**133**)—which shows that Demosthenes himself was among the infantry on

[1] On the history of Euboia in this period see Brunt *CQ* 19 (1969) 245–65. On the sequence of events in 349/8, Cawkwell's account in *CQ* 12 (1962) 127–30 is preferable to J. M. Carter's in *Historia* 20 (1971) 418–29.

[2] For convenience I call the author 'Plutarch' and the ruler of Eretria 'Ploutarkhos'.

[3] Whether Philip prompted or supported the rebellion is a disputed question, which cannot be answered with certainty. For some recent discussion of it see Cawkwell *CQ* 12 (1962) 129–30, Brunt *CQ* 19 (1969) 249–50, H.-J. Gehrke *Phokion* (1976) 33–4 n. 47, N. G. L. Hammond and G. T. Griffith *A History of Macedonia* 2.318 n. 2, Knoepfler *BCH* 105 (1981) 298–9, Burke *TAPA* 114 (1984) 111 n. 3.

[4] After the campaign was over, Hegesileos, who was a cousin of Euboulos, was prosecuted on a charge of assisting Ploutarkhos to deceive the Athenian people; and Euboulos, speaking in his support at the trial, asked the jurors to forgive himself (19.290 with Σ (513 Dilts)). This is tenuous, but sufficient, evidence that Euboulos acknowledged some responsibility for the expedition and its failure. Burke *TAPA* 114 (1984) 111–20, taking Euboulos' responsibility for granted, argues that his purpose was to prevent Philip from using Euboia as a base to interfere with Athenian commercial traffic in the Aegean; that is possible, but there is no clear evidence for it.

[5] **133** n. ἐκληροῦ.

this expedition. They left Athens before the Anthesteria (39.16), in January or February of 348. The infantry, on landing in Euboia, marched under the command of Phokion to Tamynai, while the cavalry landed further up the coast at Argoura.[1]

But then there seems to have been a pause. Plutarch, in his account of the campaign, says that Phokion did not get the support he expected from the people of Euboia, but 'found everything full of traitors, sick, and undermined by bribery'; so he simply held his position on a hill near Tamynai, apparently taking no offensive, and allowed those of his men who were 'disorderly, talkative, and bad' to go away (Plu. *Phok.* 12.2–3). Whether this vocabulary is Plutarch's own or comes from his source, it is tempting to think that 'talkative' (*λάλος*) at least may have been selected with Demosthenes in mind; for it is clear that Demosthenes did in fact return to Athens at this stage, because he was present at the Dionysia (in March). Meidias got back to Athens too. None of the cavalry ever moved from Argoura to join Phokion at Tamynai;[2] some of them sailed north to support Olynthos,[3] but Meidias returned home (**132**, **197**). Demosthenes describes this as if Meidias were a mere deserter. But more probably a decision had been taken that the cavalry was not needed in Euboia, and that part of it should go to Olynthos and the other part (including Meidias' troop, doubtless after a further drawing of lots) should return to Athens.[4] On his return Meidias made a speech in the Ekklesia criticizing the conduct of the expedition (**132**, **197**).

The news from the north continued to be bad, and the Boule agreed on another appeal for voluntary donations of triremes.

[1] On the topography and strategy see Knoepfler *BCH* 105 (1981) 289–329.

[2] The cavalry mentioned in Plutarch's account of the battle of Tamynai must have been Euboians; he distinguishes them from *οἱ Ἀθηναῖοι*, who are hoplites (Plu. *Phok.* 13.2–6).

[3] These must be the 150 cavalry who went to Khalkidike under the command of Kharidemos, as reported by Dion. Hal. *Amm.* 1.9 from Philokhoros (*F. Gr. Hist.* 328 F50). This is the only point at which I disagree with Cawkwell's account of the chronology of this campaign. He says 'presumably the cavalry that crossed from Chalcis to Olynthus went not long after the battle' of Tamynai (*CQ* 12 (1962) 129). But they must have gone earlier, because *ὅθ' ἧκεν ἐκ Χαλκίδος* (**132**) and *ὅτ' εἰς Ὄλυνθον διέβησαν* (**197**) refer to the same occasion and Meidias was in Athens before the battle.

[4] **164** *τοὺς ἐξ Ἀργούρας* shows that others besides Meidias returned from Argoura to Athens.

Demosthenes points out that Meidias, even though present at that meeting of the Boule, did not volunteer; he did, however, offer to donate a ship before the start of the next meeting of the Ekklesia (**161–2**). By that time there was also bad news from Euboia: Phokion was being besieged at Tamynai. The expectation was that the entire Athenian army would have to go out, part of it (including the cavalry) to Euboia and the rest to Olynthos, and it was possibly at this meeting that Apollodoros proposed a decree that money from the theoric fund should be diverted to military purposes (**162**, 59.4).[1] But, while the meeting was still in progress, it became clear that the situation in Euboia was less serious than had been thought, and the proposal to send reinforcements there was dropped (**163**), probably because news arrived at this time that Phokion had won the battle of Tamynai. Thus it was not until some considerable time later that the cavalry contingent which had returned from Argoura to Athens was sent out to Euboia again, to relieve the soldiers who had won the battle (**164**).[2]

Meanwhile the personal hostility between Demosthenes and Meidias came to a climax at the Dionysia in the spring of 348. For this festival Demosthenes was the chorus-producer[3] of the men's dithyrambic chorus provided by his tribe; he boasts particularly of the fact that he volunteered to perform this duty (**13–14**). As the time of the festival approached, Meidias began causing various difficulties for Demosthenes and his chorus. He raised (perhaps earlier, before the expedition to Euboia) objections to Demosthenes' choristers' being given the exemption from military service to which they were entitled for the period of the rehearsals and the festival (**15**). Then, Demosthenes alleges, he broke into the house of a goldsmith during the night and tried to destroy the gold crowns and costumes which the goldsmith had made for the chorus to wear (**16**); and he bribed the director to sabotage the training of the chorus (**17**). Demosthenes also alleges various other obstructions, intended to pre-

[1] Apollodoros' decree was stopped by a *γραφὴ παρανόμων*, and he was fined one talent (59.5–8). On this incident see M. H. Hansen *GRBS* 17 (1976) 235–46.

[2] Phokion too left Euboia, probably at the same time as the victorious soldiers. His successor, Molossos, proceeded to lose the war, so that the whole expedition was afterwards regarded as a disgrace to Athens (5.5, Plu. *Phok.* 14. 1–2).

[3] I use 'chorus-producer' to translate *χορηγός*, 'director' for *διδάσκαλος*, and 'chorister' for *χορευτής*.

vent his chorus from winning the contest; they culminated on the actual day of the performance, when Demosthenes was in the theatre, no doubt splendidly dressed as a chorus-producer, and Meidias came up and punched him in the face.

This happened in full view of everyone. Even if the reader prefers to adopt a generally sceptical attitude to the account of Meidias' misdeeds, this particular incident cannot be denied. It could be mentioned years later as a well known fact (Ais. 3.52), and it must really have happened. But why did Meidias do it? In the whole of his long speech Demosthenes never suggests any specific reason. He says a great deal about Meidias' wealth and arrogance, but he does not mention any reason that Meidias had for singling out Demosthenes as the target of his insolent conduct. Surely their dispute fourteen or fifteen years previously cannot have been the only reason. Perhaps Meidias had been provoked to renew the quarrel by some more recent incident which Demosthenes has concealed from us. But the actual punch in the face may have been unpremeditated. Meidias may have been just walking across the theatre when Demosthenes got in his way, strutting about in his splendid clothes, and perhaps made some rude remark, which made Meidias lose his temper and hit him.

Demosthenes did not hit back (and afterwards he prided himself on his self-restraint, **74**), but made a formal accusation against Meidias for committing an offence concerning the festival (ἀδικεῖν περὶ τὴν ἑορτήν). This was done by the legal procedure called *probole* at a meeting of the Ekklesia, held in the theatre on the next day after the Dionysia and the immediately ensuing festival of the Pandia were over (**8**).[1] The result was a vote condemning Meidias (καταχειροτονία). But the peculiar feature of *probole* was that no penalty followed from this. If Demosthenes wanted a penalty imposed on Meidias, he had to take further steps to prosecute; but if he thought it was shaming enough for Meidias to have had the Ekklesia vote against him, he could just leave it at that. At first, it seems, he did leave it at that. He probably thought that nothing more was necessary to get Meidias to keep his hands off him in future. Anyway Meidias was now a trierarch and was away from Athens for a time

[1] The legal procedure is considered further on pp. 13–16.

on his ship (**164–7**). However, he returned to Athens in due course and began pestering Demosthenes again.

The next events probably fall in the year 348/7. First a man named Euktemon, allegedly bribed by Meidias, accused Demosthenes of desertion from the army (*λιποτάξιον*). We are not told what grounds were given for this charge; but since we know that Demosthenes was in Athens for the Dionysia of 348, most likely he was accused of leaving the army in Euboia without permission. Of course he could have argued that it was his duty as a chorus-producer to be present at the festival; that would probably have been accepted as a good excuse for leaving the army, and in fact Euktemon in the end did not proceed with the case. Indeed Demosthenes says that the only purpose of it was to get a notice saying 'Euktemon of Lousia prosecuted Demosthenes of Paiania for desertion' put up in the Agora, bringing his name into ridicule or disrepute (**103**).

Euktemon seems to have had the support of a man named Nikodemos, and the next thing that happened was a murder: Nikodemos was violently killed by a young man named Aristarkhos. Demosthenes was a friend of this young man, and Meidias now went around saying that Demosthenes shared responsibility for the murder, and he tried to persuade the relatives of Nikodemos to prosecute Demosthenes instead of Aristarkhos. However, they did not in fact prosecute him, and we can safely assume that he really had nothing to do with the murder, although that did not stop Aiskhines and Deinarkhos raking up the story against him years later.[1]

Then Meidias made a speech, presumably in the Ekklesia, declaring that the fiasco in Euboia had been Demosthenes' fault (**110**). We are given no further information about this speech. For what, precisely, can he have blamed Demosthenes? Certainly not for the fact that the expedition was made at all, since Demosthenes opposed it (5.5); nor for its ultimate failure under Molossos, with which Demosthenes had nothing to do (Plu. *Phok.* 14.2). Meidias probably said that the difficulties of Phokion at Tamynai, when his force was blockaded (**162**), were due to the fact that Demosthenes and other soldiers had

[1] For details of this incident, **104** n.

left him and returned to Athens; perhaps he said that Demosthenes incited others to leave.[1]

After that Meidias attacked Demosthenes at a *dokimasia*. Demosthenes had been selected by lot to be a member of the Boule (**111**). We know that Demosthenes was on the Boule in 347/6;[2] so it is clear that we have now reached the summer of 347, more than a year after the punch-up at the Dionysia. The practice was that all those selected by lot for the Boule underwent a *dokimasia* or vetting by the members of the outgoing Boule, to check their eligibility (*AP* 45.3). No surviving text gives a detailed account of the *dokimasia* of the Boule, but probably it was similar to the *dokimasia* of the arkhons (described in *AP* 55.2–5): each man was asked standard questions about his parentage and his performance of public obligations; then anyone who wished could make an accusation, against which the accused man could speak in his own defence, and the Boule voted to accept or reject him. Again we do not know what Meidias said; most likely he alleged that Demosthenes was ineligible for the Boule because he was guilty of the homicide of Nikodemos, and so was polluted and unfit to perform public and religious functions.[3]

Evidently Demosthenes got the better of the argument, since we know that he did get on to the Boule: he performed inaugural rites for the Boule at the beginning of the new year 347/6 (**114**), and he was also chosen to be an arkhitheoros and a hieropoios (**115**). But he must have been highly indignant at having been virtually put on trial himself by the very man whom he was entitled to put on trial for hitting him. This was the last straw (cf. **111** τελευτῶν) which prompted him to go ahead with prosecuting Meidias for the offence committed at the Dionysia in the year before last. The text of the speech which we have was clearly written soon after this, for a trial in 347/6.[4] It includes passages (**2**, **216**) asserting that many men

[1] Demosthenes, however, blames Ploutarkhos (**110**); and Plutarch in effect agrees with him that the flight of Ploutarkhos with his mercenaries was a more serious blow to Phokion than the loss of the 'disorderly, talkative, and bad' Athenians (Plu. *Phok.* 12.3–13.2).

[2] 19.154, Ais. 2.17, 3.62.

[3] Cf. Ant. 6.45–6 on the unfitness of a man suspected of homicide to be a member of the Boule.

[4] Dionysios of Halikarnassos says (*Amm.* 1.4) that the speech was written in 349/8 (when Kallimakhos was arkhon), but that is a simple error: he has attributed it to the

have urged him to take further proceedings against Meidias and not let the matter drop; these passages are a kind of apology for reviving the accusation so long after the offence occurred and the people voted on it in the Ekklesia.

It has sometimes been thought that the personal clashes described in the oration are not sufficient explanation for Demosthenes' decision to go ahead with the prosecution, and that he must also have had a political reason for attacking Meidias at this time.[1] But it is difficult to ascertain what that reason may have been, because we have little information about Meidias' activities or affinities in politics. The only political action to which his name is firmly attached is the decision to send a force to Euboia in 349/8; as we have already seen, Demosthenes opposed that decision, and Meidias later made another speech criticizing him in connection with it (**110**), but that is hardly enough to show that they were regular political opponents, or to explain Demosthenes' decision to proceed against Meidias in 347/6. Virtually the only other information that we have about Meidias' speeches[2] is that, whenever news of a set-back reached Athens, Meidias would tell the Athenians that it was their own fault, because they were too reluctant to go out and fight and to make financial contributions (**203**). That is much the same as what Demosthenes told them in the *Olynthiacs*; on this point, at least, it appears that Demosthenes and Meidias were not in disagreement.

As for Meidias' possible political allies, a few names of men

year when Meidias assaulted Demosthenes and the *probole* was brought before the Ekklesia, overlooking the fact that two years elapsed between those events and the composition of the speech (**13** *τρίτον ἔτος τουτί*). In the nineteenth century some scholars, notably A. Schaefer *Demosthenes*[2] 2.109–18, tried to preserve Dionysios' date by adopting a different chronology, putting the assault in 351/0 and the speech in 349/8. But that is not acceptable, because the assault was in the same year as the expedition to Euboia (**163**), which was contemporary with the war at Olynthos (**197**, 59.4), which was in 349/8, as Dionysios well knew (*Amm.* 1.4, 1.9). It was refuted by Weil *Rev. Phil.* 3 (1879) 1–13, and is now universally rejected. For a recent discussion see Giugnoli *Atene e Roma* 20 (1975) 178–81.

[1] See for example Cawkwell *JHS* 83 (1963) 49. Cawkwell is properly cautious about identifying particular political motives, but the possiblity of non-political motives seems not to occur to him. A political interpretation is favoured also by Jaeger *Demosthenes* 146–7, S. Perlman *Athenaeum* 41 (1963) 351–2, Longo *Eterie* 88–97, Giugnoli *Atene e Roma* 20 (1975) 170–83.

[2] Nothing can be made of the fact that he made a speech criticizing the Boule in 355 (22.10), because we are not told what the burden of his criticism was.

expected to support him at his trial are given in the oration. Polyeuktos, Timokrates, and Euktemon were adherents who, according to Demosthenes, supported Meidias only because they were paid to do so (**139**). Philippides, Mnesarkhides, Diotimos, and Neoptolemos, on the other hand, were all apparently rich men, who would hardly have been supporting Meidias for money (**208**, **215**). If this Philippides was the man attacked in Hypereides' speech *Against Philippides*, that gives some evidence that he was a politician of anti-democratic views;[1] and Demosthenes implies that these men 'with Meidias and those like him' formed an anti-democratic group, though his suggestion that they might take control of the state is fanciful (**209**). The other named supporter of Meidias is that elusive figure Euboulos, who had declined to speak for him at the hearing of the *probole* by the Ekklesia but was expected to do so at the trial nearly two years later (**205–7**). Furthermore, if the text of **205** is sound,[2] Demosthenes alleges that the other men will speak for Meidias because of Euboulos' hostility towards Demosthenes. This implies that Euboulos was the leader of the group.

Even if it is true that Meidias belonged to a political group led by Euboulos (and the evidence for it remains fairly tenuous), that still does not prove that Demosthenes had a political reason for proceeding against him in 347/6, since we do not know of any political disagreement between Demosthenes and Euboulos at that date. The suggestion that Euboulos opposed Demosthenes' policy of extending the war against Philip in northern Greece is quite speculative.[3] In fact Demosthenes says that Euboulos' hostility to him is personal (**205** *τὴν ἰδίαν ἔχθραν*) and claims that he does not know the reason for it (**207** *ἔγωγε μὰ τοὺς θεοὺς οὐκ οἶδα ἀνθ' ὅτου*).[4]

The evidence available to us is thus not sufficient to establish that Demosthenes had any specific political reason for prosecuting Meidias in 347/6. The political arguments in the speech are of a rather general kind, to the effect that Meidias' charac-

[1] **208** n. *Φιλιππίδην*.

[2] **205** n. *οὗτος*. If Cobet's suggestion (reported there) is accepted, the link between the other supporters and Euboulos' hostility towards Demosthenes disappears.

[3] Cf. Cawkwell *JHS* 83 (1963) 51.

[4] Σ **205** (687 Dilts) says that Euboulos accused Demosthenes of being implicated in the murder of Nikodemos, but I doubt whether this information is reliable; see p. 330.

ter makes him an unsuitable man to have in Athens.[1] Their personal dislike of each other is clear, being attested by a long series of clashes extending over a period of seventeen years. With men like these, we must beware of drawing a sharp distinction between public and private conduct; action could be taken from a mixture of political and personal motives.[2] But in this case the personal motives may have predominated. Demosthenes and Meidias really hated each other.

II. PROSECUTION AND OFFENCE

The oration *Against Meidias* is virtually our only source of information about the use of the legal procedure called προβολή ('putting forward', sc. to the Ekklesia), as used for prosecution for offences at festivals; but in most respects it makes the procedure clear enough.[3] Anyone, whether Athenian or not,[4] who wished to prosecute for an offence in connection with a festival notified the prytaneis (the fifty members of the Boule responsible for arranging meetings of the Boule and the Ekklesia) and they had to place it on the agenda for the Boule and the Ekklesia immediately after the festival.[5] In the Ekklesia the prosecutor and the defendant each made a speech, and the citizens voted for one or the other; a majority vote against the defendant was called καταχειροτονία, in his favour ἀποχειροτονία. But this vote had no effect, except that it was, in the strict sense, prejudicial. If the prosecutor wished to take the matter further, there had still to be a trial with an ordinary jury in the court of the thesmothetai.[6] Neither the prosecutor

[1] See pp. 36–7.

[2] Cf. Brown *Geschäft* 163–79.

[3] Other references, adding nothing substantial to what is known from *Meidias*, may be found in *AP* 59.2, Pol. 8.46, Harp. προβαλλομένους and προβολάς, *Lex. Rhet.* 288.18–22. Quite separate, and not relevant here, is the use of *probole* to accuse a man of being a sycophant or of failing to fulfil a promise to the people (*AP* 43.5, X. *Hell.* 1.7.35, Isok. 15.314, Ais. 2.145). For modern accounts see Goodwin's edition of *Meidias* pp. 158–62, Lipsius *Recht* 211–19, Harrison *Law* 2.59–64, MacDowell *Law* 194–7, Rhodes *Comm. on AP* 526–7.

[4] In **175** there is an example of a case initiated by a Karian.

[5] The last words of the law quoted in **8**, ὅσαι ἂν μὴ ἐκτετισμέναι ὦσιν, are best interpreted as meaning that a prosecution need not be brought to the Ekklesia if it was agreed in the Boule that a small fine was sufficient penalty; see the commentary there.

[6] One themothetes presided (**3** n. τις εἰσάγει).

nor the jury was bound by the vote in the Ekklesia, but no doubt it did in practice carry much weight with the prosecutor, in deciding whether to proceed to the trial, and with the jurors, in deciding on their verdict. Presumably the main purpose of the hearing in the Ekklesia was, in the absence of any public official to rule whether a prosecution should proceed, to enable the citizens generally to express their approval of it.

Demosthenes quotes two separate laws about *probole* (**8**, **10**) and refers in passing to another (**175**). The reason why there was more than one law is evidently that the use of *probole* was extended by stages to different festivals and different offences at them. The law in **8** refers only to the city Dionysia. The law in **10** lists the Dionysia in Peiraieus, the Lenaia, the city Dionysia, and the Thargelia, and its last words (καθὰ περὶ τῶν ἄλλων τῶν ἀδικούντων γέγραπται) imply that other laws about offenders at those festivals already exist; we may infer that the Dionysia in Peiraieus, the Lenaia, and the Thargelia each had a law much like the law in **8**. There was also a law in similar terms for the Eleusinian Mysteries (**175**); whether there were similar laws for any other festivals is unknown. The law for the Mysteries was of later date than that for the city Dionysia (**175**), which may well have been the earliest of them all; yet even it did not yet exist in the days of Alkibiades (**147**). It seems that the use of *probole* for offences at festivals may not have been introduced before the end of the fifth century.

The law in **8** defines neither the word προβολή nor the offences for which the procedure could be used. Possibly the meaning of *probole* was assumed to be known from its earlier use in other connections;[1] but it is at least equally likely that the quotation of the law which we have is incomplete and some definition or explanation of the procedure was given in another sentence. But there is no reason to think that the definition of the offence would have been anything more than ἀδικεῖν περὶ τὴν ἑορτήν, which was the charge brought by Demosthenes against Meidias (**1**; cf. **9**, **26**, **28**, **175**). It was for the Ekklesia and subsequently the jury to decide whether the defendant's behaviour constituted 'wrongdoing concerning the festival'. The

[1] *Probole* may well have been used for sycophancy at an earlier date than for festival offences. Isok. 15.314 attributes its use for sycophancy to 'our ancestors', which should mean at least that it originated well before the end of the fifth century.

purpose of the law in **10** was to extend the application of the *probole* procedure to anyone who, during any of the festivals listed, seized money or other property from another person, even if it was something (such as an overdue debt) which he was otherwise legally entitled to seize. Such a seizure was not exactly 'wrongdoing concerning the festival'. But presumably the Athenians had found by experience that creditors would often hope to track down their debtors at a festival which a large part of the population attended, and that their activities were liable to cause a disturbance in the proceedings; and so the law in **10** was introduced to make them too subject to *probole*.

Demosthenes gives several examples of the use of *probole*. In one case Menippos, a Karian, complained that Euandros, a Thespian to whom he owed money, caught hold of him during the Eleusinian Mysteries; Euandros was found guilty both in the Ekklesia and in the subsequent trial, and was required to forfeit the money which Menippos owed him and pay compensation besides (**175–7**). In another case a man who had been taking a seat in the theatre to which he was not entitled complained that the arkhon's assessor (*πάρεδρος*) manhandled him; the assessor was found guilty in the Ekklesia, but died before the case reached a trial by jury (**178–9**). In a third case a man named Ktesikles, riding in a procession at a festival, caught sight of someone he did not like and struck him with his whip; the Ekklesia voted against him, and the jury condemned him to death (**180**). All these cases involved physical violence, but they do not prove that the *probole* procedure was inappropriate when violence was not involved, since Demosthenes has probably selected them for their similarity to the violent treatment which he suffered from Meidias. The other case which is recorded did not involve violence: Aristophon was accused by *probole* because he failed to supply some crowns for a festival, but then he delivered them and the charge was dropped (**218**).

When a *probole* case went for trial by a jury, the trial procedure was presumably the same as for other public cases. The case would occupy the court for a full day; hence Demosthenes has drafted a long speech for the prosecution.[1] The penalty was

[1] A full court day is what is called *διαμεμετρημένη ἡμέρα* (19.120, 53.17, Ais. 2.126, *AP* 67.3). For recent discussion of the length of time available for speeches see Rhodes *Comm. on AP* 719–28, MacDowell *CQ* 35 (1985) 525–6.

not specified by law; thus, if the defendant was found guilty, there would be further speeches and voting to assess the penalty.[1] Demosthenes says several times in the speech that Meidias deserves death, but he also envisages the possibility of a lesser penalty;[2] his final decision on what penalty to propose would not have been made until he saw how much support he had from the jury in the voting on the verdict.

It has sometimes been supposed that the name *probole* was applicable only to the hearing in the Ekklesia, and that the subsequent trial by jury, if there was one, was a trial of a separate legal action, probably a γραφὴ ἀσεβείας.[3] That is a misapprehension. Several kinds of legal action in Athens were named from the manner in which they were initiated, and the name was then applied to the whole process: for example, the word ἀπαγωγή was used strictly for the arrest of a malefactor but also more loosely for the legal proceedings which followed his arrest. In the same way the word προβολή was used strictly for the occasion when the accusation was made in the Ekklesia (e.g. **193**), but also more generally for all the ensuing proceedings. This is shown by passages in the speech in which this action is contrasted with other actions which Demosthenes might have used but did not. He predicts that Meidias will say that, instead of the present action, he ought to have brought a δίκη βλάβης for the damage to his property and a γραφὴ ὕβρεως for the alleged assault on his person; 'but I know one thing very well, and you should know it too: if I had brought a *dike* and not a *probole* against him, I should immediately have been faced with the converse argument—that, if any of these accusations was true, I ought to have brought a *probole*' (**25–6**). Later he remarks that it *would* be appropriate if Meidias were judged to be guilty of impiety (ἀσέβεια) too; this implies that impiety is not the charge on which he is actually expecting Meidias to be convicted.[4] Thus the action against Meidias is neither a γραφὴ ἀσεβείας nor a γραφὴ ὕβρεως. Formally the offence for which Meidias is being prosecuted is simply ἀδικεῖν περὶ τὴν ἑορτήν.

[1] Cf. **25** τίμημα ἐπάγειν ὅ τι χρὴ παθεῖν ἢ ἀποτεῖσαι.

[2] Cf. **152** μάλιστα μὲν θάνατος, εἰ δὲ μή, πάντα τὰ ὄντα ἀφελέσθαι.

[3] Cf. Harrison *Law* 2.62–3.

[4] **51** νῦν δέ μοι δοκεῖ κἂν ἀσέβειαν εἰ καταγιγνώσκοι τὰ προσήκοντα ποιεῖν. The use of ἄν and the optative implies that fulfilment of the hypothesis is not expected, while καί means 'even' or 'also', in addition to the charge which has been brought in fact.

Nevertheless Demosthenes keeps asserting that he is guilty of *asebeia* and *hybris*, because he thinks that will make the jury more ready to convict.

We therefore naturally want to decide whether Meidias' conduct was truly *asebeia* and *hybris*. But that is more easily said than done, because neither term is clearly defined. The text of the Athenian law of *asebeia* does not survive, but the law of *hybris* is preserved in **47**,[1] and we see that it makes no attempt at all to define *hybris*: it starts straight off 'If anyone commits *hybris* against anyone ...'. It assumes that everyone knows what *hybris* is. There is no special legal sense of *hybris*, and in cases of doubt the jurors have to decide whether a particular act amounts to *hybris* or not, according to the normal usage of the word. The law of *asebeia* was probably much the same; certainly there is no evidence that it did include a definition of the word, and I assume that it did not.

Undoubtedly *asebeia* denotes behaviour offensive to a god or gods, but what kinds of behaviour did the Athenians believe to be so? Most obviously, behaviour which contravened a law concerning a religious matter. For example the hierophant of the Eleusinian Mysteries was convicted of *asebeia* for performing a sacrifice at Eleusis on the day of the Haloa, because that contravened two rules of sacred law: the sacrifice ought to have been performed by the priestess, not by the hierophant; and sacrifices were not allowed on the day of the Haloa (59.116).[2] Prosecution for *asebeia* was not restricted to contraventions of specific written rules:[3] for example the mutilation of the Hermai in 415 was undoubtedly an ἀσέβημα even though no written law said, in so many words, that the Hermai must not be mutilated.[4] But a prosecutor would no doubt have a better chance of convincing a jury if he could point to a specific rule that had been broken. Demosthenes, in accusing Meidias of

[1] On the authenticity of this text, **47** n.

[2] For other known prosecutions for *asebeia* in the fourth century see MacDowell *Law* 197–8.

[3] J. Rudhardt *Mus. Helv.* 17 (1960) 87–105 argues that after 403/2, when a law was made that no uninscribed law was to be enforced (And. 1.85–7), no one could be prosecuted for *asebeia* unless he was accused of breaking a law forbidding some specific act or behaviour. But the argument is unsound, because there was undoubtedly an inscribed law about *asebeia* in general, which was sufficient to permit prosecution for any conduct alleged to be *asebeia*.

[4] **147**, Thuc. 6.27, Lys. 6.11–12.

asebeia (even though not formally prosecuting him for it), does his best in this direction. He points out that the celebration of festivals with choruses is ordained not merely by man-made laws but by oracles; this implies that anyone obstructing or interfering with the performance of a chorus is contravening the will of the gods (**51–5**). He also refers to a law penalizing anyone who prevented an individual chorister from performing (**56–7**). However, punching a chorus-producer was not the same thing as interfering with a chorus. Evidently there was no law or oracle giving special protection to the person of a chorus-producer, or Demosthenes would have quoted it. Probably it would have been hard to convince an Athenian jury that punching a chorus-producer was offensive to Dionysos (as asserted in **126**), and that was why Demosthenes did not formally prosecute Meidias for *asebeia*. He had a better chance of winning if he used the *probole* procedure, in which the offence was categorized as ἀδικεῖν περὶ τὴν ἑορτήν, because no one could deny that the punch-up occurred at the festival. But he still tries to involve Meidias in the opprobrium for *asebeia* too, not only by the references to laws and oracles about choruses and choristers (**51–7**) but also by using, towards the end of the speech, the expression ἀσεβεῖν περὶ τὴν ἑορτήν, which blurs the distinction between the two offences (**199**, **227**).[1]

The concept of *hybris* is even harder to pin down. It is one of the most famous elements of Greek morality, and there have been many modern attempts to define it, but none yet has been wholly successful. Some progress indeed has been made: no one, I hope, still holds the belief, prevalent until about twenty years ago, that *hybris* is an offence against the gods, leading to divine punishment of an overweening mortal. That belief was based on a very small number of passages in tragedy, and is incompatible with the vast majority of passages in Greek authors in which the word is used. But the authors use it so variously that framing a definition to fit all the instances is extremely difficult, perhaps impossible. It must be acknowledged that authors ranging in time from Homer to Aristotle, writing in different genres and in different parts of Greece, may not all have meant quite the same thing by *hybris*.

[1] He also uses ἀσέβημα and related words to refer to Meidias' conduct after the murder of Nikodemos (**104**, **114**, **120**); **104** n. ἀσέβημα.

Some years ago I published a paper in which I tried to give a more comprehensive outline of the range of the word's use than had previously been available.[1] I argued that *hybris* has several characteristic causes: youthfulness, having plenty to eat and drink, and wealth. It also has characteristic results: further eating and drinking, sexual activity, larking about, hitting and killing, taking other people's property and privileges, jeering at people, and disobeying authority both human and divine. No one of these characteristic causes and results is necessarily present in every instance of *hybris*, but its essence consists of having energy or power and misusing it self-indulgently. At the time when I wrote that paper, however, I was unaware that two other pieces of work were being done on the same subject. J. T. Hooker's paper argues that originally the word *hybris* carried no moral condemnation but meant simply 'exuberant physical strength'; only later did it take on the pejorative significance 'physical strength wrongly applied'.[2] N. R. E. Fisher has studied *hybris* in much more detail, examining virtually every instance of the word and those related to it from Homer to the fourth century.[3] He takes Aristotle's *Rhetoric* as his starting-point: Aristotle emphasizes desire to dishonour another person as characteristic of *hybris* (*Rhet.* 1374a 11–15, 1378b 23–35). Fisher considers that this is the decisive criterion: *hybris* is behaviour that causes dishonour to individuals, to groups, or to the values that hold a society together. He argues that this definition fits the usage of the word in all authors, except for certain works of Plato (especially *Phaidros*) in which its sense is deliberately extended for a philosophical purpose. In a later paper, dealing with Homer only, E. Cantarella supports Fisher and maintains that *hybris* originally meant behaviour causing dishonour in a heroic society, and invariably has this sense in the *Iliad* and *Odyssey*.[4] Her conclusion is diametrically opposite to Hooker's, though she seems to have been unacquainted with his article.[5]

[1] '*Hybris* in Athens', *G&R* 23 (1976) 14–31.

[2] 'The original meaning of *ὕβρις*', *Archiv für Begriffsgeschichte* 19 (1975) 125–37.

[3] *The concept of* hybris *in Greece from Homer to the fourth century B.C.* (D.Phil. thesis, Oxford 1976). Two articles based on the thesis have appeared in *G&R* 23 (1976) 177–93, 26 (1979) 32–47, but Fisher's full-length book on the subject has, as I write these words, not yet been published.

[4] 'Spunti di riflessione critica su *ὕβρις* e *τιμή* in Omero', *Symposion 1979* (1981) 85–96.

[5] Articles by E. Ruschenbusch and M. Gagarin are restricted to the law of *hybris*, and are considered in my commentary on that text (**47** n.).

At first sight, it seems to strengthen Fisher's and Cantarella's position that they reach the same conclusion starting from texts at opposite ends of the period, Homer and Aristotle; what is true at the beginning and at the end, one thinks, must be true throughout. However, Fisher's interpretation of Aristotle is not quite satisfactory. One of the key sentences runs: *οὐ γὰρ εἰ ἐπάταξεν πάντως ὕβρισεν, ἀλλ' εἰ ἕνεκά του, οἷον τοῦ ἀτιμάσαι ἐκεῖνον ἢ αὐτὸς ἡσθῆναι*, 'If one hits, one does not in all cases commit *hybris*, but only if it is for a purpose, such as dishonouring the man or enjoying oneself' (*Rhet.* 1374a 13–15). This does not mean, as Fisher would have it, that dishonour is the purpose of all acts of *hybris*; the word *οἷον* shows that dishonour is merely one example of the various different purposes that *hybris* may have. The other passage of Aristotle on which Fisher mainly relies (*Rhet.* 1378b 23–35) supports his case better, since it certainly emphasizes dishonour; but it too leaves a little room for doubt whether 'Doing and saying things at which the victim incurs shame is *hybris*'[1] necessarily means that nothing else is *hybris*. It is also important to keep in mind that Aristotle's *Rhetoric* is written primarily for composers of lawcourt speeches, so that he is concerned only with cases of *hybris* in which there is a victim who may want to prosecute. If there are kinds of *hybris* which do no dishonour or harm to anyone, we should not expect them necessarily to be mentioned by Aristotle.

The question whether *hybris* must have a victim becomes even more acute when we turn to Homer and consider the disagreement between Hooker and Cantarella. An extended discussion of *hybris* in Homer would not be appropriate here; so I give only one example. In *Odyssey* 4.625–7 'the suitors in front of the house of Odysseus were enjoying themselves throwing discuses and javelins on a levelled ground, as before, having *hybris*'. I interpret this as meaning that the suitors had plenty of energy, and expended it on self-indulgent pleasure. By using the word *hybris* the poet implies disapproval: they ought to have devoted their energy to some work of practical use. But Hooker considers that the word *hybris* here conveys no adverse comment on their activity, which is 'an activity wholly appropriate to noblemen'. Cantarella, on the other hand, thinks it

[1] I correct here the translation which I gave in *G&R* 23 (1976) 27. The absence of the article with *ὕβρις* shows that it is the complement, not the subject.

'hardly contestable' that their activity dishonours Odysseus. Yet Odysseus is not in Ithaka at the time, and the suitors do not even know whether he is alive or dead; it is a very strained interpretation to say that they want to dishonour him and decide that a good way of doing so is to go out of his house and throw javelins. A similar lack of agreement prevails about the interpretation of a number of other passages, in Homer and other authors, which seem to Fisher, but not to me, to involve dishonouring a victim. Among these are passages in which *hybris* is attributed to frisky animals, or to overgrown and fruitless plants;[1] these are not to be dismissed as peripheral or metaphorical, but must be given due weight in any attempt at a comprehensive definition.

In Greek literature as a whole, therefore, I still believe that it is not adequate to say simply that *hybris* is behaviour that causes dishonour. The concept is too protean for that. In particular, the notion of self-indulgence and indiscipline within the offender himself is frequently as prominent as, or more prominent than, the notion of dishonour suffered by a victim. And, to return to Demosthenes, both these notions are present in the *hybris* which he attributes to Meidias. He does link *hybris* with dishonour in several passages, especially the famous one in which he talks about the distinction between a blow struck with *hybris* and one without it:

'It wasn't the blow that made him angry, but the dishonour; nor is being hit such a serious matter to free men (though it is serious), but being hit with *hybris*. There are many things which the hitter might do, men of Athens, some of which the victim might not even be able to report to someone else—in his bearing, in his look, in his voice, when he displays *hybris*, when he displays hostility, when he strikes with the fist, when he strikes on the face. That's what rouses people, that's what makes them forget themselves, if they're not accustomed to being insulted. No one reporting this behaviour, men of Athens, could convey its seriousness to his listeners as vividly as the *hybris* is seen at the actual time by the victim and the onlookers.' (**72**)

This passage stresses the effect of *hybris* on the victim, but it stresses even more the attitude of the attacker. The feature of *hybris* to which Demosthenes directs most attention throughout

[1] Cf. M. Vogel *Chiron* (1978) 1.61–6; A. Michelini '"Ὕβρις and plants', *HSCP* 82 (1978) 35–44.

the speech is the hybristic man's state of mind. What makes an act of *hybris* is not the status of the victim (even a slave, who is a person with no honour, can suffer an act of *hybris*) but the character of the act itself (**46**). Yet it is not a matter of defining particular acts. Meidias' activity against Demosthenes took many forms: breaking into his house and using bad language (**80–1**), bribing his chorus-director (**17**), and trying to get him exiled for homicide (**115**) are all called *hybris* by Demosthenes. He also declares that Meidias behaved with *hybris* towards the whole people of Athens, not only in attacking their chorus-producer but also in other ways—for instance by making a speech in which he criticized their failure to contribute money and personal service to the navy (**203–4** ... *τοιαῦτα ὑβρίζων*). In fact Meidias displayed *hybris* towards everybody all the time (**1** *τὴν ὕβριν, ᾗ πρὸς ἅπαντας ἀεὶ χρῆται Μειδίας*). His *hybris* was caused by his wealth (**98**, **138**, etc.), which made him bold. He had a huge house at Eleusis and a carriage drawn by a pair of white horses, and he swaggered through the Agora talking about 'cups' and 'drinking-horns' and 'chalices' loudly enough for the passers-by to hear (**158**). The strongest evidence of his *hybris*, according to Demosthenes, was that he was prepared to accuse a large number of men at the same time (**135**); this statement makes self-confidence a primary element in *hybris*.

We cannot know whether an Athenian jury in the fourth century would have been willing to interpret *hybris* as broadly as Demosthenes does. The text of the law (**47**) does not define it, but leaves to the jury the task of deciding whether any particular act amounts to *hybris*. The only limitation is that, to be a legal offence, *hybris* must be directed against another human being, whether child, woman, or man, free or slave. In practice we rarely hear any mention of legal proceedings for *hybris* except in connection with physical assault or sexual violation (e.g. 45.4, 54.1, Ar. *Wasps* 1418, *Birds* 1046), and conviction for *hybris* on other grounds may not have been customary. But, unless his account of the facts is wholly misleading, Demosthenes should have been able to get Meidias convicted of *hybris* for hitting him in the theatre, if he had chosen that form of legal action.

No English translation of either *hybris* or *asebeia* is really satisfactory. Different words may seem preferable in different contexts, but for the purpose of translating the speech in the present volume I have thought it best to use the same English terms throughout, and I have chosen 'insolence' and 'impiety' as the least unsatisfactory renderings.

III. COMPOSITION AND DELIVERY

A passage of Aiskhines' speech *Against Ktesiphon* appears to imply that the trial of Meidias did not take place, and thus that the speech *Against Meidias* was never delivered. Aiskhines, attacking Demosthenes many years later, mentions the affair in a catalogue of what he claims are well known discreditable incidents in Demosthenes' past life.

τί γὰρ δεῖ νῦν ταῦτα λέγειν ..., καὶ ταῦτα δὴ τὰ περὶ Μειδίαν καὶ τοὺς κονδύλους, οὓς ἔλαβεν ἐν τῇ ὀρχήστρᾳ χορηγὸς ὤν, καὶ ὡς ἀπέδοτο τριάκοντα μνῶν ἅμα τήν τε εἰς αὑτὸν ὕβριν καὶ τὴν τοῦ δήμου καταχειροτονίαν, ἣν ἐν Διονύσου κατεχειροτόνησε Μειδίου;

'What need is there to relate these events now ..., and the incidents concerning Meidias and the punches that he received in the *orkhestra* when he was a khoregos, and how he sold for thirty mnai both the insolence to himself and the adverse vote which the people gave against Meidias in the precinct of Dionysos?' (Ais. 3.51–2)

This is most easily interpreted as meaning that Demosthenes accepted a payment of thirty mnai to drop the case before it came to trial. That is how it is interpreted by later writers who say that he gave up the case on receiving money (Plu. *Dem.* 12.4, *Eth.* 844d, Souda δ 456); for we may assume that they are just repeating the statement of Aiskhines, as they understand it, and are not to be regarded as giving independent evidence. But Aiskhines was attacking Demosthenes, and his statement must be approached with caution. Is it really credible that Demosthenes, after writing the speech we have, after his boasts that he persisted with the prosecution in spite of all attempts to dissuade him (**3**, **40**, **151**), after his insistence that a conviction was essential in order to uphold the rule of law in Athens (**221–5**), after his demands for the death penalty (**21**, **70**, **118**, **201**), accepted a bribe to drop the case? One would hardly think so; and yet Aiskhines, even after a lapse of some years, could not have told a complete falsehood about an incident which must have been notorious.

So it is worth considering whether an alternative interpretation is possible. It has been suggested that the trial was held, and Meidias was found guilty by the jury, but when the court proceeded to consideration of the penalty, Demosthenes proposed not death but a fine of thirty mnai.[1] His reason for doing this could have been, for example, that the jury had voted for conviction by only a small majority, and he judged that in the voting on the penalty he could not count on getting a majority in favour of death;[2] but he might well have been criticized by some people afterwards for giving way to this extent. Could Aiskhines use the word 'sold' (ἀπέδοτο) for a decision which resulted in payment of money to the public treasury, not to Demosthenes himself? Surely he could: he wanted to make Demosthenes' conduct sound as disreputable as possible, and if some of his listeners, not remembering the trial of Meidias, interpreted his words the other way and thought that Demosthenes had accepted a bribe, so much the better!

So the evidence of Aiskhines is not conclusive. But there is also a different kind of reason for believing that the trial never took place. Critics have pointed to a number of faults in the composition of the speech as we have it, suggesting that it lacks the author's final revision. Demosthenes would surely not have delivered the speech in this imperfect state. So, it is argued, the agreement to drop the case must have been made shortly before the trial was due, when he had done most of his work on the speech but not quite finished it; the trial never took place; the speech was never completed; nor could Demosthenes have published such a bitter attack on Meidias after making an agreement with him; but the draft was found among his possessions and documents after his death, and that was when copies of it were first made and distributed. Hints of this view

[1] This suggestion is made by G. Grote *History of Greece* (1862) 8.90 n. 1 and revived by Erbse *Hermes* 84 (1956) 135–52.

[2] In *ἀγῶνες τιμητοί* it must often have been a tricky matter for the prosecutor and the defendant to judge respectively how severe and how lenient they could afford to be in their proposals for the penalty, without provoking a majority of the jury to vote for the opponent's proposal. The size of the majority in the first vote, on the verdict, might help them to make this judgement.

may be found even among the scholia on the speech[1] and in the account of Demosthenes written by Photios in the ninth century,[2] but it was fully developed by German scholars in the nineteenth century, and prevailed unchallenged until 1956.[3] In an article published in that year, however, H. Erbse argues that there is no good reason to regard the speech as unfinished, and that it was both delivered and published by Demosthenes himself.[4]

We can agree with Erbse that the weaknesses in the composition of the speech have been overstated. Demosthenes was under no obligation to adopt a wholly logical arrangement. If he wished to impress a particular point on the jury, he could make it several times over in different parts of the speech. If he thought that some other point would damage his case, he could treat it cursorily or omit it altogether. He could deal with topics in an illogical order, if he considered that by doing so he could prevent his audience from becoming bored.[5] He could announce a plan for his speech and then not carry it out: the jurors were not likely to make notes of his plan, and then check the actual speech against it and complain of divergences. Thus, when he says that he will examine first the insults to himself, then those to the citizens of Athens, and after that the rest of Meidias' life (**21**), we must not suppose that the speech is unfinished if he does not in fact deal with those topics in that order.[6]

Nevertheless some difficulties remain. These are the ones which seem to me to be the most substantial:

[1] Σ **89** (307b Dilts) *τοῦτο τὸ μέρος ὠβέλισται παρὰ τῶν κριτικῶν καὶ ὡς ἀδιόρθωτον παραλέλειπται.* On obelized passages of the text see pp. 47–8.

[2] Phot. *Bibl.* 265.491b *καί τινες ἔφησαν ἑκάτερον λόγον* (viz. *Against Meidias* and *On the False Embassy*) *ἐν τύποις καταλειφθῆναι, ἀλλὰ μὴ πρὸς ἔκδοσιν διακεκάθαρται.*

[3] See especially Boeckh *Abh. Berlin 1818–19* 68–77, Schaefer *Demosthenes* 3.2.58–63, Blass *Beredsamkeit*[2] 3.1.338–9, C. Vielhauer *De Demosthenis Midiana* (Bratislava 1908), Sealey *REG* 68 (1955) 96–101. The most extreme proponent of this view is A. H. G. P. van den Es, who in his *Commentatio de Demosthenis Midiana* (published with *Het vierde eeuwfeest van het Stedelijk Gymnasium te Utrecht*, 1874) analyses the oration into fourteen separate fragments which were stuck together by an editor after Demosthenes' death. The view that the speech was not delivered is accepted in MacDowell *Law* 196–7, but I have subsequently become more dubious.

[4] *Hermes* 84 (1956) 135–52.

[5] For the arrangement of topics in the speech see p. 28–31.

[6] Even Erbse seems worried about this, and tries unsuccessfully to maintain that *μετὰ ταῦτα* does not mean 'after that' (p. 139). Actually the speech does in the end fulfil the plan more or less: **128–42** is about offences against other individuals, and **143–74** is about other aspects of Meidias' life.

1. When Demosthenes has a law read out to the jury, his normal practice is to give immediately afterwards some comment or explanation to ensure that the jury understands the significance of the law and its bearing on his case (**9**, **11**, **48**, and similarly in other speeches). Yet two of the laws read out in this speech, those on arbitrators and on bribery (**94**, **113**) are not followed by any such comment. In neither case is it at all obvious how the details of the law support his argument, and it is unlikely that he would have left his listeners to work it out for themselves. It seems more likely that for some reason he did not write the comments which he intended to make on these texts. A possible reason may be that, when he reached these points in writing the speech, he had not yet procured copies of the texts of the two laws, so that, though confident that they would be relevant, he did not yet know their exact wording.

2. One of the best known metaphors in Demosthenes is his comparison of life to an ἔρανος, a kind of loan which friends make to one another; but it is curious that the comparison is made twice in the speech, in passages of similar though not identical wording (**101**, **184–5**). The repetition was noticed by a scholiast, who defends it: 'One should not be surprised that he used the same phraseology also in his other passage rejecting mercy. It is customary for the ancients to do this, and he himself has done it also in the *Philippics* and *Against Timokrates*, and so has Isokrates in the *Panathenaikos* and *On the Antidosis*.'[1] The scholiast undoubtedly has in mind the passages from *On the Chersonese* which are repeated in the *Fourth Philippic* (e.g. 8.41–5 ~ 10.13–17) and those from *Against Androtion* which are repeated in *Against Timokrates* (e.g. 22.47–56 ~ 24.160–8); likewise the two Isokrates speeches mentioned repeat passages from other speeches (Isok. 3.5–9 ~ 15.253–7, 7.14 ~ 12.138). But in all those cases the second use of a passage is in a different speech from the first; not all the listeners will have been present on both occasions, and even those who were had some time to forget the earlier speech before they heard the later. Repetition within a single speech is a different matter. Erbse argues that the ἔρανος comparison is rhetorically suited to each of the parts of the speech in which it occurs.[2] So it is; but if it had been used

[1] Σ **101** (348b Dilts).

[2] *Hermes* 84 (1956) 143–4.

in both, the jury would surely have become restive. The manner of its introduction on the second occasion, 'I consider that all men contribute to loans throughout their life ...' (**184**), is clearly that of a speaker proposing a new idea to his audience for consideration, not one which is already familiar. So I do not believe that **101** and **184–5** would both have been included in a speech actually delivered. Demosthenes may perhaps have written **101** first; then on reaching **184** he thought that that might be a better place for the *ἔρανος* comparison and rewrote it to fit in there (the version in **184–5** is slightly more elaborate than that in **101**), but failed to delete it in the earlier place.[1]

3. In **208** Demosthenes says that he has heard that Philippides, Mnesarkhides, Diotimos, and other men of that sort, rich trierarchs, will plead for Meidias; and he comments on their reasons (to **212**). He then says, as if turning to a new topic, 'A large number of rich men have gathered here ..., and they will come forward to address a plea to you' (**213**). As he goes on, he names some of them, to show whom he means: 'Neoptolemos, Mnesarkhides, Philippides, and those very rich men' (**215**). Here again it is hard to believe that both passages were intended to be used together in the speech, not because Demosthenes cannot make the same point twice (he often does so) but because **213–18** has the air of introducing matters not mentioned before. So **208–12** and **213–18** should be regarded as alternatives, of which he intended to use only one.

I conclude that Demosthenes did not deliver the speech exactly in the form in which we have it. Yet, after all, that does not prove that the speech was not delivered at all. We must bear in mind the manner in which a written speech is likely to have been used. Whereas a modern speaker may bring his written text with him and read it out verbatim, an Athenian speaker seems never to have done that. He maintained a pretence, at least, of speaking extemporaneously, and to preserve that appearance he might even include in his written script a phrase like 'I nearly forgot to mention this' (**110**).[2] After writing the text in advance he would memorize it as accurately as he could. But probably no speaker ever succeeded in reciting a

[1] Batschelet-Massini *Tainia* 1.524 n.50, on the other hand, considers **101** to be an improvement on **184–5** in style and rhythm.

[2] Cf. A. P. Dorjahn *CP* 50 (1955) 193.

long speech from memory without a single change of wording, and only one with very little self-confidence would even try. Most would have the ability to improvise at least a few words if they could not remember the written text exactly, and they might need to do so in response to an unforeseen incident or development of the case in court.[1]

Demosthenes, by the time of his prosecution of Meidias, had studied and practised the art of oratory for years. He was undoubtedly capable of extemporizing at length, and of modifying a prepared speech in the course of delivering it, to suit the reception which he found he was getting from the jurors. For example, the existing text of *Meidias* includes a passage where he reads out a list of Meidias' offences and offers to speak on whichever the jurors wish for as long as they wish (**130**). Thus it was certainly not necessary for him to have every word of his speech written out in advance in its final form. The text as we have it, with incomplete sections and alternative versions, may yet be the final state which the written draft reached before delivery; from this nearly-complete draft a fairly small amount of extemporization may have enabled him to deliver a perfect speech in court. As for publication afterwards, if the speech had been a success and people were clamouring to read it, he may have allowed copies of the written draft to be made at once, without holding it back for further revision.

There is no proof, then, that the speech was not delivered. But neither is there proof that it was. The alternative scenario, that a compromise was reached before the trial came on, remains possible. Unless some further evidence about the trial is discovered, the question must remain open. We can accept, however, that Meidias atoned for his offence by paying thirty mnai, whether that payment was made to Demosthenes privately or as a legal penalty to the state.

IV. STRUCTURE AND RHETORIC

Critics have tried since ancient times to analyse the structure of Demosthenes' speeches, but the results have seldom been very satisfactory, because the speeches do not easily submit to logi-

[1] On the possibilities of extemporizing and of preparing alternative versions of a speech cf. K. J. Dover *Lysias and the Corpus Lysiacum* (1968) 150–1.

cal dissection. Often they flow on from one topic to another in no obvious systematic order and without marked divisions. A popular speech is not like an academic dissertation. Demosthenes does not wish his listeners to analyse his arguments, but to be swept along by them; and it is a part of his genius that he can switch so easily and smoothly from one topic to another. He knows that some listeners have difficulty in concentrating on the same topic for a long time, and their attention may be held more effectively by 'leapfrog' than by 'dogged' tactics. Nevertheless one can usually detect a framework of major themes. The following analysis shows how I consider the various parts of the speech *Against Meidias* may best be distinguished, but this is very much a matter of personal preference; different readers may prefer to analyse the speech in different ways.[1]

I. Introduction.
    1. This trial results from the people's own vote (**1–4**).
    2. The jurors should defend both Demosthenes and themselves (**5–8**).

II. The legal procedure of *probole* for offences at festivals (**8–12**).

III. Meidias' attack on Demosthenes at the Dionysia.
    A. Narrative (**13–18**) and evidence (**21–2**), with remarks about the arrangement of the speech (**19–21**, **23–4**).
    B. The significance of the offence.
        1. Meidias will object, unjustifiably:
            (*a*) The wrong legal procedure is being used (**25–8**).
            (*b*) The dispute is merely a personal one (**29–35**).
            (*c*) There are many precedents for what he did (**36–41**).
        2. Demosthenes argues, on the contrary:
            (*a*) The laws take a strict view of *hybris* (**42–50**).
            (*b*) The offence was also an act of impiety (**51–61**).

[1] For other analyses see Goodwin's edition pp. 127–8, Erbse *Hermes* 84 (1956) 137–9.

(*c*) Meidias' act was particularly serious (**62–76**).

IV. Other incidents between Meidias and Demosthenes.

A. Narrative and comment on specific incidents.

1. The *antidosis* (**77–82**).
2. The prosecution for slander and the disfranchisement of Straton (**83–101**).
3. The accusations of desertion and homicide (**102–22**).

B. Conclusion: these offences concern everyone (**123–7**).

V. Meidias' offences against other persons.

1. The offences are very numerous, but Demosthenes will speak about any that the jury wishes him to (**128–31**).
2. Meidias' attack on his fellow-cavalrymen (**132–5**).
3. The reluctance of victims to prosecute or testify (**136–42**).

VI. Meidias' life in general.

1. Contrast between Meidias and Alkibiades (**143–50**).
2. His claims to have performed liturgies and to have donated a trireme (**151–70**).
3. His tenure of offices (**171–4**).

VII. Various arguments for being severe to Meidias.

1. Previous offenders have been treated severely though their offences were less than his (**175–83**).
2. His previous life has not earned lenience (**184–5**).
3. Pleas from his children should be ignored (**186–8**).
4. It is irrelevant to say that Demosthenes is an orator and has prepared his speech (**189–92**).
5. Meidias' hostility to ordinary citizens (**193–204**).
6. Meidias' supporters at the trial are self-interested (**205–18**).
7. The rule of law must be upheld (**219–25**).

VIII. Peroration (**226–7**).

It will be seen that less than half of the speech is strictly relevant to the prosecution for Meidias' offence at the Dionysia in 348. A good deal is about his general character and other activities, including some which occurred many years before. This is not just because standards of relevance were more lax in ancient courts than in modern ones. It is also because personal rivalry was more prominent in Athenian courts and life. The jurors would feel that their vote was to decide not merely whether Meidias committed a particular act, but whether Meidias or Demosthenes was the better man, more deserving of their support. Demosthenes wants them not just to impose a small penalty for a single incident, but to get rid of Meidias permanently, as being a bad character who does Athens no good.

This aim makes relevant several passages which in a modern court would be irrelevant and inadmissible, especially those numbered V and VI in my analysis. These include a long contrast between the trierarchies and other liturgies which Meidias claims to have performed for Athens, and those which Demosthenes has performed himself (**154–7**). Meidias claims to have donated a trireme for the navy; but he only did it at the last minute so as to have an excuse for not going out to fight on land in the cavalry, and 'if there had been some danger at sea, of course he would have made for the land' (**160–4**). When he did go out with the cavalry, he was preposterously self-indulgent, riding on a silver mule-chair, with fine cloaks and flagons of wine (**133**). His whole life is lived in luxury, with a house so big that it overshadows the neighbourhood, a carriage drawn by a pair of white horses, and an escort of slaves when he walks (**158**). In public offices he is incompetent and ridiculous: when elected to be a cavalry commander, he had no horse of his own to ride, but had to borrow one, and then it turned out that he couldn't ride it (**171–4**). None of this has anything to do with the punch-up at the Dionysia; but it has everything to do with making the jury feel hostility and contempt for Meidias as a man.[1]

And yet, Demosthenes insists, the dispute is not just a personal one. True, he does want revenge for himself, and shows no embarrassment about admitting it.[2] But if Meidias says

[1] On the presentation of Meidias' character see Pearson *The Art of Demosthenes* 105–11. But I do not agree with Pearson that it is 'highly conventionalized'; on the contrary, it is full of details applicable only to Meidias, such as those listed above.

[2] Cf. **207** *μή μ' ἀφαιροῦ τὴν τιμωρίαν*.

'Don't deliver me up to Demosthenes! Don't destroy me because of Demosthenes! Are you going to destroy me just because I'm his enemy?' (**29**), that is an argument which Demosthenes firmly rejects. The person whom Meidias punched was not just Demosthenes the individual, but a chorus-producer performing his official function at a national festival. An attack on him, he argues, was an attack on the people of Athens through their representative, and it is in the people's own interest to impose an appropriate penalty. One of the main objects of the speech is to make the jurors feel that they too, as members of the Athenian public, will be affected by the outcome of the trial. 'Meidias has treated insolently not only me but also you and the laws and everyone else' (**7**).[1]

The two principal ingredients of the speech, as of most law-court speeches, are narrative (διήγησις) and argument (πίστις). But whereas in shorter and simpler speeches the narrative commonly comes first and the argument afterwards, in *Meidias* they are mixed. For listeners who are not very intellectual, narrative is generally more interesting than argument, and Demosthenes holds their attention by distributing it throughout the speech. The account of the events concerning the Dionysia of 348 comes, as one might expect, near the start, immediately after the introduction and the explanation of the legal procedure. It is a fairly straightforward catalogue of incidents (**13–18**); its most remarkable feature is that there is no specific mention of the punch in the face which is supposed to be the main point of the entire case.[2] Possibly Demosthenes considered that the fact was well known, and that a description of himself being punched would diminish his impressiveness in the eyes of the jury. The account of other incidents in the quarrel between Demosthenes and Meidias, though most of them occurred earlier chronologically, is placed later, nearly in the middle of the speech. This is the longest passage of narrative (extending almost continuously from **78** to **120**) and also the most effective; it includes Thrasylokhos and Meidias bursting into Demosthenes' house, and the arbitration conducted by

[1] On Demosthenes' skill at using private and public considerations to support each other cf. Brown *Geschäft* 167–8.

[2] No doubt the general expression κακὰ καὶ πράγματα ἀμύθητα (**17**) is intended to cover the blows. They are also mentioned (as πληγαί) in **1** and **12**.

the unfortunate Straton. Other incidents in the life of Meidias are scattered through the later part of the speech, especially his inglorious military and naval activities; these are shorter passages, and are clearly intended not just to inform the jurors but to entertain them too. There are also some short passages of narrative which do not involve Meidias at all. We hear how Sannion and Aristeides were allowed to take part in choral performances even when legally disqualified (**58–60**), how the arkhon's assessor treated a man taking the wrong seat in the theatre (**178–9**), and so on.

The quality of these passages varies, but some of them are among the most skilful pieces of narrative in the whole of Greek literature. The description even of complicated incidents is nearly always clear and immediately intelligible.[1] But it is not just a matter of clarity. The account is often enlivened by mention of one or two visual features. Meidias' silver mule-chair is the most obvious example (**133**). Others are the dark complexion of Sophilos the pancratiast (**71** *μέλας*), the young girl hearing the curses of the men bursting into her home (**79**), the evening shadows falling as Meidias arrives at the arkhons' office (**85** *ἑσπέρας οὔσης καὶ σκότους*),[2] Meidias giving his hand to the murderer Aristarkhos (**119** *τὴν δεξιὰν ἐμβαλών*) or swaggering through the Agora (**158** *διὰ τῆς ἀγορᾶς σοβεῖ*), Ktesikles holding a whip as he rides in a procession (**180** *σκῦτος ἔχων*), Euboulos sitting silent in the theatre when Meidias calls on him to speak (**206**), and Demosthenes' own cloak slipping from his shoulders as he tries to elude the grasp of Blepaios the banker and is left 'almost nude in my tunic' (**216** *μικροῦ γυμνὸν ἐν τῷ χιτωνίσκῳ*). Such details help the listeners to visualize each scene, and so to believe the story.

Quoted speeches and dialogue can also make a narrative passage vivid, as when we hear Meidias speaking to the Boule: 'Don't you know the facts, Boule? When you've got the perpetrator, are you still delaying and investigating? Are you out of your minds? Won't you put him to death? Won't you go to his

[1] There are rare exceptions: in the account of Euthynos and Sophilos (**71**) it is not immediately clear who killed whom.

[2] Demosthenes uses this effect again in the most famous of all his narratives, the account of the arrival at Athens of the news of the capture of Elateia: *ἑσπέρα μὲν γὰρ ἦν* ... (18.169).

house and arrest him?' (**116**). But Demosthenes makes comparatively little use of this in narrative.[1] His main use of quoted speech is in passages of argument, to propound a view which other people hold, or which Meidias is expected to put forward; Demosthenes then proceeds to demolish it. In most cases the quotation does not consist of words that Meidias has actually used, but is invented by Demosthenes and put into Meidias' mouth: 'Oh, but he donated a trireme; I know he'll drool on about that and say "I donated you a trireme!"' (**160**). An Aunt Sally set up by the orator himself makes an easy target, but it may still be an entertaining way of presenting an argument; Demosthenes may, for all we know, have been able to do a funny imitation of Meidias' voice and manner of speaking. He also has a formidable repertoire of other stylistic devices for presenting arguments in a lively and varied manner: rhetorical questions, exclamations, long periodic sentences with numerous subordinate clauses, very short sentences with asyndeton, carefully balanced antitheses, and so on.[2]

The actual arguments so presented are cleverly interwoven through the speech, but for the most part those concerned with the illegality of Meidias' conduct at the Dionysia in 348 come in the earlier part (**1–76**) and those about the rest of his life and character come later. Detailed criticism of particular arguments may be found in the commentary; here I make only some more general comments.

At the beginning comes the argument based on the *probole* procedure: because the Ekklesia has already voted against Meidias, Demosthenes argues, the jury should do likewise. This argument is not altogether cogent: Meidias could have retorted that the reason why, under this procedure, the Ekklesia's verdict was not final was precisely that its hearing of the case was superficial, and so the jury's duty was to consider carefully whether the Ekklesia's verdict was right or wrong. At this stage Demosthenes adduces two laws, but neither helps his argument much, since the one provides no definition of the offence (**8**)

[1] More effective use of dialogue in narrative can be found in the work of a generally less skilful orator half a century earlier: 'He asked whose baby it was, and they said "Kallias', son of Hipponikos". "That's me!" "Yes, and it's your baby!"' (And. 1.126).

[2] Attention is drawn in the commentary to some notable instances. On Demosthenes' style in general see G. Ronnet *Étude sur le style de Démosthène dans les discours politiques*, but she does not include *Meidias* in her study.

and the other is concerned with different offences from the one that Meidias committed (**10**).

After the narrative of events concerning the festival, better arguments follow, presented mainly in the form of answers to defences or objections which Meidias is likely to raise. This is where Demosthenes convincingly maintains that it is correct for him to use the procedure of a public case (δίκη δημοσία), such as *probole*, because the offence was committed at a public festival against a man holding an official position at it (**25–35**). He argues that it involved *hybris* and *asebeia*, both of which were proper subjects for public cases; he supports this argument by the reading of the law about *hybris* (**47**) and oracles showing the religious importance of choral festivals (**52–3**). Here some of the detailed argumentation will not quite bear the weight that he places on it. An affirmation that barbarians would think the Athenian law about *hybris* a good one if they knew of it (though they actually do not) does not really prove its importance (**48–50**); and evidence that choristers are protected from interference during a festival does not really prove that a chorus-producer should enjoy the same protection (**56–61**). Nevertheless this part of the speech contains some very effective passages, notably the famous lines explaining just why *hybris* is so offensive (**72**), and by the end of it the jurors should have been in no doubt that punching a chorus-producer at the Dionysia was an outrageous thing to do.

To portray Meidias' character in general, Demosthenes uses first the long narrative of the quarrel between the two of them. Here he does not need to add much argument or explanation, since the significance of the events is clear without it. Instead he calls for the evidence of witnesses at several points (**82**, **93**, **107**, **121**). In one instance his use of a witness is particularly clever. He calls up Straton, the man whom Meidias got disfranchised for misconduct of an arbitration. Because of his disfranchisement Straton is not allowed to speak in a lawcourt, and cannot give evidence;[1] so Demosthenes has him stand

[1] At this period the testimony of witnesses was written down in advance and read out at the trial, but the witnesses were present when their testimony was read out (cf. **82** κάλει μοι τούτων τοὺς μάρτυρας, etc.) and presumably confirmed orally that the testimony was correct. This could not be done by a man not permitted to speak, and therefore he could not testify.

before the court in silence, conveying, more vividly than any words could do, the impression that Meidias has gagged him. Thus the speech is enhanced by a striking visual, almost a theatrical, effect;[1] but Demosthenes at once builds on it verbally with a passage of pathos about poor Straton, who 'is not permitted even to tell you whether he has been treated justly or unjustly', and of indignation against Meidias and his wealth and arrogance (**95–6**).

Among the comments which Demosthenes intersperses in his account of the rest of Meidias' life two passages are especially worth noticing. One is the comparison of life to an *ἔρανος* (**101** and **184–5**), already mentioned in connection with the problem of composition (p. 26). Metaphors are not common in Greek oratory, and so extended a metaphor is most unusual. Thus the imagery here attracts special attention to the proposition that Meidias deserves to receive as little favour and consideration as he gives to others. The other striking comparison is between Meidias and Alkibiades (**143–50**). Alkibiades was the most flamboyant figure in Athenian history, whose life-style earned him much disapprobation, but who undoubtedly had great qualities too, especially as a military commander. It would do much for Meidias' case if he could persuade the jury that his own ostentatiousness was likewise counterbalanced by services to the state. But Demosthenes demolishes that argument by declaring that Meidias' offences are worse than Alkibiades', while his public services are negligible. The historical comparison, or rather contrast, is an effective device for diversifying this part of the speech.

Towards the end Demosthenes concentrates increasingly on Meidias as a member of a class antagonistic to ordinary Athenians. It is as an individual that Meidias expresses contempt for ordinary people, and annoys everybody by 'your audacious talking, your posturing, your lackeys and wealth and insolence' (**195**). But he also has rich friends who will support him, and Demosthenes paints a lurid picture of what might happen if *they* ever came to power.

'Just consider, men of the jury: if it should happen (as I hope and am sure it will not) that these men got control of the state, along with

[1] Cf. Σ **95** (321 Dilts) *βούλεται εἰσάγειν ὡς ἐπὶ σκηνῆς τὸ πρόσωπον τοῦ παθόντος*.

Meidias and those like him, and then one of you, the general public, committed an offence against one of them—not the sort of thing Meidias did to me, but anything else—and came before a court manned by these men, what pardon or what mercy do you think he would get? They'd soon do him a favour, wouldn't they, or pay careful attention to the request of an ordinary member of the public! Wouldn't they, rather, say at once "The menace! The pest! That he should be guilty of insolence, and still breathe! He ought to be satisfied if he's allowed to live"?' (**209**).

This is a flight of fancy; Demosthenes is blackening his opponents by inventing their actions and their words. But he may have been exploiting real fears in the Athenian populace. Athenian democracy was never quite secure. Even when it seemed safest, there was always some apprehension that it might be subverted, to be replaced by a tyrant or an oligarchy;[1] and Demosthenes very skilfully awakens this apprehension to support his case.

The ordinary man's protection against such calamities is the rule of law and the punishment of offenders, and this is the theme which Demosthenes chooses for the final climax. He describes each juror walking home at the end of the day, without fear that he may be assaulted on the way (**221**). Why is he so secure? Because attackers are deterred by fear of the laws. But the laws on their own have no strength; they are mere written documents. They derive their strength from the jurors who enforce them. 'The laws get their power from you, and you from the laws' (**224**). As he approaches the conclusion of his speech, Demosthenes rises above the details of his dispute with Meidias, and is telling a universal truth with both pathos and power. His defence of legal justice is as valid today as when he wrote it. Elsewhere we can analyse his cleverness and skill at arguing; at the close we can simply admire a masterpiece of eloquence.

[1] For groundless fear of subversion of the democracy in this period cf. 13.14. Similar fears were common in the fifth century; cf. especially the comment on them in Ar. *Wasps* 488–507.

V. MANUSCRIPTS AND TEXT

Ancient manuscripts

Fragments survive of eleven ancient copies of *Meidias*, and also of two ancient copies of secondary works which quote from it. None of them gives a large amount of the text.[1]

Π1 = P. Heidelberg N.S. 2.207 (Pack² no. 304), 1st century A.D.: parts of **104–5**.
Π2 = P. Oxy. 56.3850, 2nd century: parts of **131–5** and **137**.
Π3 = P. Oxy. 56.3846, early 3rd century: a passage from **6–8**.
Π4 = P. Oxy. 56.3847, 3rd century: a small part of **29–30**.
Π5 = P. Oxy. 56.3848, 3rd century: the left-hand side of a column which contained a passage from **48–51**.
Π6 = P. Oxy. 56.3849, 3rd century: small pieces of **51–6**.
Π7 = P. Oxy. 11.1378 (Pack² no. 306, Hausmann no. 32), 3rd century: parts of **151–4**.
Π8 = Membr. Berolin. 13276 (Hausmann no. 30), 3rd century, a leaf of a parchment codex. It has part of **11–12** on one side and part of **12–13** on the other side.
Π9 = P. Harris 17 (Pack² no. 305), 4th century: a few words of **147**.
Π10 = P. Rainer N.S. 1.8 + P. Whitehouse (Pack² nos. 302–3, Hausmann no. 31), 4th century. These four fragments are re-edited by J. Lenaerts in *Chronique d'Égypte* 42 (1967) 131–6; he shows that they are all from the same papyrus roll. They contain substantial parts of **33** and **39–43**.
Π11 = P. Rainer N.S. 3.47 (Pack² no. 2870), 4th or 5th century. These fragments are re-edited by J. Lenaerts *Papyrus littéraires grecs* (Brussels 1977) no. 11. He shows

[1] In this list 'Pack²' refers to R. A. Pack *The Greek and Latin Literary Texts from Greco-Roman Egypt* (2nd ed., Ann Arbor 1965). 'Hausmann' refers to B. Hausmann *Demosthenis fragmenta in papyris et membranis servata* 2 (Florence 1981). I have not collated any of these thirteen ancient manuscripts myself; I rely on the publications mentioned here, together with information about Π2, Π3, Π4, Π5, and Π6 given to me by Mr. P. J. Parsons.

that there are eighteen fragments from a papyrus codex containing *Meidias* and *Aristokrates*. The *Meidias* fragments are small pieces of **91**, **98**, **99–100**, **102**, **104**, **105**, **110**, **112**, **124**, **127**, **129**, **130**.

Π12 = P. Lit. London 179 (Pack[2] no. 307), late 1st century. The famous papyrus of Aristotle's *Athenaion Politeia* (P. London 131) has near the end of its first roll the beginning of a quite separate text, evidently written before the papyrus was taken by another scribe for *AP*. It is an introduction (ὑπόθεσις) and commentary on *Meidias*, going as far as **11**. It is published in F. G. Kenyon's edition *Aristotle on the Constitution of Athens* (3rd edition, London 1892: not in the first two editions) appendix ii. It is also printed and discussed by Blass *Jb. Cl. Ph.* 145 (1892) 29–33, and the introduction only is printed as fr. 163 in E. Ofenloch's Teubner edition of Caecilius. In the present volume, besides referring in the apparatus criticus to the quotations from the text of the oration, I reproduce the introduction as hyp. v (p. 430).

Π13 = a fragment published by C. Wessely *Studien zur Palaeographie und Papyruskunde* 4 (1905) 111–13 (Pack[2] no. 308), 4th or 5th century, one leaf of a papyrus codex which contained a lexicon to *Meidias*. The first side is occupied by a long quotation from *AP* 53, apparently adduced to explain **83** διαιτητής. The other side contains some shorter notes with lemmata from **161**, **184**, **157**, **145**, **114**, **115**.

Medieval manuscripts

The inventory compiled by Canfora[1] lists 279 manuscripts of Demosthenes, of which 64 include *Meidias* and 5 others include extracts from it. Full collation and classification of all these manuscripts is a massive task which no one has yet undertaken. No previous edition of *Meidias* has used more than about a dozen of them. More comprehensive investigations have been made by Irmer for orations 8 and 9 (*Chersonese* and *Third*

[1] L. Canfora *Inventario dei manoscritti greci di Demostene* (Padua 1968).

Philippic),[1] by Clavaud for the Budé edition of orations 60 and 61 (*Epitaphios* and *Erotikos*), by Passweg for oration 24 (*Timokrates*),[2] and by Dilts for his edition of the scholia. But since a scribe does not necessarily copy different orations from the same exemplar, we must not assume that a relationship between manuscripts established for one oration is valid for another. I have therefore made an independent investigation, in the course of which I have collated the text of *Meidias* fully or partially[3] in 47 manuscripts, including almost complete coverage of those which seem to have been written before A.D. 1400, with a selection of those of the fifteenth century. In many cases my conclusions are similar to those reached by the scholars just mentioned about other orations, but not in every case; and I emphasize here that my statements throughout this chapter, including such phrases as 'all the manuscripts', refer only to *Meidias* and only to the manuscripts which I have examined.

The following are the manuscripts which I have taken into account and the sigla used to refer to them.[4]

A = Munich: Bayerische Staatsbibliothek, gr. 485, early 10th century.[5]

Aa = Alexandria: *Πατριαρχικὴ Βιβλιοθήκη*, 124, dated A.D. 1354.[6]

Af = Milan: Biblioteca Ambrosiana, C 235 inf., 13th or 14th century.

[1] D. Irmer *Zur Genealogie der jüngeren Demostheneshandschriften: Untersuchungen an den Reden 8 und 9* (Hamburg 1972).

[2] R. Passweg *The Manuscript Tradition: Demosthenis In Timocratem, Oration 24* (dissertation, New York University 1975).

[3] Partial collation usually covers **1–55** and **202–27** and selected words or phrases in the rest of the oration. In a few cases different sections of the text have been collated; but they always include both the beginning and the end, in order to check for the possibility that a scribe switched from one exemplar to another in the course of it.

[4] My sigla follow a scheme devised by Dilts in connection with his edition of the scholia, but there are four departures from it. For Laur. plut. 59.9, I retain the siglum P which is traditional in editions of Demosthenes, not L which is used in editions of the scholia. For Urb. gr. 113, I retain the siglum U. For the Vatican manuscripts denoted by two-letter sigla, in order to avoid confusion between capital V and small v, I have substituted X for the latter, so that Dilts's ve and vg are my Xe and Xg.

[5] On the date of A, see Passweg 12.

[6] The Patriarchal Library was unable to supply photographs of Aa, and I am grateful to Mr. E. R. C. Holland of the British Consulate in Alexandria, who kindly photographed it for me.

Aq = Milan: Biblioteca Ambrosiana, Z 129 sup., 14th century.
B = Munich: Bayerische Staatsbibliothek, gr. 85, 13th century.
Bc = Bologna: Biblioteca Universitaria, 3564, 14th century.
Cd = Cesena: Biblioteca Malatestiana, plut. D 27.1, 13th century.
F = Venice: Biblioteca Nazionale Marciana, gr. 416 (=536), 10th century.[1]
Fh = Florence: Biblioteca Medicea Laurenziana, plut. 59.8, 15th century.
Fi = Florence: Biblioteca Medicea Laurenziana, plut. 59.10, 15th century.
Fk = Florence: Biblioteca Medicea Laurenziana, plut. 59.25, 15th century.
Fl = Florence: Biblioteca Medicea Laurenziana, plut. 59.29, 15th century.
Fu = Florence: Biblioteca Medicea Laurenziana, conv. supp. 168, 14th or 15th century.
Fx = Florence: Biblioteca Medicea Laurenziana, acquisti 71, 14th or 15th century.
K = Paris: Bibliothèque Nationale, gr. 2998, 13th or 14th century.
La = Leiden: Bibliotheek der Rijksuniversiteit, BPG 33, 15th century.
Lb = Leiden: Bibliotheek der Rijksuniversiteit, Periz. Q 4, dated A.D. 1457.
Lh = London: British Library, Harley 5670, 15th century.
Ln = London: British Library, Addit. 39617, 15th century.
Lp = London: Lambeth Palace Library, 1207, 13th century.[2]
Mk = Venice: Biblioteca Nazionale Marciana, gr. 417 (=839), 15th century.
Mm = Venice: Biblioteca Nazionale Marciana, gr. 420 (=860), 14th century.[3]
Mp = Venice: Biblioteca Nazionale Marciana, gr. VIII 3

[1] A description of F, with illustration, may be found in E. Mioni and M. Formentin *I codici greci in minuscola dei sec. IX e X della Biblioteca Nazionale Marciana* (Padua 1975) 47–8 with plate 31.

[2] On the date of Lp, see Passweg 96–7.

[3] On the date of Mm, see Passweg 49.

(=1193), signed and dated 13 March 1461 by the scribe George Tzangaropoulos on the last page.

Nb = Naples: Biblioteca Nazionale, II E 12, 15th century.

Nc = Naples: Biblioteca Nazionale, II E 13, 14th century.

O = Brussels: Bibliothèque Royale Albert Ier, 11294–5, 15th century.[1]

Od = Oxford: Bodleian Library, Barocci 73, 14th century.

P = Florence: Biblioteca Medicea Laurenziana, plut. 59.9, 10th century.[2]

Pq = Paris: Bibliothèque Nationale, gr. 2994, 14th century.[3]

Pr = Paris: Bibliothèque Nationale, gr. 2995, 14th century.

R = Paris: Bibliothèque Nationale, gr. 2936, 14th century.

S = Paris: Bibliothèque Nationale, gr. 2934, late 9th or early 10th century.[4]

T = Paris: Bibliothèque Nationale, gr. 2940, 13th century.

U = Rome: Biblioteca Apostolica Vaticana, Urbinas gr. 113, 11th century.

V = Paris: Bibliothèque Nationale, Coislin 339, 15th century.

Vb = Rome: Biblioteca Apostolica Vaticana, gr. 68, 14th century.[5]

Vd = Rome: Biblioteca Apostolica Vaticana, gr. 70, 14th century.

Ve = Rome: Biblioteca Apostolica Vaticana, gr. 71, 15th century.

Vf = Rome: Biblioteca Apostolica Vaticana, gr. 76, 14th century.

[1] Passweg 61, following Wittek, attributes O to a scribe named Thomas Bitzimanos, who also wrote Marc. gr. 572 and Vat. Pal. gr. 81.

[2] On the date of P, see A. Diller in *Serta Turyniana* (ed. J. L. Heller, Urbana 1974) 523. The hand is very like that of the Ravenna manuscript of Aristophanes (Rav. 429), but Diller rejects the suggestion that it was written by the same scribe.

[3] Pq has generally been attributed to the 13th century, but Irmer *Genealogie* 65–6 argues on the basis of its relationship to other manuscripts of orations 8 and 9 that it belongs to the 14th. The evidence from *Meidias* supports this: Pq is derived from Pr (see p. 68) and cannot be dated earlier than Pr. It is curious that Drerup *Sitz. Bay. Akad.* (1902) 300 says that one would regard Pr as the exemplar of Pq (he calls them β and *t* respectively) if Pq were not older, without drawing the obvious inference, that Pq is in fact not older.

[4] S is described in detail by I. T. Voemel on pp. 219–43 of his edition of orations 1–17 (*Demosthenis Contiones*, 1857). A facsimile has been published, edited by H. Omont (Paris 1892). On the date of S, see Passweg 86–7.

[5] On the date of Vb, see J. Irigoin *Scriptorium* 12 (1958) 44–50.

Vg = Rome: Biblioteca Apostolica Vaticana, gr. 927, 14th century.[1]

Vi = Rome: Biblioteca Apostolica Vaticana, gr. 1367, 15th century.

Vs = Rome: Biblioteca Apostolica Vaticana, gr. 2207, 14th century.[2]

Wd = Vienna: Österreichische Nationalbibliothek, phil. gr. 105, 14th century.

X = Florence: Biblioteca Medicea Laurenziana, plut. 59.27, 14th century.

Xe = Rome: Biblioteca Apostolica Vaticana, Palatinus gr. 113, 15th century.

Xg = Rome: Biblioteca Apostolica Vaticana, Palatinus gr. 172, 15th century.

Y = Paris: Bibliothèque Nationale, gr. 2935, 10th century.[3]

Before discussing these manuscripts individually it will be best first to consider two problems which concern the reliability of the tradition in general: the documents included in the oration, and the passages which are 'obelized'.

Documents

As is usual in forensic speeches, Demosthenes calls for documents to be read at several points: laws in five places (**8**, **10**, **47**, **94**, **113**), witnesses' statements in seven (**22**, **82**, **93**, **107**, **121**, **168**, **174**), and oracles in one (**52–3**). In all these places except one (**174**) many of the manuscripts give what purport to be the texts of those documents. But are they the genuine texts? In the nineteenth century many scholars rejected them

[1] The statement of Canfora *Inventario* 59 that Vg (his no. 195) lacks the last part of the oration is not correct. It is merely that some of the leaves have been bound in the wrong order. The sequence is: fos. 161–84, **1–149** (ἀπορρήτους); fos. 185–6, **217** (καταχειροτονήσας)–**227**; fos. 187–9, **198** (εἴτε μή)–**217** (τί οὖν; ὑμῖν); fos. 190–7, **149** (ὥσπερ)–**198** (ἄμεινον).

[2] On the date of Vs, see G. Avezzù *BIFG* 3 (1976) 190, M. L. Sosower *Palatinus Graecus 88 and the Manuscript Tradition of Lysias* (Amsterdam 1987) 16–17. *Meidias* is written by the scribe whom Sosower calls Co[5] and identifies with the scribe who wrote Naples II.F.9, the manuscript of Sophocles and Euripides containing notes attributed to Planudes.

[3] For a discussion of Y, see N. G. Wilson *CQ* 10 (1960) 200–2. He identifies the scribe of the major part, which includes *Meidias*, with the scribe of the Vatican manuscript of Plato (Vat. gr. 1).

entirely;[1] it was suggested that students of rhetoric in Hellenistic or Roman times just fabricated suitable texts, using the information provided by Demosthenes himself in the speech.

This rejection seemed to be confirmed when the study of stichometry revealed that they were not included in the ancient numbering of lines, 'a new and most convincing proof of the spuriousness of these documents'.[2] The standard length of a line (στίχος or ἔπος) of prose in antiquity was about 15 or 16 syllables, and each hundredth line was marked by a letter in the margin (α for 100, β for 200, and so on). The marginal letters were still written at the same points in the text even in copies in which the scribe had actually written longer or shorter lines, so that they could be used for reference (rather as we refer to Plato by the page-numbers of the edition by Stephanus, even though our editions have pages of different sizes).[3] The total number of lines in a work was written at the end; this will have facilitated calculation of sums due to scribes who were paid according to the length of the texts they copied. There is some evidence that the numbering of prose lines was already practised in the fourth century B.C.;[4] so it is possible that this was done for Demosthenes' speeches at the time when copies were first made for distribution and sale.[5]

By the time of the extant medieval manuscripts of Demosthenes the purpose of the marginal letters had probably been forgotten, and they are usually omitted; but some are preserved in S and F (and in B, which is derived from F). The following list gives the first words of the lines beside which the letters appear. The minor discrepancies between S and F evidently arise from differences in line-division.

[1] See especially Westermann *De litis instrumentis*.

[2] Goodwin's edition p. 178. On stichometry in Greek books generally see K. Ohly *Stichometrische Untersuchungen* (Beihefte zum Zentralblatt für Bibliothekswesen 61, Leipzig 1928).

[3] For examples of reference to passages of Greek prose by line-numbers (in multiples of 100) see Diog. Laert. 7.33, 7.187–8.

[4] Theopompos boasted that he had written more than 20,000 lines of epideictic works and more than 150,000 lines of historical works (Phot. *Bibl.* 176.120b–121a = *F. Gr. Hist.* 115 F25). Cf. also Isok. 12.136. K. Ohly *Stichometrische Untersuchungen* 92–4 argues that the practice arose in the fifth century because of the emergence of the book trade at that time.

[5] Cf. J. A. Goldstein *The Letters of Demosthenes* (1968) 9–25.

	S	F
α	**11** καὶ κατὰ	**11** καὶ κατὰ
β	missing	missing
γ	**33** -λιν γε	missing
δ	**43** -βης οὗτοι	**42** (end) -νόντων
ε	**56** -νον οὐκ	missing[1]
ζ	missing	missing
η	**74** συγγνώμην	**74** -νος πολλὴν
θ	missing	**86** -τοι τὸ πρᾶγμα
ι	missing	**98** -γής ἐστι
κ	**106** κράτιστον	**107** μὴν ὡς
λ	missing	missing
μ	**129** πάντα μὲν	missing
ν	**140** οὗπερ	missing
ξ	missing	missing
ο	**161** τόν γε	missing
π	**173** νόμους	missing
ρ	**183** ἀδικοῦντα	missing
σ	missing	missing
τ	**205** ἐπηρεάζειν	missing
υ	**216** ἕλκοντά με	missing
φ	missing	missing

At the end the total number of lines is given, not only in S and F but also in Y, as XXIII, meaning 2003. But clearly there must have been about a hundred more lines of the standard length after the point where υ (meaning 2000) appears, and so the figure must be corrupt. The correct figure may be XXHI or XXHIII, meaning 2101 or 2103 lines.[2]

When we compare the amount of text between each marginal letter and the next, it is obvious that the documents were not counted when the lines were numbered. A glance is enough to check this roughly: each hundred of the ancient lines is equivalent to about ten of the sections into which modern editors divide the speech, but the fifth hundred (ending at ε), for example, extends from **43** to **56** because three sections

[1] The report of Christ *Abh. Bay. Akad.* 16.3 (1882) 167, that B has the marginal letter ε (though F has not), is not correct. He has misread the letter β, which here means δεύτερον and is the beginning of a scholium (168b Dilts).

[2] Cf. Goodwin's edition pp. 177–9.

within it (**47**, **52**, **53**) are occupied by documents. More exact calculations have confirmed this.[1] So it appears that, when the lines were originally numbered, the text did not yet include the documents. (But otherwise it did have essentially its present form; the stichometry shows that, apart from the documents, no major insertions or omissions have occurred in the text.)

Yet this evidence is not conclusive proof that the documents are spurious. We should in any case not expect the laws, testimonies, and oracles to have been included in Demosthenes' own original draft of the speech; he needed to have them on separate sheets or tablets which could be given to the clerk of the court for reading out at the trial. Thus, when the speech was first copied for publication and its lines were numbered, the documents may still have been in a separate dossier, from which they were transferred into the text of the speech only when further copies were made at a later date. Another possibility is that an editor, seeing that Demosthenes called for a particular law to be read out, found that law in the archives, or in a collection of Athenian laws and decrees like the one formed by Krateros (*F. Gr. Hist.* 342), and inserted it in the speech. We must therefore judge each document on its merits, and not condemn them all out of hand.[2]

One of the laws included in this speech is authenticated by external evidence: Aiskhines quotes several phrases from the law of *hybris* in his speech *Against Timarkhos* (1.15), and since those phrases occur in the text given in **47** we can accept this text as genuine. Three of the other laws (**8**, **10**, **113**), though not confirmed by quotation elsewhere, contain details and phraseology which look convincing, and which are not drawn from the text of the speech, and so may be accepted; and the same may be said of the oracles (**52–3**). The remaining law (**94**), however, is open to three kinds of objection: it is about private arbitration, although Demosthenes calls for the law about public arbitration; it omits a number of matters which

[1] Figures based on lines of the Tauchnitz edition are given by F. Burger *Stichometrische Untersuchungen zu Demosthenes und Herodot* (1892) 9, and figures based on lines of S by Goodwin p. 178.

[2] Drerup *Jb. Cl. Ph.* Supp. 24 (1898) 223–47 gives a comprehensive survey of previous discussions of the authenticity of documents in Attic speeches, and rightly concludes that study should be based on the form and content of each document individually.

must have been included in the real law; and some of its terminology does not accord with the normal usage of Athenian laws. Mistakes in linguistic usage can also be found in the statements attributed to witnesses, which should probably all be rejected as spurious. For the details, see the commentary on each document.

Obelized passages

In some manuscripts certain passages of the oration are obelized (it is convenient to use this term to refer collectively to various dashes and other signs placed in the margin opposite each line). These include the documents inserted in the text, and in addition the following passages are obelized in S and F.

S	F
38–42 (... φαίνεται)	**38–41**
49 (καὶ πολλοὺς ...)	
86 (τὴν μὲν ... λαθεῖν)	**86** (τὴν μὲν ... ἀπηνέχθη)
88–92 (... ποιεῖ)	**88–92** (... ποιεῖ)
92 (εἰ γὰρ ...)	**92** (εἰ γὰρ ...)
97 (καὶ μήτε ...)	**97** (καὶ μήτε ...)
99 (ἀλλ' ἴστε ... διδόντας)	**99** (ἀλλ' ἴστε ... διδόντας)
100–1	**100–1**
133–4 (... ἤλαυνες)	**133**
139 (οὓς ...)	**139** (οὓς ...)
143–8 (... ἐνδεικνύμενος)	**143–7**
	189–92
	197–9 (ὃν γὰρ ... θεωρήσετε)
	201 (οὐδὲ γὰρ ...)
205–7	**205–7**
	210 (μὴ τοίνυν μηδ' οὗτοι ...)
217–18 (... ἥττησθε)	**217–18** (... ἡττᾶσθε)
218 (πότερ' ...)	**218** (πότερον ...)

Many of the same passages are obelized in B, which is derived from F. In P only **38–40** (... λεκτέον) and the oracles in **52–3** are obelized; probably the scribe got tired of copying symbols of which he did not know the significance. The same reason explains why other manuscripts show no obelization at all. But S and F clearly derive their obelization from the same ultimate source; the differences between them are small enough to be explained by errors and omissions in copying. Its antiquity is confirmed by a reference in the scholia at **89**, which is one of the passages obelized in S and F: *τοῦτο τὸ μέρος ὠβέλισται παρὰ τῶν κριτικῶν καὶ ὡς ἀδιόρθωτον παραλέλειπται* (307b Dilts). There is a similar reference at **95**: *ὠβέλισται δὲ καὶ ταῦτα* (323 Dilts). Actually **95** is not obelized in the surviving manuscripts,[1] but presumably it was so in earlier ones.

These scholia mean that ancient critics, like modern ones, held that Demosthenes left the speech in an unrevised condition, and they obelized the passages which they thought he would have corrected in revision. Some of the passages are ones that modern scholars have also criticized, notably **100–1**, which is the first of the two passages comparing life to an *ἔρανος* (see p. 26). Others are passages in which, to me at least, no fault is evident, though most are rhetorical sections, the removal of which would hardly interfere with the narrative and argument in the rest of the speech. This may tell us something about the criteria used by ancient critics in assessing the merits of speeches,[2] but it does not give us any additional information about Demosthenes.[3]

The older medieval manuscripts (SAFYP)

When we compare the two oldest medieval manuscripts, S and A, we find a considerable number of differences between them. Many are small discrepancies which could have arisen by errors in copying at any period, but one is very substantial: S includes all the documents, whereas A omits them. Two explanations of this fact are possible. Either the scribe of A, or of an earlier manuscript from which A is descended, had before him an exemplar which included the documents, and he left them out of his copy; or the direct line of descent from Demo-

[1] Blass (p. xxxvi of his edition) and Christ *Abh. Bay. Akad.* 16.3 (1882) 180 state that this passage is obelized in B, and this statement is followed by other scholars (Goodwin, Sykutris, Erbse), but the photographs of B in my possession do not show any obeli here.

[2] For an attempt to reconstruct the criteria see Erbse *Hermes* 84 (1956) 144–5. Cf. also Christ *Abh. Bay. Akad.* 16.3 (1882) 179–82.

[3] Blass *Beredsamkeit*[2] 3.1.338 maintains that obelization means that the passage was omitted in another manuscript; but there is no evidence that an abridged version of this oration existed.

sthenes' autograph to A never included the documents, which were first inserted in an ancestor of S after the two branches of the tradition had already diverged.

The former explanation, that a scribe deliberately jettisoned parts of the text which he had, is in my opinion very improbable. I therefore accept the alternative explanation, which implies that the division of the tradition occurred early.[1] Some of the documents are genuine (see p. 46) and their insertion in the text of the oration is not likely to have occurred later than the Hellenistic period. I postulate that in that period, perhaps in the third century B.C., there existed two copies (at least) of the oration, and the documents were added to one of those copies and not to the other.[2] S is descended from the former and A from the latter. However, that does not preclude the possibility that some contamination occurred later: some variant readings may have been copied from an ancestor of S into an ancestor of A, or vice versa.

The evidence of the ancient manuscripts, though not sufficient to be decisive, gives some support to the hypothesis that the two branches of the tradition diverged in antiquity, not merely in the middle ages; for some of them seem to agree consistently with one branch rather than the other. Notably, Π6, of the third century A.D., besides agreeing with A at other points, omits the documents in **52–3**.

In the nineteenth century, when S was first studied critically, the view became prevalent that it preserved a superior text, and that its readings should normally be given preference over those of A and other manuscripts. The culmination of this tendency may be seen in Cobet's affirmation 'Nil curo reliquos libros Demosthenis dum Parisinum S habeam'.[3] Only comparatively recently has this view been questioned by Erbse and Irmer.[4] The belief that S represents a superior tradition is

[1] As yet discussion of the problem whether the archetype (i.e. the latest common ancestor of the surviving manuscripts) of the Demosthenic corpus as a whole was ancient or medieval has not reached any generally agreed conclusion. See Drerup *Philol.* Supp. 7 (1899) 533–51, Pasquali *Storia della tradizione* 269–78, Erbse *Gesch. der Text.* 1.262–4, Irmer *Philol.* 112 (1968) 43–62.

[2] The spurious documents could, of course, have been added at a later date than the genuine ones.

[3] C. G. Cobet *Variae lectiones* (2nd edition, 1873) p. xxiv.

[4] Erbse *Gesch. der Text.* 1.263, Irmer *Philol.* 112 (1968) 43–7 and *Genealogie* 95.

based partly on scholia on two passages of *Meidias* in which the manuscripts' readings are as follows:[1]

> **133** ἀργυρᾶς τῆς ἐξ Εὐβοίας SP$_4^{\gamma\rho}$: ἐξ Ἀργούρας τῆς Εὐβοίας AF: Ἀργούρας τῆς ἐξ Εὐβοίας YP
> **147** ἱερὰ S: ἱερὰν ἐσθῆτα AFYP

A scholiast at **133** (469a Dilts) explains ἀργυρᾶς and goes on to say ἡ δὲ δημώδης "ἐξ Ἀργούρας" ἔχει. At **147** (508 Dilts) he (probably the same scholiast) remarks "ἱερὰ" μόνον ἡ ἀρχαία ἔχει. These scholia have been taken to mean that S gives us the ancient (ἀρχαία) text, while A and other manuscripts give the vulgate (δημώδης). Clearly the scholiast did have two copies of the oration before him, and maybe one was an ancestor of S and the other an ancestor of A; but in what sense was only the former 'ancient'? It may have been merely accidental that it was an older copy than the other. What the scholia do prove is that in these two passages the readings of A were current in antiquity as well as the readings of S. As it happens, this is confirmed in both cases by testimonia: the reading ἐξ Ἀργούρας τῆς Εὐβοίας is quoted by Hdn. Gr. 2.920.9 and Macr. *Sat.* 5.21.8, and the reading ἱερὰν ἐσθῆτα is quoted by a scholiast on Hermogenes (Walz 4.546.8–10). So this evidence confirms, not refutes, the hypothesis that the branches of the tradition represented for us by S and A had already diverged in ancient times. But even if S is right in these two places, that does not show that, where they differ, S is always right and A is always wrong.

Another notable difference between S and A is that in some passages one or two inessential words are included in A but omitted in S. Words included in S but omitted in A are rarer. Here are some examples (not a complete list).

> **13** τρίτον ἔτος τουτί S: τρίτον ἢ τέταρτον ἔτος τουτί A
> **18** δικαιοτάτους S: δικαιοτάτους καὶ πιστοτάτους A
> **20** πλοῦτον S: πλοῦτον καὶ ὕβριν A
> **20** ἠδίκει S: ἠδίκει τότε A
> **21** πάντων οὖν S: πάντων οὖν τούτων A
> **37** πάντας S: πάντας τοὺς ἄλλους A
> **45** ὅτι S: διὰ τί; ὅτι A

[1] Cf. Pasquali *Storia della tradizione* 280–1.

54 καὶ ἀγαθαί S: καλαὶ καὶ ἀγαθαί A
57 τὸν δὲ χορηγὸν S: τὸν δὲ χορηγὸν αὐτὸν A
97 παράδειγμα S: παράδειγμα τοῖς ἄλλοις A
106 τἀναλώμαθ' S: τὰ ἀναλώματα πάντα A
141 τάχα τοίνυν S: τάχα τοίνυν ἴσως A
158 κυμβία καὶ ῥυτὰ καὶ φιάλας S: κυμβία καὶ ῥυτὰ καὶ φιάλας καὶ τὰ τοιαῦτα A
203 φλαῦρον S: λυπηρὸν ἢ φλαῦρον A
209 οὐδ' ἔσται S: οὐδ' ἔσται ποτέ A

It appears that either a scribe somewhere in the succession of copyists leading from the archetype to S had a tendency carelessly to omit a word from time to time, or a scribe in the succession leading to A had a tendency to add extra words. At first sight one may think omission an easier and likelier fault than addition. Yet if S's omissions were simply due to hasty copying, one would expect them sometimes to remove more important or indispensable words. In fact most of the sentences concerned are quite satisfactory in S's version, while several of A's additions look like attempts to construct commonplace phrases, not all of which are quite appropriate to the context. Thus in **13**, where we expect Demosthenes to give an exact date at the beginning of his narrative, A gives 'two or three years ago', a phrase remembered perhaps from the *Third Olynthiac* (3.4). In **20** it is not appropriate to include *hybris* in a list of circumstances with which Meidias is surrounded (περὶ αὐτόν), given as a reason for not prosecuting him for acts of *hybris*. In **45** S gives us 'Because ...' and A adds 'Why?' before it; Demosthenes does indeed often say 'Why? Because ...', but that is not appropriate in this particular place where the preceding sentence already includes τί ...; meaning 'Why?' In **54** the particle καὶ in S's text links ἀγαθαί to πολλαί: A inserts καλαί, remembering that καλὸς κἀγαθός is a common phrase, but that phrase of personal commendation is not very suitable for oracles, which are the subject of this sentence. So it seems probable that most of the extra words in A are gratuitous additions, and in this respect it may be right to say that S on the whole preserves a purer text.

Where does F stand in relation to S and A? It seems to occupy an intermediate position; it sometimes agrees with S and sometimes with A, and it cannot be wholly derived from either one of them. It contains the documents, the stichometry, and the obelization, which it cannot have got from A; yet it also contains many correct readings which are found in A but lost in S. Where it differs from both S and A, it is oftener wrong than right. This evidence suggests the possibility that the exemplar from which it was copied was a copy of S (thus including the documents) to which variant readings from A had been added above the lines or in the margins; the scribe of F then copied this out, adopting the variants where he thought fit, and introducing new errors of his own. There is a good example of the conflationary character of F in **222**: where S has *περιεῖναι* and A has its synonym *ζῆν*, F has both, making nonsense.

If this were the true and complete picture, we could dismiss F as having no independent value for establishing the text of Demosthenes. However, there are some passages where F differs from both S and A and is probably or possibly right; so we must consider further whether it has a true line of inheritance from antiquity independently of S and A.

4 *παρήγγελκεν* F: *περιήγγελκεν* SA
43 *ἀποκτιννύντας* F: *ἀποκτειννύντας* SA
52 *πυκάσαντας* F: *πυκνάσαντας* S: om. A
66 *οὕτως* (ante *ἀλόγιστος*) F: *οὗτος* SA
66 *οὕτως* (ante *ἄθλιος*) F: *οὗτος* A: om. S
77 *ἂν* F: om. SA
79 *ἂν* F: om. SA
91 *ἀλλ'* (ante *οὐδὲ*) Π11 F: om. SA
104 *οὔτ' ἄλλο οὐδὲν* F: *οὔτ' οὐδὲν* S: *οὔτ' οὐδὲν ἄλλο* A
127 *ἡμῶν* F: *ὑμῶν* S: *ἐμοῦ* A
129 *ὑπομείναιτ'* F: *ὑπομειτ'* S: *ὑπομενεῖτε* A
137 *τὸν τούτου* F: om. SA
145 *καὶ στέφανοι* F: om. SA
149 *τὰ* (ante *ἐναντιώτατα*) F: om. SA
167 *ἀργύρεια* F: *ἀργύρια* SA
168 *καὶ* (ante *Μειδίας*) F: om. SA
168 *δὲ τοῦ* F: *δεκάτου* S: om. A
169 *καταλαζονεύσεται* F: *καταλαζονεύεται* SA
192 *διημάρτανε* F: *διημαρτάνει* S: *διήμαρτε* A
205 *μείζων* F: *μεῖζον* SA

208 πέπυσμαι F: πέπεισμαι SA
216 δ' (post ἐκεῖνον) om. F: δ' SA
216 δὲ (post ἐπειδὴ) F: om. SA[1]

Not all of these are impressive. More than half could easily be explained by saying that the scribe of F (or the scribe of the exemplar from which F was copied) corrected on his own initiative a small error which he found in S and A. In some of the other cases it may be that F's reading, though it looks plausible, is not in fact correct. But I find it hard to suppose that every case in this list can be explained in those ways; it is easier to believe that F did somehow manage to derive some information from a source other than S and A. On the other hand, I do not think it likely that F is the heir to a third independent transmission of the text from antiquity; its resemblances to S and A, in errors as well as in right readings, are too close for that. I still think that F's text is a conflation of the two traditions; but perhaps its exemplar was based not on S and A themselves but on manuscripts closely related to them (the exemplars from which S and A respectively were copied, or other copies of those exemplars) which had a few true readings which S and A have lost.

A similar problem arises when we turn to Y and P. About these two manuscripts one thing is clear: they are twins, two copies from the same exemplar. That is shown by the fact that some new errors (not in SAF) occur in both Y and P, some in Y only, and some in P only. (Examples may easily be found in the apparatus criticus.) What is less clear is the character of that lost exemplar. But, as in the case of F, it seems to have been essentially a conflation of the two main lines of tradition. Where the earlier manuscripts diverge, Y and P often agree with S, often with F, and less often with A; but they rarely give a good reading not found in SAF, perhaps only in the following places.

52 ἐρεχθείδαισιν YP: ἐρεχθίδεσσι S: ἐρεχθίδεσσιν F: om. A
52 λατοῖ YP: λατου SF: om. A
53 τρεῖς YP: τρὶς SF: om. A
74 οἷ YP: οὗ SAF

[1] To this list one might add the fact that in a few places F has obeli not preserved in S; see p. 47.

94 ἀπὸ YP: ὑπὸ SF: om. A
121 αὐτοῦ YP: αὐτὸν SF: om. A

It is notable that all but one of these are in the documents which A omits entirely. Perhaps that is simply because the documents, transmitted by only one line instead of two, had more passages in which the true reading had been lost and so gave more opportunities for conjectural emendation. (The oracles in **52–3** especially contain a good deal of corruption.) At any rate it seems quite probable that all these readings are due to conjecture by the scribe of the exemplar of which Y and P are copies, or else just to accident. The possibility that that scribe had access to a now-lost manuscript of independent authority, though it cannot be absolutely excluded, is more remote than in the case of F.

All that I have said so far about SAFYP refers to the text as originally written by the scribe of each manuscript. But there are also corrections and additions by various hands. In A and Y corrections are few and negligible, but in F and P they are very numerous, and there are also some in S. In some places the corrector has erased or deleted the original text, showing that he believed it to be wrong; in others he has merely added an alternative reading above the line or in the margin, with or without an abbreviation of *γράφεται*, not thereby committing himself to a decision on its correctness.

It is difficult to distinguish the correctors. In some places it is clear that the correction has been made by the original scribe, who has just noticed that he has copied his exemplar wrongly, so that we can treat the correction as the authentic text and ignore what was written at first. More often it is clear that the correction is by a different hand. We must then wonder whether the corrector has given us the reading of another manuscript or merely made a conjecture of his own; only if he adds the note *γράφεται* can we be fairly confident that it is the former. In many places it is impossible to tell whether the corrector is the original scribe or another, especially when the correction consists simply of erasure of one or more letters. Erasure also sometimes (though not always) makes it impossible to tell what the original reading was.

The corrections in S were examined in minute detail in the

nineteenth century by C. Graux and L. Duchesne, whose findings are reported by Weil. These French scholars were able to devote long study to the manuscript in Paris. Besides the original scribe, they distinguish three hands not later than the twelfth century and acknowledge the possibility of another of that period, besides other hands of later periods. I have assigned the sigla S_2 and S_3 respectively to the hands that Weil calls 'le réviseur ancien' and 'le réviseur proprement dit', while his 'quatrième main' is S_4.[1]

Corrections and variant readings in F are mostly to be attributed to the original scribe. Some are in his usual minuscule script. Others are in uncial letters, written very small, which may at first sight seem to be from a different hand. But they are not; the scribe's reason for using uncials, it seems, is merely that these letter forms are clearer when small writing is wanted. The best evidence that they are from the first hand is the fact that they were written before the obelization was added to the manuscript (as may be seen at **133**, **143**, and **189**, where the obeli are arranged around the uncial additions in the margin). But the commonest type of correction in F is simple erasure. In particular, someone has conceived the false belief that movable nu ought not to be written when the next word starts with a consonant, and he has gone through the text erasing this kind of nu systematically; he has missed a few instances, and in at least one place has misidentified a nu as movable (**79**, where he has erased the first nu of *τότε ῥηθέντων*, presumably taking *ερηθεν* as a verb!). Is there any way of telling who made these erasures? Sometimes he extends a line of the preceding letter (usually the middle bar of *ϵ*) to cover the gap left by the erasure, and since the ink with which this is done is indistinguishable from the original scribe's ink, I think it likely, though it cannot be called certain, that he was responsible for the erasures too. Yet there are a few places (e.g. **124** *πᾶσιν*) where the original scribe at first wrote a movable nu which the other manuscripts do not have. I conjecture that the scribe at first had no objection to movable nu, but after he had written

[1] See p. ii of Weil's edition. Drerup *Sitz. Bay. Akad.* (1902) 289–91 describes the hands in S generally, without specific reference to *Meidias*. In the present volume I have accepted Weil's information, with a few additions which I have made on the basis of the published facsimile; I have not studied these hands in the manuscript itself.

the text someone (perhaps his supervisor) told him that it should be removed before a consonant, and he then revised his text, not only erasing 'wrong' nus but also comparing his text with another and adding corrections and marginal variants. However, since this cannot be proved, I attribute the readings of F after an erasure merely to F^{pc} ('post correctionem') and not to a specific hand. There are few corrections and variant readings in F which cannot be attributed to the original scribe. The only important one is **53** ἐλινύειν, which is likely to be the true reading, not preserved in other manuscripts, though whether the corrector got it from a lost manuscript or by his own conjecture I cannot say. It is written by the hand which Drerup calls F_2 in his account of the various hands responsible for the scholia.[1]

P has undergone correction by several different hands. The earlier corrections are in brown ink; many of these are clearly made by the scribe who added the main body of scholia to the manuscript,[2] though some are probably by other hands. I call all the corrections in brown ink P_2. A later corrector, whom I call P_3, writes in bold black ink. Some of his interventions (which are mostly in the first half of the oration) are merely clarification of letters which he found hard to read, or addition of breathings and accents which the original scribe omitted, but in some places he does alter the text. Another corrector writes in grey ink, and is probably later again.[3] He adds some scholia, and he occasionally changes the text, but more often he just adds an alternative reading above the line of text or in the margin. A few of the corrections in grey ink may be due to yet another hand, but since these are not easily distinguished I call all the grey corrections P_4.

It is not easy to assess the importance of all these corrections. Many of them are identical with readings which we have in

[1] Drerup *Philol.* Supp. 7 (1899) 560–3 distinguishes six hands responsible for corrections and scholia in F, the first of which is the one that I identify with the original scribe. The scholia are mostly written by F_2 and F_4, and the other hands probably make little or no contribution to *Meidias*, though there are a few corrections too small for the hand to be confidently identified. Avezzù *BPEC* 27 (1979) 56–8 maintains that the hands which Drerup calls the third and fourth are actually the same, but Dilts in his edition of the scholia (vol. 1 p. vii n. 3) supports Drerup in distinguishing them.

[2] Called P^1 by Dilts *TAPA* 104 (1974) 98.

[3] Called P^2 by Dilts *TAPA* 104 (1974) 98.

other manuscripts anyway. But sometimes they offer unique readings which, especially when marked γράφεται, are evidently drawn from manuscripts now lost. Many of these variants appear to be wrong, but some could be correct; perhaps the most impressive is **132** τῷ νῦν, added above the line by an unidentified corrector in S. Although most of the good variants could be medieval conjectures, the possibility remains open that some of the correctors had access to older manuscripts not dependent on any of the manuscripts now extant.

The later medieval manuscripts

Some of the later manuscripts are derived from A, some from F, and some from Y; none, it seems, from S or P. (Hence some editors classify the manuscripts in four families. They regard S as the sole member of one family, and P as belonging to the same family as Y.) It is convenient to consider them in these three categories, though the picture is complicated by contamination: some manuscripts contain readings drawn from more than one source. We can proceed by the usual method of looking for conjunctive errors and separative errors, provided that this method is used with caution. Some kinds of confusion (e.g. between ἡμεῖς and ὑμεῖς, ἔβαλλον and ἔβαλον, ἡγεῖσθαι and ἡγῆσθε) are so common in Byzantine manuscripts that they cannot be used for this purpose, and even a less common mistake may happen to be committed by two scribes independently. Conversely, an intelligent scribe may correct an obvious error in his exemplar (as in **72**, for example, all copies of Y correct its error ἀπεγγέλλων). But most of these manuscripts have enough substantial errors to make their relationships clear.

Since the later manuscripts are not fully reported in the apparatus criticus, I provide in this chapter some samples of their errors. These are not complete lists, but a selection of about half-a-dozen readings in each case, which seem distinctive enough to be used as evidence for or against a relationship. The errors of A, F, and Y are not listed here, because they may be found in the apparatus criticus.

(*a*) Manuscripts derived mainly from Y

U is the oldest manuscript after SAFYP. It is difficult to use because of its present condition. The brown ink has faded throughout, so that none of it is easy to read; at some time water has got into the book and washed out the lines near the foot of each page, making some words entirely illegible. A corner of one leaf (fo. 129) has got torn off, removing parts of **154** and **157–8**. There are a few corrections by at least two different hands; for example, in **15** the word *ἀκούσαντες*, which is in the text in A and F, is added in the margin in U, as it is also in S and P. In general the text follows that of Y, and there are some conjunctive errors with Y alone of the older manuscripts.[1]

24 *ἀπαντᾶν*] *ἁπάντων* YU
32 *προσυβρίζει*] *προυβρίζει* YU
41 *γ' ἔνι* om. YU
70 *εἰς*] *πρὸς* Y: *προ* U
133 *συμβαλουμένους*] *συμβαλομένους* Y^{pc}U
153 *λέγειν ἐν ἁπάσαις ταῖς*] *ἐν ἁπάσαις λέγειν ταῖς* Y: *ἐν ταῖς ἁπάσαις λέγειν* U

There are also many new errors, of which the following are merely a few samples.

1 *τὴν* (ante *χορηγίαν*) om. U
2 *ὠργίσθη καὶ παρωξύνθη*] *ὠργίσθη καὶ παρωργίσθη καὶ παρωξύνθη* U
24 *τοῦτον*] *καὶ τούτων* U
44 *ἐπέταξεν*] *ἐποίησεν* U
105 *ἔλεγεν* om. U
132 *ἐγὼ πυνθάνομαι* om. U

The worst aberrations are a long duplication and a long omission. The passage from **141** *ἢ τί δὴ πάλιν* to **146** *πρόγονοι* is written out twice; the second copy has been scored out and *σφάλμα* written in the margin. The passage from **220** *ἴσως ἐμέ* to **227** *τοὺς ἄλλους σω-* is omitted; here the scribe shows no consciousness of omission, but writes *μισεῖ μειδίας φρονίσαι* and

[1] Contrast Irmer *Genealogie* 55–6. He implies that U is related to A in all the orations which it contains, but in *Meidias* I have found no conjunctive errors of U with A.

so on without, it seems, being aware that the words do not make sense. The likeliest explanation of these two blunders is that in the latter place the scribe inadvertently turned over two leaves of his exemplar together, and in the former place thought he had turned over a leaf when he had not. But these passages duplicated or omitted in U do not begin and end at points where pages of Y begin and end. This suggests that U is not copied directly from Y, but from an intermediate lost manuscript which was a copy of Y and did have pages beginning and ending at those points.

On the other hand, there are a few places where Y has an error and U the correct reading.

7 *τὰ τοιαῦτα* U: *τοιαῦτα* Y
91 *δι' ἣν* U: *δι' ἧς* Y
93 *ἀποδιαιτήσομεν* U: *ἀποδιαιτήσωμεν* Y
107 *προσεξείργασται* U: *προσεξείργασθαι* Y
139 *πολύευκτος* U: *πολύευκτον* Y
139 *ἑταιρεία* U: *ἑταιρία* Y
174 *ἱππαρχῶν* U: *ἱππάρχω* Y
188 *ἡγεῖσθε* U: *ἡγεῖσθαι* Y

And in **184**, where Y has *τοὺς δεομένους* and U is now illegible, Ve (a copy of U: see below) has the correct reading *τούσδε μόνους*, which means that U probably had it too. But all these corrections are very small and could have occurred by accident; some are very obvious in their context and could have been made by the scribe's own conjecture. They are not sufficient to prove that he had evidence independent of Y. I conclude that U is copied from a lost copy of Y.

Ve is a copy of U (made before U was damaged, for it includes words now illegible in U). It reproduces U's distinctive errors, including the long omission at **220–7**, and makes only trivial corrections. Notable among its new errors is that U's abbreviation of *ὦ ἄνδρες Ἀθηναῖοι* is persistently misinterpreted as *ὦ Ἀθηναῖοι*. I give a few other examples.

13 *τῆς φυλῆς, τῶν δ' ἐπιμελητῶν* om. Ve
17 *παρασκήνια φράττων, προσηλῶν, ἰδιώτης ὢν τὰ* om. Ve
28 *ὡς οὐ*] *υἱοῦ* Ve
38 *καὶ οὐκ ἐπὶ τούτου μόνον, ἀλλ' ἐπὶ πάντων φαίνεται προῃρημένος με ὑβρίζειν* om. Ve

158 ἐπισκοτεῖν] ἐπισκοσκεῖν Ve
216 Ἀθηναῖοι (post ποιήσεις)] ἆθλια Ve

O is derived from Y independently of U. It repeats all the substantial errors of Y, though some of them have subsequently been corrected by another hand, no doubt by reference to some other manuscript. O also has some new errors, not found elsewhere, showing that none of the other manuscripts investigated is derived from it.

6 οἶδ'] εἶδ' O
46 οὐκ ἔστιν om. O
54 καὶ αὗται om. O
211 δεήσονται] δεηθήσονται O
218 εἶναι (post ἀνεξέταστον) om. O
220 μὴ (ante τοίνυν)] καὶ O

Af is another copy of Y. It reproduces Y's errors and adds only a few more of its own.[1]

5 φυλῆς] φυλακῆς Af
10 καὶ ἐπὶ Ληναίῳ πομπὴ καὶ οἱ τραγῳδοὶ καὶ οἱ κωμῳδοί om. Af
28 and **40** κατηγόρηκα] κεκατηγόρηκα Af
203 πρῶτος] πρῶτον Af
215 προσελθόντος] προελθόντος Af

Vb is a more careless copy of Y, adding to Y's errors a substantial number of new ones.[2] The following are only a small selection.

5 νομίζων τῷ μὲν κατηγόρῳ περὶ τῶν τοιούτων προσήκειν ἐλέγχειν μόνον, τῷ δὲ φεύγοντι καὶ παραιτεῖσθαι om. Vb
18 τὸ σῶμα om. Vb
19 μιαροῦ] καθαροῦ Vb
53 θῦσαι] σῶσαι Vb
105 προσηλῶσθαι] προσκυνεῖσθαι Vb

[1] Cf. Passweg 36: 'extremely accurate in the copying of the text'.

[2] Corrections in Vb of errors in Y are ones which could easily have occurred by accident (e.g. **66** ἤ Vb: εἰ Y, **153** ὑμῖν Vb: ἡμῖν Y). They do not support the suggestion, made by Passweg 40, that another manuscript intervened between Y and Vb, though the possibility cannot be excluded.

172 *ἀνανδρίαν καὶ πονηρίαν*] *πονηρίαν καὶ ἀνανδρίαν* Vb

Vd in turn is a careless copy of Vb, adding many more errors to Vb's and thus presenting a very corrupt text.

1 *ἄλλα* om. Vd
2 *καλῶς* om. Vd
26 *ἓν μὲν ἐκεῖνο εὖ*] *ἐκεῖνο εὖ μὲν* Vd
55 *ἀσεβεῖν*] *εὐσεβεῖν* Vd
206 *ἡγεῖτο* om. Vd
211 *ἀποκαλεῖ*] *φησὶν* Vd

Vg is also derived from Y. It reproduces most of Y's errors, but it does correct a small number of them. These improvements on Y each involve only one or two letters, and it is hardly possible to say whether they are due to accident, intelligent emendation, or checking against another manuscript; in all cases they merely restore a reading which we have in other older manuscripts (e.g. **28** *ἐνέγκοι μοι*, **183** *ἀποκτενεῖτε*). There are also many new errors.

11 *γιγνόμενα*] *γιγνόμενα χρήματα* Vg
32 *κοινὸν στέφανον*] *στέφανον κοινὸν* Vg
52 *χάριν* om. Vg
166 *τινα* om. Vg
181 *εἷλεν*] *εἷλκεν* Vg
223 *δικάζοντες*] *βαδίζοντες* Vg

Vs is yet another manuscript derived from Y independently of those already mentioned. It reproduces Y's errors, but in some cases a variant reading is written above by a different hand (e.g. in **20** *λαμβάνειν* has *βεῖν* written above, and *ἠδίκει καὶ* has *τότε* written above). These variants correspond to readings of A and are presumably taken from a manuscript derived from A. Vs also has its own crop of separative errors.

11 *τοὺς* (ante *ὑπερημέρους*) om. Vs
28 *τῇ πόλει παραχωρῶ τῆς τιμωρίας*] *τῆς πόλεως παραχωρῶ τιμωρίας* Vs
43 *αἰδέσεως*] *ἀνέσεως* Vs
208 *τὴν χάριν ταύτην*] *ταύτην τὴν χάριν* Vs
218 *οὐδὲ* (ante *λαθεῖν*)] *οὐδὲ γὰρ* Vs
227 *διάγειν*] *παράγειν* Vs

Bc, **Nc**, and **T** are derived from the same hyparchetype, as numerous conjunctive errors show.

10 *ἡ πομπὴ* (post *Διονυσίοις*)] *οἱ πομποὶ* BcNcT
18 *τὸ σῶμα, τῇ φυλῇ*] *τῷ σώματι φυλῇ* Bc^{ac}NcT
59 *ὑμῶν* om. BcNcT
96 *Μειδίου καὶ* om. BcNcT
108 *καὶ θεῶν* om. BcNcT
134 *κατεσκεύαζον*] *καὶ κατεσκεύαζον* BcNcT
182 *πολλῷ τούτων*] *πολλῶν* BcNcT
211 *βοηθήσητε*] *κριθήσεται* BcNcT

As far as **207** the hyparchetype reproduced the errors of Y, but at that point the scribe must have changed to a different exemplar. The rest of the oration is derived from A or from a manuscript closely related to A. The first clear evidence of this switch is in **208**, where BcNcT all have *τοῦτον* (after *ἐξαιτήσεσθαι*), not *αὐτόν* (after *παρ' ὑμῶν*).

Bc and Nc each have a large number of separative errors, though in one place an error in Bc has by lucky accident restored a correct reading which the older manuscripts have lost (**53** *δημοτελῆ*). I list only a selection of the worst blunders.

57 *συγκόψας*] *σκώψας* Bc
72 *ὀργήν*] *πληγὴν* Bc
103 *κατεσκεύασεν*] *ἐμισθώσατο* Bc
166 *καὶ λιποτάξιον* om. Bc
172 *καὶ μὴν εἴ τις αὐτοῦ ταῦτα ἀφέλοιτο "ἱππάρχηκα, τῆς Παράλου ταμίας γέγονα"* om. Bc
197 *πάλιν νῦν μείνας πρὸς τοὺς ἐξεληλυθότας τοῦ δήμου κατηγορήσει* om. Bc

2 *σφόδρα*] *σφόδρα οὕτως* Nc
4 *ἡγῆται*] *ἡγῆται· οὐ γὰρ ἂν καταγνοίην* Nc
135 *ἔμοιγε*] *εἶναι μοι* Nc
203 *ἐμὲ οἴεσθε ὑμῖν*] *ἐμὲ οἴεσθε ἐμὲ* Nc
205 *ἤδη* om. Nc
227 *τὰ πεπραγμένα* om. Nc

A peculiarity of T is that the scholia, instead of being written in the margins in the usual way, are inserted between sentences of the text in ink of a different colour. This manuscript has fewer errors than Bc and Nc, and some of those made by the original scribe have been corrected by another hand, which also adds some variant readings; although the source of these cannot be identified, they seem all to be readings which we also have in other manuscripts. After correction T has very few and small separative errors, but they are sufficient to show that Bc and Nc are not copied from it.

29 *τῶν* (ante *λόγων*)] *τὸν* T
52 *πατρίοισι*] *πατρίῃσι* T
94 *ἕτερον*] *ἑτέρου* T
204 *θαυμάζεις εἰ κακὸς κακῶς*] *θαυμάζῃς εἰ κακῶς κακῶς* T
205 *ὑπὲρ*] *οἱπὲρ* T
207 *δύνασαι*] *δύναται* T

La, on the other hand, is a copy of T after correction, incorporating variants added to T by the second hand. Errors remaining in T are usually reproduced by La, but in certain cases the scribe has tried to correct them. Sometimes he is successful and sometimes not, as may be seen in two examples in which he plumps for the accusative in preference to the genitive.

29 *τῶν λόγων*] *τὸν λόγων* T: *τὸν λόγον* La
94 *ἕτερον δικαστήριον*] *ἑτέρου δικαστήριον* T: *ἕτερον δικαστήριον* La

La also has corrections and variants written in by a second hand, though in this case too they are readings which we also find elsewhere. But even after all this work La still has some new errors left uncorrected.

13 *χορηγοῦ*] *χοροῦ* La
47 *τῶν* (ante *ἐλευθέρων*) om. La
47 *πέμπτον*] *πέμπον* La
142 *τοῦ* (ante *δίκης*)] *τῆς* La
207 *Εὔβουλε*] *εὔβουλος* La
211 *τὴν* (ante *χάριν*) om. La

(*b*) Manuscripts derived mainly from F

B is derived from F. From the evidence of the scholia it has been argued that B is not copied directly from F but from a lost intermediate manuscript which also contained scholia from a different source.[1] That may well be correct, though I have found no evidence in the text of *Meidias* to support it. In one place a corrector in B restores, probably by conjecture, a right reading lost in other manuscripts (**10** μὴ ἐξεῖναι), but other variants in B are merely errors.

16 ἐπεβούλευσεν] ἐπεχείρησεν B
95 οὐδ' εἰ] οὐδὲν B
136 ὄντα om. B
160 ἔχειν χάριν] χάριν ἔχειν B
197 πρὸς (ante ὑμᾶς)] ὡς B
222 με (post περιόψεσθε) om. B

Fi is basically a copy of B, but has undergone much correction. Nearly all the separative errors of B are more or less perceptible in Fi, though some of them only in the form of an erasure on top of which the correct reading has been written; the only exception is **144** ἐκεῖνος, where Fi may have got the right reading by accidental miscopying of B's error ἐκεῖνον. Fi is heavily loaded with variant readings between the lines and in the margins but, in the parts which I have collated, they are readings which we already have in earlier manuscripts. I have not noticed any substantial new errors which have been left uncorrected.

Mk is essentially derived from Fi. In the parts which I have collated it reproduces those separative errors of B which Fi retains, but not those which are corrected in Fi. However, in a few places Mk shares a reading with a different branch of the tradition, and at least the following two instances can hardly be accidental.

7 καὶ εἰς τοὺς νόμους SAYPMk: om. FBFi
223 τῇ τῶν νόμων ἰσχύι SYMk: τῷ τοῖς νόμοις ἰσχύειν FBFi

These are not emendations made by the scribe's own conjecture but must have been obtained from another manuscript, probably one of those derived from Y. New errors in Mk:

10 φανερὰ] φοβερὰ Mk
22 ἐργάζομαι] κατεργάζομαι Mk

[1] Dilts in *Studia Codicologica* (ed. K. Treu) 151–8.

43 ἂν δ' ἄκων] ἂν δ' ἄκων τις Mk
46 γε om. Mk
46 τὸν τῆς ὕβρεως νόμον] τὸν νόμον τῆς ὕβρεως Mk
205 ἐπαχθεῖς] ἐπαχθὲς Mk

Lb is copied from Mk, reproducing the errors just listed and adding others.

7 εἰς (ante τοὺς νόμους) om. Lb
42 θεωρεῖθ'] θεωρεῖσθ' Lb
46 αὐτοῦ] αὐτοὺς Lb
47 ἐκτείσῃ om. Lb
203 ἀπολαύων] ἀπολαμβάνων Lb
224 τίς (ante ἐστιν)] τί Lb

Fx is another manuscript derived from F.[1] It is a careless copy, with many new errors, and readers of it have additional handicaps: water has got into the binding at some time, so that the words close to the inner margin of each page are sometimes washed out and illegible; the leaves were then rebound in the wrong order, and two (containing **128–37** and **148–58**) appear to have been lost altogether.

1 καὶ (ante βίαια) om. Fx
17 παρασκήνια] παρασχοίνια Fx
22 πάντα om. Fx
25 χορὸν] χρόνον Fx
41 ἄφνω τὸν λογισμὸν] τὸν λογισμὸν ἄφνω Fx
213 μηδενί με] μηδὲν οἶμαι Fx

Fl is partly copied from Fx; see p. 78.

The other manuscripts derived mainly from F evidently had a common hyparchetype which received corrections from another source. The core of this group consists of **X**, **Fu**, and **Pr**, which are linked by conjunctive errors.

34 ὁ (ante χορηγὸς) om. XFuPr
149 τούτου] τούτων XFuPr
167 χρηματισμὸς] χρηματισμοὺς XFuPr
197 πότερον] πρότερον XFuPr

[1] Passweg 83–4 finds evidence in *Timokrates* that Fx incorporates some corrections from S, but I see no sign of that in *Meidias*.

205 οὐδὲ (ante πεπονθὼς)] οὐ XFuPr
209 δὲ (post τοῦτον) om. XFuPr

These three manuscripts generally follow F, but sometimes they have a correct reading which is not in F.

51 καὶ (ante τοὺς ὕμνους) XFuPr: om. F
53 τρεῖς XFuPr: τρὶς F
131 τις XFuPr: τι F
137 καὶ τὴν ἀσέλγειαν XFuPr: om. F
147 γε (post χορηγῶν) XFuPr: om. F
209 ἀνθρώπων XFuPr: om. F

Further conjunctive errors link X and Fu without Pr.[1]

36 ἀπήγγελλε] ἀπήγγειλλε XFu
44 περ] περὶ XFu
62 Ἰφικράτης] ἰφιλοκράτης XFu
101 τοῦτον] τούτου XFu
125 ἀσχάλλειν] ἀσχάλειν XFu
215 ἀνεκράγετε] ἀνεκράγετο XFu

And X and Fu have separative errors, showing that neither is copied from the other.

30 τῇ (ante πόλει) om. X
42 ἤδη δεῖ σκοπεῖν] δεῖ σκοπεῖν ἤδη X
66 ἂν μίαν δραχμὴν] μίαν δραχμὴν ἂν X
99 τοῦδε] τούτου X
125 ἐξ ἀρχῆς ἀσελγὲς] ἀσελγὲς ἐξ ἀρχῆς X
170 γὰρ (post ὑπὲρ) om. X

31 ἠσέλγαινε] εἰσελγαινε Fu
32 ἴστε om. Fu
52 ἰθύνεθ'] ἰθύνεσθ' Fu
204 τῶν (ante πολλῶν) om. Fu
212 ἡγοῦνται] ἡγοῦντο Fu
214 παραμυθήσασθαι] παραμυθήσεισθαι Fu

Pr also has separative errors:

[1] Passweg 75–6 argues that in *Timokrates* the lost predecessor (which she calls π) of X and Fu (which she calls Fv) was an apograph of B. This is not true for *Meidias*, in which the separative errors of B do not occur in XFu.

40 *λέγοντι*] *λέγει* Pr
66 *οὕτω καὶ*] *καὶ οὕτω* Pr
126 *τὴν* (ante *λειτουργίαν*) om. Pr
142 *οὐ* om. Pr
151 *ἀναιδεῖ*] *ἀναιδεῖ μάλα* Pr
203 *πράγματα*] *χρήματ'* Pr

But Pr also has further correct readings which are neither in F nor in XFu:

8 *παραδιδότωσαν* Pr: *παραδότωσαν* FXFu
34 *καὶ τὸ* Pr: *καίτοι* FXFu
100 *μηδὲν* Pr: *μὲν* FXFu
132 *ἦν* Pr: om. FXFu
154 *ἦμεν* Pr: *ἦσαν* FXFu
184 *βραχέα* Pr: om. FXFu

The best hypothesis to explain this state of affairs is as follows: the lost hyparchetype of XFuPr was originally copied from F; some corrections were made in it by reference to another manuscript; a copy from it was then made, from which in turn X and Fu were copied; then further corrections were made in the hyparchetype before Pr was copied from it. The manuscript from which the corrections (or some of them) were derived was probably T; besides the correct readings already listed, Pr and T share a few readings which may have been supposed to be corrections but are actually errors.

4 *παρήγγελκεν*] *περιήγγειλεν* TPr
90 *ἕνα* om. TPr
107 *Ἀφιδναῖος*] *ἀφνιδαῖος* TPr
182 *πολλῷ*] *πολλῶν* TPr

The relationship of **Aa** to Pr is not quite clear. It shares most of Pr's errors and adds many more of its own; the following list gives only a small proportion of them.

2 *ἐφ' οἷς ἠδικημένῳ μοι συνῄδει, ὥστε πάντα ποιοῦντος τούτου καί τινων ἄλλων* om. Aa
35 *αἰκείας*] *κακίας* Aa
62 *χρήματα*] *κτήματα* Aa
93 *ἐπὶ τὴν δίαιταν, ἀλλὰ καταλιπόντα. γενομένης δὲ ἐρήμου κατὰ Μειδίου, ἐπιστάμεθα Μειδίαν* om. Aa

167 *οὐ* (ante *λειτουργία*)] *καὶ* Aa
183 *οὐδὲν*] *μηδὲν* Aa

But I have found four places where Aa is right and Pr is wrong.

42 *ἑκουσίως* Aa: *ἑκουσίους* Pr
72 *ἑτέρῳ* Aa: *ἑταίρῳ* Pr
188 *οὗτος* Aa: *οὕτως* Pr
190 *ἡγῶμαι* Aa: *ἡγοῦμαι* Pr

It is possible to explain these facts by postulating the existence of another manuscript containing all the errors common to Pr and Aa; then Pr is a remarkably accurate copy of that manuscript, with only four new errors, while Aa is a much more careless copy of it. But I prefer the explanation that Aa is a copy of Pr, and either by conjecture or by lucky accident has corrected Pr's errors in these four places.

V is certainly derived from Pr, but has undergone subsequent correction by reference to some other manuscript. Most of the errors of Pr are reproduced, but some have visibly been erased or changed; in **142**, for example, *οὐ*, omitted in Pr, is added above the line in V, and in **203**, where Pr has *χρήματ'*, V has *πράγματ'* written over an erasure. These corrections do not introduce readings not already known to us from other manuscripts (except for the accent on *προῆσθε* in **220**). New errors in V:

11 *πορίσαιτό*] *πορίσηταί* V
17 *ἐστεφανωμένον*] *στεφανούμενον* V
66 *μοι* om. V
165 *τοῦ* (ante *Νικίου*) om. V
184 *εἰπὼν*] *ἐστὶν* V
214 *προβολή*] *βουλὴ* V

Pq and **Mp** are two copies of a lost manuscript which was a copy of Pr. This is shown by the fact that they generally follow Pr but share some conjunctive errors.

8 *μὲν* (post *πρῶτον*) om. PqMp
8 *τῶν* (ante *ἄλλων*) om. PqMp
43 *τὸ βλάβος* om. PqMp
175 *ἑορτὴν*] *ἱερὰν* PqMp

190 *συμφέρειν*] *συμφέρον* PqMp
217 *καὶ* (ante *δοκοῦντα*) om. PqMp

They also have separative errors, showing that neither is copied from the other.

34 *γὰρ* (ante *εὔορκα*) om. Pq
51 *ὑμεῖς* om. Pq
83 *πένης* om. Pq
110 *μικροῦ*] *μιαροῦ* Pq
188 *διὰ* (post *οὐδὲ*) om. Pq
198 *ὡς* om. Pq

7 *ἀγωνιεῖται*] *ἀγνοεῖται* Mp
16 *χορῷ*] *χρυσῷ* Mp
41 *τοῦ* om. Mp
83 *μέν τις*] *μάντης* Mp
202 *οἷς* om. Mp
207 *εἴη ἂν*] *ἱκανὸς* Mp

Fk is a copy of Pq, repeating its errors and adding a few others. In a few places the scribe corrects, or attempts to correct, Pq by his own conjecture.

8 *μὲν* (post *πρῶτον*) om. Fk
10 *καὶ οἱ τραγῳδοί, καὶ ἐπὶ Ληναίῳ πομπὴ καὶ οἱ τραγῳδοὶ καὶ οἱ κωμῳδοί* om. Fk
22 *Παμμένους*] *μαμμένους* Fk
93 *ἐρήμου*] *ἐρήμος* (*sic*) Pq: *ἐρήμης* Fk
174 *παρεσκευάκει*] *παρεσκευάσα*[ει] Pq: *παρεσκευάσκει* Fk
174 *ἵππον, ἵππον*] *ἵππον* Pq: *ἵππαρχον* Fk

Xe is a copy of X, with many additional errors.

12 *καὶ* (post *ὥστε*) om. Xe
47 *μεταλάβῃ*] *μεταβάλῃ* Xe
185 *πολλοὺς ὑβρίζων*] *ὑβρίζων πολλοὺς* Xe
203 *ὑμῖν εἰσοίσειν, ὑμεῖς δὲ νεμεῖσθε; ἐμὲ οἴεσθε* om. Xe
225 *τέχνην*] *τεχνίτην* Xe
227 *τιμωρήσασθε*] *τιμωρήσατε* Xe

Finally in this group there are two manuscripts which change horses in mid-stream. In **Mm** the text is copied from X as far as **154** *κρίνων*. Up to that point it reproduces most of the errors of X, though the scribe has succeeded in correcting some minor ones, probably by his own conjecture (e.g. **33** *εἴπῃ*, where X has *εἴποι*), and a few more substantial ones have been corrected by a later hand, evidently by reference to some other manuscript (e.g. **12** *ἐπέσχετε*, where X has *ἔπασχε*). At **154** *οὗτος* (the beginning of a new page, fo. 182) a different scribe takes over and copies the remainder of the oration from Wd. (For the distinctive errors of Wd see p. 73–4.) Some of Wd's errors have subsequently been removed by erasure or other methods of correction, but it appears that Mm did originally have all the substantial errors of Wd in this part of the oration. New errors in Mm:

3 *λαβεῖν* om. Mm
29 *πολλάκις οἶδ' ὅτι*] *οἶδ' ὅτι πολλάκις* Mm
131 *ἡγεῖθ'*] *ἂν ἡγεῖθ'* Mm
154 *σύνδυο ἦμεν*] *δύο μὲν* Mm
204 *γὰρ* (post *τοιοῦτος*) om. Mm
206 *καὶ λιπαροῦντος*] *ἐκλιπαροῦντος* Mm

In **Aq** also the first part of the oration is copied from X, but the rest seemingly from Pq. The changeover occurs at **85**: the last conjunctive errors with X are **83** *στράτωρ* (for *Στράτων*) and **84** *καταδιήτησεν* (for *κατεδ-*), and the first with Pq against X are **85** *καταδιητήκει* (for *κατεδεδ-*) and **87** *τοὺς* om. However, I have noted a few places later in the speech where Aq is right although Pq is wrong. The most significant is **175** *ἑορτὴν* Aq: *ἱερὰν* Pq. It can hardly be the case that Aq got this reading by accident or conjecture. It therefore seems more likely that Aq is not copied directly from Pq, but from a copy of Pq in which a few corrections had been made by reference to some other manuscript. But most of the variants in Aq can be regarded as mere errors perpetrated by the scribe of Aq himself:

18 *τοίνυν τοὺς κριτὰς*] *τοὺς κρίτας τοίνυν* Aq
33 *καὶ* (ante *πάλιν*) om. Aq
171 *ὧν*] *ὧν ἂν* Aq
179 *ἔλεγε μὲν*] *ἔλεγεν* Aq
195 *τὸ* (ante *σὸν σχῆμα*) om. Aq
212 *οὗτοι*] *οὗ* Aq

There are also some corrections written in Aq by a later

hand. Most of them merely restore readings which we have in earlier manuscripts, but one gives a reading which is not found elsewhere and could possibly be correct: **45** *πράττοι*. This may be a conjecture.[1]

(*c*) Manuscripts derived mainly from A

There are many places in the text where A gives a different reading from SFYP, and thus it is not difficult to identify the later manuscripts which are derived from this branch of the tradition; but sorting out their exact relationships to one another is harder, because there is a good deal of contamination. This is particularly obvious when we look at the documents. A omits all the documents (see pp. 48–9), and naturally some of its followers do likewise (CdKLpLh). Other scribes, realizing that the documents do exist, try to supply them from elsewhere. Thus the scribe of Wd, having found them in some other manuscript, has written them all in the margins of the text. The scribe of Od has written some in the margins, in one case (**82**) under the heading *μαρτυρία εὑρεθεῖσα ἐν ἄλλῳ βιβλίῳ κειμένη*, and has included the rest in the text at the appropriate points. Other scribes have procured only some of the documents: for example, R has some in the margins (**8**, **10**, **22**, **52–3**) and one in the text (**47**) but does not have any of the documents after **53**; Ln lacks four (**93**, **94**, **121**, **168**) but has all the rest in the text. Various explanations of these permutations are possible: a scribe may have had only temporary access to a manuscript containing the documents, so that he did not have time to copy them all; or he may have overlooked some written in the margins, or have mistaken them for scholia which he was not concerned to copy. But anyway it is clear that in some cases a scribe consulted another manuscript besides the one which was his main exemplar, and we may assume that he might then use it not only for the documents but also to correct his copy of the rest of the text. Hence the difficulty of working out a clear genealogy.

Another, less important, indicator of the manuscripts derived from A is corruption of the vocatives in the oration. A regularly uses the abbreviations $\overset{\delta}{\omega}$ and $\overset{\theta}{\omega}$ respectively for ὦ

[1] The same corrector may be responsible for a good reading in the second hypothesis: hyp. ii 10 *αἰσχρουργίας*.

ἄνδρες δικασταί and ὦ ἄνδρες Ἀθηναῖοι. (In the apparatus criticus, where they need to be noted, these abbreviations are recorded by the siglum A^cp, meaning *compendium.*) Later scribes often misunderstand them; the former is commonly expanded as ὦ ἄνδρες or ὦ δικασταί, and the latter as ὦ Ἀθηναῖοι. These forms are not Demosthenic, and are marks of manuscripts derived from A.

K is a comparatively straightforward copy of A, reproducing virtually all of A's distinctive readings and errors. It omits all the documents, and it often retains A's abbreviations for the vocatives. A second hand adds a few corrections or variants, generally ones which are in SFYP, but one is not found elsewhere: in **58** the corrector has changed διδάσκων to ἐκδιδάσκων. This may come from a lost manuscript, but it can hardly be what Demosthenes wrote. The original scribe of K adds nothing but some errors of his own.

6 ἄλλος om. K
29 τῶν λόγων om. K
55 χορευτῶν] χορευόντων K
156 πλέον om. K
206 φίλον δήπου] δήπου φίλον K
210 τὴν τούτων om. K

Cd, **Wd**, and **Od** evidently descend from a lost copy of A in which some corrections had been made. They have most of A's readings, but the following errors of A do not occur in these three manuscripts.

29 μή με] μηδὲ A
30 αὐτοῖς] αὐτοὺς A
37 μόνον] νόμον A
56 μὴ (post ὅπως) om. A
64 καὶ νικῶντα om. A
102 μὲν om. A
110 αὖ om. A
115 ἱεροποιὸν] ἱεροποιῶν A
116 ἔχοντες] λέγοντες A
182 θανάτῳ] θάνατον A
183 τις om. A
216 μὴ om. A

Besides these corrections they share some conjunctive errors.

9 *τὰ* (ante *Πάνδια*) om. Cd^{ac}WdOd
45 *ἑαυτῷ*] *ἑαυτῶν* CdWdOd
46 *τὸ* (ante *γιγνόμενον*) om. Cd^{ac}WdOd
54 *καὶ ἀγαθαί*] *καλαὶ καὶ ἀγαθαὶ* A: *καὶ ἀγαθαὶ καὶ καλαὶ* CdWd: *καὶ ἀγαθαὶ καὶ πολλαὶ* Od
155 *ἧπται*] *ἥπτετο* CdWdOd
155 *πεποιήκατε*] *ἐποιήσατε* CdWdOd

Cd, which follows A in omitting all the documents, is distinguished from Wd and Od by separative errors, including one (**18** *προδιαφθείρας*) which happens by luck to restore the true text.

51 *μὴ* (ante *χορηγὸς*) om. Cd
71 *φοβερὸν*] *φανερὸν* Cd
103 *ἔγωγ'* om. Cd
133 *τοὺς* (ante *πολεμίους*) om. Cd
147 *χορηγοῦντα* om. Cd
157 *οὐδὲν*] *οὐδενὶ* Cd

Wd and Od are distinguished from Cd by a large number of conjunctive errors (many of them transposing a verb to follow its object), which must have originated in a lost manuscript intermediate between them and the ancestor which they share with Cd. These two manuscripts also have all the documents, either in the margins or in the text. They must be copies of the same exemplar, to which the documents, derived from a different source from the oration, had been added, probably in the margins.

4 *ἐλπίζω τὸ δίκαιον*] *τὸ δίκαιον ἐλπίζω* WdOd
9 *ποιεῖν τὴν ἐκκλησίαν*] *τὴν ἐκκλησίαν ποιεῖν* WdOd
26 *εἵλετο τιμωρίαν*] *τιμωρίαν εἵλετο* WdOd
47 *ἢ παῖδα* om. WdOd
161 *προσῆκεν*] *παρῆν* WdOd
192 *ἔχοι τὴν αἰτίαν*] *τὴν αἰτίαν ἔχοι* WdOd

Wd and Od each have separative errors, showing that neither is copied from the other.

4 *ἀδεῶς* om. Wd

12 ταύτας om. Wd
56 γε om. Wd
157 δίκαιον ἦν με] ἦν με δίκαιον Wd
159 πολλὰς] λαμπρὰς Wd
204 πικρίαν] πονηρίαν Wd

13 ἐγὼ χορηγήσειν] χορηγήσειν ἐγὼ Od
26 ἂν εὐθύς μοι λόγος] εὐθὺς ἂν λόγος μοι Od
49 ἀλλὰ] ἀλλὰ καὶ Od
155 συντελεῖς] τοὺς ἐντελεῖς Od
211 τοὺς (ante νόμους) om. Od
218 ἀφῆτε] ἀφανῆτε Od

The scribe of **R** (or more likely his predecessor: see below) switches from one exemplar to another about a quarter of the way through. In the first part of the oration (at least as far as **54** καὶ ἀγαθαί)[1] he uses the same exemplar as Wd and Od, and reproduces the conjunctive errors of those two manuscripts; he also has the documents up to that point. Thereafter (at the latest from **64**, where καὶ νικῶντα is omitted) he copies A, and thus has neither the documents nor the corrections which had been made in the common ancestor of CdWdOd. New errors in R:

42 δεῖ om. R
62 ποτ' ἐκεῖνον] ἐκεῖνον ποτὲ R
76 γενέσθαι] γενέσθαι τοῦτον R
135 ἔμοιγε om. R
206 δήπου om. R
211 ὑβρίζει καὶ πτωχοὺς ἀποκαλεῖ, ἃ δὲ νῦν om. R

Vf and **Fh** repeat most of R's errors, and each has many separative errors.

10 οἱ παῖδες καὶ ὁ κῶμος καὶ οἱ κωμῳδοὶ καὶ οἱ τραγῳδοί, καὶ Θαργηλίων τῇ πομπῇ καὶ τῷ ἀγῶνι om. Vf
30 ὑμᾶς ὁ παθὼν] ὁ παθὼν ὑμᾶς Vf
37 πάντας om. Vf
49 ἐζημιώκασι] ἐζημίωσαν Vf
197 πρὸς ὑμᾶς] πρὸς ὡς ὑμᾶς Vf
206 λιπαροῦντος] παρακαλοῦντος Vf

[1] Probably as far as **63**, where Od and R both have διεπράξατο. Wd's διεπράττετο may be a correction by accident.

10 προβολαὶ] προβολεῖς Fh
27 τότε om. Fh
52 ἀγυιεῖ om. Fh
53 Διὶ κτησίῳ om. Fh
217 τῆς (post ὕβρεως) om. Fh
218 νυνὶ] νῦν Fh

In some places it is clear that Vf at first carried the same error as R but was later corrected. For example, in **211**, where R omits seven words (by haplography of νῦν), Vf has an erasure on which fourteen words, including those seven, are written in a very compressed script; obviously Vf originally had the same omission as R, but it was then corrected by reference to some other manuscript. In addition, some of the documents omitted in R have been added in the margins of Vf.

From this evidence it may seem that Vf (before correction) and Fh are copies of R.[1] Yet there is also some evidence to the contrary, since there are a few errors in R which are absent (with no sign of erasure or correction) from Vf and Fh, or from Fh only. They include:

2 ἀπέβλεψεν] ὑπέβλεψεν R
37 μέλειν] μέλλειν R
44 ἐξούλην] ἐξούλης R
92 ὑμετέρους] ἡμετέρους R
209 εἰς (ante ἐμέ) om. R
209 τούτων] τίνων R

9 τὰ (ante Πάνδια) om. RVf
17 καὶ (ante πράγματα) om. RVf
41 φανερός] φανερῶς RVf
206 δέ (post νῦν) om. RVf
213 ἐφ' ἧς] ἐφ' οἷς RVf
216 μὴ (ante διαλύσει) om. RVf^{ac}

And in **211** Fh never omitted the seven words omitted in R and at first in Vf. It hardly seems possible that Vf and Fh corrected all these errors by accident or conjecture. So, despite the fact that these two manuscripts are otherwise very close to

[1] Passweg 114–16 maintains that Vf and Fh (which she calls Q) are copies of R in *Timokrates*, but she finds one place where they have a correct reading and R has an error: 24.211 ἐπαινεῖτε] ἐποιεῖτε R.

R, I think that they must be copied not from R itself but from R's immediate predecessor. We must postulate the existence of another lost manuscript which was the exemplar of R, Vf, and Fh, and it was the scribe of that manuscript, rather than the scribe of R, who changed to a different exemplar at **64**. After R and Vf were copied from it, but before Fh was, some corrections were made in it by reference to another source; that is the best explanation of the fact that some errors in R and Vf are not repeated in Fh.

Xg is a copy of Fh. It contains all the separative errors of Fh listed above and some further errors.[1]

2 *τὰς* (post *οὐσίας*) om. Xg
6 *ὑβρισμένος*] *ὑβρισμένως* Xg
46 *ὁ πάσχων*] *ὑπάσχων* Xg
221 *ὑβριεῖν*] *ὑβρίζειν* Xg
227 *ἐξελήλεγκται*] *ἐξελήλεγκεται* Xg

Lp is another manuscript derived from A. Like A it lacks all the documents. In general it reproduces the distinctive readings and errors of A, but in a dozen places a correction has been made, apparently by the original scribe. Most notably there are three passages in which words omitted by A are added in the margin of Lp: **6** *ἀγανακτήσας ... ταύτην*, **101** *οὔτε θεόν*, **179** *ἄνθρωπε, θέαν*. These words must have been obtained from a different manuscript. New errors in Lp:

5 *παραιτεῖσθαι*] *ἀξιοῦν παραιτεῖσθαι* Lp
5 *ἀδίκως ἀφαιρεθείσης*] *ἀφαιρεθείσης ἀδίκως* Lp
58 *τὰ* (ante *τοιαῦτα*) om. Lp
142 *τοιαῦτα*] *ταῦτα* Lp
162 *εἷς οὗτος*] *οὗτος εἷς* Lp
202 *δικαίως οἶμαι δίκην διδόναι*] *οἶμαι διδόναι δικαίως δίκην* Lp

Ln and **Vi** seem to be two copies of a lost manuscript which was a copy of Lp. The documents, or some of them, were evidently added to that manuscript from a different source,

[1] Xg has not been properly collated. I did not plan to include it in my investigation; but, having a spare hour and a half in the Vatican library, I noted the readings mentioned, which, though few, are sufficient to show Xg's dependence on Fh. A full collation would probably reveal far more separative errors.

probably in the margins, since the scribes of Ln and Vi sometimes miss them: Ln has some which Vi lacks, but both lack those in **93–4**, **121**, **168**. In the rest of the text most of Lp's errors are reproduced, but a few are not. This might mean that the lost exemplar of Ln and Vi was copied from an ancestor of Lp;[1] but I think it more probable that it was copied from Lp and then a few corrections were made in it by reference to the manuscript from which the documents were drawn. Conjunctive errors of Ln and Vi may also be attributed to this lost exemplar.

8 *ταῦτα* om. LnVi
18 *τοὺς κρίτας*] *καὶ τοὺς κρίτας* LnVi
34 *θῆσθε*] *χρῆσθε* LnVi
46 *τὸν τῆς ὕβρεως νόμον*] *τῆς ὕβρεως τὸν νόμον* LnVi
209 *τούτῳ*] *τούτου* LnVi
220 *ὑμῶν* om. LnVi

Ln and Vi each have separative errors.

5 *μὲν* (post *τῷ*) om. Ln
26 *τἀδικήματα*] *τὰ δικαιώματα* Ln
45 *ἱκανὴν τὴν τιμωρίαν εἶναι*] *εἶναι ἱκανὴν τὴν τιμωρίαν* Ln
66 *οὕτω* (ante *καὶ πλούσιος*) om. Ln
222 *τότε*] *τοῦτο* Ln
223 *σκοπεῖν καὶ ζητεῖν*] *ζητεῖν καὶ σκοπεῖν* Ln

16 *παρασκευάζηται*] *παρασκεβάσητε* Vi
44 *αὐτῷ*] *αὐτὰ* Vi
133 *πεντηκοστόλογοι*] *πεντηκοστολίγοι* Vi
205 *ἐπαχθής*] *ἀπηχθής* Vi
218 *ἢ* (ante *νυνὶ*) om. Vi
223 *ὅπλων*] *ὅσων* Vi

Lh and **Nb** are likewise shown by conjunctive errors to be two copies of a lost copy of Lp, but the lost copy did not have the documents added to it. Consequently Lh still lacks the documents, but in Nb they have been added in the margins by a later hand (copied from Lb, to judge from a few conjunctive errors). Conjunctive errors of Lh and Nb may be attributed to their lost exemplar.

[1] Passweg 98 states that in *Timokrates* Ln is independent of Lp.

7 μὲν (post ὕβρισμαι)] γὰρ LhNb
32 οὐδενὸς] οὐδενὰ LhNb
45 τὸν (ante ὑβρίζειν)] τὸ LhNb
46 τοῦ νόμου] τὸν νόμον LhNb
64 ὅποι] ὅτι LhNb
209 ἐᾷ] ἐᾶν LhNb

Lh and Nb each have separative errors.

2 οὕτως] οὕτωρ Lh
3 ὑπομείνας] ὑπομεῖναι Lh
28 λῆμμα] λῆμα Lh
43 ἄρξωμαι] ἄρξομαι Lh
45 γὰρ (post τὴν μὲν) om. Lh

4 λοιπὸν] λοιπῶν Nb
15 εἰς] εἰς τὰ Nb
33 στεφανηφορίαν] στεφανοφορίαν Nb
43 φονικοὶ] φονοικοὶ Nb
54 δὲ (post ἱστάναι)] καὶ Nb
222 προδῶτε] προδότε Nb

The first part of **Fl** is copied from the same exemplar as Lh and Nb, and shares their conjunctive errors as far as the end of **46**. When he reached that point, the scribe evidently realized that his exemplar was a copy omitting the documents, and he therefore switched to one including them. His decision was disastrous, for his new exemplar was Fx, a copy in which some leaves are disordered or lost and some words are washed out (see p. 65). He mindlessly copies out the text as he finds it, leaving blank spaces where words are illegible in Fx and oblivious of the absence of some leaves, so that he jumps in mid-sentence from **57** to **73** and from **128** to **158**. Later he realizes that those passages have been omitted, and tacks them on (copying them perhaps from his first exemplar) at the end of the oration, which is then repeated, so that **225–7** appears twice. I make no attempt to list individual errors in this farrago. Poor fifteenth-century reader who had to use this copy to read Demosthenes!

(*d*) Conclusion

All the later manuscripts are derived from A or F or Y. Their basic relationships to those three manuscripts are clear.

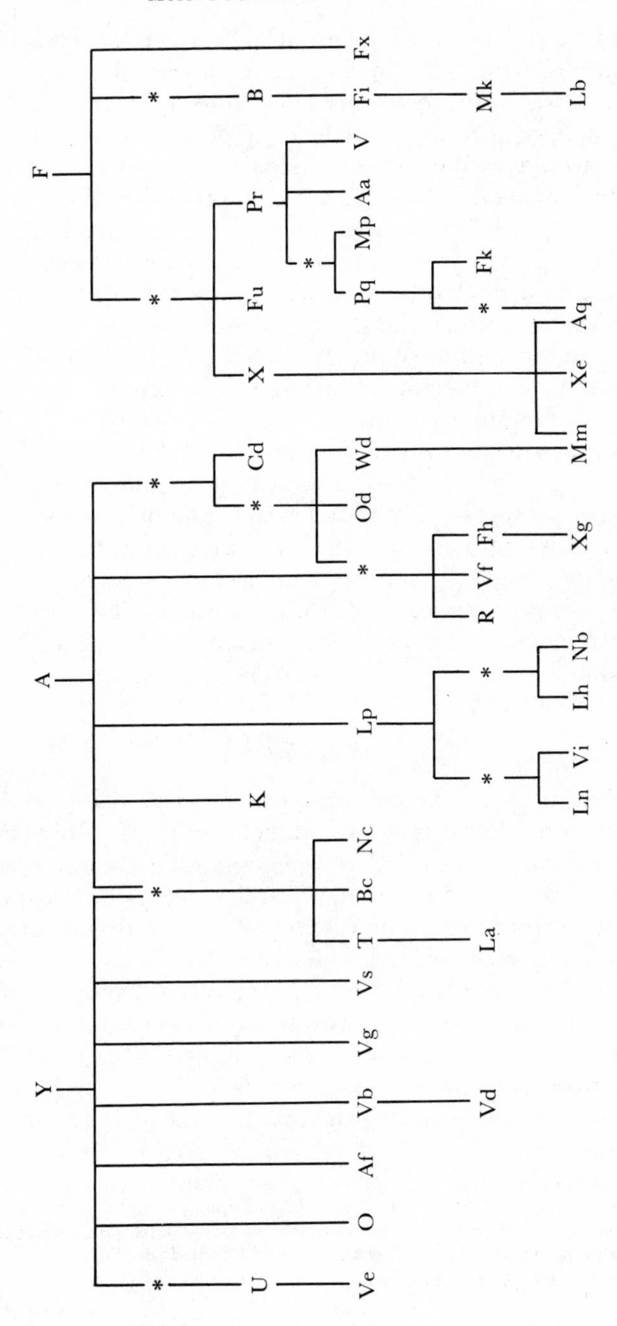
F
B
Fx
Fi
Mk
Lb
Pr
V
Aa
Mp
Fu
Pq
Fk
Aq
X
Xe
Mm
A
Cd
Od
Wd
Fh
Xg
Vf
R
Lp
Nb
Lh
Vi
Ln
K
Nc
Bc
T
La
Y
Vs
Vg
Vb
Vd
Af
O
U
Ve

though many of them incorporate corrections and variant readings of which the precise source can usually not be identified, there is no good reason to think that any of these is derived from a source independent of the older manuscripts which we have. We therefore need not treat the later manuscripts as authorities for the constitution of the text of *Meidias*. Occasionally, however, they offer a different reading which, whether it has arisen by mistake or by deliberate emendation, anticipates a modern conjecture or is worth considering as a conjecture in its own right.

The stemma on p. 79 summarizes the probable relationships of the later manuscripts (omitting the chaotic Fl). An asterisk denotes a manuscript now lost, the existence of which should be assumed to explain the relationships; there may of course have been more of these intermediate manuscripts than the diagram shows. It only shows which manuscript each scribe took as his main exemplar; it does not show the sources of corrections and alternative readings, which in some manuscripts are very numerous. Even the basic relationships are not all certain. The stemma should therefore be used with much caution.

Prose-rhythm

Rhythm in Demosthenes' prose has undergone a good deal of investigation, but it remains a difficult subject.[1] Only two features of it can be regarded as well established: that he prefers to avoid hiatus (a word ending with a vowel followed immediately by a word beginning with a vowel), and that he prefers to avoid a sequence of more than two short syllables. The two preferences were discovered in the nineteenth century by Benseler and Blass;[2] but those two scholars overstated the preferences as if they were laws (with only certain strictly defined exceptions), and it is now generally agreed that Blass went much too far in emending the text to remove violations of the

[1] S. Skimina *État actuel des études sur le rythme de la prose grecque* 1 (= *Bulletin International de l'Académie Polonaise: Classe de Philologie* Supp. 3, 1937) 106–36 gives a good account of work on Demosthenic prose-rhythm up to that date. For subsequent contributions see Batschelet-Massini *Tainia* 1.503–28, McCabe *Prose-rhythm*.

[2] Benseler *De hiatu* 62–167, Blass *Beredsamkeit*² 3.1.100–12.

supposed laws. Subsequent editors have been a little more cautious, but have still accepted numerous small emendations on these grounds. Thus, for example, where the manuscripts have τὰ τοῦδε οὐκ ἐλεηθέντα ὑπὸ τούτου (**99**) and πάνθ' ὅσα ἐστὶν ἀγαθὰ ὑμῖν (**188**), editors print τὰ τοῦδ' οὐκ ἐλεηθένθ' ὑπὸ τούτου and πάνθ' ὅσ' ἔστ' ἀγάθ' ὑμῖν to remove hiatus and sequences of three short syllables; and in making these changes they usually do not reveal in the apparatus criticus that the reading printed is not in the manuscripts.

I am reluctant to follow this practice, for several reasons:

1. Even when Demosthenes was likely to elide a vowel in speaking, he may have written the word in full in his text. Inscriptions show that in the classical period it was quite common to use *scriptio plena* even in verse where elision was required by the metre.[1] It is noteworthy that even the oldest papyrus fragment of *Meidias* (Π1 of the first century A.D.) has elided and unelided forms at the same points as the medieval manuscripts (**104** τοῦτ' ἐρῶ, **105** ἀλλ' ἕνα ὅρον ... με ἀνελεῖν); this shows that the instances of *scriptio plena* in the medieval manuscripts are not merely medieval errors.

2. In some places Demosthenes may have had rhetorical reasons, which we have not perceived, for wanting a pause between words, or a sequence of short syllables.

3. Though we may accept that he had a preference for avoiding hiatus and numerous short syllables, it has not been shown that he always avoided them on every possible occasion.

4. It is clear that *Meidias*, as we have it, is a draft rather than a finished text for publication (see pp. 23–8). Thus he may not have perfected its rhythm.

5. It is a hindrance to further study of prose-rhythm if editors do not report manuscripts' readings accurately.[2]

In editing the text the best policy is to allow avoidance of hiatus or of a sequence of short syllables to influence our choice between variant readings, but not to use them as reasons for changing the text where it is not otherwise in doubt. So in this edition readers will find τὰ τοῦδε οὐκ ἐλεηθέντα ὑπὸ τούτου printed without elisions, because that is what the manuscripts give; those who believe that Demosthenes would in fact have

[1] Cf. Threatte *Grammar* 1.424.

[2] Cf. McCabe *Prose-rhythm* 44–8, Radermacher *Gnomon* 16 (1940) 9.

elided where possible are free to make the elisions themselves as they read the speech.

The constitution of the text and the apparatus criticus

Where the manuscripts offer divergent readings, agreement of S and A may normally be assumed to indicate the reading of the archetype, having greater authority than an alternative reading given by other manuscripts. Nevertheless, even the archetype doubtless contained some errors; occasionally a reading given by both S and A is unacceptable and must be replaced by an alternative drawn either from other manuscripts or from later conjecture. In some places an explanatory gloss has got into the text of both S and A (e.g. **60** *κορυφαῖος*, **71** *ὁ τύπτων*, **86** *τὴν τοῦ Θαργηλιῶνος ἢ τοῦ Σκιροφοριῶνος γιγνομένην*, **133** *τῆς Εὐβοίας*), probably copied from an ancestor of one into an ancestor of the other by 'horizontal transmission'.[1] In the documents, because they are not preserved in A, conjecture is needed more frequently than elsewhere.

Where S and A differ, there is no rule of thumb for choosing between them. The one which is right will not necessarily be the one with which the other manuscripts agree. Even where we happen to have a fragment of an ancient manuscript agreeing with either S or A, that is not decisive, because all such fragments are later in date than the period when the ancestry of S and the ancestry of A probably diverged. But naturally some weight must be given to their evidence, and readings of FYP are always worth noticing too, because they (especially F) may include evidence drawn from manuscripts earlier than S and A.

We can also use quotations of Demosthenes by later Greek authors (generally called 'testimonia'). The rhetoricians in particular often quote Demosthenes. When the quotations are precise, they are valuable independent evidence for the text. However, authors sometimes quote carelessly, and sometimes they merely paraphrase, wishing to convey the point of Demosthenes' argument rather than his exact words. For this edition I have made a fuller collection of testimonia than has pre-

[1] One such gloss corresponds to an error of Libanios (**156** n. *αὐληταῖς ἀνδράσιν*), but that is not enough to prove that all these glosses are to be attributed to him.

viously been assembled for *Meidias*, and in a separate register above the apparatus criticus I list those which may be useful for establishing the text, but I omit those which are too brief or too corrupt or too remote from the text to be useful.

Some of the manuscripts carry many scholia, which have recently been edited by M. R. Dilts. I refer to them as Σ with the numbering of Dilts's edition; I rely on his text and have not myself collated the scholia.[1] They seem to be mainly the work of rhetoricians of the third and fourth centuries A.D. They contain little historical or other information about Demosthenes and his contemporaries, except what we can read for ourselves in the surviving speeches; the scholiasts' concern is primarily with rhetorical analysis. Thus the scholia are of interest mainly for the study of rhetorical theory in their period, and I give comparatively little attention to them in the present volume. But they also have a secondary use which is relevant here: occasionally their lemmata or quotations from the oration give a different reading from the text in the surviving manuscripts, and should be considered as additional evidence for the text. There are also a few scholia in two papyri; see p. 39.

The editors of the earliest printed editions, produced in the sixteenth century, may have been able to use some manuscripts not now extant. In particular, the editor of the Aldine edition (whose identity is uncertain) and the French editor Denys Lambin (usually called by the Latin form of his name, Lambinus) each attribute to unnamed manuscripts some readings which I have not found in any of the manuscripts that I have collated.

The apparatus criticus in this edition generally reports all variant readings found in SAFYP or in the ancient manuscripts and testimonia listed in the register above it; where any of those sources is not mentioned in the apparatus, it may be assumed that it has the reading printed in the text. But there are the following exceptions:

1. I usually ignore variation or omission of breathings, accents, punctuation, and iota subscript. But occasionally

[1] The Patmian scholia on Demosthenes, which I call Σ (Patm.), are not included in Dilts's edition. I quote them from the edition by I. Sakkelion in *BCH* 1 (1877) 14–16, reprinted in *Lexica Graeca Minora* (ed. K. Latte and H. Erbse, 1965) 144–6.

such discrepancies are noted where they affect the meaning or where the correct form is uncertain, not because the manuscripts are authoritative in these matters, but to draw the reader's attention to the alternative possibility.

2. Where the only difference among the manuscripts is between elided and unelided forms of the same word, or between inclusion and omission of movable *ν*, I have adopted a compromise to avoid expanding the apparatus excessively. If S and A (and any ancient manuscript which preserves the passage) agree, I put their reading in the text and do not record any variant found in other manuscripts. If S and A disagree, I choose between them (normally adopting the alternative which avoids hiatus or a succession of short syllables) and in the apparatus I list all the other manuscripts of the group SAFYP which give the other reading, showing only the syllable in question, e.g. -*σι* AF. Testimonia are not reported in this connection.

3. The use of an abbreviation instead of a complete word in a manuscript is not reported where the text is otherwise not in doubt.

4. Where a correction has been made in a manuscript, the readings both before and after correction are generally reported, if both are legible. If the reading before correction is illegible, that is shown by placing the siglum in square brackets, e.g. [F^{ac}]. But the reading before correction is sometimes ignored if it is clear that the original scribe has just corrected a blunder of his own, in a place where the true reading is not in doubt.

The later manuscripts are not reported except where they give a reading, not appearing in SAFYP, which either is adopted or may be seriously considered for adoption in the text. If such a reading is in many manuscripts, I give only those (up to three) which may be the earliest, not a complete list.

Readings found only in printed editions are reported only selectively. The sigla Ald^{vl} and $Lamb^{vl}$ refer to the *variae lectiones* given in the Aldine edition at the end of the volume and in the margin of Lambinus' edition. Some modern conjectures not reported in the apparatus are mentioned in the commentary.

The use of square brackets to indicate *delenda* in classical texts sometimes causes misunderstanding, because in papyri and inscriptions square brackets are used to indicate restorations. I therefore use instead the signs { } to mark words which are in the manuscripts but are in my opinion spurious. I use < > to mark words which are not in the manuscripts and are added by conjecture.

SIGLA

Π1	P. Heidelberg N.S. 2.207	saec. i
Π2	P. Oxy. 56.3850	ii
Π3	P. Oxy. 56.3846	iii
Π4	P. Oxy. 56.3847	iii
Π5	P. Oxy. 56.3848	iii
Π6	P. Oxy. 56.3849	iii
Π7	P. Oxy. 11.1378	iii
Π8	Membr. Berolin. 13276	iii
Π9	P. Harris 17	iv
Π10	P. Rainer N.S. 1.8 et P. Whitehouse	iv
Π11	P. Rainer N.S. 3.47	iv–v
Π12	P. Lit. Lond. 179	i
Π13	Wessely *St. Pal. Pap.* 4.111–13	iv–v
S	Parisinus gr. 2934	ix–x
A	Monacensis gr. 485	x
F	Marcianus gr. 416	x
Y	Parisinus gr. 2935	x
P	Laurentianus plut. 59.9	x

Sigla codicum recentiorum, qui raro memorantur, in pp. 40–3 reperiri possunt.

[ ]	periit vel legi non potest
Σ	scholium
Ald	editio Aldina (A.D. 1504)
Lamb	editio Lambini (A.D. 1570)

Notae adscriptae (e.g. S^{pc}_{3}, $F^{\gamma\rho}$)

ac	ante correctionem
cp	per compendium
mg	in margine
pc	post correctionem
sl	supra lineam
vl	varia lectio (in editione Aldina vel Lambini)
γρ	varia lectio cui scriba praescripsit γρ(άφεται) vel sim.
1	ipsius scribae manus
2, 3, 4	manus recentiores

ΚΑΤΑ ΜΕΙΔΙΟΥ
ΠΕΡΙ ΤΟΥ ΚΟΝΔΥΛΟΥ

ΚΑΤΑ ΜΕΙΔΙΟΥ
ΠΕΡΙ ΤΟΥ ΚΟΝΔΥΛΟΥ

Τὴν μὲν ἀσέλγειαν, ὦ ἄνδρες δικασταί, καὶ τὴν ὕβριν, ᾗ
πρὸς ἅπαντας ἀεὶ χρῆται Μειδίας, οὐδένα οὔθ᾽ ὑμῶν οὔτε τῶν
ἄλλων πολιτῶν ἀγνοεῖν οἴομαι. ἐγὼ δ᾽, ὅπερ ἂν καὶ ὑμῶν
ἕκαστος ὑβρισθεὶς προείλετο πρᾶξαι, τοῦτο καὶ αὐτὸς
ἐποίησα, καὶ προυβαλόμην ἀδικεῖν τουτονὶ περὶ τὴν ἑορτήν, 5
οὐ μόνον πληγὰς ὑπ᾽ αὐτοῦ λαβὼν τοῖς Διονυσίοις, ἀλλὰ καὶ
ἄλλα πολλὰ καὶ βίαια παθὼν παρὰ πᾶσαν τὴν χορηγίαν.
2 ἐπειδὴ δὲ καλῶς καὶ τὰ δίκαια ποιῶν ὁ δῆμος ἅπας οὕτως
ὠργίσθη καὶ παρωξύνθη καὶ σφόδρα ἐσπούδασεν ἐφ᾽ οἷς
ἠδικημένῳ μοι συνῄδει, ὥστε πάντα ποιοῦντος τούτου καί 10
τινων ἄλλων ὑπὲρ αὐτοῦ οὐκ ἐπείσθη οὐδ᾽ ἀπέβλεψεν εἰς τὰς
οὐσίας τὰς τούτων οὐδ᾽ εἰς τὰς ὑποσχέσεις, ἀλλὰ μιᾷ γνώμῃ
κατεχειροτόνησεν αὐτοῦ, πολλοί μοι προσιόντες, ὦ ἄνδρες
δικασταί, καὶ τῶν ἐν τῷ δικαστηρίῳ νῦν ὄντων ὑμῶν καὶ τῶν
ἄλλων πολιτῶν ἠξίουν καὶ παρεκελεύοντο ἐπεξελθεῖν καὶ 15
παραδοῦναι τοῦτον εἰς ὑμᾶς, ὡς μὲν ἐμοὶ δοκεῖ, δι᾽ ἀμφότερ᾽,
ὦ ἄνδρες Ἀθηναῖοι, νὴ τοὺς θεούς, καὶ δεινὰ πεπονθέναι
νομίζοντες ἐμὲ καὶ δίκην ἅμα βουλόμενοι λαβεῖν ὧν ἐπὶ τῶν
ἄλλων ἐτεθέαντο θρασὺν ὄντα καὶ βδελυρὸν καὶ οὐδὲ καθεκτὸν
3 ἔτι. οὕτω δὲ τούτων ἐχόντων, ὅσα μὲν παρ᾽ ἐμοῦ προσῆκεν 20

1 *τὴν μὲν … ὕβριν* Max. Plan. (Walz 5) 377.16, Sopat. (Walz 8) 111.16 1–2 *τὴν μὲν … χρῆται AB* 1.451.13 *τὴν … ὦ* et *καὶ … Μειδίας* Π12 *τὴν μὲν … Μειδίας* Greg. Cor. (Walz 7) 1152.2 1–3 *τὴν μὲν … οἴομαι* Hermog. 99.12, 418.6, Ruf. (Hammer) 400.10 3–5 *ἐγὼ … ἐποίησα* Hermog. 345.19 5 *καὶ … τουτονὶ* Π12 *καὶ … ἑορτήν* Syrian. (Rabe 2) 101.14, Max. Plan. (Walz 5) 300.27, Harp. s.v. *προβαλλομένους*, Suda π 2327 8–19 *ἐπειδὴ … βδελυρὸν* Greg. Cor. (Walz 7) 1279.8 8–9 *οὕτως … παρωξύνθη* Longin. *Fr.* (Hammer) 216.1 11–12 *εἰς … ὑποσχέσεις* Π12 12–13 *ἀλλὰ … αὐτοῦ AB* 1.151.12 16–17 *ὡς … πεπονθέναι* Hermog. 436.1

Tit.] *ΠΕΡΙ ΤΟΥ ΚΟΝΔΥΛΟΥ* om. A 1 *ὦ ἄνδρες δικασταί* om. Greg. Hermog. 418: *ὦ Ἀθηναῖοι* Ruf.: *ἀεὶ AB* *τὴν* (post *καὶ*) om. Greg. 2 *αἰεὶ* F^{ac} *οὔθ᾽* om. Ruf. 3 *οἴομαι* S_1^{pc}AYP Hermog. Ruf.: *οἶμαι* S^{ac}F *δὲ* A 5 *τουτονὶ* Π12AFY Syrian. Max.: *τοῦτον* SP Harp. Suda 7 *παρὰ*] *κατὰ* S^{sl} 9 *ἐσπούδαζεν* Greg. 10 *μοι* om. F 12 *οὐδ᾽ εἰς* Π12 Greg.: *οὐδὲ* SAFYP 14 *δικασταί* om. Greg. 16 *-ρα* AF 17 *ὦ ἄνδρες δικασταί* A^{cp}: *ὦ Ἀθηναῖοι* Hermog.: om. Greg. 17–18 *κἀμὲ νομίζοντες δεινὰ πεπονθέναι* Hermog. 17 *πολλὰ δεινὰ* Greg. 18–19 *ὄν … τεθέανται* Σ (16 Dilts) 20 *-κε* $S^{pc}AF^{pc}P^{pc}$

AGAINST MEIDIAS
ON THE PUNCH

The bullying, men of the jury, and the insolence, with which Meidias constantly treats everyone, are known to all of you and to every Athenian, I suppose. My own reaction was just what the reaction of any of you to insolence would have been: I accused him by *probole* of an offence concerning the festival. It wasn't just that he'd inflicted blows on me at the Dionysia, but I'd been subjected to a great deal of violent treatment throughout my time as a chorus-producer. The **2** people acted in the right and proper way: they were all so angry and incensed, and they supported me so strongly over the wrongs which they realized had been done to me, that despite all the efforts of Meidias and some supporters of his they weren't won over or distracted by the wealth or promises of those men, but with one accord they voted him down. After that I was approached, men of the jury, by many of you who are present in court today, and other citizens too, who asked and urged me to proceed and bring him before you. I think there were two reasons, men of Athens, certainly: they thought I'd been shockingly treated, and they wished at the same time to punish him for other cases in which they'd observed that he was audacious, disgusting, and out of control. In these circumstances I have duly pre- **3**

φυλαχθῆναι, πάντα δικαίως ὑμῖν τετήρηται, καὶ κατηγορήσων,
ἐπειδή τις εἰσάγει, πάρειμι, ὡς ὁρᾶτε, πολλὰ μέν, ὦ ἄνδρες
Ἀθηναῖοι, χρήματ', ἐξόν μοι λαβεῖν ὥστε μὴ κατηγορεῖν, οὐ
λαβών, πολλὰς δὲ δεήσεις καὶ χάριτας καὶ νὴ Δί' ἀπειλὰς
4 ὑπομείνας. ἃ δ' ἐν ὑμῖν μετὰ ταῦτά ἐστιν ὑπόλοιπα, ὅσῳ πλεί- 5
οσιν οὗτος ἠνώχληκεν καὶ παρήγγελκεν (ἑώρων γὰρ αὐτὸν
ἄρτι πρὸ τῶν δικαστηρίων οἷα ἐποίει), τοσούτῳ μᾶλλον ἐλπίζω τὸ δίκαιον ἕξειν. οὐ γὰρ ἂν καταγνοίην ὑμῶν οὐδενὸς οὔθ'
ὡς, περὶ ὧν πρὸς ἐμὲ ἐσπουδάσατε αὐτοί, τούτων ἀμελήσετε,
οὔθ' ὡς, ἵνα Μειδίας ἀδεῶς τὸ λοιπὸν ὑβρίζῃ, ψηφιεῖταί τις 10
ὑμῶν ὀμωμοκὼς ἄλλο τι πλὴν ὅ τι ἂν δίκαιον ἡγῆται.

5 Εἰ μὲν οὖν, ὦ ἄνδρες Ἀθηναῖοι, παρανόμων ἢ παραπρεσβείας ἤ τινος ἄλλης τοιαύτης αἰτίας ἔμελλον αὐτοῦ κατηγορεῖν, οὐδὲν ἂν ὑμῶν ἠξίουν δεῖσθαι, νομίζων τῷ μὲν κατηγόρῳ
περὶ τῶν τοιούτων προσήκειν ἐλέγχειν μόνον, τῷ δὲ φεύγοντι 15
καὶ παραιτεῖσθαι. ἐπειδὴ δὲ τούς τε κριτὰς διαφθείραντος
τούτου τότε καὶ διὰ τοῦτο τῆς φυλῆς ἀδίκως ἀφαιρεθείσης
6 τὸν τρίποδα, καὶ αὐτὸς πληγὰς εἰληφὼς καὶ ὑβρισμένος οἷα
οὐκ οἶδ' εἴ τις ἄλλος πώποτε χορηγὸς ὑβρίσθη, ἣν ὑπὲρ τούτων ἀγανακτήσας καὶ συνοργισθεὶς καταχειροτονίαν ὁ δῆμος 20
ἐποιήσατο, ταύτην εἰσέρχομαι, οὐκ ὀκνήσω καὶ δεῖσθαι. εἰ
γὰρ οἷόν τε τοῦτ' εἰπεῖν, ἐγὼ νῦν φεύγω, εἴπερ ὑβρισθέντα
7 μηδεμιᾶς δίκης τυχεῖν ἐστίν τις συμφορά. δέομαι οὖν ὑμῶν
ἁπάντων, ὦ ἄνδρες δικασταί, καὶ ἱκετεύω, πρῶτον μὲν εὐνοϊκῶς ἀκοῦσαί μου λέγοντος, ἔπειτ', ἐὰν ἐπιδείξω Μειδίαν 25
τουτονὶ μὴ μόνον εἰς ἐμὲ ἀλλὰ καὶ εἰς ὑμᾶς καὶ εἰς τοὺς

2–3 *ἐπειδή ... εἰσάγει* et *πολλὰ ... λαβεῖν* Π12 2–5 *πολλὰ ... ὑπομείνας* Greg. Cor. (Walz 7) 1279.22 4–5 *πολλὰς ... ὑπομείνας* Π7, Hermog. 436.3 5–6 *ὅσῳ ... ἠνώχληκεν* Longin. *Fr.* (Hammer) 216.1 12–13 *εἰ ... αὐτοῦ* Π12 12–14 *εἰ ... κατηγορεῖν* Priscian. *Inst.* 18.228 23–92.12 *-ιᾶς ... πρῶ-* Π3 25–92.1 *ἔπειτ' ... ὑβρικότα* Π12

---

1 *τετήρηνται* S^{ac} 2–3 *ὦ ἄνδρες Ἀθηναῖοι* om. Π12: *ὦ ἄνδρες δικασταί* A^{cp}: *ὦ ἄνδρες* Greg. 3 *-τα* A *μὴ οὐ κατ-* Greg. 4 *δὲ* om. S *καὶ χάριτας* om. Hermog. *νὴ δία* $Π12^{sl}$SF: om. $Π12^{ac}$ 5 *ὅσῳ γὰρ* YP^{ac} 6 *-ηκε* $S^{pc}AFP^{pc}$ *παρ-* FP_2^{pc}: *περι-* SAY: [P^{ac}] 9 *αὐτοὶ πρότερον* FYP 10 *τὸ λοιπὸν ἀδεῶς ὑβρίζει* A 11 *ἡγῆται δίκαιον* F 12 *ὦ ἄνδρες Ἀθηναῖοι* om. Π12 Priscian. 13 *ἄλλης τινος* A 13–14 *τοιαύτης* post *κατηγορεῖν* SYP 13 *αἰτίας* om. Π12 *ἤμελλον* A Priscian. 15 *ἐξε-λέγχειν* S_2^{pc}FYP 17 *τότε* AFP_2^{sl}: om. SYP_1 20–1 *ἀγανακτήσας ... ταύτην* om. A 21 *ὀκνήσων* YP^{ac} 23 *τυχεῖν* om. Y^{ac} *-τί* $S^{pc}AF^{pc}P^{pc}$: [Π3] 24 *ἁπάντων* om. F *ὦ ἄνδρες ἀθηναῖοι* F 25 *ἔπειτα ἐὰν* A: *ἐπειδὰν* Π12: [Π3] *μιδιαν* Π3 26 *τοῦτον* Π12: [Π3] 26–92.1 *καὶ εἰς τοὺς νόμους* om. Π12F

served all your rights which it was my function to protect, and now that the case is being brought into court I'm here to prosecute, as you see. A great deal of money, men of Athens, was offered me to refrain from prosecuting, but I didn't accept it, and I held out against many requests and favours —yes, and threats too. As for what it remains for you to do **4** after this, the more men he has annoyed by canvassing (I saw what he was doing in front of the courts this morning), the more I expect him to get his deserts. For I can't think that any of you will ignore the charges on which you yourselves supported me before, or that any of you will give Meidias impunity for insolence in future, by voting under oath for anything but what you consider just.

If, men of Athens, I were going to prosecute him for illegal **5** proposals or misconduct of an embassy or any other charge of that sort, I shouldn't think it right to make any appeal for favour to you. I think it proper for the prosecutor on such charges merely to prove his case; it's for the defendant to plead in addition. But as it is, this man corrupted the judges at the time, and as a result my tribe was unjustly deprived of the tripod, and I myself was subjected to blows and inso- **6** lence, such as I doubt whether any chorus-producer ever suffered before. So, since the vote which the people gave against him for those acts, in indignation and anger on my behalf, is the subject of my case, I shan't hesitate to make an appeal too. For I, so to speak, am now on trial, if it is a penalty to be a victim of insolence and not obtain justice for it. So I appeal **7** to you all, men of the jury, and I entreat you, first to listen to my speech with goodwill, and secondly, if I prove that this man Meidias has treated insolently not only me but also you

νόμους καὶ εἰς τοὺς ἄλλους ἅπαντας ὑβρικότα, βοηθῆσαι καὶ
ἐμοὶ καὶ ὑμῖν αὐτοῖς. καὶ γὰρ οὕτω πως ἔχει, ὦ ἄνδρες
Ἀθηναῖοι· ὕβρισμαι μὲν ἐγὼ καὶ προπεπηλάκισται τὸ σῶμα
τοὐμὸν τότε, ἀγωνιεῖται δὲ καὶ κριθήσεται τὸ πρᾶγμα νυνί,
πότερον ἐξεῖναι δεῖ τὰ τοιαῦτα ποιεῖν καὶ εἰς τὸν τυχόνθ' 5
8 ὑμῶν ἀδεῶς ὑβρίζειν ἢ μή. εἴ τις οὖν ὑμῶν ἄρα καὶ τὸν
ἔμπροσθεν χρόνον τῶν ἰδίων τινὸς ἕνεκα γίγνεσθαι τὸν ἀγῶνα
τόνδ' ὑπελάμβανεν, ἐνθυμηθεὶς νῦν ὅτι δημοσίᾳ συμφέρει
μηδενὶ μηδὲν ἐξεῖναι τοιοῦτο ποιεῖν, ὡς ὑπὲρ κοινοῦ τοῦ
πράγματος ὄντος καὶ προσέχων ἀκουσάτω καὶ τὰ φαινόμενα 10
αὐτῷ δικαιότατα εἶναι, ταῦτα ψηφισάσθω.

Ἀναγνώσεται δὲ πρῶτον μὲν ὑμῖν τὸν νόμον καθ' ὃν εἰσὶν αἱ
προβολαί· μετὰ δὲ ταῦτα καὶ περὶ τῶν ἄλλων πειράσομαι
διδάσκειν. λέγε τὸν νόμον.

ΝΟΜΟΣ 15

*Τοὺς πρυτάνεις ποιεῖν ἐκκλησίαν ἐν Διονύσου τῇ
ὑστεραίᾳ τῶν Πανδίων. ἐν δὲ ταύτῃ χρηματίζειν πρῶτον
μὲν περὶ ἱερῶν, ἔπειτα τὰς προβολὰς παραδιδότωσαν
τὰς γεγενημένας ἕνεκα τῆς πομπῆς ἢ τῶν ἀγώνων τῶν
ἐν τοῖς Διονυσίοις, ὅσαι ἂν μὴ ἐκτετισμέναι ὦσιν.* 20

9 Ὁ μὲν νόμος οὗτός ἐστιν, ὦ ἄνδρες Ἀθηναῖοι, καθ' ὃν αἱ
προβολαὶ γίγνονται, λέγων, ὥσπερ ἠκούσατε, ποιεῖν τὴν ἐκκλησίαν ἐν Διονύσου μετὰ τὰ Πάνδια, ἐν δὲ ταύτῃ, ἐπειδὰν
χρηματίσωσιν οἱ πρόεδροι περὶ ὧν διῴκηκεν ὁ ἄρχων, χρηματίζειν καὶ περὶ ὧν ἄν τις ἠδικηκὼς ᾖ περὶ τὴν ἑορτὴν ἢ 25
παρανενομηκώς, καλῶς, ὦ ἄνδρες Ἀθηναῖοι, καὶ συμφερόν-
τως ἔχων ὁ νόμος, ὡς τὸ πρᾶγμ' αὐτὸ μαρτυρεῖ. ὅπου γὰρ
ἐπόντος τοῦ φόβου τούτου φαίνονταί τινες οὐδὲν ἧττον ὑβρ-

2–4 καὶ γὰρ ... τότε Aristid. *Rh.* (Schmid) 10.14   21 ὁ ... ὦ Π12
23–7 ἐπειδὰν ... μαρτυρεῖ Anon. *Fig.* (Spengel 3) 122.8   26–94.1 ὀρθῶς
καὶ καλῶς ... ὕβρισται Aps. *Rh.* (Hammer) 292.19

---

2 οὕτως ἔχει Aristid.   2–3 ὦ ἄνδρες δικασταί A^{cp}: om. Aristid.   5 τὰ
om. Y   -τα A   6 ἢ] ει Π3   οὖν om. F^{ac} 7 -σθε SYP^{ac} τῶν om. A
ἕνεκά τινος A: τινὸς εἵνεκα YP^{ac}: τιν[...]νεκα Π3   γιν- A: [Π3]   8 -δε AF
9 τοιοῦτόν τι A: τοῦτο F   10 ὄντος om. Y^{ac} 11 αὑτῷ Markland
δικαιότερ' SYP   12 δὲ om. $S^{ac}YP^{ac}$ 14 διδάσκειν ὑμᾶς F   16–20 legem
om. A   17 τῶν Πανδίων Palmer: ἐν πανδίον' S: ἐν πανδίονι FYP   18 παραδότωσαν F   20 ὅσα F^{ac} 21 ὦ ἄνδρες δικασταί A^{cp} 22 γίν- A Σ (Patm.)
24 πρόοδοι S^{ac} 24–5 χρηματίζειν κελεύει $AFY^{pc}P$ *Fig.*   25 περὶ (post
ᾗ)] τι ἐπὶ *Fig.*   26–7 ὀρθῶς καὶ καλῶς ἔχων ὁ νόμος Aps.   26 ὦ ἄνδρες
δικασταί A^{cp}: ὦ θεοί *Fig.*   27 -μα AF   28 φόνου Aps.

and the laws and everyone else, to support both me and yourselves. This is the situation, men of Athens: the insolence and assault were directed at me and my person that day, but the question that will be contested and decided now is whether it should be permitted to do this sort of thing and to behave with insolence towards any one of you with impunity, or not. Therefore, if any of you did suppose previously that **8** this case arose from a private motive, you should now bear in mind that it's not beneficial to any member of the public for such an act to be permitted. Consider that the question is one that affects everyone; listen with attention; and vote for what appears to you to be the most just.

He will read to you first the law governing *probolai*. After that I shall try to explain the other matters. Read the law.

LAW

The prytaneis are to convene a meeting of the Ekklesia in the precinct of Dionysos on the day after the Pandia. At this meeting they are to deal first with sacred matters; next let them bring forward the *probolai* which have been made in connection with the procession or the contests at the Dionysia, all that have not been paid for.

That is the law, men of Athens, authorizing *probolai*. As **9** you heard, it specifies the convening of the Ekklesia in the precinct of Dionysos after the Pandia, and at this meeting the proedroi, after dealing with the arrangements made by the arkhon, are to deal with any offence or illegal act concerning the festival. That is a good and beneficial provision, men of Athens, as this very case proves; for when, in the face of this deterrent, people display insolence nonetheless, what must

ισταί, τί χρὴ τοὺς τοιούτους προσδοκᾶν ἂν ποιεῖν, εἰ μηδεὶς ἐπῆν ἀγὼν μηδὲ κίνδυνος;

10 Βούλομαι τοίνυν ὑμῖν καὶ τὸν ἑξῆς νόμον ἀναγνῶναι τούτῳ· καὶ γὰρ ἐκ τούτου φανερὰ πᾶσιν ὑμῖν ἥ τε τῶν ἄλλων ὑμῶν εὐλάβεια γενήσεται καὶ τὸ τούτου θράσος. λέγε τὸν νόμον. 5

ΝΟΜΟΣ

Εὐήγορος εἶπεν· ὅταν ἡ πομπὴ ᾖ τῷ Διονύσῳ ἐν Πειραιεῖ καὶ οἱ κωμῳδοὶ καὶ οἱ τραγῳδοί, καὶ ἐπὶ Ληναίῳ ⟨ἡ⟩ πομπὴ καὶ οἱ τραγῳδοὶ καὶ οἱ κωμῳδοί, καὶ τοῖς ἐν ἄστει Διονυσίοις ἡ πομπὴ καὶ οἱ παῖδες καὶ ὁ κῶμος καὶ 10 οἱ κωμῳδοὶ καὶ οἱ τραγῳδοί, καὶ Θαργηλίων τῇ πομπῇ καὶ τῷ ἀγῶνι, μὴ ἐξεῖναι μήτε ἐνεχυράσαι μήτε λαμβάνειν ἕτερον ἑτέρου, μηδὲ τῶν ὑπερημέρων, ἐν ταύταις ταῖς ἡμέραις. ἐὰν δέ τις τούτων τι παραβαίνῃ, ὑπόδικος ἔστω τῷ παθόντι, καὶ προβολαὶ αὐτοῦ ἔστωσαν ἐν τῇ 15 ἐκκλησίᾳ τῇ ἐν Διονύσου ὡς ἀδικοῦντος, καθὰ περὶ τῶν ἄλλων τῶν ἀδικούντων γέγραπται.

11 Ἐνθυμεῖσθε, ὦ ἄνδρες δικασταί, ὅτι ἐν τῷ προτέρῳ νόμῳ κατὰ τῶν περὶ τὴν ἑορτὴν ἀδικούντων οὔσης τῆς προβολῆς, ἐν τούτῳ καὶ κατὰ τῶν τοὺς ὑπερημέρους εἰσπραττόντων ἢ καὶ 20 ἄλλ' ὁτιοῦν τινος λαμβανόντων ἢ βιαζομένων ἐποιήσατε τὰς προβολάς. οὐ γὰρ ὅπως τὸ σῶμα ὑβρίζεσθαί τινος ἐν ταύταις ταῖς ἡμέραις, ἢ τὴν παρασκευὴν ἣν ἂν ἐκ τῶν ἰδίων πορίσαιτό τις εἰς λειτουργίαν, ᾤεσθε χρῆναι, ἀλλὰ καὶ τὰ δίκῃ καὶ ψήφῳ τῶν ἑλόντων γιγνόμενα τῶν ἑαλωκότων καὶ κεκτημένων ἐξ 25 **12** ἀρχῆς τὴν γοῦν ἑορτὴν ἀπεδώκατε εἶναι. ὑμεῖς μὲν τοίνυν, ὦ ἄνδρες Ἀθηναῖοι, πάντες εἰς τοσοῦτον ἀφῖχθε φιλανθρωπίας καὶ εὐσεβείας ὥστε καὶ τῶν πρότερον γεγενημένων ἀδικημά-

24–5 ἀλλὰ ... ἑαλωκότων Π12 26–96.2 τὴν ... ταῖς ἡμ- Π8

---

1 προσδοκᾶν τοὺς τοιούτους AF 2 ἐπῆν] ἦν A^{ac} 3 τοίνυν] δὲ A ἀναγνῶναι νόμον F τούτῳ $S^{ac}Y$: τουτονί $S^{pc}_{3}AFP^{pc}_{3}$: τουτο P^{ac} 4 ἄλλων ἁπάντων ὑμῶν AP^{pc}_{3} 7–17 legem om. A 8–11 καὶ ἐπὶ ... τραγῳδοί om. F 8 ἡ ante πομπὴ MacDowell: ante ἐπὶ Reiske 9 πομπῇ YP 11 Θαργηλίων Wolf: ὁ θαργηλιὼν SFYP: θαργηλιῶνος $BcPr^{pc}$ 12 μὴ B^{pc}: μήτε SFYP: μή τι Buttmann μήτε (ante ἐνεχ-) $S^{ac}FY^{pc}P^{ac}$: μημήτε Y^{ac}: om. $S^{pc}_{4}P^{pc}$ 13 ἕτερον] ὁτιοῦν Weil 15 αὐτῷ Y 16 τῇ ἐν Διονύσου del. Humbert 17 ἄλλων τῶν] ἄλλως πως Spalding 19 κατὰ τῶν περὶ AP^{sl}_{4}: περὶ τῶν κατὰ SFYP 22 ὅπως μὴ $S^{pc}AFYP$ 23 ἂν del. Blass πορίσηται V 24 εἰς τὴν F τὰ ἐν S^{pc} 25 γιν- Π12A 26 ἀποδεδώκατε A 26–7 ὦ ἄνδρες δικασταί A^{cp} 27 αφειχθε Π8 -πίας τε FP^{sl}_{4}

one expect such men would do if they faced no trial or risk?

Now, I want to read you also the next law to this one; for it **10** too will make clear to you all both the care that the rest of you have taken and the audacity of Meidias. Read the law.

LAW

Euegoros proposed: when the procession takes place for Dionysos in Peiraieus and the comedies and the tragedies, and the procession at the Lenaion and the tragedies and the comedies, and at the city Dionysia the procession and the boys and the revel and the comedies and the tragedies, and at the procession and the contest of the Thargelia, it is not to be permitted to distrain or to seize another thing from another person, not even from overdue debtors, during those days. If anyone transgresses any of this, let him be liable to prosecution by the victim, and let there be *probolai* of him in the Ekklesia in the precinct of Dionysos as an offender, in the same way as is laid down about the other offenders.

Notice, men of the jury, that, whereas the previous law **11** specified *probole* against those committing offences concerning the festival, in this one you created *probolai* also against those exacting overdue payments or taking anything else from a person or using violence. So far from thinking it right that insolent treatment should be inflicted during those days on anyone's person or on the preparations made by anyone out of his own resources for a liturgy, you actually decided to allow property awarded by juries to successful prosecutors to remain in the hands of the original owners who have lost their cases, at least for the duration of the festival. You, men **12** of Athens, had all displayed generosity and piety to the point even of prohibiting, for those days, exaction of a penalty for

των τὸ λαμβάνειν δίκην ἐπέσχετε ταύτας τὰς ἡμέρας· Μειδίας δ' ἐν αὐταῖς ταύταις ταῖς ἡμέραις ἄξια τοῦ δοῦναι τὴν ἐσχάτην δίκην ποιῶν δειχθήσεται. βούλομαι δ' ἕκαστον ἀπ' ἀρχῆς ὧν πέπονθα ἐπιδείξας καὶ περὶ τῶν πληγῶν εἰπεῖν ἃς τὸ τελευταῖον προσενέτεινέ μοι· ἓν γὰρ οὐδέν ἐστιν ἐφ' ᾧ τῶν (5) πεπραγμένων οὐ δίκαιος ὢν ἀπολωλέναι φανήσεται.

13 Ἐπειδὴ γὰρ οὐ καθεστηκότος χορηγοῦ τῇ Πανδιονίδι φυλῇ, τρίτον ἔτος τουτί, παρούσης δὲ τῆς ἐκκλησίας ἐν ᾗ τὸν ἄρχοντα ἐπικληροῦν ὁ νόμος τοῖς χορηγοῖς τοὺς αὐλητὰς κελεύει, λόγων καὶ λοιδορίας γιγνομένης, καὶ κατηγοροῦντος (10) τοῦ μὲν ἄρχοντος τῶν ἐπιμελητῶν τῆς φυλῆς, τῶν δ' ἐπιμελητῶν τοῦ ἄρχοντος, παρελθὼν ὑπεσχόμην ἐγὼ χορηγήσειν ἐθελοντής, καὶ κληρουμένων πρῶτος αἱρεῖσθαι **14** τὸν αὐλητὴν ἔλαχον, ὑμεῖς μέν, ὦ ἄνδρες Ἀθηναῖοι, πάντες ἀμφότερ' ὡς οἷόν τε μάλιστ' ἀπεδέξασθε, τήν τε ἐπαγγελίαν (15) τὴν ἐμὴν καὶ τὸ συμβὰν ἀπὸ τῆς τύχης, καὶ θόρυβον καὶ κρότον τοιοῦτον ὡς ἂν ἐπαινοῦντές τε καὶ συνησθέντες ἐποιήσατε, Μειδίας δ' οὑτοσὶ μόνος τῶν πάντων, ὡς ἔοικεν, ἠχθέσθη, καὶ παρηκολούθησεν παρ' ὅλην τὴν λειτουργίαν ἐπηρεάζων μοι συνεχῶς καὶ μικρὰ καὶ μείζω. (20)

15 Ὅσα μὲν οὖν τοὺς χορευτὰς ἐναντιούμενος ἡμῖν ἀφεθῆναι

3–6 *βούλομαι … φανήσεται* schol. Hermog. (Walz 7) 721.5, Io. Sard. 238.15, Ruf. (Hammer) 401.26, Syrian. (Rabe 2) 64.16 4–9 *-αὶ περὶ … νόμο-* Π8 5 *προσενέτεινέ μοι* *Rhet. Lex.* (Naoumides) 44 5–6 *ἓν … φανήσεται* *AB* 1.138.23 7 *ἐπειδὴ … χορηγοῦ* Anon. Seg. (Hammer) 371.18 7–8 *ἐπειδὴ … φυλῇ* Hermog. 288.24, schol. Hermog. (Walz 4) 489.1 *ἐπειδὴ … τουτί* schol. Hermog. (Walz 7) 409.2 7–13 *ἐπειδὴ … ἐθελοντής* Greg. Cor. (Walz 7) 1185.8, Ruf. (Hammer) 402.26 12–16 *ὑπεσχόμην … τύχης* Greg. Cor. (Walz 7) 1181.29 21–98.1 *ὅσα … ἠνώχλησεν* *AB* 1.138.12 21–98.3 *ὅσα … ἐάσω* Aristid. *Rh.* 29.15, Alex. *Fig.* (Spengel 3) 23.17, Anon. *Fig.* (Spengel 3) 142.16, Anon. *Fig. Hermog.* (Walz 3) 711.13 21–98.7 *ὅσα … φανείη* Hermog. 420.17

---

1 *ἐπέσχε* F^{ac}: *ἐπέχετε* Dobree *ταύταις ταῖς ἡμέραις* A 3–4 *δὲ καθ' ἕκαστον ὧν πέπονθα ἐξ ἀρχῆς* A: *δ' ἕκαστα ὧν πέπονθα* Ruf. Sard. 4 *ἔτι καὶ* FYP [*ε*]*ιπιν* Π8 5 *προσέτεινέ* VbV^{ac} schol. Hermog. Ruf. Sard. *ἓν γὰρ οὐδέν*] *οὐδὲν γάρ* Ruf. Sard. *-τι* A 6 *φαίνεται* schol. Hermog. 7 *ἐπεὶ* Ruf. *τοῦ χορηγοῦ* Seg. 8 *τρίτον ἢ τέταρτον* A Greg.: *τέταρτον* F$_{1}^{γρ}$ 9 *χορηγοῖς* S$_{3}^{sl}$P$_{2}^{γρ}$: *χοροῖς* SAFYP Greg. Ruf. 10 *λόγου* Ruf. *γιν-* A: *γεν-* Ruf. 12 *ἐγὼ ὑπεσχόμην* Greg. 1185 13 *ἐθελοντί* Ruf. 14 *καὶ ὑμεῖς* Greg. *ὦ ἄνδρες δικασταί* A^{cp}: *ὦ δικασταί* Greg. 15 *-ρα* AF *-τα* AYP^{ac} 18 *δὲ* A *ἁπάντων* F 19 *-σε* S^{pc}AF^{pc}P$_{3}^{pc}$ *παρ'*] *κατὰ* S^{sl} 20 *ἐμοὶ* AF 21 *ἢ τοὺς* AFP$_{2}^{sl}$ Aristid. Hermog. *ἐναντιούμενος ἡμῖν*] *κωλύων* Hermog. *μὴ ἀφεθῆναι* Lamb^{vl}

offences committed previously, whereas Meidias will be proved to have been committing acts deserving the extreme penalty on those very days. I should like to relate everything inflicted on me from the beginning, before going on to the blows in which his attacks on me culminated; there is not one of his actions that does not justify the death penalty, as you will see.

When no chorus-producer had been appointed for the Pandionis tribe, two years ago, and it was the meeting of the Ekklesia at which the law requires the arkhon to allot the pipers to the chorus-producers, there were speeches and recrimination, with the arkhon accusing the superintendents of the tribe, and the superintendents of the tribe accusing the arkhon; and so I stepped forward and volunteered to serve as chorus-producer. The allotment was made, and the first choice of piper fell to me. You, men of Athens, all gave the warmest welcome to both incidents, my offer and the lucky accident, and your cheering and applause were such as to show both approval and pleasure; but this man Meidias, it seems, was the one and only person who was annoyed, and he pursued me throughout the liturgy with continual obstruction, small and great. **13** **14**

The trouble that he gave me, opposing the release of our **15**

τῆς στρατείας ἠνώχλησεν, ἢ προβαλλόμενος καὶ κελεύων
ἑαυτὸν εἰς Διονύσια χειροτονεῖν ἐπιμελητήν, ἢ τἆλλα πάντα
ὅσα τοιαῦτα, ἐάσω· οὐ γὰρ ἀγνοῶ τοῦθ', ὅτι τῷ μὲν ἐπηρεαζομένῳ τότ' ἐμοὶ καὶ ὑβριζομένῳ τὴν αὐτὴν ὀργὴν ἕκαστον
τούτων ἥνπερ ἄλλ' ὁτιοῦν τῶν δεινοτάτων παρίστη, ὑμῖν δὲ 5
τοῖς ἄλλοις τοῖς ἔξω τοῦ πράγματος οὖσιν οὐκ ἂν ἴσως ἄξια
ταῦτα καθ' αὑτὰ ἀγῶνος φανείη· ἀλλ' ἃ πάντες ὁμοίως ἀγα-
16 νακτήσετε, ταῦτ' ἐρῶ. ἔστι δ' ὑπερβολὴ τῶν μετὰ ταῦτα, ἃ
μέλλω λέγειν, καὶ οὐδ' ἂν ἐπεχείρησα ἔγωγε κατηγορεῖν
αὐτοῦ νῦν, εἰ μὴ καὶ τότε ἐν τῷ δήμῳ παραχρῆμα ἐξήλεγξα. 10
τὴν γὰρ ἐσθῆτα τὴν ἱεράν (ἱερὰν γὰρ ἔγωγε νομίζω πᾶσαν
ὅσην ἄν τις ἕνεκα τῆς ἑορτῆς παρασκευάζηται, τέως ἂν
χρησθῇ) καὶ τοὺς στεφάνους τοὺς χρυσοῦς, οὓς ἐποιησάμην
ἐγὼ κόσμον τῷ χορῷ, ἐπεβούλευσεν, ὦ ἄνδρες Ἀθηναῖοι,
διαφθεῖραί μου νύκτωρ ἐλθὼν ἐπὶ τὴν οἰκίαν τὴν τοῦ χρυσο- 15
χόου· καὶ διέφθειρεν, οὐ μέντοι πᾶσάν γε, οὐ γὰρ ἐδυνήθη. καὶ
τοιοῦτον οὐδείς πώποτε οὐδένα φησὶν ἀκηκοέναι τολμήσαντα
17 οὐδὲ ποιήσαντα ἐν τῇ πόλει. οὐκ ἀπέχρησεν δ' αὐτῷ τοῦτο,
ἀλλὰ καὶ τὸν διδάσκαλον, ὦ ἄνδρες Ἀθηναῖοι, διέφθειρέν μου
τοῦ χοροῦ· καὶ εἰ μὴ Τηλεφάνης ὁ αὐλητὴς ἀνδρῶν βέλτιστος 20
περὶ ἐμὲ τότε ἐγένετο, καὶ τὸ πρᾶγμα αἰσθόμενος τὸν ἄνθρωπον ἀπελάσας αὐτὸς συγκροτεῖν καὶ διδάσκειν ᾤετο δεῖν
τὸν χορόν, οὐδ' ἂν ἠγωνισάμεθα, ὦ ἄνδρες Ἀθηναῖοι, ἀλλ'
ἀδίδακτος ἂν εἰσῆλθεν ὁ χορὸς καὶ πράγματα αἴσχιστ' ἂν
ἐπάθομεν. καὶ οὐδ' ἐνταῦθ' ἔστη τῆς ὕβρεως, ἀλλὰ τοσοῦτον 25

9–10 *καὶ … ἐξήλεγξα* schol. Hermog. (Walz 7) 408.22 19–20 *ἀλλὰ … χοροῦ AB* 1.132.25 20–1 *καὶ … ἐγένετο* schol. Ar. *Pax* 531 24 *ἀδίδακτος … χορὸς AB* 1.344.21, Phot. α 360

---

1 *στρατιᾶς* A *ἠνόχλησεν* S^{ac}: *ἠνώχληκεν* Hermog. *ἢ*] *καὶ* Alex. *καὶ κελεύων* om. Aristid.: *καὶ χειροτονῶν* Alex. 2 *αὐτὸν* Alex. *χειροτονεῖν* et *ἢ … τοιαῦτα* om. Alex. *ἢ*] *καὶ* Aristid. *πάντα* om. Hermog. 3 *-το* SYP 4 *τότ' … ὑβριζομένῳ* om. Hermog. *τότ' ἐμοὶ* FP: *τότε μοι* SY: *τότε ἐμοὶ* A 4–5 *τούτων ἕκαστον ὀργὴν* Hermog. 5 *ἂν ἄλλ'* Dobree 6 *τοῖς ἄλλοις* om. Hermog. *τοῖς* (post *ἄλλοις*) om. S *ἔξωθεν* A^{ac} 6–7 *αὐτὰ καθ' αὑτὰ ἄξια* A: *ἄξια αὐτὰ καθ' ἑαυτὰ* Hermog. 7 *ἃ* SFYP: *ἐφ' οἷς* $AP_1^{\gamma\rho}$ *ἅπαντες* $AP_3^{\gamma\rho}$ *ἂν ὁμοίως* $F_1^{sl}P_3^{\gamma\rho}$ *ἀκούσαντες* (post *ὁμοίως*) $S_3^{mg}F$: (post *-σετε*) AP_3^{mg} 8 *ἀγανακτήσαιτε* $S^{ac}FP$ *δὲ* AF 9 *κατηγορεῖν ἔγωγε* schol. Hermog. 12 *παρασκευάσηται* AFP_2^{pc} *τέως*] *ὡς* P^{pc} 16 *πᾶσάν* SYP: *πάντας* AF Men. Rh. ap. Σ (68b Dilts) *ἠδυνήθη* A *καὶ* SYP: *καίτοι* AFP_3^{pc} 17 *τοιοῦτον* SFYP: *τοῦτό γε* $AP_4^{\gamma\rho}$ 18 *-σε* $S^{pc}AF^{pc}$ *δὲ* SFYP 19 *ὦ θεοί AB* *-ρέ* $S^{pc}AF^{pc}P_2^{pc}$ 21 *τότε* om. schol. Ar. *ἰσθόμενος* S^{ac} 24 *-στα* AF 25 *-θα* SFYP

choristers from military service, or standing and canvassing for election as a superintendent for the Dionysia, and everything else of that sort, I will pass over. I'm well aware that although I, who suffered the obstruction and insolence at that time, felt the same anger at each of those actions as at any other really serious one, to the rest of you who weren't involved they would perhaps not appear worth a trial in themselves. But I'll tell you about what will make you just as indignant as me. **16** The subsequent events, which I'm going to relate, go beyond everything, and I shouldn't even have attempted to accuse him of them now if I hadn't convinced the Ekklesia immediately after the event. He plotted, men of Athens, to destroy my sacred clothing (I regard as sacred all clothing that one makes for the purpose of the festival, until it is used) and the gold crowns which I ordered as an adornment for the chorus, by raiding the goldsmith's house at night; and he did destroy them—though not entirely, because he wasn't able to. And everyone says they never heard of anyone in Athens venturing on or carrying out such a deed. **17** Not content with that, he also corrupted the director of my chorus, men of Athens; and if Telephanes, the piper, had not been a very good friend to me then and, when he saw what was happening, turned the man out and taken upon himself the co-ordination and direction of the chorus, we should have been out of the running, men of Athens—the chorus would have gone on stage untrained, and we should have been utterly disgraced. Nor did his insolence stop even

αὐτῷ περιῆν ὥστε τὸν ἐστεφανωμένον ἄρχοντα διέφθειρεν,
τοὺς χορηγοὺς συνῆγεν ἐπ' ἐμέ, βοῶν, ἀπειλῶν, ὀμνύουσι
παρεστηκὼς τοῖς κριταῖς, τὰ παρασκήνια φράττων,
προσηλῶν, ἰδιώτης ὢν τὰ δημόσια, κακὰ καὶ πράγματα
18 ἀμύθητά μοι παρέχων διετέλεσεν. καὶ τούτων, ὅσα γε ἐν τῷ 5
δήμῳ γέγονεν ἢ πρὸς τοῖς κριταῖς ἐν τῷ θεάτρῳ, ὑμεῖς ἐστέ
μοι μάρτυρες πάντες, ἄνδρες δικασταί. καίτοι τῶν λόγων
τούτους χρὴ δικαιοτάτους ἡγεῖσθαι, οὓς ἂν οἱ καθήμενοι τῷ
λέγοντι μαρτυρῶσιν ἀληθεῖς εἶναι. προδιαφθείρας τοίνυν τοὺς
κριτὰς τῷ ἀγῶνι τῶν ἀνδρῶν, δύο ταῦτα ὡσπερεὶ κεφάλαια 10
ἐφ' ἅπασι τοῖς ἑαυτῷ νενεανιευμένοις ἐπέθηκεν, ἐμοῦ μὲν
ὕβρισεν τὸ σῶμα, τῇ φυλῇ δὲ κρατούσῃ τὸν ἀγῶν' αἰτιώτατος
τοῦ μὴ νικῆσαι κατέστη.

19 Τὰ μὲν οὖν εἰς ἐμὲ καὶ τοὺς φυλέτας ἠσελγημένα καὶ περὶ
τὴν ἑορτὴν ἀδικήματα τούτῳ πεπραγμένα, ἐφ' οἷς αὐτὸν 15
προυβαλόμην, ταῦτ' ἐστίν, ὦ ἄνδρες Ἀθηναῖοι, καὶ πόλλ'
ἕτερα, ὧν ὅσ' ἂν οἷός τε ὦ διέξειμι πρὸς ὑμᾶς αὐτίκα δὴ
μάλα. ἔχω δὲ λέγειν καὶ πονηρίας ἑτέρας παμπληθεῖς αὐτοῦ
καὶ ὕβρεις εἰς πολλοὺς ὑμῶν καὶ τολμήματα τοῦ μιαροῦ
20 τούτου πολλὰ καὶ δεινά, ἐφ' οἷς τῶν πεπονθότων οἱ μέν, ὦ 20
ἄνδρες δικασταί, καταδείσαντες τοῦτον καὶ τὸ τούτου θράσος
καὶ τοὺς περὶ αὐτὸν ἑταίρους καὶ πλοῦτον καὶ τἆλλα ὅσα δὴ
πρόσεστι τούτῳ, ἡσυχίαν ἔσχον, οἱ δ' ἐπιχειρήσαντες δίκην
λαμβάνειν οὐκ ἠδυνήθησαν, εἰσὶ δ' οἳ διελύσαντο, ἴσως
λυσιτελεῖν ἡγούμενοι. τὴν μὲν οὖν ὑπὲρ αὐτῶν δίκην ἔχουσιν 25
οἵ γε πεισθέντες· τῆς δ' ὑπὲρ τῶν νόμων, οὓς παραβὰς οὗτος
κἀκείνους ἠδίκει καὶ νῦν ἐμὲ καὶ πάντας τοὺς ἄλλους, ὑμεῖς
21 ἐστε κληρονόμοι. πάντων οὖν ἀθρόων ἓν τίμημα ποιήσασθε, ὅ

14–16 *τὰ … ἐστίν* schol. Hermog. (Walz 4) 309.27, (Walz 7) 408.26, Sopat. (Walz 5) 119.31

---

1 -ρε $S^{pc}AF^{pc}P_2^{pc}$ 5 -σε $S_4^{pc}AF^{pc}P_3^{pc}$ γε SAYP: μὲν $FP_2^{\gamma\rho}$ 7 ὦ ἄνδρες $A^{cp}F$ 8 δικαιοτάτους καὶ πιστοτάτους A   9 προδια- CdVe: προσδια- SAFYP   10 τῷ ἀγῶνι τῶν] τῶν ἀγωνιστῶν Vd Σ (80 Dilts)   ὡσπερι S^{ac} 11 αυτῷ A   νενιανευ- S: νεανιευ- A   12 -σε A   εἰς τὸ F   14 οὖν om. schol. Hermog. (Walz 4)   ἠσελγημένα om. Sopat.: ἡμαρτημένα schol. Hermog. (Walz 4)   τὰ περὶ AP_2^{pc} schol. Hermog. Sopat.   15 τούτῳ πεπραγμένα om. schol. Hermog. (Walz 4) Sopat.   16 -λὰ A   17 ὅσα AF   ὧι A   18 αὐτοῦ παμπληθεῖς A   22 ἑτέρους S   πλοῦτον καὶ ὕβριν καὶ AFP_2^{mg} τὰ ἄλλα S   δὴ ὅσα SFYP   23 ἔσχον] ἦγον $Lamb^{vl}$ 24 λαβεῖν A   ἐδυν- FP_3^{pc} δ' οἳ καὶ A: δὲ καὶ οἳ FP_2^{pc} 25 αὐτοῖς (post -λεῖν) $AY^{sl}P^{ac}$: αὑτοῖς P_4^{pc}: αὑτοὺς F   αὑτῶν (post ὑπὲρ) P_4^{pc}: αυτῶν SA: αὐτῶν FYP^{ac} 26 δὲ A   27 ἠδίκει τότε AP_4^{sl} 28 οὖν τούτων A   ἀθρόων S_3^{pc}: ἀνθρώπων S^{ac}

there. He was so full of it that he tried to corrupt the crowned arkhon, and to get the chorus-producers to gang up against me; he shouted, he threatened, he stood beside the judges while they were taking the oath, he blocked and nailed up the side-scenes, though they are public property and he held no official position—I can't tell you how much harm and trouble he caused me, continually. And as far as concerns the **18** incidents in the Ekklesia or before the judges in the theatre, all of you are my witnesses, men of the jury. Indeed the most reliable statements are those whose truth the audience can attest for the speaker. So, corrupting the judges for the men's contest in advance, he capped all his youthful exploits, so to speak, with these two: he treated my person with insolence, and he was the man most to blame that the tribe which was best in the contest did not win.

Those, men of Athens, are the attacks on me and my **19** fellow-tribesmen and offences concerning the festival committed by this man, for which I accused him by *probole*. There are also many others, and I shall immediately relate to you all of them that I can. I am also able to tell of numerous other acts of wickedness and insolence which he has perpetrated against many of you, and many dreadful deeds which the scoundrel has ventured to commit. Some of the victims, **20** men of the jury, in fear of him and his audacity and his comrades and wealth and all the rest of the advantages you know he has, kept quiet; others attempted to obtain compensation and failed; and some made peace with him, perhaps thinking it to their advantage. Well, those who were persuaded to do that at least have compensation for themselves; but as for the compensation for the laws which he transgressed when he wronged those men, and now when he has wronged me, and all the others, you are the heirs of that. So **21** for all the crimes put together you must impose a single

τι ἂν δίκαιον ἡγῆσθε. ἐξελέγξω δὲ πρῶτον μὲν ὅσα αὐτὸς
ὑβρίσθην, ἔπειθ᾽ ὅσα ὑμεῖς· μετὰ ταῦτα δὲ καὶ τὸν ἄλλον, ὦ
ἄνδρες Ἀθηναῖοι, βίον αὐτοῦ πάντα ἐξετάσω, καὶ δείξω
πολλῶν θανάτων, οὐχ ἑνὸς ὄντα ἄξιον. λέγε μοι τὴν τοῦ
χρυσοχόου πρώτην λαβὼν μαρτυρίαν. 5

22 ΜΑΡΤΥΡΙΑ

{Παμμένης Παμμένους †ἔπερχος† ἔχων χρυσοχοεῖον ἐν
τῇ ἀγορᾷ, ἐν ᾧ καὶ καταγίγνομαι καὶ ἐργάζομαι τὴν
χρυσοχοϊκὴν τέχνην. ἐκδόντος δέ μοι Δημοσθένους, ᾧ
μαρτυρῶ, στέφανον χρυσοῦν ὥστε κατασκευάσαι καὶ 10
ἱμάτιον διάχρυσον ποιῆσαι, ὅπως πομπεύσαι ἐν αὐτοῖς
τὴν τοῦ Διονύσου πομπήν, καὶ ἐμοῦ συντελέσαντος αὐτὰ
καὶ ἔχοντος παρ᾽ ἐμαυτῷ ἕτοιμα, εἰσπηδήσας πρός με
νύκτωρ Μειδίας ὁ κρινόμενος ὑπὸ Δημοσθένους, ἔχων
μεθ᾽ ἑαυτοῦ καὶ ἄλλους, ἐπεχείρησεν διαφθείρειν τὸν 15
στέφανον καὶ τὸ ἱμάτιον, καὶ τινὰ μὲν αὐτῶν ἐλυμήνατο,
οὐ μέντοι πάντα γε ἠδυνήθη διὰ τὸ ἐπιφανέντα με
κωλῦσαι.}

23 Πολλὰ μὲν τοίνυν, ὦ ἄνδρες Ἀθηναῖοι, καὶ περὶ ὧν τοὺς
ἄλλους ἠδίκηκεν ἔχω λέγειν, ὥσπερ εἶπον ἐν ἀρχῇ τοῦ λόγου, 20
καὶ συνείλοχα ὕβρεις αὐτοῦ καὶ ἀτιμίας τοσαύτας ὅσας ἀκού-
σεσθε αὐτίκα δὴ μάλα. ἦν δ᾽ ἡ συλλογὴ ῥᾳδία· αὐτοὶ γὰρ οἱ
24 πεπονθότες προσῇσάν μοι. βούλομαι δὲ πρὸ τούτων εἰπεῖν οἷς
ἐπιχειρήσειν αὐτὸν ἀκήκοα ἐξαπατᾶν ὑμᾶς· τοὺς γὰρ ὑπὲρ
τούτων λόγους ἐμοὶ μὲν ἀναγκαιοτάτους προειπεῖν ἡγοῦμαι, 25
ὑμῖν δὲ χρησιμωτάτους ἀκοῦσαι. διὰ τί; ὅτι τοῦ δικαίαν καὶ
εὔορκον θέσθαι τὴν ψῆφον ὁ κωλύσας ἐξαπατηθῆναι λόγος
ὑμᾶς οὗτος αἴτιος ἔσται. πολὺ δὴ μάλιστα πάντων τούτῳ τῷ
λόγῳ προσέχειν ὑμᾶς δεῖ, καὶ μνημονεῦσαι τοῦτον, καὶ πρὸς
ἕκαστον ἀπαντᾶν, ὅταν οὗτος λέγῃ. 30

15–16 ἐπεχείρησεν ... στέφανον *AB* 1.132.27

---

2 -τα A ὑμεῖς ἠδίκησθε S_3^{sl}AFP_4^{mg} 7–18 testimonium om. A
7 ἔπαρχος FP_4^{sl} ἔχω Fk^{pc}LnVi 9 ᾧ καὶ YP 10 μαρτυρῶι S
11 πομπεύσαιεν αὐτοῖς Y 14 νύκτωρ καὶ S^{ac} 17 γε πάντα P 20 ἐν
ἀρχῇ εἶπον Y 21 συνείλεχα F αὐτοῦ καὶ ἀτιμίας SYP: καὶ πονηρίας αὐτοῦ
A$F_1^{γρ}$: αὐτοῦ καὶ πονηρίας F 23 προσῇσάν Dindorf: προσήεσάν SAFYP
25 ἀναγκαιοτάτους ἐμοὶ μὲν S^{ac} 27 κωλύσων A: [P^{ac}] 28 ἔσται
αἴτιος S^{ac} 30 ἁπάντων Y: [S^{ac}]

penalty, whatever you think just. I shall prove first the insolence against myself, and then that against you. After that I shall examine all the rest of his life, men of Athens, and I shall show that he deserves to die several times over, not just once. Take and read the goldsmith's testimony first.

TESTIMONY

22

{I, Pammenes son of Pammenes [...], have a goldsmith's workshop in the Agora, in which I reside and practise the craft of goldsmith. Demosthenes, for whom I testify, commissioned me to fashion a gold crown and make a cloak interwoven with gold, in order to wear them in the procession of Dionysos. After I had completed them and had them ready in my house, Meidias, who is prosecuted by Demosthenes, accompanied by other men, burst in on me at night and attempted to destroy the crown and the cloak. He ruined some of them, but was unable to ruin them all because I appeared and prevented it.}

I am in a position also to tell you a great deal, men of **23** Athens, about his offences against other men, as I said at the beginning of my speech. I've collected all his acts of insolence and dishonour that you'll hear about immediately. The collecting was easy, because the victims approached me of their own accord. But before that, I want to tell you the means by **24** which I've heard he will attempt to deceive you. I consider it very necessary for me to speak to you about these in advance, and very useful for you to hear about them. Why? Because the part of my speech which prevents your being deceived will enable you to vote justly and in accordance with your oath. You ought to pay most attention to this part of my speech, and remember it, and resist each point when he brings it forward.

25 Ἔστι δὲ πρῶτον μὲν ἐκεῖνο οὐκ ἄδηλος ἐρῶν, ἐξ ὧν ἰδίᾳ
πρός τινας αὐτὸς διεξιὼν ἀπηγγέλλετό μοι, ὡς εἴπερ ἀληθῶς
ἐπεπόνθειν ταῦτα ἃ λέγω, δίκας ἰδίας μοι προσῆκεν αὐτῷ
λαχεῖν, τῶν μὲν ἱματίων καὶ τῶν χρυσῶν στεφάνων τῆς διαφθορᾶς καὶ τῆς περὶ τὸν χορὸν πάσης ἐπηρείας, βλάβης, ὧν δ' 5
εἰς τὸ σῶμα ὑβρίσθαι φημί, ὕβρεως, οὐ μὰ Δί' οὐχὶ δημοσίᾳ
κρίνειν αὐτὸν καὶ τίμημα ἐπάγειν ὅ τι χρὴ παθεῖν ἢ ἀποτεῖ-
26 σαι. ἐγὼ δὲ ἓν μὲν ἐκεῖνο εὖ οἶδα, καὶ ὑμᾶς δὲ εἰδέναι χρή, ὅτι
εἰ μὴ προυβαλόμην αὐτὸν ἀλλ' ἐδικαζόμην, οὑναντίος ἧκεν ἂν
εὐθύς μοι λόγος, ὡς εἴπερ ἦν τι τούτων ἀληθές, προβαλέσθαι 10
μ' ἔδει καὶ παρ' αὐτὰ τἀδικήματα τὴν τιμωρίαν ποιεῖσθαι· ὅ
τε γὰρ χορὸς ἦν τῆς πόλεως, ἥ τε ἐσθὴς τῆς ἑορτῆς ἕνεκα
πᾶσα παρεσκευάζετο, ἐγώ τε ὁ πεπονθὼς ταῦτα χορηγὸς ἦν·
τίς ἂν οὖν ἑτέραν μᾶλλον εἵλετο τιμωρίαν ἢ τὴν ἐκ τοῦ νόμου
27 κατὰ τῶν περὶ τὴν ἑορτὴν ἀδικούντων οὖσαν; ταῦτ' εὖ οἶδ' ὅτι 15
πάντ' ἂν ἔλεγεν οὗτος τότε. φεύγοντος μὲν γάρ, οἶμαι, καὶ
ἠδικηκότος ἐστὶν τὸ τὸν παρόντα τρόπον τοῦ μὴ δοῦναι δίκην
διακρουόμενον τὸν οὐκ ὄνθ' ὡς ἔδει γενέσθαι λέγειν, δικαστῶν δέ γε σωφρόνων τούτοις τε μὴ προσέχειν καὶ ὃν ἂν
28 λάβωσιν ἀσελγαίνοντα κολάζειν. μὴ δὴ τοῦτο λέγειν αὐτὸν 20
ἐᾶτε, ὅτι καὶ δίκας ἰδίας δίδωσιν ὁ νόμος μοι καὶ γραφὴν
ὕβρεως. δίδωσι γάρ, ἀλλ' ὡς οὐ πεποίηκεν ἃ κατηγόρηκα, ἢ
πεποιηκὼς οὐ περὶ τὴν ἑορτὴν ἀδικεῖ, τοῦτο δεικνύτω· τοῦτο
γὰρ αὐτὸν ἐγὼ προυβαλόμην, καὶ περὶ τούτου τὴν ψῆφον

1–2 ἔστι ... μοι Hermog. 439.14 1–6 ἔστι ... ὕβρεως schol. Hermog. (Walz 4) 500.16 1–8 ἔστι ... ἀποτεῖσαι Greg. Cor. (Walz 7) 1291.11, schol. Hermog. (Walz 7) 417.11 20–3 μὴ ... δεικνύτω schol. Hermog. (Walz 4) 500.26, (Walz 7) 418.1

---

2 αὐτὸς om. Greg. schol. Hermog. (Walz 7) ἀπηγγείλατό schol. Hermog. (Walz 7) Lamb^{vl} ἀληθὲς SY: [P^{ac}] 3 ταῦτα ἐπεπόνθειν schol. Hermog. δίκην ἰδίως schol. Hermog. (Walz 4) 4 λαβεῖν schol. Hermog. (Walz 4) ἕνεκα (post στεφάνων) P$^{sl}_{3}$ 5 παρὰ schol. Hermog. (Walz 4) ἁπάσης F καὶ βλάβης Greg. schol. Hermog. (Walz 7) 6 ὑβρίζεσθαί φημι SYP^{ac}: φημι ὑβρίσθαι Greg.: ὑβρίσθαι φαμὲν schol. Hermog. (Walz 4) ὕβρεων schol. Hermog. (Walz 7) μὰ τὸν Greg. δία A 7 ἑαυτὸν S^{ac}YP^{ac}: αὐτὸν Bc παθεῖν αὐτὸν schol. Hermog. ἀποτεῖσαι Blass: ἀποτίσαι SAFYP Greg. schol. Hermog. 9 προυβαλλόμην BcVbWd ἂν ἧκεν F 10 προβάλλεσθαί S 11 με AF τα δικαια S^{ac} 13 παρεσκεύαστο A 14 μᾶλλον om. S τῶν νόμων A 16 ἂν SFP$^{pc}_{2}$: ἐὰν A: om. YP^{ac} μὲν om. S^{ac} γὰρ ἂν οἶμαι S 17 -τὶ AFP^{pc} μὴ om. A 18 -τα A 21 μοι δίδωσιν ὁ νόμος schol. Hermog. (Walz 4) 21–2 καὶ γραφὴν ὕβρεως om. schol. Hermog. (Walz 4) 22 κατηγόρηκεν schol. Hermog. (Walz 7) 23 οὐ om. schol. Hermog. (Walz 4)

First, one thing that it is clear he will say, to judge from **25** what he himself was reported to me as telling some people privately, is that, if I had really suffered the treatment I allege, I ought to have brought private cases against him—for the destruction of the cloaks and the gold crowns, and for all the obstruction concerning the chorus, a case of damage; and for the insolent treatment of my person which I allege, a case of *hybris*—certainly not to subject him to a public prosecution and assessment of the penalty which he should suffer or pay. But I know one thing very well, and you should know **26** it too: if I had brought a private case and not a *probole* against him, I should immediately have been faced with the converse argument, that, if any of these accusations was true, I ought to have brought a *probole* and taken steps to punish him at the actual time of the offences. The chorus belonged to the city, and all the clothing was being made for the festival, and I, the victim of all this, was a chorus-producer; so who would have thought any other punishment preferable to the one prescribed by the law against persons committing offences concerning the festival? That would have been his tale then, **27** I know very well. An offender on trial, I suppose, will try to evade the method of being brought to justice which is being used and say that the one which is not ought to have been; yes, but sensible jurors will pay no attention to that, and will punish anyone they catch bullying. So don't let him say that **28** the law permits me private cases and a *graphe* for *hybris*. It does permit them, but what he has to show is that he hasn't done the things of which I've accused him, or that, though he has done them, he's not an offender concerning the festival; that's what I've prosecuted him for by *probole*, and on that

οἴσετε νῦν ὑμεῖς. εἰ δ' ἐγὼ τὴν ἐπὶ τῶν ἰδίων δικῶν πλεονεξίαν ἀφεὶς τῇ πόλει παραχωρῶ τῆς τιμωρίας, καὶ τοῦτον
εἱλόμην τὸν ἀγῶνα ἀφ' οὗ μηδὲν ἔστι λῆμμα λαβεῖν ἐμοί,
χάριν, οὐ βλάβην δήπου τοῦτ' ἂν εἰκότως ἐνέγκοι μοι παρ'
ὑμῶν. 5

29 Οἶδα τοίνυν ὅτι καὶ τούτῳ πολλῷ χρήσεται τῷ λόγῳ, "μή
με Δημοσθένει παραδῶτε, μηδὲ διὰ Δημοσθένην με ἀνέλητε·
ὅτι τούτῳ πολεμῶ, διὰ τοῦτό με ἀναιρήσετε;" τὰ τοιαῦτα
πολλάκις οἶδ' ὅτι φθέγξεται, βουλόμενος φθόνον τινὰ ἐμοὶ διὰ
30 τούτων τῶν λόγων συνάγειν. ἔχει δ' οὐχ οὕτω ταῦτα, οὐδ' 10
ἐγγύς. οὐδένα γὰρ τῶν ἀδικούντων ὑμεῖς οὐδενὶ τῶν
κατηγόρων ἐκδίδοτε· οὐδὲ γάρ, ἐπειδὰν ἀδικηθῇ τις, ὡς ἂν
ἕκαστος ὑμᾶς ὁ παθὼν πείσῃ ποιεῖσθε τὴν τιμωρίαν, ἀλλὰ
τοὐναντίον νόμους ἔθεσθε πρὸ τῶν ἀδικημάτων, ἐπ' ἀδήλοις
μὲν τοῖς ἀδικήσουσιν, ἀδήλοις δὲ τοῖς ἀδικησομένοις. οὗτοι δὲ 15
τί ποιοῦσιν οἱ νόμοι; πᾶσιν ὑπισχνοῦνται τοῖς ἐν τῇ πόλει
δίκην, ἂν ἀδικηθῇ τις, ἔσεσθαι δι' αὑτῶν λαβεῖν. ὅταν τοίνυν
τῶν παραβαινόντων τινὰ τοὺς νόμους κολάζητε, οὐ τοῖς
κατηγόροις τοῦτον ἐκδίδοτε, ἀλλὰ τοὺς νόμους ὑμῖν αὐτοῖς
31 βεβαιοῦτε. ἀλλὰ μὴν πρός γε τὸ τοιοῦτον, ὅτι "Δημοσθένης" 20
φησὶν "ὕβρισται", δίκαιος καὶ κοινὸς καὶ ὑπὲρ ἁπάντων ἔσθ'
ὁ λόγος. οὐ γὰρ εἰς Δημοσθένην ὄντα με ἠσέλγαινε μόνον
ταύτην τὴν ἡμέραν, ἀλλὰ καὶ εἰς χορηγὸν ὑμέτερον· τοῦτο δ'
32 ὅσον δύναται γνοίητ' ἂν ἐκ τωνδί. ἴστε δήπου τοῦθ', ὅτι τῶν
θεσμοθετῶν τούτων οὐδενὶ θεσμοθέτης ἔστ' ὄνομα, ἀλλ' 25
ὁτιδήποθ' ἑκάστῳ. ἂν μὲν τοίνυν ἰδιώτην ὄντα τινὰ αὐτῶν
ὑβρίσῃ τις ἢ κακῶς εἴπῃ, γραφὴν ὕβρεως καὶ δίκην κακηγορίας ἰδίαν φεύξεται, ἐὰν δὲ θεσμοθέτην, ἄτιμος ἔσται καθάπαξ.
διὰ τί; ὅτι τοὺς νόμους ἤδη ὁ τοῦτο ποιῶν προσυβρίζει καὶ
τὸν ὑμέτερον κοινὸν στέφανον καὶ τὸ τῆς πόλεως ὄνομα· ὁ γὰρ 30

6–7 μή ... ἀνέλητε Max. (Rabe *PS*) 447.7 10–14 -γων ... ἐπ' ἀδη- Π4 22–3 οὐ ... ὑμέτερον Minuc. (Hammer) 345.8

---

1 δὲ τὴν A 4 ἐνέγκοι μοι S: ἐνέγκαι μοι AFP$_2^{pc}$: ἐνέγκοιμι F$_1^{sl}$YP^{ac}: ἐνεγκαίμην Lamb^{vl} 6–7 μή με] μηδε A 8 τούτῳ AFP$_2^{γρ}$Σ (98c Dilts): ἐκείνῳ SYP 10 συλλέγειν S^{mg}F$_1^{γρ}$ δὲ S ουτως Π4 12 οὐδὲ γάρ SFYP: οὐδ' A: ⟦ο⟧ουδ[ε]γαρ Π4 ἀδικηθείς Y: [P^{ac}] 13 ὑμῶν Mm:]ν Π4 ποιεῖσθαι Π4 S^{ac} 15 τοῖς ... δὲ om. P^{ac} 16 ποιήσουσιν SF 17 ἐὰν A αὐτῶν Buttmann 18 οὐχὶ A 19 αὐτοὺς A 20 τοιοῦτο F 21 ὑβρίσθαι S$_1^{pc}$ 22 μόνον ὄντα με ἠσέλγαινεν Minuc. 23 ταύτην τὴν ἡμέραν om. Minuc. 24 δύνατε A -ητε A τῶνδε AFP$_3^{pc}$ -το SYP 25 -τιν A 26 -ποτ' A: -ποτε F 27–8 κατηγορίας S$_4^{sl}$ 29 προυβρίζει Y

you will cast your votes today. If I, forgoing the gain arising in private cases, allow the city to receive the penalty, and if I chose this action from which it's impossible for me to obtain any profit, surely that should win me favour from you, not disadvantage.

I know he'll also make much use of this argument: 'Don't **29** deliver me up to Demosthenes! Don't destroy me because of Demosthenes! Are you going to destroy me just because I'm his enemy?' That's the sort of thing he'll keep uttering, I know, in the hope of building up some resentment against me by this talk. But that's not the position; far from it. You do **30** not hand any offender over to any prosecutor. You do not even impose the penalty in whatever way each victim of an offence persuades you to do it. On the contrary, you have made laws in advance of the offences, for offenders not yet known and for victims not yet known. And what do those laws do? They promise to all in the city that, if anyone is wronged, it will be possible to obtain justice through them. So when you punish someone who transgresses the laws, you're not handing him over to the prosecutors: you're confirming the laws in your own interest. And as for the **31** assertion that he makes to the effect 'It's Demosthenes who is the victim', the answer is one that is just and fair and concerns everybody: it wasn't just me, Demosthenes, that he was bullying during that day, but also your chorus-producer. Let me explain the significance of that. You know of course that **32** none of the thesmothetai here has the name Thesmothetes, but whatever name each one has. Well then, if one treats insolently or slanders any of them as a private individual, one will be prosecuted in a *graphe* for *hybris* or a private case for slander; if as a thesmothetes, one will be permanently disfranchised. Why? Because the man who does that is using insolence also against the laws, and against the crown that belongs to you all, and against the city's name; for the name

θεσμοθέτης οὐδενὸς ἀνθρώπων ἔστ' ὄνομα, ἀλλὰ τῆς πόλεως.
33 καὶ πάλιν γε τὸν ἄρχοντα, ταὐτὸ τοῦτο, ἐὰν μὲν ἐστεφανωμένον πατάξῃ τις ἢ κακῶς εἴπῃ, ἄτιμος, ἐὰν δὲ ἰδιώτην, ἰδίᾳ ὑπόδικος. καὶ οὐ μόνον περὶ τούτων οὕτω ταῦτ' ἔχει, ἀλλὰ καὶ περὶ πάντων οἷς ἂν ἡ πόλις τινὰ {ἄδειαν ἢ} στεφανηφορίαν 5
ἤ τινα τιμὴν δῷ. οὕτω τοίνυν καὶ ἐμέ, εἰ μὲν ἐν ἄλλαις τισὶν ἡμέραις ἠδίκησέν τι τούτων Μειδίας ἰδιώτην ὄντα, ἰδίᾳ καὶ
34 δίκην προσῆκεν αὐτῷ διδόναι· εἰ δὲ χορηγὸν ὄντα ὑμέτερον ἱερομηνίας οὔσης πάνθ' ὅσα ἠδίκηκεν ὑβρίσας φαίνεται, δημοσίας ὀργῆς καὶ τιμωρίας δίκαιός ἐστι τυγχάνειν· ἅμα 10
γὰρ τῷ Δημοσθένει καὶ ὁ χορηγὸς ὑβρίζετο, τοῦτο δ' ἐστὶ τῆς πόλεως, καὶ τὸ ταύταις ταῖς ἡμέραις, αἷς οὐκ ἐῶσιν οἱ νόμοι. χρὴ δέ, ὅταν μὲν τιθῆσθε τοὺς νόμους, ὁποῖοί τινές εἰσιν σκοπεῖν, ἐπειδὰν δὲ θῆσθε, φυλάττειν καὶ χρῆσθαι· καὶ γὰρ
35 εὔορκα ταῦθ' ὑμῖν ἐστιν καὶ ἄλλως δίκαια. ἦν ὁ τῆς βλάβης 15
ὑμῖν νόμος πάλαι, ἦν ὁ τῆς αἰκείας, ἦν ὁ τῆς ὕβρεως. εἰ τοίνυν ἀπέχρη τοὺς τοῖς Διονυσίοις τι ποιοῦντας τούτων κατὰ τούτους τοὺς νόμους δίκην διδόναι, οὐδὲν ἂν προσέδει τοῦδε τοῦ νόμου. ἀλλ' οὐκ ἀπέχρη. σημεῖον δέ· ἔθεσθε ἱερὸν νόμον αὐτῷ τῷ θεῷ περὶ τῆς ἱερομηνίας. εἴ τις οὖν κἀκείνοις τοῖς 20
προϋπάρχουσι νόμοις καὶ τούτῳ τῷ μετ' ἐκείνους τεθέντι καὶ πᾶσι τοῖς λοιποῖς ἔστ' ἔνοχος, ὁ τοιοῦτος πότερα μὴ δῷ διὰ τοῦτο δίκην ἢ ⟨κἂν⟩ μείζω δοίη δικαίως; ἐγὼ μὲν οἶμαι μείζω.

36 Ἀπήγγελλε τοίνυν τίς μοι περιιόντ' αὐτὸν συλλέγειν καὶ 25
πυνθάνεσθαι τίσιν πώποτε συμβέβηκεν ὑβρισθῆναι, καὶ λέγειν τούτους καὶ διηγεῖσθαι πρὸς ὑμᾶς μέλλειν, οἷον, ὦ ἄνδρες Ἀθηναῖοι, τὸν πρόεδρον ὅν ποτέ φασιν ἐν ὑμῖν ὑπὸ Πολυζήλου πληγῆναι, καὶ τὸν θεσμοθέτην ὃς ἔναγχος ἐπλήγη τὴν αὐλητρίδα ἀφαιρούμενος, καὶ τοιούτους τινάς, ὡς, ἐὰν 30

3–7 τις ... ἰδιώτην Π10

3 πατάξῃ τις ... εἴπῃ $S_3^{\gamma\rho}$AFYP:]ις ... ειπηι Π10: πατάξῃς ... εἴπῃς S 4 οὕτω SAFYP: τουτ[Π10 5 ἄδειαν ἢ del. MacDowell 6 ἐν om. FuPrVb 7 -σέ AFP^{pc} τι SAFYP: ετ[Π10 9 ἱερονομίας Y: [P^{ac}] 10 δίκαιός SFYP: ἄξιός A 11 ὁ om. FuPrX χορὸς S^{ac} 12 πόλεως ὄνομα $S_3^{\gamma\rho}$AFP_2^{mg} καίτοι ταύταις S_4^{pc}FP_2^{pc}: καὶ ταῦτ' αὐταῖς Reiske 13 -σι AF^{pc}YP 15 ἔνορκα AF_1^{sl} -τι AF^{pc}YP 16 αἰκίας S_3^{sl}AFYP 18 ἂν ὑμῖν AFP_2^{mg} τοῦδε SFYP: τούτου A 22 λοιποῖς SFYP: ἄλλοις A 23 κἂν add. G. H. Schaefer 25 τίς μοι om. Y^{ac} περιόν- SAYP^{ac} -τα AF 26 -σι AF^{pc}YP

Thesmothetes does not belong to any person, but to the city. And again in the case of the arkhon, the same thing is true: if **33** one strikes or slanders him when he's wearing his crown, one is disfranchised; if as a private individual, one is liable to private prosecution. And this applies not only to them, but to all on whom the city confers any crown or any honour. So in my case too, if Meidias had committed any of these offences against me as a private individual on any other days, it would be proper for him to be punished by private prosecution; but if it's plain that all his insolent offences were com- **34** mitted against me while I was your chorus-producer and at a sacred season, he deserves public anger and punishment. For not only Demosthenes but also the chorus-producer suffered insolence; and this act, and the fact that it was committed on those days on which the laws forbid it, is the concern of the city. You should scrutinize the quality of laws at the time when you're making them, but once you've made them you should observe and enforce them; for that is in accordance with your oath, and also just. Your law of damage existed **35** long ago; so did the law of battery; so did the law of *hybris*. So, if it had been satisfactory that persons committing any of those offences at the Dionysia should be punished according to those laws, there would have been no need for this law in addition. But it was not satisfactory; witness the fact that you made a sacred law for this god in particular concerning the sacred season. So, if a man is liable under those laws which existed before and also under the subsequent one and also under all the others, is such a man to escape punishment for that reason, or does he deserve an even heavier one? A heavier one, I think.

Someone reported to me that Meidias was going around **36** listing and enquiring about persons who had ever been victims of insolence, and that he was intending to mention these men and tell their stories to you, men of Athens; for example, the proedros who is said to have been hit by Polyzelos at a meeting of the Ekklesia, and the thesmothetes who was hit recently when removing the woman piper, and some men of

πολλοὺς ἑτέρους δεινὰ καὶ πολλὰ πεπονθότας ἐπιδείξῃ, ἧττον
37 ὑμᾶς ἐφ' οἷς ἐγὼ πέπονθα ὀργιουμένους. ἐμοὶ δ' αὖ τοὐναντίον, ὦ ἄνδρες Ἀθηναῖοι, δοκεῖτε ποιεῖν ἂν εἰκότως, εἴπερ
ὑπὲρ τοῦ κοινῇ βελτίστου δεῖ μέλειν ὑμῖν. τίς γὰρ οὐκ οἶδεν
ὑμῶν τοῦ μὲν πολλὰ τοιαῦτα γίγνεσθαι τὸ μὴ κολάζεσθαι 5
τοὺς ἐξαμαρτάνοντας αἴτιον ὄν, τοῦ δὲ μηδένα ὑβρίζειν τὸ
λοιπὸν τὸ δίκην τὸν ἀεὶ ληφθέντα ἣν προσήκει διδόναι μόνον
αἴτιον ἂν γενόμενον; εἰ μὲν τοίνυν ἀποτρέψαι συμφέρει τοὺς
ἄλλους, τοῦτον καὶ δι' ἐκεῖνα κολαστέον, καὶ μᾶλλόν γε ὅσῳ-
περ ἂν ᾖ πλείω καὶ μείζω· εἰ δὲ παροξῦναι καὶ τοῦτον καὶ 10
38 πάντας, ἐατέον. ἔτι τοίνυν οὐδ' ὁμοίαν οὖσαν τούτῳ κἀκείνοις
συγγνώμην εὑρήσομεν. πρῶτον μὲν γὰρ ὁ τὸν θεσμοθέτην
πατάξας τρεῖς εἶχεν προφάσεις, μέθην, ἔρωτα, ἄγνοιαν διὰ τὸ
σκότους καὶ νυκτὸς τὸ πρᾶγμα γενέσθαι. ἔπειθ' ὁ Πολύζηλος
ὀργῇ καὶ τρόπου προπετείᾳ φθάσας τὸν λογισμὸν ἁμαρτὼν 15
ἔπαισεν· οὐ γὰρ ἐχθρός γε ὑπῆρχεν ὤν, οὐδ' ἐφ' ὕβρει τοῦτ'
ἐποίησεν. ἀλλ' οὐ Μειδίᾳ τούτων οὐδὲν ἔστ' εἰπεῖν· καὶ γὰρ
ἐχθρὸς ἦν, καὶ μεθ' ἡμέραν εἰδὼς ὕβριζεν, καὶ οὐκ ἐπὶ τούτου
μόνον, ἀλλ' ἐπὶ πάντων φαίνεται προῃρημένος με ὑβρίζειν.
39 καὶ μὴν οὐδὲ τῶν πεπραγμένων ἐμοὶ καὶ τούτοις οὐδὲν ὅμοιον 20
ὁρῶ. πρῶτον μὲν γὰρ ὁ θεσμοθέτης οὐχ ὑπὲρ ὑμῶν οὐδὲ τῶν
νόμων φροντίσας οὐδ' ἀγανακτήσας φανήσεται, ἀλλ' ἰδίᾳ πει-
σθεὶς ὁπόσῳ δήποτε ἀργυρίῳ καθυφεὶς τὸν ἀγῶνα. ἔπειθ' ὁ
πληγεὶς ἐκεῖνος ὑπὸ τοῦ Πολυζήλου, ταὐτὸ τοῦτο, ἰδίᾳ
διαλυσάμενος, ἐρρῶσθαι πολλὰ τοῖς νόμοις εἰπὼν καὶ ὑμῖν, 25
40 οὐδ' εἰσήγαγε τὸν Πολύζηλον. εἰ μὲν τοίνυν ἐκείνων
κατηγορεῖν βούλεταί τις ἐν τῷ παρόντι, δεῖ λέγειν ταῦτα· εἰ δ'
ὑπὲρ ὧν ἐγὼ τούτου κατηγόρηκα ἀπολογεῖσθαι, πάντα

4–8 *τίς … γενόμενον* Theon (Spengel 2) 64.5 11–14 *ἔτι … γενέσθαι* schol. Hermog. (Walz 4) 529.30 12–13 *ὁ … ἄγνοιαν* schol. Hermog. (Walz 4) 250.2 15–16 *ὀργῇ … ἔπαισεν* schol. Hermog. (Walz 4) 250.4 23–112.22 *ἔπειθ' … τούτων* Π10

---

1 *πολλὰ καὶ δεινὰ* F 4 *ὑπὲρ* om. SYP *μέλλειν* A 4–5 *ἡμῶν οὐκ οἶδε* Theon 5 *γίν-* A: *γεν-* Theon 6 *ὄν* om. A 7 *νόμον* A 8 *ἂν* om. Theon 10 *ᾖ* om. A *πλειωι* S 11 *πάντες* S^{ac}: *πάντας τοὺς ἄλλους* A *-δὲ* A 12 *εὑρήσετε* schol. Hermog. 13 *-χε* AF *ἔρωτα, ἄνοιαν* schol. Hermog. 250.2: *ἄγνοιαν, ἔρωτα* schol. Hermog. 529.30 14 *γεγενῆσθαι* A *-ζηλος ἐκεῖνος* A 15 *φθάσας* $AF_1^{sl}P$ schol. Hermog.: *φθάσαι* $SFYP_4^{sl}$: *προφθάσας* $S_3^{γρ}P_4^{γρ}$ *ἁμαρτὼν* om. schol. Hermog. 16 *ἔπαισεν* $AF_1^{mg}P_2^{γρ}$ schol. Hermog.: *ἔφησεν* SFYP: *ἔπεσεν* $S_3^{γρ}$ *-το* A 17 *ἐστιν* AF 18 *ὕβριζε* $AF^{pc}P_2^{pc}$: *ὑβρίζειν* YP^{ac} *τούτων* F

that sort. He assumes that, if he shows that many other men have been victims of many serious incidents, you will be less angry at what has been done to me. But it seems to me, men **37** of Athens, that it would be reasonable for you to do just the reverse, if the common good should be your consideration. For every one of you knows that the reason why many such offences are committed is that the offenders are not punished, whereas the only way to ensure that no one acts insolently in future would be for every man caught to pay the proper penalty. So if deterrence of other offenders is desirable, their offences are an additional reason for punishing Meidias, and all the more so, the more numerous and serious they are; but if it's desirable to spur him and all of them on, you should let him off. Besides, we shall find that he doesn't **38** even have as good a defence as those others had. First, the man who hit the thesmothetes had three excuses: drink, love, and inability to recognize him because the incident happened at night and in the dark. Secondly, Polyzelos, from anger and impetuosity of character, acted without thinking and struck by mistake; for he wasn't on hostile terms, and didn't intend his action to be insolent. But Meidias can't say any of these things. He was hostile; he acted insolently in daylight, knowing who I was; and it's clear that he was deliberately insolent to me not only on this but on all occasions. Furthermore, I see no resemblance between my con- **39** duct and theirs. First, it will be clear that the thesmothetes showed no concern or resentment on behalf of the public interest and the laws, but dropped the case because he was privately induced to do so, by whatever amount of money. Secondly, in the case of the man hit by Polyzelos, the same thing is true: he made a private agreement, said goodbye to the laws and the public interest, and didn't even take Polyzelos to court. So, if one wants to make accusations against **40** those men today, this evidence is relevant; but if what is wanted is a defence concerning my accusations against

μᾶλλον ἢ ταῦτα λεκτέα. πᾶν γὰρ τοὐναντίον ἐκείνοις αὐτὸς
μὲν οὔτε λαβὼν οὐδὲν οὔτ' ἐπιχειρήσας λαβεῖν φανήσομαι,
τὴν δ' ὑπὲρ τῶν νόμων καὶ τὴν ὑπὲρ τοῦ θεοῦ καὶ τὴν ὑπὲρ
ὑμῶν τιμωρίαν δικαίως φυλάξας καὶ νῦν ἀποδεδωκὼς ὑμῖν.
μὴ τοίνυν ἐᾶτε ταῦτ' αὐτὸν λέγειν, μηδ', ἂν βιάζηται, πείθ- 5
41 εσθε ὡς δίκαιόν τι λέγοντι. ἂν γὰρ ταῦθ' οὕτως ἐγνωσμένα
ὑπάρχῃ παρ' ὑμῖν, οὐκ ἔνεστ' αὐτῷ λόγος οὐδείς. ποία γὰρ
πρόφασις, τίς ἀνθρωπίνη καὶ μετρία σκῆψις φανεῖται τῶν
πεπραγμένων αὐτῷ; ὀργὴ νὴ Δία· καὶ γὰρ τοῦτο τυχὸν λέξει.
ἀλλ' ἃ μὲν ἄν τις ἄφνω τὸν λογισμὸν φθάσας ἐξαχθῇ πρᾶξαι, 10
κἂν ὑβριστικῶς ποιήσῃ, δι' ὀργήν γ' ἔνι φῆσαι πεποιηκέναι· ἃ
δ' ἂν ἐκ πολλοῦ συνεχῶς ἐπὶ πολλὰς ἡμέρας παρὰ τοὺς νόμους
πράττων τις φωρᾶται, οὐ μόνον δήπου τοῦ μὴ μετ' ὀργῆς
ἀπέχει, ἀλλὰ καὶ βεβουλευμένως ὁ τοιοῦτος ὑβρίζων ἐστὶν
ἤδη φανερός. 15

42 Ἀλλὰ μὴν ὁπηνίκα καὶ πεποιηκὼς ἃ κατηγορῶ καὶ ὕβρει
πεποιηκὼς φαίνεται, τοὺς νόμους ἤδη δεῖ σκοπεῖν, ὦ ἄνδρες
δικασταί· κατὰ γὰρ τούτους δικάσειν ὀμωμόκατε. καὶ θεωρεῖθ' ὅσῳ μείζονος ὀργῆς καὶ ζημίας ἀξιοῦσι τοὺς ἑκουσίως καὶ
δι' ὕβριν πλημμελοῦντας τῶν ἄλλως πως ἐξαμαρτανόντων. 20
43 πρῶτον μὲν τοίνυν οἱ περὶ τῆς βλάβης οὗτοι νόμοι πάντες, ἵν'
ἐκ τούτων ἄρξωμαι, ἂν μὲν ἑκὼν βλάψῃ, διπλοῦν, ἂν δ' ἄκων,
ἁπλοῦν τὸ βλάβος κελεύουσιν ἐκτίνειν. εἰκότως· ὁ μὲν γὰρ
παθὼν πανταχοῦ βοηθείας δίκαιός ἐστι τυγχάνειν, τῷ δρά-
σαντι δ' οὐκ ἴσην τὴν ὀργήν, ἄν θ' ἑκὼν ἄν τ' ἄκων, ἔταξεν ὁ 25
νόμος. ἔπειθ' οἱ φονικοὶ τοὺς μὲν ἐκ προνοίας ἀποκτιννύντας
θανάτῳ καὶ ἀειφυγίᾳ καὶ δημεύσει τῶν ὑπαρχόντων ζημιοῦ-

10–11 ἀλλ' ... πεποιηκέναι Gal. *Plac. Hipp. Pl.* 5.7.86

---

1 λεκτέον FYP: [Π10] πᾶν S^{pc}AFP$_2^{pc}$: πάντα S^{ac}YP^{ac}: [Π10] 2 οὐδὲν om. A:]δεν Π10 5 αὐτὸν ταῦτα A: ταυ[τ αυτον] Π10 πείθεσθαι S^{ac}: πείθεσθ' F: [Π10] 6 ἐὰν A: [Π10] 7 οὐκ ἐνέσται FYP: [Π10]: οὐκέτ' ἔσται T οὐδ' εἷς S: οὐδὲ εἷς FYP: [Π10] 10 ἀλλ' ἃ μὲν S$_3^{γρ}$AFYP: ἀλλὰ μὴν S Gal.: αλ[...]εν Π10 τί (post ἐξαχθῇ) A: τι FP^{sl}: τα Π10 10–11 πράξεις ὑβριστικῶς ποιήσει Gal. 11 τι (post ὑβριστικῶς) A: τοῦτο F δι' ὀργήν om. S (add. S$_2^{mg}$) γ' ἔνι S^{pc}AFP$_2^{pc}$: γενυ S^{ac}: πάνυ P^{ac}: [Π10]: om. Y: δ' ὁ Πλάτων Gal. φῆσαι Y: ἔφησε Gal. 13 μὴ om. O^{pc} 14 [απεχε]ιν Π10 βεβουλευμένος AF:]ευμεν[Π10 15 φανερῶς A: [Π10] 16–17 ἃ ... πεποιηκὼς om. P^{ac} 16 ὕβριν Y: [Π10]: δι' ὕβριν Lamb^{vl} 17 σκοπεῖν δεῖ SYP ὦ ἄνδρες Ἀθηναῖοι A^{cp}: [Π10] 18 [ομωμο]κοτε Π10^{ac} 19 -τε A 20 τι (post ὕβριν) A 22 ἐὰν A ἑκών τις AP^{pc} ἐὰν δὲ A 24 ἐστι om. S 25 δὲ A τε (ante ἄκων) A 26 ἀποκτειννύντας SA

Meidias, it's the last thing that is appropriate. Unlike those men, it will be clear that I have neither received nor attempted to get any payment; I have justly preserved the opportunity for retribution, on behalf of the laws and the god and yourselves, and now I have delivered it to you. So don't let him say this; and if he insists, don't believe that he is making a fair point. If this is your decision on this question, **41** he has no possible argument. What kind of excuse, what human or reasonable pretext will be produced for his behaviour? 'Anger'; perhaps he'll say that too. But acts that one is suddenly carried away to commit without thinking, even if done insolently, may be said to have been due to anger; acts that one is discovered to have been committing illegally over a long period, for many days in succession, are not only far from being committed in anger, but it becomes clear that such a man is guilty of deliberate insolence.

Now, since it's clear that he has committed the acts of **42** which I accuse him, and has committed them with insolence, we should turn to consideration of the laws, men of the jury; for you have sworn to judge in accordance with them. Observe how much more severely they punish those who offend intentionally and from insolence than those who err in some other way. First, then, all those laws about damage (to begin **43** with them) require payment of double the amount if damage is intentional, but the single amount if unintentional. Quite reasonably: the victim deserves a remedy in every case, but the law has not ordained equal severity for the perpetrator whether he acted intentionally or unintentionally. Next, the homicide laws punish those killing deliberately with death, perpetual exile, and confiscation of property, but those kill-

σι, τοὺς δ' ἀκουσίως αἰδέσεως καὶ φιλανθρωπίας πολλῆς
44 ἠξίωσαν. οὐ μόνον δ' ἐπὶ τούτων τοῖς ἐκ προαιρέσεως ὑβρ-
ισταῖς χαλεποὺς ὄντας ἰδεῖν ἐστιν τοὺς νόμους, ἀλλὰ καὶ ἐφ'
ἁπάντων. τί γὰρ δή ποτ', ἄν τις ὀφλὼν δίκην μὴ ἐκτίνῃ,
οὐκέτ' ἐποίησεν ὁ νόμος τὴν ἐξούλην ἰδίαν, ἀλλὰ προστιμᾶν 5
ἐπέταξεν τῷ δημοσίῳ; καὶ πάλιν τί δή ποτ', ἂν μὲν ἑκὼν παρ'
ἑκόντος τις λάβῃ τάλαντον ἓν ἢ δύο ἢ δέκα καὶ ταῦτ' ἀποστερήσῃ, οὐδὲν αὐτῷ πρὸς τὴν πόλιν ἐστίν, ἂν δὲ μικροῦ
πάνυ τι τιμήματος ἄξιόν τις λάβῃ, βίᾳ δὲ τοῦτο ἀφέληται, τὸ
ἴσον τῷ δημοσίῳ προστιμᾶν οἱ νόμοι κελεύουσιν ὅσον περ δὴ 10
45 τῷ ἰδιώτῃ; ὅτι πάνθ' ὅσα τις βιαζόμενος πράττει κοινὰ
ἀδικήματα καὶ κατὰ τῶν ἔξω τοῦ πράγματος ὄντων ἡγεῖτο
ὁ νομοθέτης· τὴν μὲν γὰρ ἰσχὺν ὀλίγων, τοὺς δὲ νόμους ἁπάντων εἶναι, καὶ τὸν μὲν πεισθέντα ἰδίας, τὸν δὲ βιασθέντα
δημοσίας δεῖσθαι βοηθείας. διόπερ καὶ τῆς ὕβρεως αὐτῆς τὰς 15
μὲν γραφὰς ἔδωκεν ἅπαντι τῷ βουλομένῳ, τὸ δὲ τίμημα
ἐποίησεν ὅλον δημόσιον· τὴν γὰρ πόλιν ἡγεῖτο ἀδικεῖν, οὐχὶ
τὸν παθόντα μόνον, τὸν ὑβρίζειν ἐπιχειροῦντα, καὶ δίκην
ἱκανὴν τὴν τιμωρίαν εἶναι τῷ παθόντι, χρήματα δ' οὐ προσήκ-
46 ειν τῶν τοιούτων ἐφ' ἑαυτῷ λαμβάνειν. καὶ τοσαύτῃ γε ἐχρή- 20
σατο ὑπερβολῇ, ὥστε, κἂν εἰς δοῦλον ὑβρίζῃ τις, ὁμοίως
ἔδωκεν ὑπὲρ τούτου γραφήν. οὐ γὰρ ὅστις ὁ πάσχων ᾤετο
δεῖν σκοπεῖν, ἀλλὰ τὸ πρᾶγμα ὁποῖόν τι τὸ γιγνόμενον· ἐπειδὴ
δ' εὗρεν οὐκ ἐπιτήδειον, μήτε πρὸς δοῦλον μήθ' ὅλως ἐξεῖναι
πράττειν ἐπέταξεν. οὐ γάρ ἐστιν, οὐκ ἔστιν, ὦ ἄνδρες Ἀθην- 25
αῖοι, τῶν πάντων οὐδὲν ὕβρεως ἀφορητότερον, οὐδ' ἐφ' ὅτῳ
μᾶλλον ὑμῖν ὀργίζεσθαι προσήκει. ἀνάγνωθι δ' αὐτόν μοι τὸν
τῆς ὕβρεως νόμον· οὐδὲν γὰρ οἷον ἀκούειν αὐτοῦ τοῦ νόμου.

1–2 *τοὺς ... ἠξίωσαν* *AB* 1.361.32, Phot. α 550 28 *οὐδὲν ... νόμου* Phot. s.vv. *οὐδὲν ...*, *οἷον*, Suda ο 784, οι 147

---

1 *τοῦ δὲ AB* Phot. *ἀνέσεως* VsVeT^{sl} 2 *ἠξίωσεν AB* Phot. 3 *-τι* AFP^{pc} 4 *γὰρ δή* S$^{pc}_{4}$AFY^{pc}: *δὴ γάρ* S^{ac}Y^{ac}P *-τε* SF *ἐάν* YP *ὀφείλων* A *ἐκτίσῃ* AFY^{sl}P$^{sl}_{2}$ 5 *ἐξούλης* YP 6 *-τε* A *ἐὰν* YP 7 *ἢ καὶ δέκα καὶ* F *-τα* SFYP 8 *ἐὰν* A 9 *τι* Aa: om. SAFYP 10 *δὴ* SYP: *ἂν* A: om. F 11 *διὰ τί; ὅτι* AFP$^{sl}_{2}$ *-τα* SYP *ὅσα ἂν* A *πράττῃ* WdRVf: *πράττοι* Aq^{pc} 17 *οὐχὶ* AFYP: *οὐ* S 19 *τὴν* om. S 20 *τὸν τοιοῦτον* P$^{sl}_{4}$: *τῶν τοιούτων ἕνεκα* T^{pc} *ἑαυτῶν* CdOdWd 23 *γιν-* A *ἐπεὶ* A 24 *δὲ* AF *-τε* A 25 *ταῦτα πράττειν* A *ἐπέτρεψεν* SF$^{γρ}_{1}$YP 27 *μοι λαβὼν* FP$^{mg}_{4}$ 28 *τὸ ἀκούειν* Suda οι 147

ing unintentionally they have deemed worthy of pardon and great clemency. Not only in these cases is it possible to see **44** that the laws are strict towards those who are wilfully insolent, but in all cases. Why is it that, if a man loses a private case and doesn't pay up, the law has then not made the ejectment a private matter but required the payment of an additional penalty to the treasury? And again why is it that, if a man gets one or two or ten talents from someone by mutual agreement and then misappropriates it, the city isn't concerned with him, but if he gets something which is of very small value but takes it away by force, the laws order an additional payment of the same amount to the treasury as to the individual? Because the legislator considered all violent **45** actions to be offences against everyone and harmful to those not involved; for, he thought, strength is possessed by a few men but the laws belong to all, and the man who is persuaded requires a private remedy, but the man who is compelled, a public one. Consequently, for *hybris* too he permitted everyone who wished to prosecute, and he made the penalty entirely payable to the state; for, he considered, the man who turns to insolence wrongs not only the victim but the city, and revenge is sufficient compensation for the victim, who ought not to make money for himself for matters of this kind. And in fact he went so far as to permit similar **46** prosecution even for a slave treated insolently. He thought it right not to consider the identity of the victim but the character of the act committed; and since he found the act unacceptable, he ordered that it should not be permitted, either against a slave or at all. For there is nothing, nothing at all, men of Athens, more intolerable than insolence, or more deserving your anger. Please read the law of *hybris*; there is nothing like hearing the actual law.

47

ΝΟΜΟΣ

Ἐάν τις ὑβρίζῃ εἴς τινα, ἢ παῖδα ἢ γυναῖκα ἢ ἄνδρα, τῶν ἐλευθέρων ἢ τῶν δούλων, ἢ παράνομόν τι ποιήσῃ εἰς τούτων τινά, γραφέσθω πρὸς τοὺς θεσμοθέτας ὁ βουλόμενος Ἀθηναίων οἷς ἔξεστιν, οἱ δὲ θεσμοθέται εἰσ- 5
αγόντων εἰς τὴν ἡλιαίαν τριάκοντα ἡμερῶν ἀφ᾽ ἧς ἂν ἡ γραφή, ἐὰν μή τι δημόσιον κωλύῃ, εἰ δὲ μή, ὅταν ᾖ πρῶτον οἷόν τε. ὅτου δ᾽ ἂν καταγνῷ ἡ ἡλιαία, τιμάτω περὶ αὐτοῦ παραχρῆμα, ὅτου ἂν δοκῇ ἄξιος εἶναι παθεῖν ἢ ἀποτεῖσαι. ὅσοι δ᾽ ἂν γράφωνται {γραφὰς ἰδίας} κατὰ 10
τὸν νόμον, ἐάν τις μὴ ἐπεξέλθῃ ἢ ἐπεξιὼν μὴ μεταλάβῃ τὸ πέμπτον μέρος τῶν ψήφων, ἀποτεισάτω χιλίας δραχμὰς τῷ δημοσίῳ. ἐὰν δὲ ἀργυρίου τιμηθῇ τῆς ὕβρεως, δεδέσθω, ἐὰν {δὲ} ἐλεύθερον ὑβρίσῃ, μέχρι ἂν ἐκτείσῃ.

48 Ἀκούετε, ὦ ἄνδρες Ἀθηναῖοι, τοῦ νόμου τῆς φιλανθρωπίας, 15
ὃς οὐδὲ τοὺς δούλους ὑβρίζεσθαι ἀξιοῖ. τί οὖν, πρὸς θεῶν; εἴ τις εἰς τοὺς βαρβάρους ἐνεγκὼν τὸν νόμον τοῦτον, παρ᾽ ὧν τὰ ἀνδράποδα εἰς τοὺς Ἕλληνας κομίζεται, ἐπαινῶν ὑμᾶς καὶ
49 διεξιὼν περὶ τῆς πόλεως εἴποι πρὸς αὐτοὺς ὅτι "εἰσὶν Ἕλληνές τινες ἄνθρωποι οὕτως ἥμεροι καὶ φιλάνθρωποι τοὺς 20
τρόπους ὥστε πολλὰ ὑφ᾽ ὑμῶν ἠδικημένοι, καὶ φύσει τῆς πρὸς ὑμᾶς ἔχθρας αὐτοῖς ὑπαρχούσης πατρικῆς, ὅμως οὐδ᾽ ὅσων ἂν τιμὴν καταθέντες δούλους κτήσωνται, οὐδὲ τούτους ὑβρίζειν ἀξιοῦσιν, ἀλλὰ νόμον δημοσίᾳ τὸν ταῦτα κωλύσοντα τέθεινται τουτονί, καὶ πολλοὺς ἤδη παραβάντας τὸν νόμον 25
50 τοῦτον ἐζημιώκασι θανάτῳ", εἰ ταῦτ᾽ ἀκούσαιεν καὶ συνεῖεν οἱ βάρβαροι, οὐκ ἂν οἴεσθε δημοσίᾳ πάντας ὑμᾶς προξένους αὑτῶν ποιήσασθαι; τὸν τοίνυν οὐ παρὰ τοῖς Ἕλλησι μόνον εὐδοκιμοῦντα νόμον, ἀλλὰ καὶ παρὰ τοῖς βαρβάροις εὖ

2 *ἐάν ... ἄνδρα* *AB* 1.176.26 16–17 *τί ... τοῦτον* Io. Sard. 211.23
18–118.5 *καὶ ... νῦν* Π5 28–9 *τὸν ... βαρβάροις* Aps. *Rh.* (Hammer) 292.22

2–14 legem om. A 2 *ὑβρίσῃ* Lamb[vl] 6 *ἡλιαίαν* FP_3^{pc} *ἂν ᾖ ἡ* Markland 8 *καταγνῷ* Lamb[vl]: *καταγνῶτε* SFYP: *καταγνωσθῇ*, Wolf *ἡ* $S_3^{pc}FP_3^{pc}$: om. $S^{ac}YP^{ac}$ *ἡλιαία* FP_3^{pc} *τιμάτωι* S 10 *ἀποτεῖσαι* Blass: *ἀποτίσαι* SYP^{ac}: *ἀποτῖσαι* FP_4^{pc} *γραφὰς ἰδίας* del. MacDowell 12 *ἀποτεισάτω* Blass: *ἀποτισάτω* SFYP 14 *δὲ* del. Oporinus *ἐκτείσῃ* Blass: *ἐκτίσῃ* SFYP 16 *ἀξιοῖ;* FY 17 *τοῦτον* om. Y^{ac} 18 *νομίζεται* Y: [P^{ac}] 19 *περὶ*] *ὑπὲρ* P 21 *ἡμῶν* Af: [Π5] 22 *ἡμᾶς* Σ (151 Dilts): [Π5] 25 *τέθειναι* S^{ac}: *τεθῆναι* Y: [Π5P^{ac}] *ἤδη* om. Π5S 26 *ἀκούσειαν* $AP_4^{γρ}$: [Π5] *συνιεν* Π5S^{ac} 27 *ἅπαντας* A 28 *αὑτῶν* P: *αὐτῶν* $S_3^{pc}FY$: *αὐτῷ* S^{ac}: *αυτῶν* ante *προξένους* A: [Π5] *οὐ*] *μὴ* Lamb[vl]: [Π5]

LAW **47**

If anyone treats with *hybris* any person, either child or woman or man, free or slave, or does anything unlawful against any of these, let anyone who wishes, of those Athenians who are entitled, submit a *graphe* to the thesmothetai. Let the thesmothetai bring the case to the Eliaia within thirty days of the submission of the *graphe*, if no public business prevents it, or otherwise as soon as possible. Whoever the Eliaia finds guilty, let it immediately assess whatever penalty it thinks right for him to suffer or pay. Of those who submit *graphai* according to the law, if anyone does not proceed, or when proceeding does not get one-fifth of the votes, let him pay one thousand drachmas to the public treasury. If he is assessed to pay money for his *hybris*, let him be imprisoned, if the *hybris* is against a free person, until he pays it.

You hear how considerate the law is, men of Athens: it **48** doesn't even authorize insolence to slaves. Well then! If someone took this law to the barbarians from whom slaves are imported to Greece, and praised you and gave them a description of Athens, saying 'There are some people in **49** Greece of such a civilized and considerate character that, although you have often done them wrong and they naturally have an inherited hostility towards you, nevertheless they don't think it right to treat insolently even the slaves whom they acquire by paying a price for them, but have publicly made this law to prevent it, and have before now imposed the death penalty on many who transgressed it'—if **50** the barbarians heard and understood this, don't you think they'd appoint all of you to be their official representatives? Well then, if the law not only has a high reputation in Greece

δόξαντ' ἂν ἔχειν, σκοπεῖσθ' ὁ παραβὰς ἥντινα δοὺς δίκην ἀξίαν ἔσται δεδωκώς.

51 Εἰ μὲν τοίνυν, ὦ ἄνδρες Ἀθηναῖοι, μὴ χορηγὸς ὢν ταῦτ' ἐπεπόνθειν ὑπὸ Μειδίου, ὕβριν ἄν τις μόνον κατέγνω τῶν πεπραγμένων αὐτῷ· νῦν δέ μοι δοκεῖ κἂν ἀσέβειαν εἰ καταγιγνώσκοι τὰ προσήκοντα ποιεῖν. ἴστε γὰρ δήπου τοῦθ', ὅτι τοὺς χοροὺς ὑμεῖς ἅπαντας τούτους καὶ τοὺς ὕμνους τῷ θεῷ ποιεῖτε, οὐ μόνον κατὰ τοὺς νόμους τοὺς περὶ τῶν Διονυσίων, ἀλλὰ καὶ κατὰ τὰς μαντείας, ἐν αἷς ἁπάσαις ἀνῃρημένον εὑρήσετε τῇ πόλει, ὁμοίως ἐκ Δελφῶν καὶ ἐκ Δωδώνης, χοροὺς ἱστάναι κατὰ τὰ πάτρια καὶ κνισᾶν ἀγυιὰς καὶ στεφανη-**52** φορεῖν. ἀνάγνωθι δέ μοι λαβὼν αὐτὰς τὰς μαντείας.

ΜΑΝΤΕΙΑΙ

Αὐδῶ Ἐρεχθείδαισιν, ὅσοι Πανδίονος ἄστυ
ναίετε καὶ πατρίοισι νόμοις ἰθύνεθ' ἑορτάς,
μεμνῆσθαι Βάκχοιο, καὶ εὐρυχόρους κατ' ἀγυιὰς
ἱστάναι ὡραίων Βρομίῳ χάριν ἄμμιγα πάντας,
καὶ κνισᾶν βωμοῖσι κάρη στεφάνοις πυκάσαντας.

Περὶ ὑγιείας θύειν καὶ εὔχεσθαι Διὶ ὑπάτῳ, {καὶ} Ἡρακλεῖ, Ἀπόλλωνι προστατηρίῳ· περὶ τύχας ἀγαθᾶς Ἀπόλλωνι ἀγυιεῖ, Λατοῖ, Ἀρτέμιδι. καὶ κατ' ἀγυιὰς κρατῆρας ἱστάμεν καὶ χορούς, καὶ στεφανηφορεῖν, καὶ κατὰ τὰ πάτρια θεοῖς Ὀλυμπίοις πάντεσσι καὶ πάσαις {ἰδίας} δεξιὰς καὶ ἀριστερὰς ἀνίσχοντας {καὶ} μνασιδωρεῖν.

3–6 εἰ … ποιεῖν schol. Hermog. (Walz 4) 524.16, (Walz 7) 433.26
3–120.13 Ἀθηναῖοι … Ἀθηναῖοι Π6   11 κατὰ … ἀγυιὰς Harp. s.v. ἀγυιᾶς

---

1 ἂν om. A: [Π5]   -σθε A   2 δεδοκὼς S: [Π5]   3 ἄνδρες om. schol. Hermog. (Walz 7)   3–4 ἐπεπόνθειν ταῦτα Π5Π6A   4 μόνον om. schol. Hermog. (Walz 4): post κατέγνω F   5–6 καταγινώσκοι A schol. Hermog. (Walz 7): καταγιγνώσκοι τις FY: καταγιγνώσκοι τις αὐτοῦ schol. Hermog. (Walz 4): [Π6P^{ac}]   6 -το SFYP   7 καὶ τοὺς S: καὶ YP: om. AF: [Π6]   ὕμνους S^{pc}_{1}AF$P^{γρ}_{4}$: ὕμνους οὓς S^{ac}YP: [Π6]   8 ποιεῖσθε F   11 ἀγυιᾶς Harp.   12 μαντίας $S^{ac}F^{ac}$: [Π6]   13 ΜΑΝΤΕΙΑ A: [Π6]   14–120.12 oracula om. Π6A   14 ἐρεχθίδεσσι S: ἐρεχθίδεσσιν F   ὅσσοι F   πανδειονος S^{ac} 15 ναίοιτε F^{sl}_{1}YP^{ac} -νετ' εορ- S: [P^{ac}]   16 βάκχοο S^{ac} 18 κνισᾶιν S   πυκνάσαντας S   19 διει S^{ac} καὶ om. $Lamb^{vl}$ 20 ἀγαθὰς Y   21 ἀγυιεῖ UAfPr: ἀγυεῖ SFYP   λατου SF: λατω S^{sl}_{3} 22 καὶ (ante κατὰ) TBcNc: om. SFYP   23 τὰ SFP^{sl}_{4}: om. YP   24 ἰδίας del. Taylor   ἀνίσχοντας Ln: ἀνίσχοντες S^{pc}_{1}FYP: [S^{ac}]   καὶ del. Buttmann

but would meet with the barbarians' approval too, just think what will be a fair penalty for the transgressor to pay.

51 If, men of Athens, I had not been a chorus-producer when Meidias treated me in this way, one would have condemned his actions only for insolence. As it is, I think it would be proper to condemn them for impiety too. You know of course that you hold all these performances of choruses and hymns for the god, not only in accordance with the laws about the Dionysia, but also in accordance with the oracles, in all of which you will find it ordained for the city, from Delphi and from Dodona alike, to establish choruses in accordance with tradition, to make streets smell of sacrifice, and to wear crowns. Please take and read the actual oracles. 52

ORACLES

I declare to the sons of Erekhtheus, all you who dwell in Pandion's town and direct festivals by inherited laws, to remember Bakkhos, and all together to establish a thanksgiving to Bromios for ripe crops along the broad-spaced streets, and to make a smell of sacrifice on the altars, covering your heads with crowns.

For health, sacrifice and pray to Zeus the highest, Herakles, and Apollo the protector; for good fortune, to Apollo of streets, Leto, and Artemis. Along the streets establish bowls of wine and choruses, and wear crowns, and in accordance with tradition raise your right and left hands to all the Olympian gods and goddesses and remember their gifts.

53 ## ΕΚ ΔΩΔΩΝΗΣ ΜΑΝΤΕΙΑΙ

Τῷ δήμῳ τῷ Ἀθηναίων ὁ τοῦ Διὸς σημαίνει, ὅτι τὰς ὥρας παρηνέγκατε τῆς θυσίας καὶ τῆς θεωρίας, αἱρετοὺς πέμπειν {κελεύει} θεωροὺς ἐννέα· καὶ τούτους διὰ ταχέων τῷ Διὶ τῷ ναΐῳ τρεῖς βοῦς καὶ πρὸς ἑκάστῳ δύο βοῒ σῦς, τῇ δὲ Διώνῃ βοῦν καλλιερεῖν· καὶ τράπεζαν χαλκῆν καθιστάναι πρὸς τὸ ἀνάθημα ὃ ἀνέθηκεν ὁ δῆμος ὁ Ἀθηναίων.

Ὁ τοῦ Διὸς σημαίνει ἐν Δωδώνῃ Διονύσῳ δημοτελῆ ἱερὰ τελεῖν, καὶ κρατῆρας κεράσαι, καὶ χοροὺς ἱστάναι, καὶ στεφανηφορεῖν ἐλευθέρους καὶ δούλους, καὶ ἐλινύειν μίαν ἡμέραν. Ἀπόλλωνι ἀποτροπαίῳ βοῦν θῦσαι, Διὶ κτησίῳ βοῦν λευκόν.

54 Εἰσίν, ὦ ἄνδρες Ἀθηναῖοι, καὶ αὗται καὶ ἄλλαι πολλαὶ μαντεῖαι τῇ πόλει καὶ ἀγαθαί. τί οὖν ἐκ τούτων ὑμᾶς ἐνθυμεῖσθαι δεῖ; ὅτι τὰς μὲν ἄλλας θυσίας τοῖς ἐφ' ἑκάστης μαντείας προφαινομένοις θεοῖς προστάττουσι θύειν, ἱστάναι δὲ χοροὺς καὶ στεφανηφορεῖν κατὰ τὰ πάτρια πρὸς ἁπάσαις ἀεὶ ταῖς **55** ἀφικνουμέναις μαντείαις προσαναιροῦσιν ὑμῖν. οἱ τοίνυν χοροὶ πάντες οἱ γιγνόμενοι καὶ οἱ χορηγοὶ δῆλον ὅτι τὰς μὲν ἡμέρας ἐκείνας, ἃς συνερχόμεθα ἐπὶ τὸν ἀγῶνα, κατὰ τὰς μαντείας ταύτας ὑπὲρ ὑμῶν ἐστεφανώμεθα ὁμοίως ὅ τε μέλλων νικᾶν καὶ ὁ πάντων ὕστατος γενήσεσθαι, τὴν δὲ τῶν ἐπινικίων ὑπὲρ αὑτοῦ τότ' ἤδη στεφανοῦται ὁ νικῶν. τὸν οὖν εἴς τινα τούτων τῶν χορευτῶν ἢ τῶν χορηγῶν ὑβρίζοντα ἐπ' ἔχθρᾳ, καὶ ταῦτα ἐν αὐτῷ τῷ ἀγῶνι καὶ ἐν τῷ τοῦ θεοῦ ἱερῷ, τοῦτον ἄλλο τι πλὴν ἀσεβεῖν φήσομεν;

23–122.5 *οὖν ... ἐστεφανωμένον* Π6

---

3 *παρήκατε* Spalding   *θεωρείας* S^{ac} 4 *κελεύει* om. Vd^{ac} *τούτους* P^{pc}: *τούτου τοὺς* $SFYP^{ac}$ 5 *τῷ ναΐῳ τρεῖς* Buttmann: *τῶν ἀρωτρὶς* SF: *τῶν ἀρωτρεῖς* $F_4^{mg}YP$: *τῶν ἀροτρεῖς* TBcNc: *τῷ Δωδωναίῳ ἀροτρεῖς* Felicianus   *βοῒ σῦς* Radermacher: *βοιήσεις* $S_1^{pc}FYP$: *βοήσεις* S^{ac} 6 *δὲ* Y^{pc}: om. $SFY^{ac}P$ *καλλιερεῖν* Sauppe: *καὶ ἄλλα ἱερεῖα* SFYP   6–7 *καθιστάναι* MacDowell: *καὶ* SFYP: om. Vf   8 *σημαίνει* Vf: *σημαίν'* SFYP   *δημοτελῆ* $BcVg^{ac}$: *δημοτελεῖ* SFYP   9 *ἱερὰ τελεῖν* Buttmann: *ἱερεῖον τέλειον* SFYP   *κρατῆρας* Humbert: *κρατῆρα* SFYP   9–11 *Ἀπόλλωνι ... θῦσαι* post *ἡμέραν* Weil: post *ἱστάναι* SFYP   10 *ἐλινύειν* $F_4^{γρ}$: *ελεεινύειν* S: *ἐλεεῖν θύειν* FYP   11 *θῦσαι* om. Y^{ac} 13 *εἴσι μὲν* A: [Π6]   *πολλαὶ* bis S^{ac} 14 *ἐν τῇ* FP_3^{pc} *πόλει καλαὶ* $AFY^{pc}P_4^{mg}$ *κἀγαθαί* FYP   17 *ἀεὶ* om. SFYP   18 *προσανεροῦσιν* A   20 *ἐπὶ* SAFY: *εἰς* $Y^{sl}P$ 21 *ὑμῶν* AFP: *αυτῶν* S: *ὑμῶν αὐτῶν* Y   *ἐστεφανούμεθα* SFYP   23 *αὑτοῦ* $LpT^{pc}O^{pc}$: *αυτοῦ* SA: *αὐτοῦ* FYP   24 *τῶν χορευτῶν ἢ* om. Π6 ut vid.

ORACLES FROM DODONA

53

The oracle of Zeus gives an indication to the people of Athens, because you let pass the times of the sacrifice and of the sacred embassy, to send nine elected men as envoys; and these with all speed are to sacrifice with good omens to Zeus Naios three oxen, and besides each ox two pigs, and to Dione an ox; and they are to set up a bronze table for the dedication which the people of Athens dedicated.

The oracle of Zeus gives an indication at Dodona to perform rites for Dionysos at public expense, and to mix bowls of wine, and to establish choruses, and to wear crowns, free men and slaves alike, and to take one day's holiday. Sacrifice to Apollo the averter of evil an ox, to Zeus the protector of property a white ox.

54 The city has, men of Athens, both these and many other good oracles. What inference should you draw from them? That, besides instructing you to make the sacrifices to the gods specified in the case of each oracle, they also command you to establish choruses and to wear crowns in accordance with tradition, in addition to all the oracles which come on every occasion. **55** So it's plain that all the appointed choruses and chorus-producers throughout those days when we assemble for the contest wear crowns for Athens, in accordance with these oracles—all of us alike, both the one that is going to be the victor and the one that is going to come last of all; it is only for the day of the victory celebrations that the victor puts on a crown for himself. So when a man treats any of these choristers or chorus-producers with insolence, as an enemy, and that while the contest is actually in progress and in the precinct of the god, can we deny that he is guilty of impiety?

56 Καὶ μὴν ἴστε γε τοῦθ', ὅτι βουλόμενοι μηδέν' ἀγωνίζεσθαι ξένον οὐκ ἐδώκατε ἁπλῶς τῶν χορηγῶν οὐδενὶ προσκαλέσαντι τοὺς χορευτὰς σκοπεῖν, ἀλλ' ἐὰν μὲν καλέσῃ, πεντήκοντα δραχμάς, ἐὰν δὲ καθίζεσθαι κελεύσῃ, χιλίας ἀποτίνειν ἐτάξατε. τίνος ἕνεκα; ὅπως μὴ τὸν ἐστεφανωμένον καὶ 5 λειτουργοῦντα τῷ θεῷ ταύτην τὴν ἡμέραν καλῇ μηδ' ἐπηρ- **57** εάζῃ μηδ' ὑβρίζῃ μηδεὶς ἐξεπίτηδες. εἶτα τὸν μὲν χορευτὴν οὐδ' ὁ προσκαλέσας κατὰ τὸν νόμον ἀζήμιος ἔσται, τὸν δὲ χορηγὸν αὐτὸν οὐδ' ὁ συγκόψας παρὰ πάντας τοὺς νόμους οὕτω φανερῶς δώσει δίκην; ἀλλὰ μὴν οὐδέν ἐστ' ὄφελος κα- 10 λῶς καὶ φιλανθρώπως τοὺς νόμους ὑπὲρ τῶν πολλῶν κεῖσθαι, εἰ τοῖς ἀπειθοῦσι καὶ βιαζομένοις αὐτοὺς ἡ παρ' ὑμῶν ὀργὴ τῶν ἀεὶ κυρίων μὴ γενήσεται.

58 Φέρε δὴ πρὸς θεῶν κἀκεῖνο σκέψασθε. παραιτήσομαι δ' ὑμᾶς μηδὲν ἀχθεσθῆναί μοι, ἐὰν ἐπὶ συμφοραῖς τινων γεγονό- 15 των ὀνομαστὶ μνησθῶ· οὐ γὰρ ὀνειδίσαι μὰ τοὺς θεοὺς οὐδενὶ δυσχερὲς οὐδὲν βουλόμενος τοῦτο ποιήσω, ἀλλὰ δεῖξαι τὸ βιάζεσθαι καὶ ὑβρίζειν καὶ τὰ τοιαῦτα ποιεῖν ὡς ἅπαντες ὑμεῖς οἱ ἄλλοι φεύγετε. Σαννίων ἐστὶ δήπου τις ὁ τοὺς τραγικοὺς χοροὺς διδάσκων· οὗτος ἀστρατείας ἑάλω καὶ κέχρηται 20 **59** συμφορᾷ. τοῦτον μετὰ τὴν ἀτυχίαν ταύτην ἐμισθώσατό τις φιλονικῶν χορηγὸς τραγῳδῶν, οἶμαι, Θεοζοτίδης. τὸ μὲν οὖν πρῶτον ἠγανάκτουν οἱ ἀντιχορηγοὶ καὶ κωλύσειν ἔφασαν, ὡς δ' ἐπληρώθη τὸ θέατρον καὶ τὸν ὄχλον συνειλεγμένον εἶδον ἐπὶ τὸν ἀγῶνα, ὤκνησαν, εἴασαν, οὐδεὶς ἥψατο, ἀλλὰ τοσοῦ- 25 τον τῆς εὐσεβείας ἐν ἑκάστῳ τις ἂν ὑμῶν ἴδοι τὸ συγκεχωρηκὸς ὥστε πάντα τὸν μετὰ ταῦτα χρόνον διδάσκει τοὺς

7–9 εἶτα … νόμους Anon. Seg. (Hammer) 395.21 7–10 εἶτα … δίκην Minuc. (Hammer) 346.12, Max. Plan. (Walz 5) 506.10 14–15 παραιτήσομαι … μοι *AB* 1.163.14 19–20 Σαννίων … χορούς Io. Sic. (Walz 6) 141.11 Σαννίων … διδάσκων Max. Plan. (Walz 5) 454.4 19–21 Σαννίων … συμφορᾷ Hermog. 225.16, 227.6, 233.9, 268.16, 289.3 20–1 οὗτος … συμφορᾷ Castor (Walz 3) 713.12, schol. Hermog. (Walz 7) 935.20

---

1 -το SFYP: [Π6] μηδένα AFP$^{sl}_{3}$:]νạ Π6: μηδὲν P 2 οὐκ ἐδώκατε bis F: ọụκ εδ[ω]ḳαθ[Π6 τῶν χορηγῶν om. Π6 ut vid. 4]ẹζεσθạị Π6 5 εἵνεκα Π6A μὴ om. A 8 προκαλέσας F κατὰ τὸν νόμον om. Seg.: post ἔσται Minuc. 9 αὐτὸν AFP$^{mg}_{4}$ Max.: om. SYP Minuc. Seg. 9–10 παρὰ … φανερῶς om. Minuc. πάντα τὸν νόμον Seg. 10 δώσει S Max.: οὐ δώσει AFYP: ὑφέξει Minuc. 12 -σιν S 13 αἰεὶ F^{ac} 16 ὀνομαστὶ Reiske: ὀνόματι SAFYP 18 καὶ τὸ ὑβρίζειν καὶ τὸ τὰ A 19 -τιν S 21 τοῦτο S^{ac} 22 φιλονεικῶν AF$^{pc}_{1}$YP θεοσδοτίδης AFYP: [S^{ac}] 24 δὲ A 26 τὸ AP$^{sl}_{2}$: om. SFYP

Furthermore, as you know, although you don't want any **56** alien to take part in the contest, you don't permit any of the chorus-producers simply to accost the choristers and make an investigation: you have laid down that, if he does accost one, he is to pay fifty drachmas, and if he orders him to be seated, a thousand drachmas. What is the purpose? To ensure that a man who is wearing a crown and serving the god during that day is not accosted or subjected to interference or insolence by anyone for his own reasons. So since, in the case of the **57** chorister, not even a man who accosts him legally will get off without penalty, in the case of the chorus-producer shall not even a man who thrashes him, quite illegally and so openly, be punished? After all, there's no point in having good humane laws to protect ordinary people, if those who disobey and violate them escape the anger of you who have authority to enforce them on each occasion.

Please consider the following facts too. (I shall ask you not **58** to be annoyed with me if I mention by name some men who have come to grief. My object in doing this will certainly not be to find any fault with anyone, but to show how all the rest of you eschew violence and insolence and conduct of that sort.) There is, as you know, a man called Sannion, who directs tragic choruses. He was convicted of not doing military service, and has suffered for it. After this misfortune he **59** was employed by a chorus-producer who was eager to win the tragedy contest—Theozotides I think it was. At first the competing producers were indignant and said they would put a stop to it. But when the theatre was full, and they saw the crowd assembled for the contest, they shrank from it; they let it go; no one laid a finger on him. So great is the forbearance arising from piety which may be seen in every one of you that he has been directing his choruses ever since,

χοροὺς καὶ οὐδὲ τῶν ἰδίων ἐχθρῶν οὐδεὶς κωλύει· τοσοῦτ'
60 ἀπέχει τῶν χορηγῶν. ἄλλος ἐστὶν Ἀριστείδης Οἰνηίδος
φυλῆς, ἠτυχηκώς τι καὶ οὗτος τοιοῦτον, ὃς νῦν μὲν καὶ γέρων
ἐστὶν ἤδη καὶ ἴσως ἥττων χορευτής, ἦν δέ ποθ' ἡγεμὼν τῆς
φυλῆς {κορυφαῖος}. ἴστε δὲ δήπου τοῦθ', ὅτι τὸν ἡγεμόνα ἂν 5
ἀφέλῃ τις, οἴχεται ὁ λοιπὸς χορός. ἀλλ' ὅμως πολλῶν
χορηγῶν φιλονικησάντων οὐδεὶς πώποτε τοῦτ' εἶδεν τὸ πλεον-
έκτημα, οὐδ' ἐτόλμησε τοῦτον ἐξαγαγεῖν οὐδὲ κωλῦσαι· διὰ
γὰρ τὸ δεῖν αὐτὸν ἐπιλαβόμενον τῇ χειρὶ τοῦτο ποιῆσαι καὶ
μὴ προσκαλέσασθαι πρὸς τὸν ἄρχοντα ἐξεῖναι, ὥσπερ ἂν εἰ 10
ξένον τις ἐξαγαγεῖν ἠβούλετο, ἅπας τις ὤκνει τῆς ἀσελγείας
61 ταύτης αὐτόχειρ ὀφθῆναι γιγνόμενος. οὔκουν δεινόν, ὦ ἄνδρες
δικασταί, καὶ σχέτλιον τῶν μὲν νικᾶν ἂν παρὰ τοῦτ' οἰομένων
χορηγῶν, τῶν ἀνηλωκότων πολλάκις πάντα τὰ ὄντα εἰς τὰς
λειτουργίας, μηδένα τολμῆσαι πώποτε μηδ' ὧν οἱ νόμοι 15
διδόασιν ἅψασθαι, ἀλλ' οὕτως εὐλαβῶς, οὕτως εὐσεβῶς,
οὕτω μετρίως διακεῖσθαι ὥστε ἀναλίσκοντας, ἀγωνιῶντας
ὅμως ἀπέχεσθαι καὶ προορᾶσθαι τὰς ὑμετέρας βουλήσεις καὶ
τὴν περὶ τὴν ἑορτὴν σπουδήν, Μειδίαν δὲ ἰδιώτην ὄντα, μηδὲν
ἀνηλωκότα, ὅτι τῳ προσέκρουσεν καὶ ἐχθρὸς ὑπῆρχεν, τοῦτον 20
ἀναλίσκοντα, χορηγοῦντα, ἐπίτιμον ὄντα προπηλακίζειν καὶ
τύπτειν, καὶ μήτε τῆς ἑορτῆς μήτε τῶν νόμων μήτε τί ὑμεῖς
ἐρεῖτε μήτε τοῦ θεοῦ φροντίζειν;

62 Πολλῶν τοίνυν, ὦ ἄνδρες Ἀθηναῖοι, γεγενημένων ἐχθρῶν
ἀλλήλοις, οὐ μόνον ἐξ ἰδίων ἀλλὰ καὶ ἐκ κοινῶν πραγμάτων, 25
οὐδεὶς πώποτ' εἰς τοσοῦτ' ἀναιδείας ἀφίκετο ὥστε τοιοῦτόν
τι τολμῆσαι ποιεῖν. καίτοι φασὶν Ἰφικράτην ποτ' ἐκεῖνον
Διοκλεῖ τῷ Πιθεῖ τὰ μάλιστα ἐλθεῖν εἰς ἔχθραν, καὶ ἔτι

27–126.3 καίτοι ... Ἰφικράτης Alex. *Fig.* (Spengel 3) 24.13

---

1 οὐδεὶς ἐχθρὸν ἐκώλυσε Σ (187 Dilts)   τοσοῦτον $S_3^{\gamma\rho}$ Σ (187 Dilts)   2 τῶν χορηγῶν S: τοῦ χορηγόν τινος ἅψασθαι A Σ (187 Dilts): τῶν χορηγῶν τινος ἅψασθαι FP_3^{sl}: τοῦ χορηγῶν τινος ἅψασθαι F_1^{sl}YP: τοῦ τὸν χορηγόν τινος ἅπτεσθαι $S_3^{\gamma\rho}$ 3 ἠτυχηκός S^{ac} 5 κορυφαῖος del. Reiske   δὲ om. AFP_4^{pc} ἐὰν A   6 ἀφέληταί AP_2^{sl} 7 φιλονικησάντων Baiter: φιλονεικησάντων SAFYP   οὐδε εἰς A: οὐδὲ εἷς RVfWd   -δε $AF^{pc}YP^{pc}$ 11 ἐβούλετο Cd^{pc} 12 γιν- A   οὔκουν Buttmann: οὐκοῦν SAFYP   ὦ ανδρες, ἀθη, δικασται S: ὦ ἄνδρες Ἀθηναῖοι A^{cp} 13 ἂν ταυτὶ SYP   ταῦτ' A   14 ἀναλωκότων F_1^{sl} 16 οὕτως εὐσεβῶς S_3^{mg}AFYP: om. S   17 οὕτως μετρίως F   18 ἀπέχθεσθαι F   βοήσεις Y^{sl}P   20 ἀναλωκότα F_1^{sl} -σε $AF^{pc}P^{pc}$ -χε $AF^{pc}P^{pc}$ 21 καὶ χορηγοῦντα S_3^{pc} 22 μήθ' ἑορτῆς F   26 -τε AFY^{pc} -τον A   27 -τὲ AF   28 πιτθεῖ $S_1^{pc}AF_1^{pc}P_2^{pc}$ Alex.: πειθεῖ Y: [P^{ac}]

and is not prevented even by any of his personal enemies, much less the chorus-producers. There's also Aristeides, a **60** member of the Oineis tribe. He too has suffered the same kind of misfortune. By now he's an old man and perhaps not such a good chorister, but at one time he was his tribe's leader. As of course you know, if one takes away the leader, the rest of the chorus is done for. And yet, though there were many chorus-producers eager for victory, none of them ever took advantage of this or ventured to remove him or stop him. Because this had to be done by taking hold of him oneself, manually, and it wasn't possible to summon him to appear before the arkhon, as one would if one wished to remove an alien, everyone shrank from being seen to perpetrate this outrage. Isn't it a terrible and dreadful thing, men **61** of the jury? On the one hand, of all the chorus-producers who thought this would make them victorious, and who had in many cases spent all their money on liturgies, not one ever ventured to touch even those whom the laws permit them to; their attitude was so cautious, so reverent, so restrained that, despite their expenditure and their rivalry, they still held off, and took notice of your wishes and your enthusiasm for the festival. Meidias, on the other hand, a private individual who had spent no money, just because he had a quarrel with someone and was hostile to him—someone who was spending money, and was a chorus-producer, and was a fully enfranchised citizen—Meidias abused him and hit him, and paid no regard to the festival or to the laws or to popular opinion or to the god.

Many men have quarrelled with one another, men of Ath- **62** ens, not only on personal but also on political grounds, but none of them ever sank to such depths of shamelessness as to perpetrate anything like this. Yet they say the famous Iphikrates at one time was a bitter enemy of Diokles of Pithos,

πρὸς τούτῳ συμβῆναι Τεισίαν τὸν Ἰφικράτους ἀδελφὸν ἀντιχορηγῆσαι τῷ Διοκλεῖ· ἀλλ' ὅμως πολλοὺς μὲν ἔχων φίλους Ἰφικράτης, πολλὰ δὲ χρήματα κεκτημένος, φρονῶν δ' ἐφ' αὑτῷ τηλικοῦτον ἡλίκον εἰκὸς ἄνδρα καὶ δόξης καὶ τιμῶν
63 τετυχηκότα ὧν ἐκεῖνος ἠξίωτο παρ' ὑμῶν, οὐκ ἐβάδιζεν ἐπὶ 5
τὰς τῶν χρυσοχόων οἰκίας νύκτωρ, οὐδὲ κατερρήγνυε τὰ παρασκευαζόμενα ἱμάτια εἰς τὴν ἑορτήν, οὐδὲ διέφθειρεν διδάσκαλον, οὐδὲ χορὸν μανθάνειν ἐκώλυεν, οὐδὲ τῶν ἄλλων οὐδὲν ὧν οὗτος διεπράττετο ἐποίει, ἀλλὰ τοῖς νόμοις καὶ τῇ τῶν ἄλλων βουλήσει συγχωρῶν ἠνείχετο καὶ νικῶντα καὶ 10
στεφανούμενον τὸν ἐχθρὸν ὁρῶν. εἰκότως· ἐν ᾗ γὰρ αὐτὸς εὐδαίμων ᾔδει γεγονὼς πολιτείᾳ, ταύτῃ συγχωρεῖν τὰ τοιαῦ-
64 τα ἠξίου. πάλιν Φιλόστρατον πάντες ἴσμεν τὸν Κολωνῆθεν Χαβρίου κατηγοροῦντα, ὅτ' ἐκρίνετο τὴν περὶ Ὠρωποῦ κρίσιν θανάτου, καὶ πάντων τῶν κατηγόρων πικρότατον 15
γενόμενον, καὶ μετὰ ταῦτα χορηγοῦντα παισὶν Διονύσια καὶ νικῶντα, καὶ Χαβρίαν οὔτε τύπτοντα οὔτε ἀφαρπάζοντα τὸν στέφανον οὔθ' ὅλως προσιόνθ' ὅποι μὴ προσῆκεν αὐτῷ.
65 πολλοὺς δ' ἂν ἔχων εἰπεῖν ἔτι καὶ διὰ πολλὰς προφάσεις ἐχθροὺς γεγενημένους ἀλλήλοις, οὐδένα πώποτε οὔτ' ἀκήκοα 20
οὔτε ἑόρακα ὅστις εἰς τοσοῦτον ἐλήλυθεν ὕβρεως ὥστε τοιοῦτόν τι ποιεῖν. οὐδέ γε ἐκεῖνο οὐδεὶς ὑμῶν οἶδ' ὅτι μνημονεύει πρότερον, τῶν ἐπὶ τοῖς ἰδίοις ἢ καὶ τοῖς κοινοῖς ἐχθρῶν ἀλλήλοις οὐδένα οὔτε καλουμένων τῶν κριτῶν παρεστηκότα, οὔθ' ὅταν ὀμνύωσιν ἐξορκοῦντα, οὔθ' ὅλως ἐπ' 25
66 οὐδενὶ τῶν τοιούτων ἐχθρὸν ἐξεταζόμενον. ταῦτα γὰρ πάντα καὶ τὰ τοιαῦτα, ὦ ἄνδρες Ἀθηναῖοι, φιλονικίᾳ μὲν ὑπαχθέντα χορηγὸν ὄντα ποιεῖν ἔχει τινὰ συγγνώμην· ἔχθρᾳ δ' ἐλαύνοντά τινα ἐκ προαιρέσεως ἐφ' ἅπασι, καὶ τὴν ἰδίαν δύναμιν

5–7 οὐκ ... ἑορτήν Alex. *Fig.* (Spengel 3) 24.18 5–8 οὐκ ... ἐκώλυεν Tib. *Fig.* (Ballaira) 15.6 17–18 καὶ Χαβρίαν ... στέφανον *AB* 1.122.21

---

1 Τεισίαν Blass: τισίαν S_1^{pc}AFYP Alex.: [S^{ac}] τὸν ἀδελφὸν τὸν Ἰφικράτους Alex. 2 μὲν om. A Alex. 3 φίλους om. Alex. 4 ἑαυτῷ AFP_3^{pc} τιμῶν καὶ δόξης A 6 κατερρήγνυ Alex. 7 κατεσκευασμένα Alex. -ρε A$F^{pc}$$P^{pc}$ 10 ἄλλων SFYP: πολλῶν A 12 ἤδη A^{ac} 13 ἅπαντες AF 14 -τε A ὠροποῦ S 16 γεγενημένον A -σὶ AFYP 16–17 καὶ νικῶντα om. A 18 -τε A -τα AF 20 -τε AF 21 ἑώρακα AF_1^{sl}P 24 ἀλλήλοις γενομένων A: γενομένων ἀλλήλοις P_2^{mg}: γεγενημένων ἀλλήλοις P_3^{mg} 26 ἐχθρῶν P_3^{pc} πάντα γὰρ ταῦτα A 27 φιλονικίᾳ O: φιλονεικίᾳ SAFYP 28 ἔχειν A δὲ AF

and on top of that Teisias, Iphikrates' brother, competed with Diokles as a chorus-producer; nevertheless, although Iphikrates had many friends and possessed much money, and was as proud of himself as one would expect of a man who had received the honour and distinctions that Iphikrates had been given by you, still he didn't break into the goldsmiths' **63** houses at night, and he didn't rip up the cloaks that were being prepared for the festival, and he didn't corrupt a director, and he didn't prevent a chorus from rehearsing, and he didn't do any of the other things in which Meidias was engaged. He yielded to the laws and the wishes of other people, and he put up with seeing his enemy victorious and crowned. Of course he did; he thought it right to yield in such matters to the society to which he owed his own success. Again, we all **64** know that Philostratos of Kolonai prosecuted Khabrias, when he was being tried on a capital charge concerning Oropos, and showed himself the most hostile of all the prosecutors. Subsequently he won a victory as the producer of a chorus of boys at the Dionysia; and Khabrias neither hit him nor snatched away his crown nor went anywhere at all that he shouldn't have gone. I could mention plenty of other men **65** who have quarrelled with one another on various grounds, but I have never heard of or seen anyone who has reached such a pitch of insolence as to do anything like this. And there's another thing which I'm sure none of you remembers happening before: anyone, a personal or indeed a political adversary, either standing by while the judges were being called up, or dictating an oath to them, or displaying hostility in any such proceedings at all. To do all these and similar **66** things, men of Athens, because one is led on by eagerness for victory when one is a chorus-producer, is excusable; but to do them because one is deliberately harassing someone as an enemy at every opportunity, and demonstrating that one's

κρείττω τῶν νόμων οὖσαν ἐνδεικνύμενον, Ἡράκλεις, βαρὺ
καὶ οὐχὶ δίκαιόν ἐστιν οὐδὲ συμφέρον ὑμῖν. εἰ γὰρ ἑκάστῳ
τῶν χορηγούντων τοῦτο πρόδηλον γένοιτο, ὅτι, ἂν ὁ δεῖνα
ἐχθρὸς ᾖ μοι, Μειδίας ἤ τις ἄλλος θρασὺς οὕτω καὶ πλουσίος,
πρῶτον μὲν ἀφαιρεθήσομαι τὴν νίκην, κἂν ἄμεινον ἀγωνίσω- 5
μαί τινος, ἔπειτα ἐφ᾽ ἅπασιν ἐλαττωθήσομαι καὶ προπηλακι-
ζόμενος διατελέσω, τίς οὕτως ἀλόγιστος ἢ τίς οὕτως ἄθλιός
ἐστιν, ὅστις ἑκὼν ἂν μίαν δραχμὴν ἐθελήσειεν ἀναλῶσαι; οὐ-
67 δεὶς δήπου. ἀλλ᾽, οἶμαι, τὸ πάντας ποιοῦν καὶ φιλοτιμεῖσθαι
καὶ ἀναλίσκειν ἐθέλειν ἐκεῖν᾽ ἐστίν, ὅτι τῶν ἴσων καὶ τῶν 10
δικαίων ἕκαστος ἡγεῖται ἑαυτῷ μετεῖναι ἐν δημοκρατίᾳ. ἐγὼ
τοίνυν, ὦ ἄνδρες Ἀθηναῖοι, τούτων οὐκ ἔτυχον διὰ τοῦτον,
ἀλλὰ χωρὶς ὧν ὑβρίσθην, καὶ τῆς νίκης προσαπεστερήθην.
καίτοι πᾶσιν ὑμῖν ἐγὼ τοῦτο δείξω σαφῶς, ὅτι μηδὲν ἀσελγὲς
ἐξῆν ποιοῦντι Μειδίᾳ μηδ᾽ ὑβρίζοντι μηδὲ τύπτοντι καὶ λυπ- 15
εῖν ἐμὲ καὶ κατὰ τοὺς νόμους αὑτῷ φιλοτιμεῖσθαι πρὸς ὑμᾶς,
68 καὶ μηδὲ διᾶραι τὸ στόμα περὶ αὑτοῦ νῦν ἔχειν ἐμέ. ἐχρῆν γὰρ
αὐτόν, ὦ ἄνδρες Ἀθηναῖοι, ὅτ᾽ ἐγὼ τῆς Πανδιονίδος χορηγὸς
ὑπέστην ἐν τῷ δήμῳ, τότε τῆς Ἐρεχθηίδος ἀναστάντα, τῆς
ἑαυτοῦ φυλῆς, ἀνθυποστῆναι, καὶ καταστήσανθ᾽ ἑαυτὸν ἐξ 20
ἴσου καὶ τὰ ὄντα ἀναλίσκοντα ὥσπερ ἐγώ, οὕτω μ᾽ ἀφαιρεῖσθαι τὴν νίκην, ὑβρίζειν δὲ τοιαῦτα καὶ τύπτειν μηδὲ τότε.
69 νῦν δὲ τοῦτο μὲν οὐκ ἐποίησεν ἐν ᾧ τὸν δῆμον ἐτίμησεν ἄν,
οὐδ᾽ ἐνεανιεύσατο τοιοῦτον οὐδέν· ἐμοὶ δ᾽, ὅς, εἴτε τις, ὦ ἄνδρες Ἀθηναῖοι, βούλεται νομίσαι μανείς (μανία γὰρ ἴσως ἐστὶν 25

9–11 *ἀλλ᾽ … δημοκρατίᾳ* *AB* 1.156.15 17–19 *ἐχρῆν … τότε* schol. Hermog. (Walz 4) 523.5 23–130.4 *νῦν … μου* Dion. Hal. *Dem.* 9 24–130.1 *ἐμοὶ … ποιεῖν* Hermog. 281.11, 348.9, 364.23, 366.6, Syrian. (Rabe 1) 74.1, Max. Plan. (Walz 5) 528.6 *ἐμοὶ … φιλοτιμίᾳ* Alex. *Fig.* (Spengel 3) 13.26 *ἐμοὶ … ὑπέστην* Tib. *Fig.* (Ballaira) 3.9

---

1 *καὶ βίαν κρείττω* AP: *κρείττω καὶ βίαν* F 2 *-τι* A 3 *χορηγῶν* A 4 *ἤ*] *εἰ* Y: [P^{ac}] 7 *οὕτως* (ante *ἀλόγιστος*) FYP: *οὗτος* SA *οὕτως* (ante *ἄθλιός*) F: *οὗτος* A: om. SYP 8 *ἐθελήσῃ* A 10 *-νό* AF 13 *χωρὶς ὧν* $S_3^{\gamma\rho}$AFYP: *χωρὶς ὢν* S^{ac}: *χορηγὸς ὢν* $S_2^{pc}P_4^{\gamma\rho}$ 15 *μήτε … μήτε* A 17 *διαῖραι* S *περὶ αὐτοῦ τὸ στόμα* A *νῦν* om. SY^{ac} 18 *ὦ ἄνδρες Ἀθηναῖοι*] *ἔσθ᾽* schol. Hermog. 19 *ἐν τῷ δήμῳ* om. schol. Hermog. 20 *αυτοῦ* A: *αὑτοῦ* Cd *-τα* A *αυτὸν* A: *αὑτὸν* Cd 21 *με* AF: *μὲν* schol. Hermog.: *μὴ* Ln 22 *μηδὲ*] *οὐδὲ* P_3^{sl} 24 *-δε* A *τοιοῦτον οὐδέν* $S_3^{\gamma\rho}$AFYP Dion.: *τοῦτο* S *ἐμοὶ*] *ἐγὼ* Tib.: *ἐμὲ* Syrian. *δὲ* $S_3^{\gamma\rho}$A *ὡς* Tib. Syrian.: *ὅσον* Alex. 24–5 *ὦ Ἀθηναῖοι* Dion. Hermog. Syrian. Max.: *θεῶν* Tib.: om. Alex. 25 *βούλεται* om. $S_3^{\gamma\rho}$ *μανείς* S: *μανίᾳ* A Syrian.: *μανίαν* $S_3^{\gamma\rho}$FYP Dion. Hermog. Alex. Tib. Max. 25–130.1 *μανία … ποιεῖν* om. Alex. *ἐστὶ τὸ ὑπὲρ* Hermog. 348.9, 364.23, 366.6, Tib. Max.: *ἐστὶ τὸ παρὰ* Hermog. 281.11

personal power is stronger than the laws—good heavens! That is oppressive, and quite wrong, and bad for Athens. If it became clear to every chorus-producer that, if someone or other were hostile to me—Meidias or any other such audacious and wealthy man—I should first be deprived of my victory even if my performance were better than anyone else's, and besides I should be put at a disadvantage at every opportunity and continually abused, who is so improvident or so foolish that he would be willing voluntarily to spend a single drachma? Surely no one. The thing that makes them **67** all emulous and willing to spend money, I think, is that each of them believes he enjoys equality and fairness in a democracy. Now I, men of Athens, was prevented by Meidias from obtaining those; apart from the insolence, I was also robbed of my victory. Yet I shall prove to you all that it was possible for Meidias, without any bullying or insolence or hitting, to vex me and seek honour himself from you quite legally, and I shouldn't have been able to say a word about him now. What he ought to have done, men of Athens, at the time **68** when I undertook in the assembly to be chorus-producer of Pandionis, was to stand up and undertake the same duty for Erekhtheis, his own tribe; putting himself on equal terms and spending his money like me, he should have tried to take the victory from me in that way, though even then he shouldn't have treated me with such insolence and blows. As it is, he **69** didn't do what would have implied a compliment to the people, nor did he show any such youthful enterprise; but me, who—whether one wishes, men of Athens, to think because I was mad (for perhaps it is madness to tackle

ὑπὲρ δύναμίν τι ποιεῖν) εἴτε καὶ φιλοτιμίᾳ, χορηγὸς ὑπέστην,
οὕτω φανερῶς καὶ μιαρῶς ἐπηρεάζων παρηκολούθησεν ὥστε
μηδὲ τῶν ἱερῶν ἱματίων μηδὲ τοῦ χοροῦ μηδὲ τοῦ σώματος
τὼ χεῖρε τελευτῶν ἀποσχέσθαι μου.

70 Εἰ τοίνυν τις ὑμῶν, ὦ ἄνδρες Ἀθηναῖοι, ἄλλως πως ἔχει 5
τὴν ὀργὴν ἐπὶ Μειδίαν ἢ ὡς οὐ δέον αὐτὸν τεθνάναι, οὐκ
ὀρθῶς ἔχει. οὐ γάρ ἐστι δίκαιον οὐδὲ προσῆκον τὴν τοῦ παθόντος εὐλάβειαν τῷ μηδὲν ὑποστειλαμένῳ πρὸς ὕβριν μερίδα
εἰς σωτηρίαν ὑπάρχειν· ἀλλὰ τὸν μὲν ὡς ἁπάντων τῶν ἀν-
ηκέστων αἴτιον κολάζειν προσήκει, τῷ δ' ἐπὶ τοῦ βοηθεῖν 10
71 ἀποδιδόναι τὴν χάριν. οὐδὲ γὰρ αὖ τοῦτ' ἔστιν εἰπεῖν, ὡς οὐ
γεγενημένου πώποτ' οὐδενὸς ἐκ τῶν τοιούτων δεινοῦ τῷ
λόγῳ τὸ πρᾶγμ' ἐγὼ νῦν αἴρω καὶ φοβερὸν ποιῶ. πολλοῦ γε
καὶ δεῖ· ἀλλ' ἴσασιν ἅπαντες, εἰ δὲ μή, πολλοί γε, Εὔθυνον τὸν
παλαίσαντά ποτ' ἐκεῖνον, τὸν νεανίσκον, καὶ Σώφιλον τὸν 15
παγκρατιαστήν (ἰσχυρός τις ἦν, μέλας, εὖ οἶδ' ὅτι γιγνώσκουσίν τινες ὑμῶν ὃν λέγω), τοῦτον ἐν Σάμῳ ἐν συνουσίᾳ
τινὶ καὶ διατριβῇ οὕτως ἰδίᾳ, ὅτι {ὁ τύπτων} αὐτὸν ὑβρίζειν
ᾤετο, ἀμυνάμενον οὕτως ὥστε καὶ ἀποκτεῖναι. ἴσασιν
Εὐαίωνα πολλοί, τὸν Λεωδάμαντος ἀδελφόν, ἀποκτείναντα 20
72 Βοιωτὸν ἐν δείπνῳ καὶ συνόδῳ κοινῇ διὰ πληγὴν μίαν. οὐ γὰρ
ἡ πληγὴ παρέστησε τὴν ὀργήν, ἀλλ' ἡ ἀτιμία· οὐδὲ τὸ
τύπτεσθαι τοῖς ἐλευθέροις ἐστὶ δεινόν, καίπερ ὂν δεινόν, ἀλλὰ

16 *ἰσχυρός ... μέλας* Anon. Seg. (Hammer) 369.4, Io. Sard. 23.19, Doxop. (Walz 2) 234.18, Max. Plan. (Walz 5) 516.8 19–20 *ἴσασιν ... ἀδελφόν* Tib. *Fig.* (Ballaira) 40.5 19–21 *ἴσασιν ... δείπνῳ* Theon *Prog.* (Spengel 2) 82.31, Doxop. (Walz 2) 225.6 *ἴσασιν ... μίαν* Syrian. (Rabe 2) 112.9 19–132.1 *ἴσασιν ... ὕβρει* schol. Hermog. (Walz 4) 529.7 21–132.1 *οὐ ... ὕβρει* Syrian. (Rabe 2) 105.6

---

1 *καὶ* om. Tib. *φιλοτιμίαν* Dion. Alex. Tib. 2 *καὶ μιαρῶς* om. Dion. *ἐπαρηκολούθησεν* S: *ἐπηκολούθησεν* YP 5 *ὦ ἄνδρες δικασταί* A[cp] 6 *οὐ* om. S 7 *ἔχει*] *γινώσκει* A 9 *εἰς* SAFY[sl]P: *πρὸς* Y 10 *τῷ*] *τὸ* A[ac] 11 *οὐδὲν* F[ac] *ἐν εστίν* A: *ἔνεστιν* Cd 11–12 *οὐ γενημένου* S[ac]: *οὐδὲ γεγενημένου* S_3^{pc} 13 *-μα* AF 15 *τὸν Νεμεόνικον* Weil: *ὄντα νεανίσκον* MacDowell *καὶ* del. Reiske 16 *τις* om. Seg. Sard. Doxop. *ἦν* om. Sard. Doxop. *μέγας* Seg. Sard. Doxop. *γιν-* A 17 *-σί* AF 18 *ὁ τύπτων* del. Boeckh 19 *ἀμυνόμενον* F 19–20 *ἴσασι τὸν Εὐαίωνα πολλὰ* Doxop. 20 *Λαοδάμαντος* A Tib. Doxop. 21 *τὸν Βοιωτὸν* Doxop. *ἐν ... κοινῇ* om. schol. Hermog. *κοινῇ*] *οἰκείων* Weil 22 *παρίστησι* Syrian. 23 *τοὺς ἐλευθέρους* schol. Hermog. *ἐστὶ* om. schol. Hermog.

something beyond one's ability) or from love of honour—undertook to be a chorus-producer, he pursued with such open and vicious obstruction that he didn't keep away from my sacred cloaks or my chorus, and in the end didn't keep his hands off my person.

If the feeling of anger which any of you has against Meidias, men of Athens, is anything less than that he ought to be put to death, you are not right. It's not just or proper that the victim's discretion should be a contribution towards acquittal for a man who didn't restrain himself from insolence; it's proper to punish the latter as being responsible for every disaster, and to show your gratitude to the former in supporting him. For it can't be maintained that no serious consequence has ever followed from such behaviour and that I'm just exaggerating the matter now and making it formidable by what I'm saying. Far from it. Everyone knows—or if not everyone, many people—that on one occasion Euthynos the famous wrestler, the young lad, defended himself even against Sophilos the pancratiast. (He was a strong man, dark—I'm sure some of you know the man I mean.) They were in Samos, just passing the time privately with some friends; and because he thought him insolent, he defended himself so vigorously that he actually killed him. Many people know that Euaion, the brother of Leodamas, killed Boiotos at a dinner party because of one blow. It wasn't the blow that made him angry, but the dishonour; nor is being hit such a serious matter to free men (though it is serious), but being hit **70** **71** **72**

τὸ ἐφ' ὕβρει. πολλὰ γὰρ ἂν ποιήσειεν ὁ τύπτων, ὦ ἄνδρες
Ἀθηναῖοι, ὧν ὁ παθὼν ἔνια οὐδ' ἂν ἀπαγγεῖλαι δύναιθ' ἑτέρῳ,
τῷ σχήματι, τῷ βλέμματι, τῇ φωνῇ, ὅταν ὡς ὑβρίζων, ὅταν
ὡς ἐχθρὸς ὑπάρχων, ὅταν κονδύλοις, ὅταν ἐπὶ κόρρης. ταῦτα
κινεῖ, ταῦτ' ἐξίστησιν ἀνθρώπους αὑτῶν, ἀήθεις ὄντας τοῦ 5
προπηλακίζεσθαι. οὐδεὶς ἄν, ὦ ἄνδρες Ἀθηναῖοι, ταῦτ' ἀπαγ-
γέλλων δύναιτο τὸ δεινὸν παραστῆσαι τοῖς ἀκούουσιν οὕτως
ὡς ἐπὶ τῆς ἀληθείας καὶ τοῦ πράγματος τῷ πάσχοντι καὶ τοῖς
73 ὁρῶσιν ἐναργὴς ἡ ὕβρις φαίνεται. σκέψασθε δὴ πρὸς Διὸς καὶ
θεῶν, ὦ ἄνδρες Ἀθηναῖοι, καὶ λογίσασθε παρ' ὑμῖν αὐτοῖς, 10
ὅσῳ πλείονα ὀργὴν ἐμοὶ προσῆκε παραστῆναι πάσχοντι τοιαῦτα ὑπὸ Μειδίου ἢ τότε ἐκείνῳ τῷ Εὐαίωνι τῷ τὸν Βοιωτὸν
ἀποκτείναντι. ὁ μέν γε ὑπὸ γνωρίμου, καὶ τούτου μεθύοντος,
ἐναντίον ἓξ ἢ ἑπτὰ ἀνθρώπων ἐπλήγη, καὶ τούτων γνωρίμων,
οἳ τὸν μὲν κακιεῖν οἷς ἔπραξε, τὸν δ' ἐπαινέσεσθαι μετὰ ταῦτα 15
ἀνασχόμενον καὶ κατασχόνθ' ἑαυτὸν ἤμελλον, καὶ ταῦτ' εἰς
74 οἰκίαν ἐλθὼν ἐπὶ δεῖπνον, οἷ μηδὲ βαδίζειν ἐξῆν αὐτῷ· ἐγὼ δ'
ὑπ' ἐχθροῦ, νήφοντος, ἕωθεν, ὕβρει καὶ οὐκ οἴνῳ τοῦτο ποιοῦντος, ἐναντίον πολλῶν καὶ ξένων καὶ πολιτῶν ὑβριζόμην,
καὶ ταῦτ' ἐν ἱερῷ καὶ οἷ πολλή μοι ἦν ἀνάγκη βαδίζειν χορηγ- 20
οῦντι. καὶ ἐμαυτὸν μέν γε, ὦ ἄνδρες Ἀθηναῖοι, σωφρόνως,
μᾶλλον δ' εὐτυχῶς οἶμαι βεβουλεῦσθαι, ἀνασχόμενον τότε καὶ
οὐδὲν ἀνήκεστον ἐξαχθέντα πρᾶξαι· τῷ δ' Εὐαίωνι καὶ πᾶσιν,

1–6 πολλὰ ... προπηλακίζεσθαι Plu. *Mor.* 1010e, schol. Hermog. (Walz 4) 530.15 1–7 πολλὰ ... παραστῆσαι Longin. *Subl.* 20 3–4 ὅταν ὡς ὑβρίζων ... κόρρης Minuc. (Hammer) 345.20 4–5 ὅταν κονδύλοις ... ἐξίστησιν Tib. *Fig.* (Ballaira) 29.2, 40.7 9–14 σκέψασθε ... γνωρίμων schol. Hermog. (Walz 4) 529.12 17–18 ἐγὼ ... οἴνῳ schol. Hermog. (Walz 4) 529.18 17–19 ἐγὼ ... ὑβριζόμην Alex. *Fig.* (Spengel 3) 18.6

---

1 ἂν om. schol. Hermog. ὦ ἄνδρες Ἀθηναῖοι om. Plu. Longin. schol. Hermog. 2 ἔνια ὁ παθὼν schol. Hermog. δύναται schol. Hermog. ἑταίρῳ F 3–4 ὡς utrumque om. Plu. ὅταν ὡς ἐχθρὸς ὑπάρχων om. Minuc. 4 ὑπάρχων om. Longin.: ὤν schol. Hermog. ὅταν κονδύλοις ante ὅταν ὡς ἐχθρὸς schol. Hermog.: post κόρρης Tib. 5 -τα AF αὑτῶν FP_3^{pc}: αυτῶν A: αὐτῶν SYP^{ac}: αὐτῶν ante ἀνθρώπους Plu.: om. Longin. schol. Hermog. ὄντας om. Plu. 6 ἄν om. A Longin. ὦ ἄνδρες Ἀθηναῖοι om. Longin. ἀπαγλων S^{ac}: ἀπεγγέλλων Y 7 δύναιτο ἂν A 9 σκέψασθαι S^{ac} δὲ schol. Hermog. 10 ὦ ... αὐτοῖς om. schol. Hermog. ὦ ἄνδρες δικασταί A^{cp} λογίσασθε FYP: λογίσασθαι S: λογίζεσθε A 11 παραστῆναι] γενέσθαι schol. Hermog. πάσχοντι ... Μειδίου om. schol. Hermog. 12 ευωνι S^{ac} 13 γε] γὰρ schol. Hermog. 15 ἐφ' οἷς A δὲ A ἐπαινεσθαι S^{ac} 16 -τα A ἔμελλον P_3^{sl} ταῦτ' ἔπαθεν AP_2^{sl} 17 τὸ δεῖπνον S^{ac} μη S^{ac} ἐξὸν F_1^{sl} δὲ A 18 ταῦτα A Alex. 20 οἷ YP: οὗ SAF$P_3^{\gamma\rho}$ 22 δὲ A 23 μηδὲν AF

insolently. There are many things which the hitter might do, men of Athens, some of which the victim might not even be able to report to someone else—in his bearing, in his look, in his voice, when he displays insolence, when he displays hostility, when he strikes with the fist, when he strikes on the face. That's what rouses people, that's what makes them forget themselves, if they're not accustomed to being insulted. No one reporting this behaviour, men of Athens, could convey its seriousness to his listeners as vividly as the insolence is seen at the actual time by the victim and the onlookers. **73** Please consider, men of Athens, and calculate to yourselves how much more angry it was proper for me to be when I suffered such treatment from Meidias than it was on the earlier occasion for that man Euaion, who killed Boiotos. He was hit by a friend, and that friend was drunk, and it was in the presence of six or seven people, who were also friends, and who were likely to criticize Boiotos for what he did and praise Euaion afterwards if he acquiesced and restrained himself; and it was when he had gone to a house for dinner, where he need not have gone at all. **74** I, on the other hand, was assaulted by an enemy, who was sober, in the morning, doing it from insolence and not from wine, in the presence of a large number of both foreigners and citizens; and it was in a sacred place, where it was very necessary for me to go because I was a chorus-producer. I think my decision was prudent, men of Athens, or rather it was fortunate, when I acquiesced at the time and wasn't induced to do anything disastrous—though

εἴ τις αὐτῷ βεβοήθηκεν ἀτιμαζόμενος, πολλὴν συγγνώμην
75 ἔχω. δοκοῦσι δέ μοι καὶ τῶν δικασάντων τότε πολλοί· ἀκούω
γὰρ αὐτὸν ἔγωγε μιᾷ μόνον ἁλῶναι ψήφῳ, καὶ ταῦτα οὔτε
κλαύσαντα οὔτε δεηθέντα τῶν δικαστῶν οὐδενός, οὔτε φιλάνθρωπον οὔτε μικρὸν οὔτε μέγα οὐδ' ὁτιοῦν πρὸς τοὺς 5
δικαστὰς ποιήσαντα. θῶμεν τοίνυν οὑτωσί, τοὺς μὲν κατα-
γνόντας αὐτοῦ μὴ ὅτι ἠμύνατο, διὰ τοῦτο καταψηφίσασθαι,
ἀλλ' ὅτι τοῦτον τὸν τρόπον ὥστε καὶ ἀποκτεῖναι, τοὺς δ'
ἀπογνόντας καὶ ταύτην τὴν ὑπερβολὴν τῆς τιμωρίας τῷ γε τὸ
76 σῶμα ὑβρισμένῳ δεδωκέναι. τί οὖν; ἐμοὶ τῷ τοσαύτῃ κεχρη- 10
μένῳ προνοίᾳ τοῦ μηδὲν ἀνήκεστον γενέσθαι, ὥστε μηδ' ἀμύνασθαι, παρὰ τοῦ τὴν τιμωρίαν ὧν πέπονθα ἀποδοθῆναι
προσήκει; ἐγὼ μὲν οἶμαι παρ' ὑμῶν καὶ τῶν νόμων, καὶ παράδειγμά γε πᾶσι γενέσθαι τοῖς ἄλλοις, ὅτι τοὺς ὑβρίζοντας
ἅπαντας καὶ τοὺς ἀσελγεῖς οὐκ αὐτὸν ἀμύνεσθαι μετὰ τῆς 15
ὀργῆς, ἀλλ' ἐφ' ὑμᾶς ἄγειν δεῖ, ὡς βεβαιούντων ὑμῶν καὶ
φυλαττόντων τὰς ἐν τοῖς νόμοις τοῖς παθοῦσι βοηθείας.

77 Οἶμαι τοίνυν τινὰς ὑμῶν, ὦ ἄνδρες δικασταί, ποθεῖν ἀκοῦσαι τὴν ἔχθραν, ἥτις ἦν ἡμῖν πρὸς ἀλλήλους· νομίζειν γὰρ
οὐδένα ἂν ἀνθρώπων οὕτως ἀσελγῶς καὶ βιαίως οὐδενὶ τῶν 20
πολιτῶν χρήσασθαι, μὴ μεγάλου τινὸς ὄντος ὃ αὐτῷ προωφείλετο. βούλομαι δὴ καὶ περὶ ταύτης ὑμῖν ἐξ ἀρχῆς εἰπεῖν καὶ
διηγήσασθαι, ἵν' εἰδῆθ' ὅτι καὶ τούτων ὀφείλων δίκην φανήσεται. ἔσται δὲ βραχὺς περὶ αὐτῶν ὁ λόγος, κἂν ἄνωθεν ἄρχε-
78 σθαι δοκῶ. ἡνίκα τὰς δίκας ἔλαχον τῶν πατρῴων τοῖς 25
ἐπιτρόποις, μειρακύλλιον ὢν κομιδῇ καὶ τοῦτον οὐδ' εἰ γέγονεν εἰδὼς οὐδὲ γιγνώσκων (ὡς μηδὲ νῦν ὤφελον), τότε μοι

1–2 πολλὴν ... τότε Tib. *Fig.* (Ballaira) 37.2 18–19 οἶμαι ... ἡμῖν Io. Sard. (Rabe *PS*) 359.20 25–7 ἡνίκα ... ὤφελον Hermog. 354.24 25–6 τὰς ... ἐπιτρόποις *Rhet. Lex.* (Naoumides) 61 26–7 καὶ ... ὤφελον Syrian. (Rabe 1) 56.14, schol. Hermog. (Walz 7) 1021.10

1 εἴ τις om. S^{ac} ἑαυτῷ A ἀτιμαζομένῳ YP 2 δικαστῶν A 3 αὐτὸν post μόνον A 7 καὶ ψηφίσασθαι YP^{ac} 8 τὸν bis S δὲ A 11 -δε A 12 παρά του P κατὰ τὴν P$_3^{sl}$ 13 ὑμῖν καὶ τοῖς νόμοις P$_3^{sl}$ καὶ (ante παράδειγμα) om. P$_3^{pc}$ 14 γε πᾶσι SFY: τοῦτον A: με πᾶσι P 15 ἀμύνασθαι A 16 ἄγει A^{ac} 17 νόμοις κατὰ τῶν ἀδικούντων FYP βοηθείας SFYP: τιμωρίας AP$_2^{γρ}$ 18 ὑμῶν τινας A ὦ ανδρες, ἀθη'ναιοι', δικασταὶ S: ὦ Ἀθηναῖοι Sard. 20 ἂν om. SA 21 ἂν χρῆσθαι A 23 -τε AF τούτων τὴν μεγίστην AF: περὶ ταύτης P$_3^{sl}$ δοῦναι (post ὀφείλων) AFYP: om. S Σ (245 Dilts) 24 περὶ αὐτῶν βραχὺς SFYP κἂν S$_3^{pc}$AFYP: καὶ S^{ac} 25–6 δίκας τοῖς ἐπιτρόποις ἐλάγχανον Hermog. 26 μειρακύλλυον F 27 γιν- A Hermog. ὄφελον AY: [P^{ac}]

I fully sympathize with Euaion and anyone else who has defended himself when dishonoured. So did many of the jury **75** in that case, I think; for I'm told that he was convicted by only one vote, even though he neither shed tears, nor addressed a plea to any of the jurors, nor did the jurors any favour whatever, large or small. So let's make the assumption that those who convicted him voted against him not because of the fact that he defended himself but because he did it in such a way as to cause death, while those who voted for acquittal conceded even that excessive degree of revenge to a man who had suffered insolent treatment of his person. Well **76** then! When I exercised so much care to prevent any disastrous result that I didn't defend myself at all, from whom ought I to obtain atonement for what was done to me? From you and the laws, I think; and an example ought to be set, to show everyone else that all insolent bullies should not be fought off at the moment of anger, but referred to you, in the knowledge that you are the guarantors and guardians of legal protection for victims.

I expect some of you, men of the jury, want to hear what **77** quarrel we had with each other, and think that no human being would treat one of his fellow-citizens so aggressively and violently unless the atonement owed to him were a substantial one. I should like to relate this to you also from the beginning, to let you understand that he will obviously deserve punishment for these actions too. The account of them will be short, even if I seem to be starting from a long way back. When I initiated the cases for my inheritance **78** against my guardians—I was a very young lad then, not knowing that Meidias had even been born, and not being acquainted with him (I wish I weren't now either!)—at that

μελλουσῶν εἰσιέναι τῶν δικῶν εἰς ἡμέραν ὡσπερεὶ τετάρτην ἢ
πέμπτην εἰσεπήδησαν ἀδελφὸς ὁ τούτου καὶ οὗτος εἰς τὴν
οἰκίαν ἀντιδιδόντες τριηραρχίαν. τοὔνομα μὲν δὴ παρέσχεν
ἐκεῖνος, καὶ ἦν ὁ ἀντιδιδοὺς Θρασύλοχος· τὰ δ' ἔργα πάντ' ἦν
79 καὶ τὰ πραττόμενα ὑπὸ τούτου. καὶ πρῶτον μὲν κατέσχισαν 5
τὰς θύρας τῶν οἰκημάτων, ὡς αὑτῶν ἤδη γιγνομένας κατὰ
τὴν ἀντίδοσιν· εἶτα τῆς ἀδελφῆς, ἔτ' ἔνδον οὔσης τότε καὶ
παιδὸς οὔσης κόρης, ἐναντίον ἐφθέγγοντο αἰσχρὰ καὶ τοιαῦτα
οἷ' ἂν ἄνθρωποι τοιοῦτοι φθέγξαιντο (οὐ γὰρ ἔγωγε προαχθ-
είην ἂν εἰπεῖν πρὸς ὑμᾶς τῶν τότε ῥηθέντων οὐδέν), καὶ τὴν 10
μητέρα κἀμὲ καὶ πάντας ἡμᾶς ῥητὰ καὶ ἄρρητα κακὰ ἐξεῖπον·
ὃ δ' οὖν δεινότατον καὶ οὐ λόγος ἀλλ' ἔργον ἤδη, τὰς δίκας ὡς
80 αὑτῶν οὔσας ἀφίεσαν τοῖς ἐπιτρόποις. καὶ ταῦτ' ἐστὶ μὲν
παλαιά, ὅμως δέ τινας ὑμῶν μνημονεύειν οἴομαι· ὅλη γὰρ ἡ
πόλις τὴν ἀντίδοσιν καὶ τὴν ἐπιβουλὴν τότε ταύτην καὶ τὴν 15
ἀσέλγειαν ᾔσθετο. κἀγὼ τότε παντάπασιν ἔρημος ὢν καὶ νέος
κομιδῇ, ἵνα μὴ τῶν παρὰ τοῖς ἐπιτρόποις ἀποστερηθείην, οὐχ
ὅσα ἐδυνήθην ἀνακομίσασθαι προσδοκῶν εἰσπράξειν, ἀλλ'
ὅσων ἐμαυτῷ συνῄδειν ἀπεστερημένῳ, δίδωμι εἴκοσι μνᾶς
τούτοις, ὅσου τὴν τριηραρχίαν ἦσαν μεμισθωκότες. τὰ μὲν δὴ 20
81 τότε ὑβρίσματα τούτων εἰς ἐμὲ ταῦτ' ἐστίν. δίκην δὲ τούτῳ
λαχὼν ὕστερον τῆς κακηγορίας εἷλον ἐρήμην· οὐ γὰρ ἀπήντα.
λαβὼν δ' ὑπερήμερον καὶ ἔχων οὐδενὸς ἡψάμην πώποτε τῶν
τούτου, ἀλλὰ λαχὼν ἐξούλης πάλιν οὐδέπω καὶ τήμερον εἰσ-
ελθεῖν δεδύνημαι· τοσαύτας τέχνας καὶ σκήψεις οὗτος εὑρί- 25
σκων ἐκκρούει. κἀγὼ μὲν οὕτως εὐλαβῶς τῇ δίκῃ, τοῖς νόμοις
ἅπαντα πράττειν ἀξιῶ· ὁ δ', ὡς ὑμεῖς ἀκούετε, ἀσελγῶς οὐ
μόνον εἰς ἐμὲ καὶ τοὺς ἐμοὺς ᾤετο δεῖν ὑβρίζειν, ἀλλὰ καὶ εἰς
82 τοὺς φυλέτας δι' ἐμέ. ὡς οὖν ταῦτ' ἀληθῆ λέγω, κάλει μοι

11 ῥητὰ ... κακὰ Phot. (Theodoridis) α 2870 14–16 ὅλη ... ᾔσθετο AB 1.122.23 21–2 δίκην ... ἀπήντα Suda ε 1815

---

1 ὡσπερὶ F: ὡς περὶ Σ (252 Dilts) 2 εἰσεπήδησεν A ἀδελφὸς Bekker: αδελφὸς SA: ἀδελφὸς YP: ὁ ἀδελφὸς F 3 ἀντιδιδόντος A 4 ἐκεῖνος del. Taylor δὲ AF ἔργ' ἅπαντ' S: ἔργα πάντα AF 4–5 ἦν post πραττόμενα A 6 αὑτῶν Taylor: αυτῶν SAY: αὐτῶν FP γιν- AYP 7 οὔσης del. Weil 9 οἷ' ἂν YP: οἱ S: οἷα A: οἷα ἂν F 10 ῥηθέ των F^{pc} 11 καὶ ἐμὲ AF 13 ἠφίεσαν FYP -τα A 14 μνημονεύειν ὑμῶν A οἶμαι F 15 ταύτην τότε S^{ac} 18 ἠδυνήθην AF κομίσασθαι AFP$_4^{\gamma\rho}$ 21 τούτων ὑβρίσματα A: ἠσελγημένα F$_1^{\gamma\rho}$ -τί AF^{pc}P^{pc} τούτῳ] τούτων S$_2^{pc}$FYP^{ac}P$_4^{\gamma\rho}$ 22 κατηγορίας Σ (Patm.) Suda ἄπαντα S^{ac} 23 δὲ A ἐξὸν Herwerden πω A 26 καὶ ἐγὼ A 26–7 τοῖς νόμοις τῇ δίκῃ πάντα A 27 δὲ A

time, when my cases were due to come into court in about three or four days, his brother and he burst into my house because they were proposing an exchange for a trierarchy. Nominally it was done by the former, and the man proposing the exchange was Thrasylokhos; but all the actions and proceedings were Meidias' doing. First they broke down the **79** doors of the rooms, as being already their own property in accordance with the exchange. Next, in the presence of my sister, who was still at home then and was a young girl, they used bad language of the kind that men of that kind would use (for nothing would induce me to repeat to you any of what was said on that occasion), and hurled abuse, decent and indecent, at my mother and me and all of us. But what was most dreadful, and not merely words but deeds, they proceeded to drop the cases against my guardians, as being their own property. This happened a long time ago, but still **80** I expect some of you remember it; for at the time the whole city knew about the proposal for exchange and this plot and the bullying. I was entirely on my own at that time, and very young; and, to avoid being deprived of the property which was in the hands of my guardians, because I was expecting to be paid not merely as much as I was in fact able to recover but as much as I knew I had been robbed of, I offered twenty mnai to these men, the sum for which they had contracted out the trierarchy. So that was their insolent behaviour to me at that time. Later I initiated a case against Meidias for the **81** slander, and won it by default; for he didn't attend. He became and remained my overdue debtor, yet I never touched any of his property; I initiated another case, for ejectment, but to this day I've never been able to get it tried, he finds so many devices and excuses for delaying it. I take such care to do everything rightly and legally, while he, as you hear, thought fit to be an insolent bully, not only to me and my family but even to the members of my tribe on my account. To prove that what I say is true, please call the **82**

τούτων τοὺς μάρτυρας, ἵν' εἰδῆθ' ὅτι, πρὶν κατὰ τοὺς νόμους
δίκην ὧν πρότερον ἠδικήθην λαβεῖν, πάλιν τοιαῦτα οἷα ἀκηκόατε ὕβρισμαι.

ΜΑΡΤΥΡΙΑ

{Καλλισθένης Σφήττιος, Διόγνητος Θορίκιος, Μνη- 5
σίθεος Ἀλωπεκῆθεν οἴδαμεν Δημοσθένη, ᾧ μαρτυροῦμεν, κρίσιν λελογχότα Μειδίᾳ ἐξούλης, τῷ καὶ νῦν
ὑπ' αὐτοῦ κρινομένῳ δημοσίᾳ, καὶ ἤδη τῇ κρίσει ἐκείνῃ
διαγεγονότα ἔτη ὀκτώ, καὶ τοῦ χρόνου γεγενημένον
παντὸς αἴτιον Μειδίαν ἀεὶ προφασιζόμενον καὶ ἀνα- 10
βαλλόμενον.}

83 Ὃ τοίνυν πεποίηκεν, ὦ ἄνδρες Ἀθηναῖοι, περὶ τῆς δίκης,
ἀκούσατε, καὶ θεωρεῖτε ἐφ' ἑκάστου τὴν ὕβριν καὶ τὴν ὑπερηφανίαν αὐτοῦ. τῆς γὰρ δίκης, ταύτης λέγω ἧς εἷλον αὐτόν,
γίγνεταί μοι διαιτητὴς Στράτων Φαληρεύς, ἄνθρωπος πένης 15
μέν τις καὶ ἀπράγμων, ἄλλως δ' οὐ πονηρὸς ἀλλὰ καὶ πάνυ
χρηστός· ὅπερ τὸν ταλαίπωρον οὐκ ὀρθῶς οὐδὲ δικαίως ἀλλὰ
84 καὶ πάνυ αἰσχρῶς ἀπολώλεκεν. οὗτος διαιτῶν ἡμῖν ὁ Στράτων, ἐπειδή ποθ' ἧκεν ἡ κυρία, πάντα δ' ἤδη διεξεληλύθει τἀκ
τῶν νόμων, ὑπωμοσίαι καὶ παραγραφαί, καὶ οὐδὲν ἔτ' ἦν 20
ὑπόλοιπον, τὸ μὲν πρῶτον ἐπισχεῖν ἐδεῖτό μου τὴν δίαιταν,
ἔπειτα εἰς τὴν ὑστεραίαν ἀναβαλέσθαι· τὸ τελευταῖον δ', ὡς
οὔτ' ἐγὼ συνεχώρουν οὔθ' οὗτος ἀπήντα, τῆς δ' ὥρας
85 ἐγίγνετ' ὀψέ, κατεδιῄτησεν. ἤδη δ' ἑσπέρας οὔσης καὶ σκότους
ἔρχεται Μειδίας οὑτοσὶ πρὸς τὸ τῶν ἀρχόντων οἴκημα, καὶ 25
καταλαμβάνει τοὺς ἄρχοντας ἐξιόντας καὶ τὸν Στράτωνα ἀπ-
ιόντ' ἤδη, τὴν ἔρημον δεδωκότα, ὡς ἐγὼ τῶν παραγενομένων
τινὸς ἐπυνθανόμην. τὸ μὲν οὖν πρῶτον οἷός τε ἦν πείθειν
αὐτόν, ἣν κατεδεδιῃτήκει, ταύτην ἀποδεδιῃτημένην ἀποφαίνειν, καὶ τοὺς ἄρχοντας μεταγράφειν, καὶ πεντήκοντα δραχ- 30

1 -τε AYP 4 ΜΑΡΤΥΡΕΣ YP 5–11 testimonium om. A
5 Θορίκιος Palmer: θεωρίσκος SFYP 6 δημοσθένην FP_4^{pc} 8 αὐτοῦ
FP_4^{sl}: αὐτῷ SYP κρινομένου S^{ac} 10 δ' ἀεὶ Y^{pc} 12 πεποίηκε
κακὸν FP_3^{pc} 14 ἧς λέγω ταύτης ἣν $S_3^{γρ}$ 15 γίν- A 17–18 οὐδὲ δικαίως
post αἰσχρῶς SFYP 18 οὑτοσὶ AF^{pc}: οὑτωσὶ F^{ac} 19 δὲ A ταῦτα
τἀκ S: τὰ ἐκ AF 20 -τι A 22 τὸ τελευταῖον] τελευτῶν A 23 -τε ἐ- A
24 ἐγίν- A -το AF καταδι- F δὲ AF 27 -τα AF 28 τε
om. $S^{ac}P_4^{mg}$ 29 αὐτὸν τὴν δίκην A κατεδε- Dindorf: καταδε- SAFYP
29–30 ἀποφαίνειν $AF_1^{γρ}Y^{sl}P_2^{γρ}$: ἀποφέρειν SFYP

witnesses of it—to show you that I've suffered further insolence, in the way you've heard, before obtaining legal redress for the previous offences against me.

TESTIMONY

{We, Kallisthenes of Sphettos, Diognetos of Thorikos, and Mnesitheos of Alopeke, know that Demosthenes, for whom we testify, has initiated a case for ejectment against Meidias, who now also is publicly prosecuted by him, and that eight years have already passed in that case, and that Meidias has been responsible for all the delay, making excuses and putting it off.}

Now, listen to what he's done about the case, men of Athens, and observe his insolence and arrogance at every point. **83** As arbitrator in the case, I mean the one in which I got him convicted, I got Straton of Phaleron, a poor man without experience of affairs, but in other ways not bad, in fact very respectable; and that's what has ruined the unfortunate man, in a manner not right or just but quite disgraceful. This man **84** Straton, acting as arbitrator for us, when the day for judgement eventually arrived and all the legal possibilities were now exhausted, *hypomosiai* and *paragraphai*, and there was nothing else left, first asked me to put off the arbitration, and next to postpone it to the following day; but finally, as I wouldn't agree and Meidias didn't appear, and it was getting late in the day, he gave a verdict against him. And when **85** it was already evening and dark, this man Meidias came to the magistrates' office, and found the magistrates coming out and Straton already leaving after delivering the verdict given in the deserted case (as I was told by one of the bystanders). At first he had the face to urge him to change his verdict from condemnation to acquittal, and the magistrates to alter the

86 μὰς αὐτοῖς ἐδίδου· ὡς δ' ἐδυσχέραινον οὗτοι τὸ πρᾶγμα καὶ
οὐδετέρους ἔπειθεν, ἀπειλήσας καὶ διαλοιδορηθεὶς ἀπελθὼν τί
ποιεῖ; καὶ θεάσασθε τὴν κακοήθειαν· τὴν μὲν δίαιταν ἀντιλαχὼν οὐκ ὤμοσεν, ἀλλ' εἴασε καθ' ἑαυτοῦ κυρίαν γενέσθαι, καὶ
ἀνώμοτος ἀπηνέχθη· βουλόμενος δὲ τὸ μέλλον λαθεῖν, φυλά- 5
ξας τὴν τελευταίαν ἡμέραν τῶν διαιτητῶν {τὴν τοῦ
Θαργηλιῶνος ἢ τοῦ Σκιροφοριῶνος γιγνομένην}, εἰς ἣν ὁ μὲν
87 ἦλθεν τῶν διαιτητῶν, ὁ δ' οὐκ ἦλθεν, πείσας τὸν πρυτανεύοντα δοῦναι τὴν ψῆφον παρὰ πάντας τοὺς νόμους, κλητῆρα
οὐδ' ὀντινοῦν ἐπιγραψάμενος, κατηγορῶν ἔρημον, οὐδενὸς 10
παρόντος, ἐκβάλλει καὶ ἀτιμοῖ τὸν διαιτητήν. καὶ νῦν εἷς
Ἀθηναίων, ὅτι Μειδίας ἔρημον ὦφλεν δίκην, ἁπάντων
ἀπεστέρηται τῶν ἐν τῇ πόλει καὶ καθάπαξ ἄτιμος γέγονεν· καὶ
οὔτε λαχεῖν ἀδικηθέντα οὔτε διαιτητὴν γενέσθαι Μειδίᾳ οὔθ'
ὅλως τὴν αὐτὴν ὁδὸν βαδίζειν, ὡς ἔοικεν, ἔστ' ἀσφαλές. 15

88 Δεῖ δὴ τοῦτο τὸ πρᾶγμα ὑμᾶς οὑτωσὶ σκέψασθαι, καὶ λογίσασθαι τί ποτ' ἐστὶν ὃ παθὼν Μειδίας οὕτως ὠμὸν τηλικαύτην ἐπεβούλευσε λαβεῖν τῶν πεπραγμένων παρ' ἀνδρὸς
πολίτου δίκην, κἂν μὲν ᾖ τι δεινὸν ὡς ἀληθῶς καὶ ὑπερφυές,
συγγνώμην ἔχειν, ἐὰν δὲ μηδέν, θεάσασθε τὴν ἀσέλγειαν καὶ 20
τὴν ὠμότητα, ᾗ καθ' ἁπάντων χρῆται τῶν ἐντυγχανόντων. τί
οὖν ἔσθ' ὃ πέπονθεν; μεγάλην νὴ Δί' ὦφλε δίκην καὶ τοσαύτην
ὥστε ἀποστερεῖσθαι τῶν ὄντων. ἀλλὰ χιλίων ἡ δίκη μόνον ἦν
89 δραχμῶν. πάνυ γε, ἀλλὰ δάκνει καὶ τοῦτο, φαίη τις ἄν, ὅταν
ἐκτίνειν ἀδίκως δέῃ· συνέβη δὲ ὑπερημέρῳ γενομένῳ λαθεῖν 25
αὐτῷ διὰ τὸ ἀδικηθῆναι. ἀλλ' αὐθημερὸν μὲν ᾔσθετο, ὃ καὶ
μέγιστόν ἐστι τεκμήριον τοῦ μηδὲν ἠδικηκέναι τὸν ἄνθρωπον,
90 δραχμὴν δ' οὐδέπω μίαν ἐκτέτεικεν. ἀλλὰ μήπω τοῦτο. ἀλλὰ
τὴν μὴ οὖσαν ἀντιλαχεῖν ἐξῆν αὐτῷ δήπου, καὶ πρὸς ἐμὲ τὸ
πρᾶγμα καταστήσασθαι, πρὸς ὅνπερ ἐξ ἀρχῆς ἦν ἡ δίκη. ἀλλ' 30
οὐκ ἐβούλετο· ἀλλ' ἵνα μὴ Μειδίας ἀτίμητον ἀγωνίσηται δέκα
μνῶν δίκην, πρὸς ἣν οὐκ ἀπήντα δέον, καὶ εἰ μὲν ἠδίκηκεν,

1 δὲ A 3 καὶ om. A κακοήθιαν S^{ac} 4 αὐτοῦ YP^{ac}
6–7 τὴν … γιγνομένην del. G. H. Schaefer 7 ἢ τὴν τοῦ A Σ (Patm.)
γιν- A Σ (Patm.) 8 -θε AFP^{pc} -θε $AF^{pc}P^{pc}$ 9 τὴν om. YP
10 ὅντιν' οὖν S^{pc}AF Σ (Patm.): όντι S^{ac}: ὅντιν' YP^{ac} κατηγόρων FYP ἐρήμην
$S_3^{\gamma\rho}F_1^{\gamma\rho}P_4^{\gamma\rho}$ 12 -λε AFP^{pc} 13 -νε $AF^{pc}P^{pc}$ 14 λαχεῖν δίκην S^{sl}F:
δίκην λαχεῖν AP_3^{pc} 15 -τιν A 17 -τέ A 18 ἐπεβούλευσαι S:
ὥστε ἐπιβουλευσαι (ante τηλικαύτην) $S_3^{\gamma\rho}$ 20 ἔχει $S^{ac}YP^{ac}$ θεάσασθαι
Spalding 21 κατὰ πάντων ἀεὶ A 22 -θε $AF^{pc}P^{pc}$ δία A
23 ὥστ' A^{pc} 26 καὶ ὃ P^{ac} 28 ἐκτέτεικεν Blass: ἐκτέτικεν SAFYP
29 αὐτῷ ἐξῆν SYP 31 ἠβούλετο A μὴ om. A 32 -κε $AF^{pc}P^{pc}$

record, and he offered them fifty drachmas. But as they got **86** annoyed at his doing this and he wasn't managing to persuade either of them, after threatening and abusing them he went away and—does what? Just look at the nasty way he behaved. After applying to have the arbitration annulled he didn't take the oath, but he let the condemnation become final and was reported as having failed to take the oath. Wishing his intention to escape notice, he waited for the arbitrators' last day, on which some arbitrators came and some didn't; he persuaded the chairman to take a vote, quite **87** illegally; and without putting down the name of a single summons-witness, accusing in a deserted case, when nobody was there, he expelled and disfranchised the arbitrator. And now one Athenian, just because Meidias was convicted in a deserted case, has been deprived of all the rights of a citizen and permanently disfranchised. It isn't safe, it appears, to prosecute Meidias when he wrongs you, or be appointed arbitrator for him, or even walk along the same street!

You should consider the matter this way, and work out **88** what treatment Meidias had suffered that was harsh enough for him to plot the exaction of such a severe penalty for it from an Athenian citizen. If it really is something terrible and monstrous, you should forgive him; but if it's nothing of the sort, look at the aggressiveness and harshness with which he treats everyone he meets. What is it that he's suffered, then? Oh, he's incurred a heavy fine, large enough to deprive him of all his property. But the fine was only a thousand drachmas. Yes indeed, but even that is annoying, someone **89** may say, when it has to be paid unjustly; and the injustice caused the payment to become overdue without his knowing it. But he learned of it on the very day that it was imposed, which is also very strong evidence that the man had done him no injustice; and yet he still hasn't paid a single drachma. More of that later; but you'd think he could have **90** applied for annulment of the arbitration and raised his action against me, his original opponent in the case. Yet he wouldn't do that; to save Meidias from contesting a case with a fixed penalty of ten mnai, which he didn't attend at the time when he ought to have, and from paying a penalty if

δίκην δῷ, εἰ δὲ μή, ἀποφύγῃ, ἄτιμον Ἀθηναίων ἕνα εἶναι δεῖ καὶ μήτε συγγνώμης μήτε λόγου μήτε ἐπιεικείας μηδεμιᾶς
91 τυχεῖν, ἃ καὶ τοῖς ὄντως ἀδικοῦσιν ἅπανθ' ὑπάρχει. ἀλλ' ἐπειδή γε ἠτίμωσεν ὃν ἠβουλήθη καὶ τοῦτ' ἐχαρίσασθε αὐτῷ καὶ τὴν ἀναιδῆ γνώμην, ᾗ ταῦτα προαιρεῖται ποιεῖν, ἐνέπλησεν 5 αὑτοῦ, ἐκεῖν' ἐποίησε, τὴν καταδίκην ἐκτέτεικε, δι' ἣν τὸν ἄνθρωπον ἀπώλεσεν; οὐδὲ χαλκοῦν οὐδέπω καὶ τήμερον, ἀλλὰ δίκην ἐξούλης ὑπομένει φεύγειν. οὐκοῦν ὁ μὲν ἠτίμωται καὶ παραπόλωλεν, ὁ δ' οὐδ' ὁτιοῦν πέπονθεν, ἀλλ' ἄνω κάτω τοὺς νόμους, τοὺς διαιτητάς, πάνθ' ὅσ' ἂν βούληται στρέφει. 10
92 καὶ τὴν μὲν κατὰ τοῦ διαιτητοῦ γνῶσιν, ἣν ἀπρόσκλητον κατεσκεύασεν, αὐτὸς κυρίαν αὑτῷ πεποίηται· ἣν δ' αὐτὸς ὦφλεν ἐμοὶ προσκληθείς, εἰδώς, οὐκ ἀπαντῶν, ἄκυρον ποιεῖ. καίτοι εἰ παρὰ τῶν ἔρημον καταδιαιτησάντων αὐτοῦ τηλικαύτην δίκην οὗτος ἀξιοῖ λαμβάνειν, τίν' ὑμῖν προσήκει παρὰ 15 τούτου λαβεῖν, τοῦ φανερῶς τοὺς ὑμετέρους νόμους ἐφ' ὕβρει παραβαίνοντος; εἰ γὰρ ἀτιμία καὶ νόμων καὶ δικῶν καὶ πάντων στέρησις ἐκείνου τἀδικήματος προσήκουσά ἐστι δίκη,
93 τῆς γε ὕβρεως μικρὰ θάνατος φαίνεται. ἀλλὰ μὴν ὡς ἀληθῆ λέγω, κάλει μοι τούτων τοὺς μάρτυρας, καὶ τὸν τῶν 20 διαιτητῶν ἀνάγνωθι νόμον.

ΜΑΡΤΥΡΙΑ

{Νικόστρατος Μυρρινούσιος, Φανίας Ἀφιδναῖος οἴδαμεν Δημοσθένην, ᾧ μαρτυροῦμεν, καὶ Μειδίαν τὸν κρινόμενον ὑπὸ Δημοσθένους, ὅτ' αὐτῷ Δημοσθένης ἔλαχεν 25 τὴν τοῦ κακηγορίου δίκην, ἑλομένους διαιτητὴν Στράτωνα, καὶ ἐπεὶ ἧκεν ἡ κυρία τοῦ νόμου, οὐκ ἀπ-

6–9 τὴν ... δ' Π11 7 οὐδὲ ... τήμερον Pol. 9.92 14–15 καίτοι ... λαμβάνειν AB 1.150.7 23–5 οἴδαμεν ... κρινόμενον AB 1.150.11

2 λόγου] ἐλέου $S_3^{γρ}$ επ' είκειας S 2–3 τυχεῖν μηδεμιᾶς F 4 ἐβούλετο A 5 ἢ A 6 αὐτοῦ P: αὑτοῦ SFY: αυτοῦ A -νο AFP ἐκτέτεικε Blass: ἐκτέτικε SAFYP: ε]ξετι[σ]ε Π11 ἧς YP^{ac} 7 ἀλλ' (ante οὐδὲ) $Π11FP_3^{pc}$ χαλκὸν F τήμερον οὐδένα Pol. 8 φεύγην S^{ac}: φεύγων Y^{sl}: [Π11] 9 ἄνω καὶ κάτω AFP_3^{pc} 10 ἅπανθ' ὅσα βούλεται A 11 μὲν om. A 12 αὑτῷ Bc^{pc}: αυτῷ SY: αὐτῷ FP: ἑαυτῷ A πεποίηκεν F 14 παρὰ τῶν] πάντες AB ἐρήμην AB 15 -να AF 16 φανερῶς οὕτως FP_3^{pc} 18 τοῦ ἀδ- AF -τιν S 19 ἔμοιγε φαίνεται FP_4^{mg} 20 τοὺς AFP_3^{pc}: om. SYP^{ac} 22 ΜΑΡΤΥΡΙΑ ΝΟΜΟΣ S 23–144.9 testimonium om. A 23 ἀφνιδαῖος F 26 κακηγορίου Ald: κατηγορίου $SFP_3^{γρ}$: κατηγόρου YP 27 ἐπεὶ ἧκεν FYP: επικεν S^{ac}: ἐπ' ἧκεν S_3^{pc}

he'd done wrong and being acquitted if he hadn't, an Athenian citizen must be disfranchised and must be granted neither pardon nor an opportunity to speak nor any fair treatment—all of which are given even to real offenders. Well, after he'd disfranchised the man he wanted to, and **91** you'd done him that favour, and he'd satiated his own shameless inclination which prompted him to do that, did he do the other thing—has he paid the damages which were his reason for destroying the man? Not a penny to this day! Instead he lets a case of ejectment be brought against him. So Straton is disfranchised and destroyed, while Meidias hasn't suffered a thing, but embroils the laws, the arbitrators, and everything he wishes. The verdict against the arbitrator, **92** which he engineered without a summons, he treats as valid for his own benefit; but the verdict given against himself in my favour, when he had been summoned and was well aware of it and didn't attend, he nullifies. Yet, if he thinks fit to exact so severe a penalty from people who decide arbitrations against him in his absence, what is it proper for you to exact from him, who flouts your laws openly and insolently? If disfranchisement, and loss of legal rights and everything else, is the appropriate penalty for that offence, for insolence death is obviously too little. But please call the witnesses to **93** confirm my statements, and read the law of arbitration.

TESTIMONY

{We, Nikostratos of Myrrhinous and Phanias of Aphidna, know that Demosthenes, for whom we testify, and Meidias, who is prosecuted by Demosthenes, at the time when Demosthenes brought the case of slander against him, chose Straton as arbitrator; and when the

αντήσαντα Μειδίαν ἐπὶ τὴν δίαιταν, ἀλλὰ καταλιπόντα. γενομένης δὲ ἐρήμου κατὰ Μειδίου, ἐπιστάμεθα Μειδίαν πείθοντα τόν τε Στράτωνα τὸν διαιτητὴν καὶ ἡμᾶς, ὄντας ἐκείνοις τοῖς χρόνοις ἄρχοντας, ὅπως τὴν δίαιταν αὐτῷ ἀποδιαιτήσομεν, καὶ διδόντα δραχμὰς πεντήκ- (5) οντα, καὶ ἐπειδὴ οὐχ ὑπεμείναμεν, προσαπειλήσαντα ἡμῖν καὶ οὕτως ἀπαλλαγέντα. καὶ διὰ ταύτην τὴν αἰτίαν ἐπιστάμεθα Στράτωνα ὑπὸ Μειδίου καταβραβευθέντα καὶ παρὰ πάντα τὰ δίκαια ἀτιμωθέντα.}

94 *Λέγε δὴ καὶ τὸν τῶν διαιτητῶν νόμον.* (10)

ΝΟΜΟΣ

{Ἐὰν δέ τινες περὶ συμβολαίων ἰδίων πρὸς ἀλλήλους ἀμφισβητῶσιν καὶ βούλωνται διαιτητὴν ἑλέσθαι ὁντινοῦν, ἐξέστω αὐτοῖς αἱρεῖσθαι ὃν ἂν βούλωνται {διαιτητὴν ἑλέσθαι}. ἐπειδὰν δ' ἕλωνται κατὰ κοινόν, (15) μενέτωσαν ἐν τοῖς ὑπὸ τούτου διαγνωσθεῖσι, καὶ μηκέτι μεταφερέτωσαν ἀπὸ τούτου ἐφ' ἕτερον δικαστήριον ταὐτὰ ἐγκλήματα, ἀλλ' ἔστω τὰ κριθέντα ὑπὸ τοῦ διαιτητοῦ κύρια.}

95 *Κάλει δὴ καὶ τὸν Στράτωνα αὐτὸν τὸν τὰ τοιαῦτα πεπον- (20) θότα· ἑστάναι γὰρ ἐξέσται δήπουθεν αὐτῷ.*

Οὗτος, ὦ ἄνδρες Ἀθηναῖοι, πένης μὲν ἴσως ἐστίν, οὐ πονηρὸς δέ γε. οὗτος μέντοι πολίτης ὤν, ἐστρατευμένος ἁπάσας τὰς ἐν ⟨τῇ⟩ ἡλικίᾳ στρατείας καὶ δεινὸν οὐδὲν εἰργασμένος, ἕστηκε νυνὶ σιωπῇ, οὐ μόνον τῶν ἄλλων ἀγαθῶν τῶν κοινῶν (25) ἀπεστερημένος, ἀλλὰ καὶ τοῦ φθέγξασθαι καὶ ὀδύρασθαι· καὶ οὐδ' εἰ δίκαια ἢ ἄδικα πέπονθεν, οὐδὲ ταῦτ' ἔξεστιν αὐτῷ **96** *πρὸς ὑμᾶς εἰπεῖν. καὶ ταῦτα πέπονθεν ὑπὸ Μειδίου καὶ τοῦ Μειδίου πλούτου καὶ τῆς ὑπερηφανίας παρὰ τὴν πενίαν καὶ ἐρημίαν καὶ τὸ τῶν πολλῶν εἷς εἶναι. καὶ εἰ μὲν παραβὰς τοὺς* (30)

28–9 *καὶ ταῦτα ... πλούτου* Aps. *Rh.* (Hammer) 328.12 28–30 *καὶ ταῦτα ... εἶναι* *AB* 1.163.16

---

1 *μειδείαν* Y *καταλειπόντα* S 4 *πως* S^{ac} 5 *ἀποδιαιτήσωμεν* F^{pc}YP$^{pc}_{3}$ 10 *λέγε ... νόμον* om. A *τῶν* om. Y^{ac} 12–19 legem om. A 14 *ερεισθαι* S^{ac} 15 *ἑλέσθαι* om. Ald: *διαιτητὴν ἑλέσθαι* del. Sauppe *δ' ἕλωνται* Reiske: *βούλωνται* SFYP 17 *μετα-* O^{sl}: *κατα-* SFYP: *ἀνα-* Lamb^{vl} *ἀπὸ* YP: *ὑπὸ* SF 18 *ἐκκλήματα* SF 20 *τὰ* om. S 24 *τῇ* add. MacDowell 26 *καὶ* (ante *ὀδύρασθαι*) A: *ἢ* SFYP 27 *οὔτ'* A *ταῦτα* A: *ταῦθ'* F 30 *τὴν ἐρημίαν* A

legal day came, Meidias did not attend the arbitration but deserted it. After it was decided against Meidias in absence, we know that Meidias urged both Straton, the arbitrator, and us, who were magistrates at that time, to decide the arbitration in his favour, and offered fifty drachmas. When we refused, he also threatened us and so made off. For this reason we know that Straton was condemned by Meidias and disfranchised contrary to all justice.}

Read also the law of arbitration. **94**

LAW

{If any men dispute with each other about private contracts and wish to choose anyone as arbitrator, let them be permitted to choose whomever they wish. When they have made an agreed choice, let them abide by the decisions of that man, and let them not thereafter remit the same charges from him to another court, but let the judgements of the arbitrator be valid.}

Call Straton himself too, the victim of this treatment. He **95** will be permitted to stand here, I suppose.

This man, men of Athens, is poor, perhaps; but he's not a bad man. He's an Athenian citizen; he has served on all the campaigns while he was of military age; and he has done nothing wicked. But now he stands in silence, deprived not only of the other benefits that everyone enjoys, but even of the chance to utter a word of complaint; he's not permitted even to tell you whether he has been treated justly or unjustly. And he has suffered this treatment from Meidias and **96** from Meidias' wealth and arrogance, because he was poor, and friendless, and just an ordinary person. If he'd trans-

νόμους ἔλαβεν τὰς πεντήκοντα δραχμὰς παρ' αὐτοῦ, καὶ τὴν
δίκην ἣν κατεδιῄτησεν ἀποδεδιῃτημένην ἀπέφηνεν, ἐπίτιμος
ἂν ἦν καὶ οὐδὲν ἔχων κακὸν τῶν ἴσων μετεῖχεν τοῖς ἄλλοις
ἡμῖν· ἐπειδὴ δὲ παρεῖδε πρὸς τὰ δίκαια Μειδίαν, καὶ τοὺς
νόμους μᾶλλον ἔδεισεν τῶν ἀπειλῶν τῶν τούτου, τηνικαῦτα 5
τηλικαύτῃ καὶ τοιαύτῃ συμφορᾷ περιπέπτωκεν ὑπὸ τούτου.
97 εἶθ' ὑμεῖς τὸν οὕτως ὠμόν, τὸν οὕτως ἀγνώμονα, τὸν τηλικ-
αύτας δίκας λαμβάνοντα ὧν αὐτὸς ἠδικῆσθαί φησι μόνον (οὐ
γὰρ ἠδίκητό γε), τοῦτον ὑβρίζοντα λαβόντες εἴς τινα τῶν
πολιτῶν ἀφήσετε; καὶ μήτε ἑορτῆς μήτε ἱερῶν μήτε νόμου 10
μήτε ἄλλου μηδενὸς πρόνοιαν ποιούμενον, οὐ καταψηφιεῖσθε;
98 οὐ παράδειγμα ποιήσετε; καὶ τί φήσετε, ὦ ἄνδρες δικασταί;
καὶ τίνα, ὢ πρὸς τῶν θεῶν, ἕξετ' εἰπεῖν πρόφασιν δικαίαν ἢ
καλήν; ὅτι νὴ Δί' ἀσελγής ἐστιν καὶ βδελυρός· ταῦτα γάρ
ἐστιν τἀληθῆ. ἀλλὰ μισεῖν ὀφείλετ', ἄνδρες Ἀθηναῖοι, δήπου 15
τοὺς τοιούτους μᾶλλον ἢ σῴζειν. ἀλλ' ὅτι πλούσιός ἐστιν.
ἀλλὰ τοῦτό γε τῆς ὕβρεως αὐτοῦ σχεδὸν αἴτιον εὑρήσετε ὄν,
ὥστ' ἀφελεῖν τὴν ἀφορμήν, δι' ἣν ὑβρίζει, προσῆκε μᾶλλον ἢ
σῶσαι διὰ ταύτην· τὸ γὰρ χρημάτων πολλῶν θρασὺν καὶ
βδελυρὸν καὶ τοιοῦτον ἄνθρωπον ἐᾶν εἶναι κύριον, ἀφορμήν 20
99 ἐστιν ἐφ' ὑμᾶς αὐτοὺς δεδωκέναι. τί οὖν ὑπόλοιπον; ἐλεῆσαι
νὴ Δία· παιδία γὰρ παραστήσεται καὶ κλαιήσει καὶ τούτοις
αὐτὸν ἐξαιτήσεται· τοῦτο γὰρ λοιπόν. ἀλλ' ἴστε δήπου τοῦθ',
ὅτι τοὺς ἀδίκως τι πάσχοντας, ὃ μὴ δυνήσονται φέρειν, ἐλεεῖν
προσήκει, οὐ τοὺς ὧν πεποιήκασι δεινῶν δίκην διδόντας. καὶ 25
τίς ἂν ταῦτ' ἐλεήσειεν δικαίως, ὁρῶν τὰ τοῦδε οὐκ ἐλεηθέντα
ὑπὸ τούτου; ἃ τῇ τοῦ πατρὸς συμφορᾷ χωρὶς τῶν ἄλλων

17–20 σχεδὸν ... τοιοῦτον Π11 22–7 παιδία ... ἄλλων schol. Hermog. (Walz 7) 336.4 26 τίς ... ἐλεηθέντα Io. Sard. 103.17

---

1 -βε $AF^{pc}P^{pc}$ 2 καταδιηίτησεν F^{pc} 3 -χε AF^{pc} 5 -σε $AF^{pc}P^{pc}$ 5–6 τηνικαῦτα ... τούτου om. F, add. F_1^{mg} 6 περι-πέπτωκεν ἀδίκως AP_4^{mg} 9 λαμβάνοντες A 10 νόμων A: νόμον F_1^{sl} 11 ποιούμενον om. S^{ac}: ποιουμένου Buttmann 12 παράδειγμα τοῖς ἄλ-λοις A 13 καὶ om. AFP_4^{pc} -τε AF 14 δία AF -τι AFP^{pc} βδελλυρός S 15 ἔσται A: ἐστι FYP -τε AF ὦ ἄνδρες Ἀθηναῖοι $A^{cp}F$ 15–16 δήπου post τοιούτους A: om. F 18 ὥστ' Π11F: ὥστε SAYP ὑβρίζειν A^{pc}: υβρι[Π11 προσήκει A: [Π11] 19 διαύτην S^{ac}: [Π11] 20 ἐᾶν] ἀν A 22 τὰ παιδία P_2^{sl} κλαύσει schol. Hermog. 23 ὑπό-λοιπον A schol. Hermog. που A τοῦθ' om. SYP^{ac} 24 τι κακὸν FP_3^{pc} δύναιντο schol. Hermog. ἀφαιρεῖν $S_3^{γρ}$ προσήκει ἐλεεῖν schol. Hermog. 25 -σιν S δεῖν schol. Hermog. 26 -τα A -ειε AP^{pc} 27 ὑπὸ τούτου; ἃ] ἂν schol. Hermog.

gressed the laws and accepted the fifty drachmas which Meidias offered him, and declared a verdict for him after deciding the arbitration against him, he wouldn't be disfranchised, and without coming to any harm he would possess the same rights as the rest of us. But since he ignored Meidias in favour of justice, and feared the laws more than Meidias' threats, the result is this great and terrible disaster which Meidias has brought upon him. And then, when a man is so **97** cruel and so heartless, when he exacts such heavy penalties for the wrongs which he merely says that he had suffered (for really he had not), on catching him treating a citizen with insolence will you let him off? When he shows no concern for festival, sacred precincts, law, or anything else, won't you vote against him? Won't you make an example of him? What **98** answer have you, men of the jury? What just or good excuse can you give, in heaven's name? You can say he's a loathsome bully. That is the truth. But you surely ought to hate men like that, men of Athens, not protect them. You can say he's rich. Yes, but you'll find that's the main thing that makes him insolent, so that it would be more appropriate to remove the basis of his insolence than to protect him because of it; to let an audacious and loathsome person like that remain in possession of a great deal of money is to have given him means of attacking yourselves. What other defence is **99** left? Pity, you may say; he'll bring forward children and weep and ask you to let him off for their sake—that's what's left. But surely you realize that pity is appropriate for those who suffer some injustice which they will be unable to bear, not for those who are being punished for their crimes. And who would justly pity those children, when he sees that the children of Straton here were not pitied by Meidias? These, besides their other troubles, can also see that no remedy for

κακῶν οὐδ' ἐπικουρίαν ἐνοῦσαν ὁρᾷ· οὐ γάρ ἐστιν ὄφλημα ὅ τι
χρὴ καταθέντα ἐπίτιμον γενέσθαι τουτονί, ἀλλ' ἁπλῶς οὕτως
100 ἠτίμωται τῇ ῥύμῃ τῆς ὀργῆς καὶ τῆς ὕβρεως τῆς Μειδίου. τίς
οὖν ὑβρίζων παύσεται καὶ δι' ἃ ταῦτα ποιεῖ χρήματα ἀφαι-
ρεθήσεται, εἰ τοῦτον ὥσπερ δεινὰ πάσχοντα ἐλεήσετε, εἰ δέ 5
τις πένης μηδὲν ἠδικηκὼς ταῖς ἐσχάταις συμφοραῖς ἀδίκως
ὑπὸ τούτου περιπέπτωκεν, τούτῳ δ' οὐδὲ συνοργισθήσεσθε;
μηδαμῶς· οὐδεὶς γάρ ἐστιν δίκαιος τυγχάνειν ἐλέου τῶν
101 μηδένα ἐλεούντων, οὐδὲ συγγνώμης τῶν ἀσυγγνωμόνων. ἐγὼ
γὰρ οἶμαι πάντας ἀνθρώπους φέρειν ἀξιοῦν παρ' αὑτῶν εἰς 10
τὸν βίον αὑτοῖς ἔρανον παρὰ πάνθ' ὅσα πράττουσιν· οἷον ἐγώ
τις οὑτοσὶ μέτριος πρὸς ἅπαντάς εἰμι, ἐλεήμων, εὖ ποιῶν πολ-
λούς· ἅπασι προσήκει τῷ τοιούτῳ ταὐτὰ εἰσφέρειν, ἐάν που
καιρὸς ἢ χρεία παραστῇ. ἕτερος οὑτοσί τις βίαιος, οὐδένα οὔτ'
ἐλεῶν οὔθ' ὅλως ἄνθρωπον ἡγούμενος· τούτῳ τὰς ὁμοίας 15
φορὰς παρ' ἑκάστου δίκαιον ὑπάρχειν. σὺ δή, πληρωτὴς τοιούτου γεγονὼς ἐράνου σεαυτῷ, τοῦτον δίκαιος εἶ συλλέξασθαι.
102 Ἡγοῦμαι μὲν τοίνυν, ὦ ἄνδρες Ἀθηναῖοι, καὶ εἰ μηδὲν ἔτι
ἄλλο εἶχον κατηγορεῖν Μειδίου, μηδὲ δεινότερα ἦν ἃ μέλλω
λέγειν ὧν εἴρηκα, δικαίως ἂν ὑμᾶς ἐκ τῶν εἰρημένων καὶ 20
καταψηφίσασθαι καὶ τιμᾶν αὐτῷ τῶν ἐσχάτων. οὐ μὴν
ἐνταῦθ' ἕστηκεν τὸ πρᾶγμα, οὐδ' ἀπορήσειν μοι δοκῶ τῶν
μετὰ ταῦτα· τοσαύτην ἀφθονίαν οὗτος πεποίηκε κατηγοριῶν.
103 ὅτι μὲν δὴ λιποταξίου γραφὴν κατεσκεύασεν κατ' ἐμοῦ καὶ

2–3 καταθέντα ... ἠτίμωται Π11 ἀλλ' ... Μειδίου Alex. *Fig.* (Spengel 3) 32.10 3 ἠτίμωται ... Μειδίου Tib. *Fig.* (Ballaira) 35.4 16–17 σὺ ... συλλέξασθαι Aristid. *Rh.* (Schmid) 14.15 20–1 δικαίως ... ἐσχάτων *AB* 1.175.4 21–3 ἐσχάτων ... ἀφθονίαν Π11 24–150.2 ὅτι ... ἐάσω Aristid. *Rh.* (Schmid) 29.18

---

1 ὁρῶ $S_3^{γρ}$ 2 τοῦτον F: του[Π11 οὗτος SA: [Π11] 3 τῆς (post ὕβρεως)] τῇ Tib. 4 διὰ ταῦθ' ἃ A 6 μηδὲν SYP^{ac}: ὧν μηδὲν AP_3^{pc}: μὲν F 7 -κε $AF^{pc}P^{pc}$ τούτων A δ' οὐδὲ SYP: δὲ οὐ AF 8 -τι AFP^{pc} 10 οἶμαι SFYP: ἡγοῦμαι A ἀξιοῦν del. Reiske: ἄξιον εἶναι $Lamb^{vl}$ 10–11 παρ' ... πράττουσιν] αὐτοῖς ἐράνους οὐ τούτους μόνους οὓς οὗτοι συλλέγουσιν ἀλλὰ καὶ ἄλλους $S_3^{γρ}P_4^{γρ}$ 11 post πράττουσιν] οὐ τοῦτον μόνον ὃν συλλέγουσιν καὶ ὧν πληρωταὶ γίγνονταί τινες ἀλλὰ καὶ ἄλλων $FY^{mg}P$ 13 ταὐτὰ Cd^{pc}: ταυτα S: ταῦτα $AFYP^{ac}$: τοιαῦτα P_3^{pc} που AFP: του S: ποι Y 14 οὗτος S βίαιος ὠμός A οὐδένα δ' SYP^{ac} 15 ἡγούμενος ἄνθρωπον F ὁμοίας SAYP: αὐτὰς F 16–17 τοῦ τοιούτου F 17 τουτονὶ Aristid. 18 μὲν om. A μηδὲν] μὲν Y^{ac} ἔτι om. A: ἔτ' F 19 κατηγορεῖν εἶχον A 20 καὶ om. A 22 -θα A: [Π11] -κε $AF^{pc}YP$: [Π11] 23 κατηγορῶν Y^{ac} 24 δὴ om. Aristid. λειπο- $F_4^{sl}P^{ac}$ γραφὴν λειποταξίου Aristid. -σε $AF^{pc}P^{pc}$

their father's misfortune is possible: there's no payment that Straton must make to be re-enfranchised, but he has quite simply been disfranchised by the force of Meidias' anger and insolence. Who will ever cease from being insolent, and be **100** deprived of the money that makes him so, if you show pity to Meidias as a victim of ill treatment, while refusing even to show indignation in support of a poor man on whom, though he's done nothing wrong, Meidias has unjustly inflicted the utmost penalties? Do not do it. No one deserves pity if he pities no one, and no one unmerciful deserves mercy. For I **101** suppose all men think it right to contribute from their own resources to a loan, for the benefit of their own life, throughout all their activities. Suppose I, for example, am well-behaved to everyone, sympathetic, helpful to many: it's proper for everyone to give a man of that sort similar contributions, wherever opportunity or need arises. Suppose someone else is violent, not sympathetic to any other man or regarding him as human at all: it's right for him to get like payments from each person. Well, Meidias, that's the sort of loan to which you've been a contributor for yourself, and that's what you deserve to collect.

Now, I consider, men of Athens, that even if I had no **102** further accusation to make against Meidias, and what I'm going to say were not more serious than what I've said so far, what has been said would justify you in voting against him and condemning him to the severest penalties. But it doesn't stop there, and I don't think I shall be short of things to say next; he has provided such an abundance of material for accusation. Now, I shall say nothing of the fact that he pro- **103** cured a prosecution for desertion against me, and hired the

τὸν τοῦτο ποιήσοντα ἐμισθώσατο, τὸν μιαρὸν καὶ λίαν εὐχερῆ, τὸν κονιορτὸν Εὐκτήμονα, ἐάσω. καὶ γὰρ οὔτ' ἀνεκρίνετο
ταύτην ὁ συκοφάντης ἐκεῖνος, οὔθ' οὗτος οὐδενὸς ἕνεκα αὐτὸν
ἐμισθώσατο πλὴν ἵν' ἐκκέοιτο πρὸ τῶν ἐπωνύμων καὶ πάντες
ὁρῷεν "Εὐκτήμων Λουσιεὺς ἐγράψατο Δημοσθένην Παιανιέα 5
λιποταξίου"· καί μοι δοκεῖ κἂν προσγράψασθαι τοῦθ' ἡδέως,
εἴ πως ἐνῆν, ὅτι Μειδίου μισθωσαμένου γέγραπται. ἀλλ' ἐῶ
τοῦτο· ἐφ' ᾗ γὰρ ἐκεῖνος ἠτίμωκεν αὑτὸν οὐκ ἐπεξελθών,
104 οὐδεμιᾶς ἔγωγ' ἔτι προσδέομαι δίκης, ἀλλ' ἱκανὴν ἔχω. ἀλλ'
ὃ καὶ δεινόν, ὦ ἄνδρες Ἀθηναῖοι, καὶ σχέτλιον καὶ κοινὸν 10
ἔμοιγ' ἀσέβημα, οὐκ ἀδίκημα μόνον, τούτῳ πεπρᾶχθαι δοκεῖ,
τοῦτ' ἐρῶ. τῷ γὰρ ἀθλίῳ καὶ ταλαιπώρῳ κακῆς καὶ χαλεπῆς
συμβάσης αἰτίας Ἀριστάρχῳ τῷ Μόσχου, τὸ μὲν πρῶτον, ὦ
ἄνδρες Ἀθηναῖοι, κατὰ τὴν ἀγορὰν περιιὼν ἀσεβεῖς καὶ δει-
νοὺς λόγους ἐτόλμα περὶ ἐμοῦ λέγειν, ὡς ἐγὼ τὸ πρᾶγμ' εἰμὶ 15
τοῦτο δεδρακώς· ὡς δ' οὐδὲν ἤνυεν τούτοις, προσελθὼν τοῖς
ἐπ' ἐκεῖνον ἄγουσιν τὴν αἰτίαν τοῦ φόνου, τοῖς τοῦ
τετελευτηκότος οἰκείοις, χρήμαθ' ὑπισχνεῖτο δώσειν εἰ τοῦ
πράγματος αἰτιῷντο ἐμέ, καὶ οὔτε θεοὺς οὔθ' ὁσίαν οὔτ' ἄλλο
105 οὐδὲν ἐποιήσατ' ἐμποδὼν τοιούτῳ λόγῳ, οὐδ' ὤκνησεν. ἀλλ' 20
οὐδὲ πρὸς οὓς ἔλεγεν αὐτοὺς ᾐσχύνθη, εἰ τοιοῦτο κακὸν καὶ
τηλικοῦτο ἀδίκως ἐπάγει τῳ, ἀλλ' ἕνα ὅρον θέμενος παντὶ
τρόπῳ με ἀνελεῖν, οὐδὲν ἐλλείπειν ᾤετο δεῖν, ὡς δέον, εἴ τις
ὑβρισθεὶς ὑπὸ τούτου δίκης ἀξιοῖ τυχεῖν καὶ μὴ σιωπᾷ, τοῦτον ἐξόριστον ἀνῃρῆσθαι καὶ μηδαμῇ παρεθῆναι, ἀλλὰ καὶ 25
λιποταξίου γραφὴν ἡλωκέναι καὶ ἐφ' αἵματι φεύγειν καὶ
μόνον οὐ προσηλῶσθαι. καίτοι ταῦθ' ὅταν ἐξελεγχθῇ ποιῶν

9–11 ἀλλ' ... δοκεῖ Aristid. *Rh.* (Schmid) 47.19 10–13 ὦ ... συμβάσης Π1 12–13 τῷ ... αἰτίας Aristid. *Rh.* (Schmid) 38.1 τῷ ... Μόσχου schol. Hermog. (Walz 4) 459.8 12–14 χαλεπῆς ... περιιὼν Π11 21–4 αὐτοὺς ... ἀξιοῖ Π1 23–4 ὡς ... σιωπᾷ Π11, Aristid. *Rh.* (Schmid) 25.22

---

1 ποιήσαντα SYP^{ac} Aristid. 2 ἀνεκρίνατο S 4 πρωτῶν S^{ac} 5 ἐγράψατο om. Y^{ac} -σθένη Σ (Patm.) δημοσθένους (post -σθένην) S^{sl} 6 λειπο- F$_4^{sl}$P^{ac} 8 ᾗ] ᾧ S$_3^{\gamma\rho}$F$_1^{sl}$ κἀκεῖνος Σ (360 Dilts) ἑαυτὸν ἠτίμωκεν A 9 -γε A 10 καὶ κοινὸν om. Aristid. 11 -γε A: [Π1] τούτῳ] τοῦτο Aristid. 12 κακῆς] καὶ δεινῆς Aristid. 14 περιιὼν S^{sl}AFYP$_4^{pc}$: περιὼν SP^{ac}: περ[Π11 15 μου S^{ac} -μά AF εἴην Cd^{pc} 16 δεδρακὼς τοῦτο F δὲ A -νε AFP^{pc} 17 -σι AF^{pc}P^{pc} 18 -τα A ὑφισχνειτο S^{ac} 19–20 ἄλλο ante οὐδὲν F: post οὐδὲν AP$_2^{sl}$: om. SYP 20 τῷ τοιούτῳ FP$_4^{sl}$ 21 τοιοῦτον S 22 τηλικοῦτον AFP^{pc} ἀδίκως om. A 24 ὑβρισθεὶς] ὀργισθεὶς Σ (369 Dilts): [Π11] σιωπᾶν Aristid.: σι[Π11 26 λειπο- P^{ac} 27 οὐ om. Y^{ac} ἐξελέγχηται F

man to do that, that over-pliable scoundrel 'dusty' Euktemon. That sycophant didn't even proceed to the preliminary inquiry in the case, nor did Meidias hire him for any purpose except to get posted up in front of the eponymous heroes for all to see 'Euktemon of Lousia prosecuted Demosthenes of Paiania for desertion'—and I think he would gladly have had added, if it had somehow been possible, that Meidias hired him to prosecute! But I say nothing about that; for a prosecution for which the man has disfranchised himself by not proceeding, I need no further compensation, I have sufficient. But the thing done by Meidias which I consider really **104** dreadful, men of Athens, and shocking and of public concern, an act of impiety, not only of injustice, I will tell you. When a serious and painful charge was made against that poor miserable man Aristarkhos son of Moskhos, first of all, men of Athens, Meidias went all round the Agora and had the effrontery to say impious and dreadful things about me, alleging that I was the man who had done that deed. As he didn't achieve anything by this talk, he approached the people who were bringing the charge of homicide against Aristarkhos, the relatives of the deceased, and promised to pay them money if they accused me of the act; he considered neither gods nor divine law nor anything else an obstacle to speaking in that way, and he had no scruple. Not even the **105** particular people whom he was addressing made him ashamed to bring such terrible trouble upon someone unjustly. The one end he set for himself was to destroy me by any means, and he thought he should leave nothing untried, regarding it as right that, if a man treated insolently by Meidias asks for justice and doesn't keep silent, he should be destroyed by exile and not be let off in any way, but be convicted of desertion and banished for bloodshed and all but nailed up! But when it's proved that he was doing all this

πρὸς οἷς ὕβριζέ με χορηγοῦντα, τίνος συγγνώμης ἢ τίνος
106 ἐλέου δικαίως τεύξεται παρ' ὑμῶν; ἐγὼ μὲν γὰρ αὐτόν, ὦ
ἄνδρες Ἀθηναῖοι, νομίζω αὐτόχειρά μου γεγενῆσθαι τούτοις
τοῖς ἔργοις, καὶ τότε μὲν τοῖς Διονυσίοις τὴν παρασκευὴν καὶ
τὸ σῶμα καὶ τἀναλώμαθ' ὑβρίζειν, νῦν δὲ τούτοις οἷς ἐποίει 5
καὶ διεπράττετο ἐκεῖνά τε καὶ τὰ λοιπὰ πάντα, τὴν πόλιν, τὸ
γένος, τὴν ἐπιτιμίαν, τὰς ἐλπίδας· εἰ γὰρ ἓν ὧν ἐπεβούλευσε
κατώρθωσεν, ἁπάντων ἂν ἀπεστερήμην ἐγὼ καὶ μηδὲ ταφῆναι προσυπῆρχεν οἴκοι μοι. διὰ τί, ἄνδρες δικασταί; εἰ γάρ,
ἐάν τις παρὰ πάντας τοὺς νόμους ὑβρισθεὶς ὑπὸ Μειδίου βο- 10
ηθεῖν αὐτῷ πειρᾶται, ταῦτα καὶ τοιαῦτα ἕτερ' αὐτῷ παθεῖν
ὑπάρξει, προσκυνεῖν τοὺς ὑβρίζοντας ὥσπερ ἐν τοῖς βαρ-
107 βάροις, οὐκ ἀμύνεσθαι κράτιστον ἔσται. ἀλλὰ μὴν ὡς ἀληθῆ
λέγω καὶ προσεξείργασται ταῦτα τῷ βδελυρῷ τούτῳ καὶ
ἀναιδεῖ, κάλει μοι καὶ τούτων τοὺς μάρτυρας. 15

ΜΑΡΤΥΡΕΣ

{Διονύσιος Ἀφιδναῖος, Ἀντίφιλος Παιανιεὺς διαφθαρ-
έντος Νικοδήμου τοῦ οἰκείου ἡμῶν βιαίῳ θανάτῳ ὑπὸ
Ἀριστάρχου τοῦ Μόσχου ἐπεξῄμεν τοῦ φόνου τὸν
Ἀρίσταρχον. αἰσθόμενος δὲ ταῦτα Μειδίας ὁ νῦν κρινό- 20
μενος ὑπὸ Δημοσθένους, ᾧ μαρτυροῦμεν, ἔπειθεν ἡμᾶς
διδοὺς κέρματα τὸν μὲν Ἀρίσταρχον ἀθῷον ἀφεῖναι,
Δημοσθένει δὲ τὴν γραφὴν τοῦ φόνου παραγράψασθαι.}

Λαβὲ δή μοι καὶ τὸν περὶ τῶν δώρων νόμον.

108 Ἐν ὅσῳ δὲ τὸν νόμον, ὦ ἄνδρες Ἀθηναῖοι, λαμβάνει, βού- 25
λομαι μικρὰ πρὸς ὑμᾶς εἰπεῖν, δεηθεὶς ὑμῶν ἁπάντων πρὸς
Διὸς καὶ θεῶν, ὦ ἄνδρες δικασταί· περὶ πάντων ὧν ἂν
ἀκούητε, τοῦθ' ὑποθέντες ἀκούετε τῇ γνώμῃ, τί ἄν, εἴ τις
ἔπασχε ταῦθ' ὑμῶν, ἐποίει, καὶ τίν' ἂν εἶχεν ὀργὴν ὑπὲρ
αὑτοῦ πρὸς τὸν ποιοῦντα. ἐγὼ γὰρ ἐνηνοχὼς χαλεπῶς ἐφ' οἷς 30

12–13 προσκυνεῖν ... ἔσται Aristid. *Rh.* (Schmid) 46.5 19–20 ἐπεξῄμεν ... Ἀρίσταρχον *AB* 1.141.33

2–3 νομίζω ante ὦ A 5 τὰ ἀνα- AF -ματα πάντα A 7 τὰς ἐλπίδας, τὴν ἐπιτιμίαν A 8 οὐδὲ A 9 προσυπῆρξεν A ὦ ἄνδρες Ἀθηναῖοι A^{cp} 11 -ρα AF 13 ἀμύνασθαι AF ἔστιν A 14 προσεξείργασθαι $S^{ac}YP^{ac}$ 15 καὶ om. F 17–23 testimonium om. A 19 ἐπέξιμεν *AB* 24 καὶ om. S 28 ἀκούσητε SYP^{ac}: ἀκούητέ μου AP_4^{pc} -το A 29 τοιαῦθ' FP_4^{pc} -να AF 30 αὑτοῦ FP_3^{pc}: αυτοῦ SA: αὐτοῦ YP^{ac}

in addition to his insolence to me when I was a chorus-producer, what pardon, or what sympathy, will he deserve to get from you? For I consider, men of Athens, that by these **106** actions he has been destroying me. Previously, at the Dionysia, his insolence was directed at my preparations and my person and my expenditures, and now, in these recent doings and activities, not only at those but at everything else too, my city, my family, my enfranchisement, my prospects: for if he had carried out just one of his plots successfully, I should have been deprived of all those, and I should have had the additional misfortune of not even being buried at home. For what reason, men of the jury? If anyone who tries to stand up for himself when quite illegally assaulted by Meidias is going to suffer this and similar treatment, it will be best to kowtow to assailants, as they do in foreign parts, not to resist them. Well, to show that I'm telling the truth and these acts also **107** have been committed by this loathsome brute, please call the witnesses of these too.

WITNESSES

{We, Dionysios of Aphidna and Antiphilos of Paiania, after the violent killing of our relative Nikodemos by Aristarkhos son of Moskhos, proceeded against Aristarkhos for the homicide. Observing this, Meidias, who is prosecuted in the present case by Demosthenes, for whom we testify, tried to persuade us, by offering cash, to let Aristarkhos off unpunished and bring the prosecution for homicide against Demosthenes instead.}

Now please take the law about bribery too.

While he's getting out the law, men of Athens, I want to **108** say a few words to you, after making an earnest request of you all, men of the jury. Concerning everything you hear, bear this question in mind as you listen: what would any one of you have done if you'd suffered this treatment, and how angry would you have been with the doer on your own behalf? I've been annoyed at the insolent treatment I re-

περὶ τὴν λειτουργίαν ὑβρίσθην, ἔτι πολλῷ χαλεπώτερον, ὦ ἄνδρες Ἀθηναῖοι, τούτοις τοῖς μετὰ ταῦτα ἐνήνοχα καὶ μᾶλλον **109** ἠγανάκτηκα. τί γὰρ ὡς ἀληθῶς πέρας ἂν φήσειέ τις εἶναι κακίας καὶ τίν' ὑπερβολὴν ἀναιδείας καὶ ὠμότητος καὶ ὕβρεως, ἄνθρωπος εἰ ποιήσας δεινὰ νὴ Δία καὶ πολλὰ ἀδίκως τινά, ἀντὶ 5 τοῦ ταῦτ' ἀναλαμβάνειν καὶ μεταγιγνώσκειν, ἔτι πολλῷ δεινότερα ὕστερον ἄλλα προσεξεργάζοιτο, καὶ χρῷτο τῷ πλουτεῖν μὴ ἐπὶ ταῦτα ἐν οἷς μηδένα βλάπτων αὐτὸς ἄμεινόν τι τῶν ἰδίων θήσεται, ἀλλ' ἐπὶ τἀναντία, ἐν οἷς ἀδίκως ἐκβαλών τινα καὶ προπηλακίσας αὐτὸν εὐδαιμονιεῖ τῆς περιουσίας; 10 **110** ταῦτα τοίνυν, ὦ ἄνδρες Ἀθηναῖοι, πάντα τούτῳ πέπρακται κατ' ἐμοῦ. καὶ γὰρ αἰτίαν ἐπήγαγέ μοι φόνου ψευδῆ καὶ οὐδὲν ἐμοὶ προσήκουσαν, ὡς τὸ πρᾶγμα αὐτὸ ἐδήλωσεν, καὶ γραφὴν λιποταξίου με ἐγράψατο τρεῖς αὐτὸς τάξεις λελοιπώς, καὶ τῶν ἐν Εὐβοίᾳ πραγμάτων (τουτὶ γὰρ αὖ μικροῦ παρῆλθέ με 15 εἰπεῖν), ἃ Πλούταρχος ὁ τούτου ξένος καὶ φίλος διεπράξατο, ὡς ἐγὼ αἴτιός εἰμι κατεσκεύαζεν πρὸ τοῦ τὸ πρᾶγμα γενέσθαι **111** πᾶσι φανερὸν διὰ Πλουτάρχου γεγονός. καὶ τελευτῶν βουλεύειν μου λαχόντος δοκιμαζομένου κατηγόρει, καὶ τὸ πρᾶγμα εἰς ὑπέρδεινόν μοι περιέστη· ἀντὶ γὰρ τοῦ δίκην ὑπὲρ ὧν 20 ἐπεπόνθειν λαβεῖν, δοῦναι πραγμάτων ὧν οὐδὲν ἐμοὶ προσῆκεν ἐκινδύνευον. καὶ ταῦτα πάσχων ἐγὼ καὶ τοῦτον τὸν τρόπον ὃν διεξέρχομαι νυνὶ πρὸς ὑμᾶς ἐλαυνόμενος, οὐκ ὢν οὔτε τῶν ἐρημοτάτων οὔτε τῶν ἀπόρων κομιδῇ, οὐκ ἔχω, ὦ ἄνδρες **112** Ἀθηναῖοι, τί χρὴ ποιῆσαι. εἰ γὰρ εἰπεῖν τι καὶ περὶ τούτων ἤδη 25 δεῖ, οὐ μέτεστι τῶν ἴσων οὐδὲ τῶν νόμων, ὦ ἄνδρες Ἀθηναῖοι, πρὸς τοὺς πλουσίους τοῖς λοιποῖς ἡμῖν, οὐ μέτεστιν, οὔ· ἀλλὰ καὶ χρόνοι τούτοις τοῦ τὴν δίκην ὑποσχεῖν, οὓς ἂν αὐτοὶ

11–12 *πέπρακται ... ψευδῆ* Π11 15 *τουτὶ ... με* Tib. *Fig.* (Ballaira) 17.4 15–16 *τουτὶ ... εἰπεῖν* Alex. *Fig.* (Spengel 3) 14.9 20–2 *ἀντὶ ... ἐκινδύνευον* *AB* 1.132.1 22–3 *καὶ ... ἐλαυνόμενος* Aristid. *Rh.* (Schmid) 45.8 27–8 *λοιποῖς ... δίκην* Π11

---

2 *μεταυτα* S^{ac} 3 *ἠγανάκτησα* A 4 *κακίας ἢ καὶ* F *-να* AF 5 *νὴ δία δεινὰ* A *τινά* om. S 6 *-τα* A *μεταγιν-* A *ἔτι* om. S 7 *ὕστερον* om. F 9 *ἐκβάλλων* S 13 *αὐτὸ τὸ πρᾶγμα* A *-σε* AF^{pc}P^{pc} 14 *λειπο-* P^{ac} 15 *γὰρ* om. Alex. *αὖ* om. A Alex. *με παρῆλθεν* Alex. Tib. 17 *κατεσκεύασε* A 19 *λαχόντος καὶ* A 20 *ὑπὲρ* om. AFP$_4^{\gamma\rho}$ 21 *προσῆκεν δίκην* SYP^{ac} *AB* 22–3 *τοῦτον ὃν διεξέρχομαι τρόπον* Aristid. 23 *νῦν* F Aristid. 25 *τί* P: *τὶ* SY: *ὅτι* A: *ὅ,τι* F 26 *νόμων* SYP: *ὁμοίων* AFP$_4^{\gamma\rho}$ *ὦ ἄνδρες δικασταί* A^{cp} 27 *πλουσίοις* A^{ac} *λοιποῖς* SYP: *πολλοῖς* AFP$_4^{\gamma\rho}$:]*λοις* Π11 *ἡμῖν* Taylor: *ἡμῶν* SFYP: *ὑμῶν* A: *η*[Π11 *μέστιν* S: [Π11] 28 *τὴν* om. AFP$_4^{pc}$: [Π11] *ἐὰν* F

ceived in connection with the liturgy, but I've been still more annoyed and more indignant, men of Athens, at these subsequent events. For what could one say was really the limit of **109** wickedness, and what shamelessness, cruelty, and insolence could one say go beyond this: if a person who did wrong to another quite terribly and repeatedly, instead of offering atonement and regret, were to go on to commit yet other much worse crimes in addition, and were to employ his wealth, not for activities by which he would improve something of his own while doing no harm to anyone, but on the contrary for ones by which he would expel and vilify someone unjustly, and then congratulate himself on his affluence? Well, all those things, men of Athens, have been done to me **110** by Meidias. He brought against me an accusation of homicide which was false and had nothing to do with me, as the event itself made clear; and he got me prosecuted for desertion, though he has deserted his post three times himself; and for the events in Euboia (I nearly forgot to mention this), which were effected by Meidias' host and friend Ploutarkhos, he tried to make out that I was responsible, before it became plain to everyone that the affair was Ploutarkhos' doing. And finally, after I was selected by lot to be a member **111** of the Boule, he made an accusation against me at the examination of my status, and my situation turned very alarming: instead of exacting a penalty for what I'd suffered, I was in danger of paying one for acts which had nothing to do with me. And while suffering this treatment and being harassed in this manner which I've just been relating to you, although I'm neither a very friendless person nor a particularly poor one, I'm at a loss to know what I should do, men of Athens. If it's right to say a word on that subject too at this stage, we, **112** the rest of the population, men of Athens, have not a bit of equality or of legal protection when we face the rich men, we really haven't. They are given whatever dates they wish for

βούλωνται, δίδονται, καὶ τἀδικήματα ἕωλα τὰ τούτων ὡς ὑμᾶς καὶ ψυχρὰ ἀφικνεῖται, τῶν δ' ἄλλων ἡμῶν ἕκαστος, ἄν τι συμβῇ, πρόσφατος κρίνεται. καὶ μάρτυρές εἰσιν ἕτοιμοι τούτοις καὶ συνήγοροι πάντες καθ' ἡμῶν εὐτρεπεῖς· ἐμοὶ δὲ **113** οὐδὲ τἀληθῆ μαρτυρεῖν ἐθέλοντας ὁρᾶτ' ἐνίους. ταῦτα μὲν οὖν 5 ἀπείποι τις ἄν, οἶμαι, θρηνῶν. τὸν δὲ νόμον μοι λέγε ἐφεξῆς, ὥσπερ ἠρξάμην. λέγε.

ΝΟΜΟΣ

Ἐάν τις Ἀθηναίων λαμβάνῃ παρά τινος, ἢ αὐτὸς διδῷ ἑτέρῳ, ἢ διαφθείρῃ τινὰς ἐπαγγελλόμενος, ἐπὶ βλάβῃ 10 τοῦ δήμου ἢ ἰδίᾳ τινὸς τῶν πολιτῶν, τρόπῳ ἢ μηχανῇ ᾑτινιοῦν, ἄτιμος ἔστω καὶ παῖδες καὶ τὰ ἐκείνου.

114 Οὕτω τοίνυν οὗτός ἐστιν ἀσεβὴς καὶ μιαρὸς καὶ πᾶν ἂν ὑποστὰς εἰπεῖν καὶ πρᾶξαι, εἰ δ' ἀληθὲς ἢ ψεῦδος ἢ πρὸς ἐχθρὸν ἢ φίλον ἢ τὰ τοιαῦτα, ἀλλ' οὐδ' ὁτιοῦν διορίζων, ὥστ' 15 ἐπαιτιασάμενός με φόνου καὶ τοιοῦτο πρᾶγμα ἐπαγαγών, εἴασε μέν με εἰσιτητήρια ὑπὲρ τῆς βουλῆς ἱεροποιῆσαι καὶ θῦσαι καὶ κατάρξασθαι τῶν ἱερῶν ὑπὲρ ὑμῶν καὶ ὅλης τῆς **115** πόλεως, εἴασε δ' ἀρχιθεωροῦντα ἀγαγεῖν τῷ Διὶ τῷ Νεμείῳ τὴν κοινὴν ὑπὲρ τῆς πόλεως θεωρίαν, περιεῖδε δὲ ταῖς Σεμναῖς 20 θεαῖς ἱεροποιὸν αἱρεθέντα ἐξ Ἀθηναίων ἁπάντων τρίτον αὐτὸν καὶ καταρξάμενον τῶν ἱερῶν. ἆρ' ἄν, εἴ γ' εἶχε στιγμὴν ἢ σκιὰν τούτων ὧν κατεσκεύαζεν κατ' ἐμοῦ, ταῦτ' ἂν εἴασεν; ἐγὼ μὲν οὐκ οἶμαι. οὐκοῦν ἐξελέγχεται τούτοις ἐναργῶς ὕβρει ζητῶν με ἐκβαλεῖν ἐκ τῆς πατρίδος. 25

116 Ἐπειδὴ τοίνυν τοῦτο τὸ πρᾶγμα οὐδὲ καθ' ἕν, πανταχῇ στρέφων, οἷός τ' ἦν ἀγαγεῖν ἐπ' ἐμέ, φανερῶς ἤδη δι' ἐμὲ τὸν Ἀρίσταρχον ἐσυκοφάντει. καὶ τὰ μὲν ἄλλα σιωπῶ· τῆς δὲ

27–8 φανερῶς ... ἐσυκοφάντει schol. Hermog. (Walz 4) 459.10

---

2 δὲ A ὑμῶν A ἐάν A 4 ὑμῶν A εὐπρεπεῖς A: εὐτρεπεῖς ante καθ' F 5 -τε AF 6 ἀπείποι S: εἴποι AF: ἃ εἴποι YP θρήνων Y 7 λέγων S 9–12 legem om. A 10 διαφθείρει Y: [P^ac] 11 ἢ (post δήμου) Iurinus: καὶ SFYP ἰδίᾳ Reiske: διά SFYP 13 -τι S ἀσεβὴς ἄνθρωπος AFYP 14 διαπρᾶξαι YP^ac δ'] τ' S^sl ἢ (ante πρὸς)] εἰ Meier: ἢ εἰ Dobree 15 ἢ (ante τὰ)] καὶ Bc: del. Spalding 15–16 ὥστε αἰτιασάμενός A 17 εἰσιτητήρια Herwerden: εἰσητήρια S: εἰσιτήρια S^sl AFYP:]ιτηρια Π13 19 δὲ A νεμίωι F 20 περιεῖδε S^{sl}_3AFYP: περιεῖλε S 21 ἱεροποιῶν A αιρηθ[Π13: ἀναρρηθέντα P 22 ἆρα A γε A 23 κατεσκεύασε A ταῦτα A 25 ἐκβάλλειν SYP 27 ἐμὲ B^sl T^sl schol. Hermog.: ἐμοῦ SAFYP

undergoing trial, and their offences are stale and cold when they come before you; but the rest of us, if anything happens to us, are each tried when fresh. And they have witnesses ready, and supporters all prepared to speak against us; yet for me, as you see, some men are unwilling to give even true evidence! Still, I suppose it would wear one out to bewail **113** that. Please read out the law next, to resume what I began. Read it.

LAW

> If any Athenian accepts from anyone, or himself gives to another, or corrupts any persons by promises, to the detriment of the people or of any of the citizens individually, by any manner or means whatever, let him and his children and his property be deprived of rights.

This man, then, is such an impious and wicked person, so **114** ready to say and do anything, not distinguishing in the slightest whether it is true or false, or is done to an enemy or a friend, or any question of that sort, that, after accusing me of homicide and bringing such a serious charge against me, he allowed me to hold inaugural rites on behalf of the Boule and to sacrifice and to initiate the rites for you and the whole city; he allowed me, as arkhitheoros, to lead the city's festival **115** representatives for Zeus of Nemea; he stood by when I was elected from the whole of Athens as one of three hieropoioi for the Solemn Goddesses and initiated the rites. If he'd had a scrap or shadow of evidence for the charges he got up against me, would he have allowed that? I think not. So these facts clearly prove that insolence was his reason for seeking to expel me from my country.

So since, twisting it every way, he was totally unable to **116** bring this charge against me, he now started publicly denouncing Aristarkhos, to get at me. I shall pass over the rest;

βουλῆς περὶ τούτων καθημένης καὶ σκοπουμένης, παρελθὼν
οὗτος "ἀγνοεῖτ'" ἔφη, "ὦ βουλή, τὸ πρᾶγμα; καὶ τὸν αὐτόχειρα ἔχοντες" λέγων τὸν Ἀρίσταρχον "μέλλετε καὶ ζητεῖτε
καὶ τετύφωσθε; οὐκ ἀποκτενεῖτε; οὐκ ἐπὶ τὴν οἰκίαν βαδι-
117 εῖσθε; οὐχὶ συλλήψεσθε;" καὶ ταῦτ' ἔλεγεν ἡ μιαρὰ καὶ ἀναιδὴς 5
αὕτη κεφαλή, ἐξεληλυθὼς τῇ προτεραίᾳ παρ' Ἀριστάρχου,
καὶ χρώμενος ὥσπερ ἂν ἄλλος τις αὐτῷ τὰ πρὸ τούτου, καί,
ὅτ' ηὐτύχει, πλεῖστα παρεσχηκότος πάντων ἐκείνου πράγ-
ματά μοι περὶ τῶν πρὸς τοῦτον διαλλαγῶν. εἰ μὲν οὖν εἰργά-
σθαι τι τούτων ἐφ' οἷς ἀπόλωλεν ἡγούμενος τὸν Ἀρίσταρχον 10
καὶ πεπιστευκὼς τοῖς τῶν αἰτιασαμένων λόγοις ταῦτ' ἔλεγεν,
118 χρῆν μὲν οὐδ' οὕτως· μετρία γὰρ δίκη παρὰ τῶν φίλων ἐστίν,
ἄν τι δοκῶσιν πεποιηκέναι δεινόν, μηκέτι τῆς λοιπῆς φιλίας
κοινωνεῖν, τὸ δὲ τιμωρεῖσθαι καὶ ἐπεξιέναι τοῖς πεπονθόσιν
καὶ τοῖς ἐχθροῖς παραλείπεται· ὅμως δ' ἔστω τούτῳ γε συγ- 15
γνώμη. εἰ δὲ λαλῶν μὲν καὶ ὁμωρόφιος γιγνόμενος ὡς οὐδὲν
εἰργασμένῳ φανήσεται, λέγων δὲ καὶ καταιτιώμενος ταῦθ'
ἕνεκα τοῦ συκοφαντεῖν ἐμέ, πῶς οὐ δεκάκις, μᾶλλον δὲ μυρ-
119 ιάκις δίκαιός ἐστ' ἀπολωλέναι; ἀλλὰ μὴν ὡς ἀληθῆ λέγω, καὶ
τῇ προτεραίᾳ ὅτε ταῦτ' ἔλεγεν εἰσεληλύθει καὶ διείλεκτο ἐκ- 20
είνῳ, τῇ δ' ὑστεραίᾳ πάλιν (τοῦτο γάρ, τοῦτο οὐκ ἔχον ἐστὶν
ὑπερβολὴν ἀκαθαρσίας, ἄνδρες Ἀθηναῖοι) εἰσελθὼν οἴκαδε ὡς
ἐκεῖνον καὶ ἐφεξῆς οὑτωσὶ καθιζόμενος, τὴν δεξιὰν ἐμβαλών,
παρόντων πολλῶν, μετὰ τοὺς ἐν τῇ βουλῇ τούτους λόγους, ἐν
οἷς αὐτόχειρα καὶ τὰ δεινότατ' εἰρήκει τὸν Ἀρίσταρχον, ὤμ- 25
νυε μὲν κατ' ἐξωλείας μηδὲν εἰρηκέναι περὶ αὐτοῦ φλαῦρον,
καὶ οὐδὲν ἐφρόντιζεν ἐπιορκῶν, καὶ ταῦτα παρόντων τῶν

21–2 τοῦτο ... ἀκαθαρσίας Aristid. *Rh.* (Schmid) 44.14

---

2 -τε A τὸ πρᾶγμα ὦ βουλὴ AF 3 ἔχοντες SFYP: λέγοντες A
4 τετυφωσθαι S^{ac} καὶ οὐκ ἀπο- YP^{ac} 5 συλλήμψεσθε S^{ac} 6 αυτηι
κεφαληι S^{ac} -ρὰ SFYP 7 αὐτῷ AFP_2^{pc}: om. SYP^{ac} 8 ὅτε A
εὐτύχει SFYP παρεσχηκότας A^{ac} πάντων $S_3^{\gamma\rho}$AFYP: om. S ἐκείνου]
τούτου $S_3^{\gamma\rho}$ 9 διαλλαγῶν $S_3^{\gamma\rho}$AF: ἀπαλλαγῶν SP^{pc}: διὰ ἀπαλλαγῶν Y: δι'
ἀπαλλαγῶν P^{ac} εἴργασταί SY 11 -τα SYP -γε AF 12 ἐχρῆν S^{pc}
οὕτω SYP 13 -σι AFYP τοῦ λοιποῦ Dobree 14 -σι $AF^{pc}YP$
15 ὑπολείπεται A: καταλείπεται FP_4^{sl} δὲ A 16 λαλῶν SFP_3^{pc}: αλων
A^{pc}: ἄλλων YP^{ac}: [A^{ac}] ἁλῶν μὲν κοινωνήσας Stephanus γενόμενος A
17 εἰργασμένων Y: [P^{ac}] -τα A 19 δίκαιος post ἀπολωλέναι F -τιν A
20 τῇ μὲν AP_4^{pc} 21 δὲ A πάλιν αὖ A τοῦτο μὲν γὰρ οὐκ Aristid.
ἔχων A Aristid. 22 ὑπερβολὴ Aristid. ὦ ἄνδρες Ἀθηναῖοι A^{cp}
23 καθεζόμενος AF 25 -τα AF 26 περὶ SFYP: κατ' A φλαῦρον
AFP_3^{pc}: φαῦλον SYP^{ac}

but when the Boule was in session about this affair and enquiring into it, this man stepped forward and said 'Don't you know the facts, Boule? When you've got the perpetrator,' meaning Aristarkhos, 'are you still delaying and investigating? Are you out of your minds? Won't you put him to death? Won't you go to his house and arrest him?' And he said all **117** this, the foul brute, although he'd been in Aristarkhos' house the day before, and until then was as friendly with him as anyone else, and although before his misfortune Aristarkhos pestered me more than anyone about settling my quarrel with Meidias. Now, if he said this because he considered that Aristarkhos had committed any of the deeds which brought about his downfall, and because he believed the accusers' statements, even so he ought to have kept his mouth shut. Friends who are thought to have done something dreadful **118** are punished moderately, by having the friendship broken off; revenge and legal proceedings are left to their victims and enemies. Still, it could be forgiven in this case. But if it's shown that he chatted under the same roof with him as if he were an innocent man, and then made these statements and accusations for the purpose of incriminating me, surely he deserves death ten times over, or rather ten thousand times! To show that I'm telling the truth, and that on the day **119** before he made these statements he'd gone in and conversed with Aristarkhos, and on the day after once again (now this really is the supreme example of impurity, men of Athens) he went into his house, sat down next to him, as close as this, gave him his hand, and in the presence of a large number of people, after that speech in the Boule in which he'd called Aristarkhos 'the perpetrator' and the most dreadful things, he swore, invoking destruction on himself, that he had made no allegation against him, and he didn't care that he was perjuring himself, and doing so in the presence of people who

συνειδότων, ἠξίου δὲ καὶ πρὸς ἐμὲ αὐτῷ δι' ἐκείνου γίγνεσθαι τὰς διαλύσεις, καὶ τούτων τοὺς παρόντας ὑμῖν καλῶ

120 μάρτυρας. καίτοι πῶς οὐ δεινόν, ὦ ἄνδρες Ἀθηναῖοι, μᾶλλον δὲ ἀσεβές, λέγειν ὡς φονεύς, καὶ πάλιν ὡς οὐκ εἴρηκεν ταῦτα ἀπομνύναι, καὶ φόνον μὲν ὀνειδίζειν, τούτῳ δ' ὁμωρόφιον 5 γίγνεσθαι; κἂν μὲν ἀφῶ τοῦτον ἐγὼ καὶ προδῶ τὴν ὑμετέραν καταχειροτονίαν, οὐδέν, ὡς ἔοικ', ἀδικῶ· ἂν δ' ἐπεξίω, λέλοιπα τὴν τάξιν, φόνου κοινωνῶ, δεῖ με ἀνηρπάσθαι. ἐγὼ δ' αὖ τοὐναντίον οἶμαι, εἰ τοῦτον ἀφῆκα, λελοιπέναι μὲν ⟨ἄν⟩, ὦ ἄνδρες Ἀθηναῖοι, τὴν τοῦ δικαίου τάξιν, φόνου δ' ἂν εἰκότως 10 ἐμαυτῷ λαχεῖν· οὐ γὰρ ἦν μοι δήπου βιωτὸν τοῦτο ποιήσαντι.

121 ὅτι τοίνυν καὶ ταῦτ' ἀληθῆ λέγω, κάλει μοι καὶ τούτων τοὺς μάρτυρας.

ΜΑΡΤΥΡΙΑ

{Λυσίμαχος Ἀλωπεκῆθεν, Δημέας Σουνιεύς, Χάρης 15 Θορίκιος, Φιλήμων Σφήττιος, Μόσχος Παιανιεύς, καθ' οὓς καιροὺς ἡ εἰσαγγελία ἐδόθη {ἡ} εἰς τὴν βουλὴν ὑπὲρ Ἀριστάρχου τοῦ Μόσχου, ὅτι εἴη Νικόδημον ἀπεκτονώς, οἴδαμεν Μειδίαν τὸν κρινόμενον ὑπὸ Δημοσθένους, ᾧ μαρτυροῦμεν, ἐλθόντα πρὸς τὴν βουλὴν καὶ 20 λέγοντα μηδένα ἕτερον εἶναι τὸν Νικοδήμου φονέα ἀλλ' Ἀρίσταρχον, καὶ τοῦτον αὐτοῦ γεγονέναι αὐτόχειρα, καὶ συμβουλεύοντα τῇ βουλῇ βαδίζειν ἐπὶ τὴν οἰκίαν τὴν Ἀριστάρχου καὶ συλλαμβάνειν αὐτόν. ταῦτα δ' ἔλεγεν πρὸς τὴν βουλὴν τῇ προτεραίᾳ μετ' Ἀριστάρχου καὶ 25 μεθ' ἡμῶν συνδεδειπνηκώς. οἴδαμεν δὲ καὶ Μειδίαν, ὡς ἀπῆλθεν ἀπὸ τῆς βουλῆς τούτους τοὺς λόγους εἰρηκώς, εἰσεληλυθότα πάλιν ὡς Ἀρίσταρχον καὶ τὴν δεξιὰν {ὡς} ἐμβεβληκότα καὶ ὀμνύοντα κατ' ἐξωλείας μηδὲν κατ'

16–19 *καθ' ... ἀπεκτονώς AB* 1.176.30

---

1 *αὐτῷ* ed. Lond. 1586: *αυτῷ* SA: *αὐτῷ* FYP *γίν-* A 2 *καὶ* om. S 3 *ΜΑΡΤΥΡΙΑ* (post *μάρτυρας*) SY: *ΜΑΡΤΥΡΕΣ* AF: om. P 4 *-κε* AF[pc] 6 *γίν-* A *προδῶι* F 7 *-κεν* AF *ἐὰν* A 9 *αὖ* AFYP: *αὐτὸ* S *ἂν* add. Blass 9–10 *ὦ ἄνδρες δικασταί* A[cp] 11 *δήπου μοι* A 14 *ΜΑΡΤΥΡΙΑ* S: *ΜΑΡΤΥΡΕΣ* AF: *ΜΑΡΤΥΡΙΑΙ* YP 15–162.2 testimonium om. A 15 *σουνιεὺς* BX: *σουνεὺς* SF[ac]YP: *σουννιεὺς* F[pc] *Χάρης* Reiske: *χιάρης* SFYP 16 *θόρκιος* YP[ac] 17 *ἡ* om. Aa 18 *Μόσχου*] *Νικοδήμου AB* 22 *αὐτοῦ* YP: *αὐτὸν* SFP$^{sl}_{4}$ 23 *οικίειαν* S[ac] 25 *προτεραίᾳ* F$^{pc}_{1}$: *προτέρᾳ* SYP: [F[ac]] 26 *δὲ* om. P[ac] 28 *ὡς* del. Taylor: *οἱ* G. H. Schaefer

were aware of that, but even asked Aristarkhos to be the intermediary for a settlement between himself and me—I shall call those who were present to give you evidence of all this too. But isn't it a terrible thing, men of Athens, or rather **120** impious, to call a man a murderer and then again to deny on oath having said that, to accuse a man of homicide and then to go under the same roof with him? If I let him off and betray your vote against him, I'm not guilty of anything, it seems; but if I take proceedings, I'm a deserter from the ranks, I'm an accomplice in homicide, I have to be annihilated! But my own opinion is the opposite: if I'd let him off, I should have been a deserter, men of Athens, from the ranks of justice, and I could reasonably have prosecuted myself for homicide; for my life surely wouldn't have been worth living if I'd done that. Well, to show that these statements of mine **121** are also true, please call the witnesses for them too.

TESTIMONY

{We, Lysimakhos of Alopeke, Demeas of Sounion, Khares of Thorikos, Philemon of Sphettos, and Moskhos of Paiania, know that, at the time when the denunciation was made to the Boule concerning Aristarkhos son of Moskhos, alleging that he was the killer of Nikodemos, Meidias, who is prosecuted by Demosthenes, for whom we testify, went to the Boule and said that the murderer of Nikodemos was none other than Aristarkhos and that it was he who was the perpetrator; and he advised the Boule to go to Aristarkhos' house and arrest him. He said this to the Boule after dining with Aristarkhos and with us on the previous day. We also know that, when Meidias left the Boule after making those statements, he went into Aristarkhos' house again, gave him his hand, and swore, invoking destruction on

αὐτοῦ πρὸς τὴν βουλὴν εἰρηκέναι φαῦλον, καὶ ἀξιοῦντα
Ἀρίσταρχον ὅπως ἂν διαλλάξῃ αὐτῷ Δημοσθένην.}

122 Τίς οὖν ὑπερβολή; τίς ὁμοία τῇ τούτου γέγονεν ἢ γένοιτ' ἂν
πονηρία; ὃς ἄνδρα ἀτυχοῦντα, οὐδὲν αὑτὸν ἠδικηκότα (ἐῶ
γὰρ εἰ φίλον), ἅμα συκοφαντεῖν ᾤετο δεῖν καὶ πρὸς ἐμὲ αὑτὸν 5
διαλύειν ἠξίου· καὶ ταῦτ' ἔπραττεν καὶ χρήματ' ἀνήλισκεν ἐπὶ
τῷ μετ' ἐκείνου κἀμὲ προσεκβαλεῖν ἀδίκως.

123 Τοῦτο μέντοι τὸ τοιοῦτον ἔθος καὶ τὸ κατασκεύασμα, ὦ
ἄνδρες Ἀθηναῖοι, τὸ τοῖς ὑπὲρ αὐτῶν ἐπεξιοῦσι δικαίως ἔτι
πλείω περιιστάναι κακά, οὐκ ἐμοὶ μὲν ἄξιόν ἐστ' ἀγανακτεῖν 10
καὶ βαρέως φέρειν, ὑμῖν δὲ τοῖς ἄλλοις παριδεῖν. πολλοῦ γε
καὶ δεῖ· ἀλλὰ πᾶσιν ὁμοίως ὀργιστέον, ἐκλογιζομένοις καὶ
θεωροῦσιν ὅτι τοῦ μέν, ὦ ἄνδρες Ἀθηναῖοι, ῥᾳδίως κακῶς
παθεῖν ἐγγύτατα ὑμῶν εἰσιν οἱ πενέστατοι καὶ ἀσθενέστατοι,
τοῦ δ' ὑβρίσαι καὶ τοῦ ποιήσαντας μὴ δοῦναι δίκην ἀλλὰ τοὺς 15
ἀντιπαρέξοντας πράγματα μισθώσασθαι οἱ βδελυροὶ καὶ χρή-

124 ματ' ἔχοντες {εἰσιν ἐγγυτάτω}. οὐ δὴ δεῖ παρορᾶν τὰ τοιαῦτα,
οὐδὲ τὸν ἐξείργοντα δέει καὶ φόβῳ τὸ δίκην ὧν ἂν ἡμῶν
ἀδικηθῇ τις λαμβάνειν παρ' αὑτοῦ ἄλλο τι χρὴ νομίζειν ποιεῖν
ἢ τὰς τῆς ἰσηγορίας καὶ τὰς τῆς ἐλευθερίας ἡμῶν μετουσίας 20
ἀφαιρεῖσθαι. ἐγὼ μὲν γὰρ ἴσως διεωσάμην, καὶ ἄλλος τις
ἄν, ψευδῆ λόγον καὶ συκοφαντίαν, καὶ οὐκ ἀνήρπασμαι· οἱ
δὲ πολλοὶ τί ποιήσετε, ἂν μὴ δημοσίᾳ πᾶσι φοβερὸν

125 καταστήσητε τὸ εἰς ταῦτα ἀποχρῆσθαι τῷ πλουτεῖν; δόντα
λόγον καὶ ὑποσχόντα κρίσιν περὶ ὧν ἄν τις ἐγκαλῇ, τότ' 25
ἀμύνεσθαι τοὺς ἀδίκως ἐφ' αὑτὸν ἐλθόντας χρή, {καὶ τότ', ἂν
ἀδικοῦντας ὁρᾷ τις,} οὐ προαναρπάζειν, οὐδ' ἐπάγοντ' αἰτίας
ψευδεῖς ἄκριτον ζητεῖν ἀποφεύγειν, οὐδ' ἐπὶ τῷ διδόναι δίκην
ἀσχάλλειν, ἀλλὰ μὴ ποιεῖν ἐξ ἀρχῆς ἀσελγὲς μηδέν.

20–2 ἐλευθερίας ... ψευδῆ Π11

---

1 φλαῦρον Dindorf   4 πονηρίᾳ S^{ac}Y   αὑτὸν Bekker: αυτὸν SA: αὐτὸν FYP   5 εἰ] εἰπεῖν $K^{pc}Lp^{sl}$ αὑτὸν Markland: αυτὸν SA: αὐτὸν FYP   6 διαλῦσαι Σ (Patm.)   -ατα A   10 περιστάναι $S^{ac}YP^{ac}$ -τιν A   12 ἐγλογ- S^{ac} 13 ὦ ἄνδρες δικασταί A^{cp} 14 παθεῖν] πάλιν $AF_1^{γρ}$ ἡμῶν F   15 ποιήσαντάς τι AP_4^{pc} 16 οἱ δὲ A   βδελλυροὶ S   17 εἰσιν ἐγγυτάτωι SA^{ac}: del. Reiske   παρορῶντα τοιαῦτα A   19 αὐτοῦ Bekker   20 τῆς ἐλευθερίας καὶ τῆς ἰσηγορίας A   εισηγορίας S^{ac} ὑμῶν A: [Π11]   21 διεσωσάμην S: διεω[Π11   23 ἐὰν A   μὴ om. P^{pc} φοβερὸν πᾶσι A: πᾶσι φανερὸν YP   24 καταστήσεται A   25 ἐγκαλέσῃ A   26 ἀμύνασθαι A   ἐπ' αὐτὸν A   26–7 καὶ ... τις del. Sykutris   -τε SFYP   27 ἀδικοῦντα SYP^{ac} -τα A

himself, that he had made no allegation against him in the presence of the Boule; and he asked Aristarkhos to reconcile Demosthenes to him.}

Can you beat it? What villainy has there ever been, or could there be, to equal his? He thought fit to denounce a man when he was down, who had done him no wrong (I say nothing of whether he was a friend), and at the same time asked him to make a settlement between himself and me; and while doing this he was also spending money for the purpose of expelling me along with Aristarkhos, quite unjustly. **122**

But this proclivity, men of Athens, and this device of entangling in yet more trouble those who take just proceedings in their own defence, is not a thing which it's right only for me to resent indignantly and the rest of you to overlook. Far from it. You should all be equally angry, in view of the fact that the likeliest of you to suffer easy maltreatment are the poorest and weakest, whereas the likeliest to be insolent, and then to avoid punishment for it and hire men to get up actions in retaliation, are the filthy rich. So such conduct mustn't be overlooked; consider that the man who by intimidation prevents the imposition of a penalty on him for his offences against any of us is simply taking away our enjoyment of free speech and liberty. In my own case, no doubt, I repulsed lies and accusations, and so might one or two others; I haven't been annihilated. But what will you people in general do, unless you publicly frighten everyone away from the misuse of wealth for this purpose? One should render an account and stand trial for the charges which anyone makes, and only then retaliate against those who proceed against one unjustly; one shouldn't destroy them in advance, nor seek to get away without trial by bringing false accusations, nor fret at being punished, but refrain from bullying in the first place. **123** **124** **125**

126 Ὅσα μὲν τοίνυν εἴς τε τὴν λειτουργίαν καὶ τὸ σῶμ᾽ ὑβρίσθην, καὶ πάντ᾽ ἐπιβουλευόμενος τρόπον καὶ πάσχων κακῶς
ἐκπέφευγα, ἀκηκόατε, ὦ ἄνδρες Ἀθηναῖοι. καὶ παραλείπω
δὲ πολλά· οὐ γὰρ ἴσως ῥᾴδιον πάντ᾽ εἰπεῖν. ἔχει δ᾽ οὕτως· οὐκ
ἔστ᾽ ἐφ᾽ ὅτῳ τῶν πεπραγμένων ἐγὼ μόνος ἠδίκημαι, ἀλλ᾽ ἐπὶ 5
μὲν τοῖς εἰς τὸν χορὸν γεγενημένοις ἀδικήμασιν ἡ φυλή, τὸ
δέκατον μέρος ὑμῶν, συνηδίκηται, ἐπὶ δ᾽ οἷς ἐμὲ ὕβρισεν καὶ
ἐπεβούλευσεν, οἱ νόμοι, δι᾽ οὓς εἷς ἕκαστος ὑμῶν σῶς ἐστιν,
ἐφ᾽ ἅπασι δὲ τούτοις ὁ θεός, ᾧ χορηγὸς ἐγὼ καθειστήκειν,
καὶ τὸ τῆς ὁσίας, ὁτιδήποτ᾽ ἐστίν, τὸ σεμνὸν καὶ τὸ δαιμόνιον 10
127 {συνηδίκηται}. δεῖ δὴ τούς γε βουλομένους ὀρθῶς τὴν κατ᾽
ἀξίαν τῶν πεπραγμένων παρὰ τούτου δίκην λαβεῖν, οὐχ ὡς
ὑπὲρ ἡμῶν ὄντος μόνον τοῦ λόγου τὴν ὀργὴν ἔχειν, ἀλλ᾽ ὡς ἐν
ταὐτῷ τῶν νόμων, τοῦ θεοῦ, τῆς πόλεως, ὁμοῦ πάντων
ἠδικημένων, οὕτω ποιεῖσθαι τὴν τιμωρίαν, καὶ τοὺς βοηθ- 15
οῦντας καὶ τοὺς συνεξεταζομένους μετὰ τούτου μὴ συνηγόρους μόνον, ἀλλὰ καὶ δοκιμαστὰς τῶν τούτῳ πεπραγμένων
ὑπολαμβάνειν εἶναι.

128 Εἰ μὲν τοίνυν, ὦ ἄνδρες Ἀθηναῖοι, σώφρονα καὶ μέτριον
πρὸς τἆλλα παρεσχηκὼς αὑτὸν Μειδίας καὶ μηδένα τῶν 20
ἄλλων πολιτῶν ἠδικηκὼς εἰς ἐμὲ ἀσελγὴς μόνον οὕτω καὶ
βίαιος ἐγεγόνει, πρῶτον μὲν ἔγωγ᾽ ἀτύχημ᾽ ἂν ἐμαυτοῦ τοῦτο
ἡγούμην, ἔπειτ᾽ ἐφοβούμην ἂν μὴ τὸν ἄλλον ἑαυτοῦ βίον
οὗτος μέτριον δεικνύων καὶ φιλάνθρωπον διακρούσηται
129 τούτῳ τὸ δίκην ὧν ἐμὲ ὕβρικεν δοῦναι. νυνὶ δὲ τοσαῦτ᾽ ἐστὶν 25
τἆλλα ἃ πολλοὺς ὑμῶν ἠδίκηκεν καὶ τοιαῦτα, ὥστε τούτου
μὲν τοῦ δέους ἀπήλλαγμαι, φοβοῦμαι δὲ πάλιν τοὐναντίον
μή, ἐπειδὰν πολλὰ καὶ δεινὰ ἑτέρους ἀκούηθ᾽ ὑπ᾽ αὐτοῦ πεπονθότας, τοιοῦτός τις ὑμῖν λογισμὸς ἐμπέσῃ· "τί οὖν; σὺ

12–14 ὡς … νόμων Π11 28–166.1 δεινὰ … ἄλλων Π11

---

1 εἴς τε] ἴστε A -μα AF 4 πατ᾽ S^{ac}: ἅπαντ᾽ A 5 ἔσται A 6 τὸ om. S 7 δὲ A -ισε AF^{pc} 8 εἰς om. S^{ac}AY 10 -τί AF 11 συνηδίκηται del. Blass 12 τὴν δίκην S^{ac} λαμβάνειν A 13 ἡμῶν F: ὑμῶν SF_1^{sl}YP: ἐμοῦ $AF_1^{\gamma\rho}$: [Π11] ὄντος μόνον SFYP:]ν Π11: μόνον ὄντος A αλλα Π11 14 ὁμοῦ A: ἐμοῦ SFYP 16 τοὺς om. F 18 ὑπο-λαμβάνετ᾽ SFYP 20 τὰ ἄλλα AY: [P^{ac}] ἑαυτὸν A 21 πολιτῶν μηδὲν AFP_4^{mg} μόνον ἀσελγὴς FYP 22 -μα A 23 αυτοῦ $S_3^{\gamma\rho}$ 24 δεικνύων μέτριον $S_3^{\gamma\rho}$ 24–5 οὕτω τὸ δίκην ὧν ἐμὲ ὕβρικεν δοῦναι διακρ-ούσηται $S_3^{\gamma\rho}$: τούτῳ τὸ δίκην δοῦναι ὧν ἐμὲ ὕβρικε διακρούσηται A: τούτῳ τὸ δίκην ὧν ἐμὲ ὕβρικε δοῦναι διακρούσηται F 25 νῦν A ἐστι δὴ F 26 τἆλλα om. A 28 ἀκούητε SYP: ἀκούσητε F: [Π11] 29 τις om. Π11 τί οὖν; om. S^{ac} σὺ] οὐ Reiske

Now, all the insults which I received with regard to my **126** liturgy and my person, and my escapes from every sort of plot and ill treatment, you have heard about, men of Athens. There are also many which I pass over; for it isn't easy, I think, to mention them all. But the fact is, there's not one of his actions which is an offence against me alone. The offences done to the chorus are also offences against the tribe, one tenth of yourselves; the insults and plots which he directed at me are against the laws that preserve every one of you; and all of them are against the god, for whom I had been appointed a chorus-producer, and the majesty and divine power of holiness, whatsoever it is. So if you wish to sentence **127** him correctly to the penalty which is appropriate for his behaviour, you mustn't base your anger on the assumption that I'm speaking only for myself. Considering that at the same time the laws, the god, and the city have all been wronged together, you must impose the penalty on that basis, and regard those who support him and those who take his side as not merely advocates but also approvers of his behaviour.

Now, if, men of Athens, Meidias had shown himself other- **128** wise restrained and well-behaved, and had not offended against any other citizen, and had been bullying and violent like this towards me alone, I should in the first place have thought this a misfortune for myself; but besides that, I should have been afraid that, by pointing out that the rest of his life was well-behaved and kind, he might thereby evade punishment for his insolence towards me. But as it is, the **129** other offences he's committed against many of you are so numerous and so serious that I'm freed from that fear; but I'm afraid on the contrary that, when you hear about his frequent ill treatment of other people, you may start arguing like this: 'Well then! Is it because you've suffered worse treat-

δεινότερα ἢ τῶν ἄλλων εἷς ἕκαστος πεπονθὼς ἀγανακτεῖς;" πάντα μὲν δὴ τὰ τούτῳ πεπραγμένα οὔτ' ἂν ἐγὼ δυναίμην πρὸς ὑμᾶς εἰπεῖν, οὔτ' ἂν ὑμεῖς ὑπομείναιτ' ἀκούειν, οὐδ', εἰ τὸ παρ' ἀμφοτέρων ἡμῶν ὕδωρ ὑπάρξειεν πρὸς τὸ λοιπόν, πᾶν τό τ' ἐμὸν καὶ τὸ τούτου προστεθέν, οὐκ ἂν ἐξαρκέσειεν· 5
130 ἃ δ' ἐστὶ μέγιστα καὶ φανερώτατα, ταῦτ' ἐρῶ. μᾶλλον δ' ἐκεῖνο ποιήσω· ἀναγνώσομαι μὲν ὑμῖν, ὡς ἐμαυτῷ γέγραμμαι, πάντα τὰ ὑπομνήματα, λέξω δ' ὅ τι ἂν πρῶτον ἀκούειν βουλομένοις ὑμῖν ᾖ, τοῦτο πρῶτον, εἶθ' ἕτερον, καὶ τἄλλα τὸν αὐτὸν τρόπον, ἕως ἂν ἀκούειν βούλησθε. ἔστι δὲ ταῦτα 10 παντοδαπά, καὶ ὕβρεις πολλαὶ καὶ περὶ τοὺς οἰκείους κακουργήματα καὶ περὶ τοὺς θεοὺς ἀσεβήματα, καὶ τόπος οὐδείς ἐστιν ἐν ᾧ τοῦτον οὐ θανάτου πεποιηκότα ἄξια πολλὰ εὑρήσετε.

ΥΠΟΜΝΗΜΑΤΑ ΤΩΝ ΜΕΙΔΙΟΥ ΑΔΙΚΗΜΑΤΩΝ 15

131 Ὅσα μὲν τοίνυν, ὦ ἄνδρες δικασταί, τὸν ἀεὶ προστυχόντ' αὐτῷ πεποίηκεν, ταῦτ' ἐστίν. καὶ παραλέλοιφ' ἕτερα· οὐ γὰρ ἂν δύναιτ' οὐδεὶς εἰσάπαξ εἰπεῖν ἃ πολὺν χρόνον οὗτος ὑβρίζων συνεχῶς ἅπαντα τὸν βίον εἴργασται. ἄξιον δ' ἰδεῖν ἐφ' 20 ὅσον φρονήματος ἤδη προελήλυθεν τῷ τούτων δίκην μηδενὸς δεδωκέναι. οὐ γὰρ ἡγεῖθ', ὡς ἐμοὶ δοκεῖ, λαμπρὸν οὐδὲ νεανικὸν οὐδ' ἄξιον θαύματος ὅ τι ἄν τις πρὸς ἕνα εἷς διαπράττηται, ἀλλ' εἰ μὴ φυλὴν ὅλην καὶ βουλὴν καὶ ἔθνος προπηλακιεῖ καὶ πολλοὺς ἀθρόους ὑμῶν ἅμα ἐλᾷ, ἀβίωτον 25
132 ᾤετ' ἔσεσθαι τὸν βίον αὑτῷ. καὶ τὰ μὲν ἄλλα σιωπῶ, μυρία ἂν εἰπεῖν ἔχων· περὶ δὲ τῶν συστρατευσαμένων ἱππέων εἰς Ἀρ-

2–3 *πάντα ... ἀκούειν* Aristid. *Rh.* (Schmid) 29.23 8–10 *ἀκούειν ... ταῦτα* Π11 21–168.1 *φρονήματος ... πάντες* Π2

---

2 *δὴ ... πεπραγμένα* om. Aristid. *ἔγωγε* F Aristid. 3 *ὑπομείναιτ'* FP^{mg}_{4}: *ὑπομειτ'* S^{ac}: *ὑπομείνετ'* S^{pc}_{4}: *ὑπομενεῖτε* A: *ὑπομενεῖτ'* YP: *βούλοισθε* Aristid. *ἂν ἀκούειν* F 4 *ὑμῶν* SY *-ειε* AF^{pc} 5 *-ειε* A 6 *ἔστιν* S *δὲ* A 8 *τὰ ὑπομνήματα πάντα* A 9 *βουλομένοις ... πρῶτον* om. Π11 ut vid. *τοῦτο* om. F 15–16 titulum SFYP^{mg}_{2}: om. AP 17 *ὦ ἄνδρες ἀθηναῖοι* F *-ντα* SFYP 18 *-κε* AF^{pc} *-τί* AF^{pc} *-πα* A 19 *δύναιτ'* Cd^{pc}K: *δύναιτο* SFYP: *δύνοιτ'* A *ἅπαξ* S^{ac} 20 *δὲ* A 21 *-θε* AF^{pc}: [Π2] 22 *ἡγεῖ θ'* SYP: *ἡγεῖσθ'* TBcNc: [Π2] 23 *οὐδ'* SP: *οὐδέν· οὐδε* A: *οὐδὲ* F: *οὐδὲν* Y: [Π2] *ἄξιον εἶναι* FP^{mg}_{4}: [Π2] *θαύματος* Herwerden: *θανάτου* SAFYP: [Π2]: *αὐτοῦ* Buttmann *ὁτι τις ạ[ν* Π2: *ὅ τι ἄν τι* F: *ἐάν τι* $P^{γρ}_{4}$ 24 *]τεται* Π2 26 *-το* AF: [Π2] *ἑαυτῷ* A: [Π2] *ἂν* AFP^{mg}_{4}: *]ν* Π2: om. SYP 27 *ἔχων εἰπεῖν* AF

ment than any other individual that you're indignant?' I can't relate to you everything he's done, nor would you have the patience to listen; and if our time on both sides, all mine and his added together, were available for the rest of it, it wouldn't be enough. But I'll tell you the things which are the most serious and blatant. Or rather, this is what I'll do: I'll **130** read you all my notes, as I've written them for myself; and then I'll relate first whichever you wish to hear first, then another, and so on, as long as you wish to listen. They're of every sort—many acts of insolence, acts of improbity to his family and impiety to the gods; there's no area in which you won't find that he's committed many capital offences.

NOTES OF MEIDIAS' OFFENCES

So, men of the jury, that's a list of the things he's done to **131** anyone who came into contact with him. And I've left others out, because no one could recount in one speech all that he has perpetrated over a long period, continually behaving with insolence throughout his life. It's worth noticing what heights of arrogance he has now reached, from not having paid a penalty for any of these acts. I suppose he considered any transaction between single individuals not distinguished or macho or worthy of admiration; unless he abused a whole tribe and Boule and class, and harassed large numbers of you at once, he thought his life would not be worth living. I pass **132** over the rest, though there are thousands of things I could mention. But concerning the cavalry who served with him at

γουραν ἴστε δήπου πάντες οἷα ἐδημηγόρησεν παρ' ὑμῖν, ὅθ'
ἧκεν ἐκ Χαλκίδος, κατηγορῶν καὶ φάσκων ὄνειδος ἐξελθεῖν
τὴν στρατιὰν ταύτην τῇ πόλει· καὶ τὴν λοιδορίαν ἣν ἐλοιδορήθη Κρατίνῳ περὶ τούτων, τῷ νῦν, ὡς ἐγὼ πυνθάνομαι,
μέλλοντι βοηθεῖν αὐτῷ, μέμνησθε. τὸν δὴ τοσούτοις ἀθρόοις 5
τῶν πολιτῶν ἔχθραν ἐπ' οὐδενὶ τηλικαύτην ἀράμενον πόσῃ
133 πονηρίᾳ καὶ θρασύτητι ταῦτα χρὴ νομίζειν πράττειν; καίτοι
πότερ' εἰσὶν ὄνειδος, ὦ Μειδία, τῇ πόλει οἱ διαβάντες ἐν τάξει
καὶ τὴν σκευὴν ἔχοντες ἣν προσῆκε τοὺς ἐπὶ τοὺς πολεμίους
ἐξιόντας καὶ συμβαλουμένους τοῖς συμμάχοις, ἢ σὺ ὁ μηδὲ 10
λαχεῖν εὐχόμενος τῶν ἐξιόντων ὅτ' ἐκληροῦ, τὸν θώρακα δὲ
οὐδεπώποτ' ἐνδύς, ἐπ' ἀστράβης δὲ ὀχούμενος ἀργυρᾶς,
χλανίδας δὲ καὶ κυμβία καὶ κάδους ἔχων, ὧν ἐπελαμβάνοντο
οἱ πεντηκοστολόγοι; ταῦτα γὰρ εἰς τοὺς ὁπλίτας ἡμᾶς
134 ἀπηγγέλλετο· οὐ γὰρ εἰς ταὐτὸν ἡμεῖς τούτοις διέβημεν. εἶτα, 15
εἴ σ' ἐπὶ τούτοις ἔσκωψεν Ἀρχετίων ἤ τις ἄλλος, πάντας
ἤλαυνες; εἰ μὲν γὰρ ἐποίεις ταῦτα, ὦ Μειδία, ἅ σέ φασιν οἱ
συνιππεῖς καὶ κατηγόρεις ὡς λέγοιεν περὶ σοῦ, δικαίως κακῶς ἤκουες· καὶ γὰρ ἐκείνους καὶ τουτουσὶ καὶ ὅλην τὴν πόλιν
ἠδίκεις καὶ κατῄσχυνες. εἰ δὲ μὴ ποιοῦντός σου κατεσκεύαζόν 20
τινες καταψευδόμενοί σου, οἱ δὲ λοιποὶ τῶν στρατιωτῶν οὐκ
ἐκείνοις ἐπετίμων ἀλλὰ σοὶ ἐπέχαιρον, δῆλον ὅτι ἐκ τῶν
ἄλλων ὧν ἔζης ἄξιος αὐτοῖς ἐδόκεις εἶναι τοῦ τοιαῦτ' ἀκούειν·
σαυτὸν οὖν μετριώτερον ἐχρῆν παρέχειν, οὐκ ἐκείνους δια-
135 βάλλειν. σὺ δ' ἀπειλεῖς πᾶσιν, ἐλαύνεις πάντας· τοὺς ἄλλους 25
ἀξιοῖς ὅ τι σὺ βούλει σκοπεῖν, οὐκ αὐτὸς σκοπεῖς ὅ τι μὴ

3–10 *πόλει ... συμβαλουμένους* Π2 12 *ἐπ' ... ἀργυρᾶς* Helladius ap. Phot. *Bibl.* 533a 12–14 *ἐπ' ... πεντηκοστολόγοι* Macr. *Sat.* 5.21.8 12 *ἀργυρᾶς* Men. rhet. ap. Σ (470a Dilts), Hdn. Gr. 2.920.9 13–25 *καὶ κάδους ... ἄλλους* Π2 25 *σὺ ... πᾶσιν* *AB* 1.120.29

---

3 *στρατείαν* $A^{ac}YP^{ac}$ *ἣν* om. F 4 *τῷ νῦν* S^{sl}: om. Π2 SAFYP 5 *αὐτῷ* S 6 *εκχθρ[* Π2 *αἱράμενον* SP: [Π2] 7 *]νηρια και π[οση* Π2 8 *πότεροι* A: [Π2] 9 *-κεν* $SF^{ac}YP$ 10 *ἐξιόντας καὶ συμβαλομένους* Y^{pc}: om. Y^{ac} *μη* S^{ac} 12 *ἀργυρᾶς* Men. Helladius: *ἀργυρᾶς τῆς ἐξ Εὐβοίας* $SP_4^{\gamma\rho}$: *ἐξ Ἀργούρας τῆς Εὐβοίας* AF Hdn. Macr.: *Ἀργούρας τῆς ἐξ Εὐβοίας* YP 13 *δὲ καὶ* SFYP: *καὶ* Σ (Patm.) Macr.: om. A *κυμβεῖα* A Σ (Patm.) *καὶ κάδους* om. Macr. 14 *]λειτας* Π2 15 *ἐπηγγέλλετο* A: [Π2] *ταυτὸ* SYP: [Π2] 16 *εἴ σ'* S: *εἰς* $S_3^{\gamma\rho}$: *εἴ σε* AFYP: [Π2] *τούτου τοις* $S_3^{\gamma\rho}$ *ἔσκωψεν* $S_3^{\gamma\rho}$AFYP: *εἰσκώψομεν* S: [Π2] *ἤ*] *εἴ* F 17 *ταῦτα*] *πάντα* $Lamb^{vl}$ *ω μιδια ταυ[* Π2 18 *]ν[ι]ππης* Π2 19 *τούτους* F: [Π2] 22 *ἐπεχείρουν* S: [Π2] 23 *ἄξιος ἐδόκεις αὐτοῖς* A: *αὐτοῖς ἄξιος ἐδόκεις* YP: *αξιος[* Π2 *τοῦ τὰ* A: *τοῦ[* Π2 *-τα* AYP: [Π2] 26 *ὅ* utrumque om. S^{ac} *σοι* A

Argoura, I expect you all know what a speech he made in the Ekklesia on his return from Khalkis, criticizing it and alleging that that expeditionary force was a disgrace to Athens; and you remember the abuse he hurled at Kratinos on this subject—who today, as I am told, is going to support him! How much wickedness and rashness must we suppose prompted a man who made such a large number of his fellow-citizens all at once so intensely hostile to him, for no reason? Indeed, which is really a disgrace to Athens, Meidias: men who made the crossing in good order and with the appropriate equipment for facing their enemies and supporting their allies; or you, who prayed when you were drawing lots that you wouldn't get on to the expedition at all, who never put on your breastplate, who rode on a silver mule-chair, and who had fine cloaks and cups and flagons of wine which the two-per-cent tax collectors tried to seize? That's the information which we hoplites were given; for we didn't cross to the same point as they did. And then, because you were mocked for this by Arkhetion or someone else, you attacked the whole lot of them, did you? But if you did what your fellow-cavalrymen say you did and you accused them of saying about you, Meidias, you deserved the bad names; for you were wronging and disgracing both them and these gentlemen and the city as a whole. But if you didn't do it and people were fabricating lies against you, and the rest of the soldiers didn't object to what they said but enjoyed your discomfiture, it's obvious that you seemed to them, from the rest of your life, to deserve that sort of comment; so you ought to have behaved with more restraint, not to have accused them. But you threaten them all, you harass them all; you expect the others to consider your wishes, but you yourself

133 **134** **135**

λυπήσεις τοὺς ἄλλους ποιῶν. καὶ τὸ δὴ σχετλιώτατον καὶ μέγιστον ἔμοιγε δοκοῦν ὕβρεως εἶναι σημεῖον, τοσούτων ἀνθρώπων, ὦ μιαρὰ κεφαλή, σὺ παρελθὼν ἀθρόων κατηγόρεις· ὃ τίς οὐκ ἂν ἔφριξεν ποιῆσαι τῶν ἄλλων;

136 Τοῖς μὲν τοίνυν ἄλλοις ἅπασιν ἀνθρώποις ὁρῶ τοῖς κρινομένοις, ὦ ἄνδρες δικασταί, ἓν μὲν ἢ δύο ὄντα τἀδικήματα ἃ κατηγορεῖται, λόγους δ' ἀφθόνους τοιούτους ὑπάρχοντας "τίς ὑμῶν ἐμοί τι σύνοιδεν τοιοῦτον; τίς ὑμῶν ἐμὲ ταῦθ' ἑόρακε ποιοῦντα; οὐκ ἔστιν, ἀλλ' οὗτοι δι' ἔχθραν καταψεύδονταί μου, καταψευδομαρτυροῦμαι", τὰ τοιαῦτα· τούτῳ δ' αὖ τἀναντία τούτων. **137** πάντας γὰρ ὑμᾶς εἰδέναι νομίζω τὸν τρόπον τὸν τούτου καὶ τὴν ἀσέλγειαν καὶ τὴν ὑπερηφανίαν τοῦ βίου, καὶ πάλαι θαυμάζειν ἐνίους οἶμαι ὧν αὐτοὶ μὲν ἴσασιν, οὐκ ἀκηκόασι δὲ νῦν ἐμοῦ. πολλοὺς δὲ τῶν πεπονθότων οὐδὲ πάνθ' ὅσα ἠδίκηνται μαρτυρεῖν ἐθέλοντας ὁρῶ, τὴν βίαν καὶ τὴν φιλοπραγμοσύνην ὀρρωδοῦντας τὴν τούτου καὶ τὴν ἀφορμήν, ἥπερ ἰσχυρὸν ποιεῖ καὶ φοβερὸν τὸν κατάπτυστον τουτονί. **138** τὸ γὰρ ἐπ' ἐξουσίας καὶ πλούτου πονηρὸν εἶναι καὶ ὑβριστὴν τεῖχός ἐστι πρὸς τὸ μηδὲν ἂν αὐτὸν ἐξ ἐπιδρομῆς παθεῖν. ἐπεὶ περιαιρεθεὶς οὗτος τὰ ὄντα ἴσως μὲν οὐκ ἂν ὑβρίζοι, εἰ δ' ἄρα, ἐλάττονος ἄξιος ἔσται τοῦ μικροτάτου παρ' ὑμῖν· μάτην γὰρ λοιδορήσεται καὶ βοήσεται, δίκην δ', ἂν ἀσελγαίνῃ τι, τοῖς ἄλλοις ἡμῖν ἐξ ἴσου δώσει. **139** νῦν δ', οἶμαι, τούτου προβέβληται Πολύευκτος, Τιμοκράτης, Εὐκτήμων ὁ κονιορτός· τοιοῦτοί τινές εἰσι μισθοφόροι περὶ αὐτόν. καὶ πρὸς ἔτι ἕτεροι τούτοις, μαρτύρων συνεστῶσα ἑταιρεία, φανερῶς μὲν οὐκ ἐνοχλούντων ὑμῖν, σιγῇ δὲ τὰ ψευδῆ ῥᾳστ' ἐπινευόντων· οὓς μὰ τοὺς θεοὺς οὐδὲν ὠφελεῖσθαι νομίζω παρὰ τούτου, ἀλλὰ δεινοί τινές εἰσιν, ὦ ἄνδρες Ἀθηναῖοι,

10 καταψευδομαρτυροῦμαι Poll. 6.153, 8.31 12–17 τὸν ... ἀφορμήν Π2

3 ἀθρόον AF: ἀθρόως Σ (476 Dilts) κατηγορεῖς AF 4 -ξε AF^{pc} τῶν ἄλλων ποιῆσαι A 5 ἀνθρώποις om. A 7 κατηγοροῦνται FYP δὲ SFYP 8 -δε AF^{pc} -το $S^{pc}YP$ ἢ τίς A ἑώρακε $SAF_1^{sl}P$ 11 εἰδέναι ὑμᾶς A 12 τὸν τούτου om. SA τούτου καὶ om. P καὶ τὴν ἀσέλγειαν om. F τὴν τοῦ βίου FP^{pc} 13 οἴομαι SFYP 14 -σιν Π2 16 ὀρρωδοῦντας Taylor: ὁρῶντας SFYP: δεδιότας A: δεδοικότας $P_4^{γρ}$: δεδ̣[Π2 17 καὶ φοβερὸν ποιεῖ AF 18 τοῦτον YP πονηρὸν SFYP: θρασὺν A 19 -τιν S ἂν om. A 24 προβέβληνται AF πολύευκτος $S^{mg}AFP$: πολύευκτον Y: om. S 26 πρὸς ἔτι T: προσέτι $S_1^{pc}AFYP$: προς S^{ac} τούτοις ἕτεροι F ἑταιρεία S: ἑταιρία $S^{sl}AFYP$ 27 ἐποχλουντων S -τα AF 29 δεῖν οἵτινές F

don't consider what behaviour of your own will cause no annoyance to them. And the most outrageous thing, the one which seems to me to be the greatest evidence of insolence —you came forward, you scoundrel, and accused all that number of people at once! Who else would not have flinched from doing that?

I notice, men of the jury, that all other persons who are on **136** trial, though the offences alleged against them are only one or two, have an abundance of arguments of this kind: 'Which of you is aware of my doing anything like that? Which of you has seen me doing that? It's not so; these men, because they are my enemies, are telling lies against me, I'm the victim of false evidence'—that sort of thing. But with Meidias it's just the opposite. You all know, I think, the character of this **137** man, and the aggressiveness and arrogance of his life, and I expect some of you have been surprised not to have heard from me today facts of which you have personal knowledge. But I notice that many of the victims are unwilling even to give evidence of all the offences against them, for fear of his violence and intrusiveness, and the resources which make this abominable man strong and formidable. Unscrupulous inso- **138** lence, based on licence and wealth, is a bulwark for avoiding any injury from an attack oneself. For although, if Meidias is stripped of his property, he may perhaps not be insolent, or if in fact he is, he will not carry the slightest weight with you, will gain nothing by abuse and shouting, and will be punished equally with the rest of us for any aggressive act, at **139** present I suppose he's protected by Polyeuktos, Timokrates, and 'dusty' Euktemon; people of that sort are his paid henchmen. And there are others besides them too, a company of witnesses got together, who give you no trouble openly, but without a word readily nod their heads at his lies. Not that I think for a moment that those men make anything out of him; but some people are dreadful, men of Athens, for going

φθείρεσθαι πρὸς τοὺς πλουσίους καὶ παρεῖναι καὶ μαρτυρεῖν.
140 πάντα δὲ ταῦτ', οἶμαι, φοβερά ἐστιν τῶν ἄλλων ὑμῶν ἑκάστῳ
καθ' ἑαυτὸν ὅπως δύναται ζῶντι. οὗπερ ἕνεκα συλλέγεσθε
ὑμεῖς, ἵνα, ὧν καθ' ἕνα ἐστὶν ἕκαστος ὑμῶν ἐλάττων ἢ φίλοις
ἢ τοῖς οὖσιν ἢ τῶν ἄλλων τινί, τούτων συλλεγέντες ἑκάστου 5
κρείττους τε γίγνησθε καὶ παύητε τὴν ὕβριν.
141 Τάχα τοίνυν καὶ τοιοῦτός τις ἥξει πρὸς ὑμᾶς λόγος, "τί δὴ
τὰ καὶ τὰ πεπονθὼς ὁ δεῖνα οὐκ ἐλάμβανεν δίκην παρ' ἐμοῦ;"
ἢ "τί δὴ" πάλιν ἄλλον ἴσως τινὰ τῶν ἠδικημένων ὀνομάζων.
ἐγὼ δέ, δι' ἃς μὲν προφάσεις ἕκαστος ἀφίσταται τοῦ βοηθεῖν 10
αὑτῷ, πάντας ὑμᾶς εἰδέναι νομίζω· καὶ γὰρ ἀσχολία καὶ
ἀπραγμοσύνη καὶ τὸ μὴ δύνασθαι λέγειν καὶ ἀπορία καὶ μυρία
142 ἐστὶν αἴτια. προσήκειν μέντοι τούτῳ μὴ ταῦτα λέγειν ἡγοῦμαι νυνί, ἀλλ' ὡς οὐ πεποίηκέν τι τούτων ὧν αὐτοῦ κατηγόρηκα διδάσκειν, ἐὰν δὲ μὴ δύνηται, διὰ ταῦτ' ἀπολωλέναι πολὺ 15
μᾶλλον. εἰ γὰρ τηλικοῦτός τίς ἐστιν ὥστε τοιαῦτα ποιῶν
δύνασθαι καθ' ἕνα ἕκαστον ἡμῶν ἀποστερεῖν τοῦ δίκης παρ'
αὐτοῦ τυχεῖν, κοινῇ νῦν, ἐπειδήπερ εἴληπται, πᾶσιν ὑπὲρ
ἁπάντων ἐστὶ τιμωρητέος ὡς κοινὸς ἐχθρὸς τῇ πολιτείᾳ.
143 Λέγεται τοίνυν ποτὲ ἐν τῇ πόλει κατὰ τὴν παλαιὰν ἐκείνην 20
εὐδαιμονίαν Ἀλκιβιάδης γενέσθαι, ᾧ σκέψασθε τίνων εὐεργεσιῶν ὑπαρχουσῶν καὶ ποίων τινῶν πρὸς τὸν δῆμον πῶς
ἐχρήσανθ' ὑμῶν οἱ πρόγονοι, ἐπειδὴ βδελυρὸς καὶ ὑβριστὴς
ᾤετο δεῖν εἶναι. καὶ οὐκ ἀπεικάσαι δήπου Μειδίαν Ἀλκιβιάδῃ
βουλόμενος τούτου μέμνημαι τοῦ λόγου (οὐχ οὕτως εἰμὶ 25
ἄφρων οὐδ' ἀπόπληκτος ἐγώ), ἀλλ' ἵν' εἰδῆθ' ὑμεῖς, ὦ ἄνδρες

20–1 λέγεται ... Ἀλκιβιάδης Hermog. 268.17 20–2 λέγεται ... ὑπαρχ-ουσῶν Aps. *Rh.* (Hammer) 280.5, 282.1, 282.9 20–174.3 λέγεται ... φέρειν schol. Hermog. (Walz 4) 539.1

---

2 -τα A -τι AF ἑκάστῳ ἀνθρώπων P^{ac} 3 συλλέγεσθαι Y^{ac}
5 συλλεγέντας A 6 τε om. A γίνησθε A: γίγνεσθε F 7 τοίνυν ἴσως A
8 -νε AF 11 ἑαυτῷ F 13 αἴτια SFYP: ἕτερα $AF_1^{\gamma\rho}P_4^{\gamma\rho}$ μὲν Y^{ac}
14 νῦν F -κέ $AF^{pc}F_1^{\gamma\rho}P_4^{\gamma\rho}$ τι τούτων ὧν SFYP: ταῦτ' ἐφ' οἷς $AP_4^{\gamma\rho}$: ταῦθ' ἐφ' οἷς $F_1^{\gamma\rho}$ 16 μᾶλλόν ἐστι δίκαιος $S_3^{sl}AFP_4^{\gamma\rho}$ 17 ἡμῶν ἕκ-αστον F 18 τυγχάνειν $AF_1^{\gamma\rho}P_4^{\gamma\rho}$ 19 ἐστὶ om. Y 20 λέγε S^{ac}
τοίνυν om. Hermog. Aps. 280.5, 282.9 ποτὲ om. Aps. ἐν τῇ πόλει om. Hermog. Aps. schol. Hermog. 20–1 ἐπὶ τῆς παλαιᾶς ἐκείνης εὐδαιμονίας Hermog. 21 Ἀλκιβιάδην schol. Hermog. 21–2 εὐεργεσιῶν ὑπαρχουσῶν] ὑπαρχόντων Aps. 22 ποίων τιμῶν $AF_1^{\gamma\rho}P_4^{\gamma\rho}$: οἵων δὴ τότε schol. Hermog.
24 δήπου om. schol. Hermog. 25 μέμνηται schol. Hermog. οὐχ] οὐδ' F
26 ἐγώ om. schol. Hermog. -να A -τε SFYP 26–174.1 ὑμεῖς et καὶ γνῶθ' et οὔτ' et οὔτ' ἔσται om. schol. Hermog.

to the rich to be corrupted and giving them attendance and testimony. All this, I suppose, is frightening to each one of the **140** rest of you, living individually as best you can. That's why you should unite: individually each of you is weaker than they are, either in friends or in resources or in something else; but united you'll be stronger than each of them and you'll put a stop to their insolence.

Perhaps you'll also be given this sort of argument: 'Why, **141** pray, did so-and-so, who has suffered such-and-such, not prosecute me?' or 'Why, pray ...', giving next perhaps the name of another of the victims. I think you all know the excuses which an individual has for refraining from standing up for himself: lack of time, aversion to public business, inability to speak, lack of means—thousands of things are responsible. But that's not what he should be saying today, in **142** my opinion; he should be demonstrating that he hasn't done any of the things of which I've accused him, and if he can't, that's all the more reason for him to go down. If a man is so mighty that, when he behaves in this way, he can prevent each of us singly from getting justice from him, now, since he is in our grasp, he must be punished jointly by all for all, as a common enemy of the state.

We are told that there was in Athens during its golden age **143** a man named Alkibiades. Just consider what public services he had to his credit and of what kind they were, and how your ancestors dealt with him when he thought fit to be unpleasant and insolent. Not that I wish to compare Meidias to Alkibiades, of course; I'm not as senseless or demented as that! But I want you to be fully aware, men of Athens, that

Ἀθηναῖοι, καὶ γνῶθ' ὅτι οὐδὲν οὔτ' ἔστιν οὔτ' ἔσται, οὐ γένος, οὐ πλοῦτος, οὐ δύναμις, ὅ τι τοῖς πολλοῖς ὑμῖν, ἂν ὕβρις **144** προσῇ, προσήκει φέρειν. ἐκεῖνος γάρ, ὦ ἄνδρες Ἀθηναῖοι, λέγεται πρὸς πατρὸς μὲν Ἀλκμεωνιδῶν εἶναι (τούτους δέ φασιν ὑπὸ τῶν τυράννων ὑπὲρ τοῦ δήμου στασιάζοντας ἐκπε- 5 σεῖν, καὶ δανεισαμένους χρήματ' ἐκ Δελφῶν ἐλευθερῶσαι τὴν πόλιν καὶ τοὺς Πεισιστράτου παῖδας ἐκβαλεῖν), πρὸς δὲ μητρὸς Ἱππονίκου καὶ ταύτης τῆς οἰκίας, οἷς ὑπάρχουσιν **145** πολλαὶ καὶ μεγάλαι πρὸς τὸν δῆμον εὐεργεσίαι. οὐ μόνον δὲ ταῦθ' ὑπῆρχεν αὐτῷ, ἀλλὰ καὶ αὐτὸς ὑπὲρ τοῦ δήμου θέμενος 10 τὰ ὅπλα δὶς μὲν ἐν Σάμῳ, τρίτον δ' ἐν αὐτῇ τῇ πόλει, τῷ σώματι τὴν εὔνοιαν, οὐ χρήμασιν οὐδὲ λόγοις ἐνεδείξατο τῇ πατρίδι. ἔτι δὲ ἵππων Ὀλυμπίασιν ἀγῶνες ὑπῆρχον αὐτῷ καὶ νῖκαι καὶ στέφανοι· καὶ στρατηγὸς ἄριστος καὶ λέγειν ἐδόκει **146** πάντων, ὥς φασιν, εἶναι δεινότατος. ἀλλ' ὅμως οἱ κατ' ἐκεῖ- 15 νον ὑμέτεροι πρόγονοι οὐδενὸς τούτων αὐτῷ συνεχώρησαν ὑβρίζειν αὐτούς, ἀλλὰ ποιήσαντες φυγάδα ἐξέβαλον· καὶ Λακεδαιμονίων ὄντων ἰσχυρῶν τότε, καὶ Δεκέλειαν ἑαυτοῖς ἐπιτειχισθῆναι καὶ τὰς ναῦς ἁλῶναι καὶ πάντα ὑπέμειναν, ὁτιοῦν ἄκοντες παθεῖν κάλλιον εἶναι νομίζοντες ἢ ἑκόντες 20 **147** ὑβρίζεσθαι συγχωρῆσαι. καίτοι τί τοσοῦτον ἐκεῖνος ὕβρισεν, ἡλίκον οὗτος νῦν ἐξελήλεγκται; Ταυρέαν ἐπάταξεν χορηγοῦντα ἐπὶ κόρρης. ἔστω ταῦτα, ἀλλὰ χορηγῶν γε χορηγοῦντα τοῦτ' ἐποίησεν, οὔπω τόνδε τὸν νόμον παραβαίνων· οὐ γὰρ ἔκειτό πω. εἷρξεν Ἀγάθαρχον τὸν γραφέα· καὶ γὰρ ταῦτα 25 λέγουσιν. λαβών γέ τι πλημμελοῦντα, ὥς φασιν· ὅπερ οὐδ' ὀνειδίζειν ἄξιον. τοὺς Ἑρμᾶς περιέκοπτεν. ἅπαντα μέν, οἶμαι, τἀσεβήματα τῆς αὐτῆς ὀργῆς δίκαιον ἀξιοῦν· τὸ δ'

10–11 *θέμενος τὰ ὅπλα* Harp. s.v. 25–6 *πω ... ὅπερ* Π9 25 *εἷρξεν ... γραφέα* *Lex. Cant.* s.v. *εἰργμοῦ* 28–176.2 *τὸ ... διαφέρει* schol. Hermog. (Walz 4) 546.8

---

1 *οὔτ'* (ante *ἔστιν*) AFP$_4^{\gamma\rho}$: *ἔτ'* SYP 2 *ὅ τι* SYP: *ὃ* AF schol. Hermog. *ὑμῖν* SF$_1^{sl}$: *ὑμῶν* AFYP schol. Hermog. 3 *προσῇ*] *ἢ̂* F 4 *ἀλκμαιονιδῶν* A 5 *ὑπὲρ*] *ὑπὸ* F *στασιάσαντας* FP$_4^{\gamma\rho}$ 6 *-τα* SFYP 7 *τοὺς* SAYP: *τοῦ* F 8 *ταύτης δὴ* A *συγγενὴς* (post *οἰκίας*) S^{sl}: *συγγενεῖς* F$_1^{\gamma\rho}$ *οἷς* S: *ἧς* AFYP *-σι* AF 11 *δὲ* A 12 *-σι* S 13 *ἱππέων* FYP^{ac} 14 *καὶ στέφανοι* S$_2^{mg}$FP$_4^{mg}$: om. SAYP 15–16 *ἐκεῖνον τὸν χρόνον ἡμέτεροι* FYP 17 *αὑτούς* Lamb: *αυτούς* SA: *αὐτούς* FYP *ἄτιμον καὶ* (ante *φυγάδα*) S$_2^{sl}$F$_1^{\gamma\rho}$P$_4^{\gamma\rho}$ 18 *ἰσχυρῶν ὄντων* F *αυτοῖς* A: *αὐτοῖς* F 19 *καὶ* (ante *τὰς*) bis S 21 *συγχωρῆσαι* del. Cobet *ὕβρισεν ἐκεῖνος* A 22 *νῦν οὗτος* A *-ξε* AF^{pc} 23 *γε* om. F 25 *ἀγάθαργον* A 26 *γ' ἔτι* Y *φα[............]τεροιοπε̣[* Π9 *-δὲ* SFYP

there is not and never will be anything—not birth, nor wealth, nor influence—which you, the majority, should tolerate if insolence is combined with it. Alkibiades, we are told, **144** on his father's side was descended from the Alkmeonids, who are said to have been exiled by the tyrants for championing the people in civil strife, and to have borrowed money from Delphi, liberated the city, and expelled the sons of Peisistratos. On his mother's side he was descended from Hipponikos and that family, who have to their credit many substantial public services. In addition to those distinctions, he himself **145** bore arms in defence of the people, twice at Samos and a third time at Athens itself, and so displayed his patriotism by personal service, not just by money or speeches. There were also his horse-races at Olympia and victories and crowns; and he was considered an excellent general and the cleverest of all speakers, they say. Nevertheless, your ancestors in his **146** time did not concede, in return for any of these things, that he should treat them insolently; they turned him out and sent him into exile. Despite the power of the Spartans at that time, they put up with the fortification of Dekeleia against them, the capture of the navy, and everything else, because they considered it more honourable to suffer anything whatever involuntarily, than voluntarily to tolerate insolence. Yet **147** what insolence did Alkibiades commit that was as serious as what Meidias has now been proved to have committed? He struck Taureas, a chorus-producer, on the face. All right, but that was one chorus-producer striking another, and not in contravention of this law, which had not yet been passed. He imprisoned Agatharkhos the painter; they say he did that too. Yes, but it was because he caught him in some transgression, they say; that was not even reprehensible. He mutilated the Hermai. Well, all acts of impiety, I suppose, justify the

ὅλως ἀφανίζειν ἱερὰ ἔσθ' ὅ τι τοῦ κόπτειν {τοὺς Ἑρμᾶς}
διαφέρει· οὐκοῦν οὗτος ἐξελήλεγκται τοῦτο ποιῶν.

148 Ἀντιθῶμεν δὴ τίς ὢν καὶ τίσι ταῦτ' ἐποίησεν δακνόμενος.
μὴ τοίνυν ὑμῖν, πρὸς τῷ μὴ καλόν, μηδὲ θεμιτὸν νομίζετε,
ἄνδρες δικασταί, μηδ' ὅσιον εἶναι τοιούτων ἀνδρῶν οὖσιν 5
ἀπογόνοις, πονηρὸν καὶ βίαιον καὶ ὑβριστὴν λαβοῦσιν
ἄνθρωπον καὶ μηδένα μηδαμόθεν, συγγνώμης ἢ φιλανθρωπίας
ἢ χάριτός τινος ἀξιῶσαι. τίνος γὰρ ἕνεκα; τῶν στρατηγιῶν.
ἀλλ' οὐδὲ καθ' αὑτὸν στρατιώτης οὗτός γε οὐδενός ἐστ' ἄξιος,
μή τί γε τῶν ἄλλων ἡγεμών. ἀλλὰ τῶν λόγων. ἐν οἷς κοινῇ μὲν 10
οὐδὲν πώποτ' εἶπεν ἀγαθόν, κακῶς δὲ ἰδίᾳ πάντας ἀνθρώπους

149 λέγει. γένους ἕνεκα νὴ Δία. καὶ τίς οὐκ οἶδεν ὑμῶν τὰς
ἀπορρήτους, ὥσπερ ἐν τραγῳδίᾳ, τὰς τούτου γονάς; ᾧ δύο τὰ
ἐναντιώτατα συμβέβηκεν εἶναι· ἡ μὲν γὰρ ὡς ἀληθῶς μήτηρ,
ἡ τεκοῦσα αὐτόν, πλεῖστον ἁπάντων ἀνθρώπων εἶχε νοῦν, ἡ 15
δὲ δοκοῦσα καὶ ὑποβαλομένη πασῶν ἦν ἀνοητοτάτη γυναι-
κῶν. σημεῖον δέ· ἡ μὲν γὰρ ἀπέδοτο εὐθὺς γενόμενον, ἡ δ'
ἐξὸν αὐτῇ βελτίω πρίασθαι ταύτης τῆς τιμῆς τοῦτον ἠγόρα-

150 σεν. καὶ γάρ τοι διὰ τοῦτο τῶν οὐ προσηκόντων ἀγαθῶν
κύριος γεγονώς, καὶ πατρίδος τετυχηκὼς ἣ νόμοις τῶν 20
ἁπασῶν πόλεων μάλιστ' οἰκεῖσθαι δοκεῖ, οὐδένα οἶμαι τρόπον
φέρειν οὐδὲ χρῆσθαι τούτοις δύναται, ἀλλὰ τὸ τῆς φύσεως
βάρβαρον ἀληθῶς καὶ θεοῖς ἐχθρὸν ἕλκει καὶ βιάζεται, καὶ
φανερὸν ποιεῖ τοῖς παροῦσιν ὥσπερ ἀλλοτρίοις, ὅπερ ἐστίν,
αὐτὸν χρώμενον. 25

151 Τοσούτων τοίνυν καὶ τοιούτων ὄντων ἃ τῷ βδελυρῷ τούτῳ
καὶ ἀναιδεῖ βεβίωται, ἔνιοί μοι προσιόντες, ὦ ἄνδρες δικασταί, τῶν χρωμένων αὐτῷ, παραινοῦντες ἀπαλλαγῆναι καὶ
καθυφεῖναι τὸν ἀγῶνα τουτονί, ἐπειδή με μὴ πείθοιεν, ὡς μὲν

1 *ἱερὰ* S: *ἱερὰν ἐσθῆτα* S$^{sl}_{2}$AFYP schol. Hermog. *ὅτε* S$^{sl}_{2}$ *κόπτειν* S: *περικόπτειν* AFYP schol. Hermog. *τοὺς Ἑρμᾶς* del. Dobree 2 *οὔκουν* Y 3 *δὲ* A -*τα* S *ἐποίησεν δακνόμενος* MacDowell: *ἐνδεικνύμενος* SAFYP *ἐπήρθη ταῦτα ποιεῖν* (post *ἐνδεικνύμενος*) S$^{mg}_{2}$ 4 *μήτ' οὖν ὑμεῖς* S$^{\gamma\rho}_{3}$ *ὑμῖν* post *καλόν* A: om. F *τῷ*] *τὸ* Σ (511 Dilts) 5 *ὦ ἄνδρες δικασταί* A^{cp} 9 *γε* om. F 10 *γε* om. SYP 13 *δύ'* S *τὰ* om. SA 14 *γὰρ* om. Y^{ac} 15 *νῦν* F^{ac}: [P^{ac}] 16 *ὑπολαμβανομένη* A: *ὑποβαλλομένη* YP^{ac} *ἀνοητάτη* S^{ac}Y^{ac} 17 *γενομενη δ'* S^{ac}: *γενομένωι· ἡ δ'* YP^{ac} 18 *ταύτης τῆς* SYP: *τῆς ἴσης* AFP$^{\gamma\rho}_{4}$ 19 -*σε* AF^{pc} *τῶν* om. AFP$^{\gamma\rho}_{4}$ 20 *τετυχηκὼς* AFP$^{pc}_{4}$: *τετευχὼς* SY: [P^{ac}] 21 -*τα* AF 22 *οὐδὲ*] *οὐδένα* S^{ac} 23 *ὡς ἀληθῶς βάρβαρον* AF: *βάρβαρον ὡς ἀληθῶς* P$^{pc}_{4}$: *ἀληθῶς βάρβαρον* Σ (519 Dilts) 26 *ἃ* om. SF^{pc}YP 27 *ὧν βεβίωται* SFYP 27–8 *ὦ ἄνδρες Ἀθηναῖοι* A^{cp} 29 *τοῦτον* F

same anger; but total destruction of sacred things is rather different from mutilation, and that's what Meidias has been proved to have been doing.

148 Let's compare him with Alkibiades, then: who is he, and what provoked him to do these things? Men of the jury, you must consider that, besides not being praiseworthy, it's not even in accord with morality or religion for you, descended from such distinguished forefathers, when you have in your hands a person who is bad, violent, and insolent, a nobody and son of nobody, to grant him any pardon, kindness, or favour. What reason for it is there? His services as a general? But the man isn't even any use as a private soldier, much less as a commander of others. His speeches? Speeches in which he never said anything for the public good, but abuses everyone individually! **149** His high birth, do you say? Well, every one of you knows the secret; it's like something in a tragedy, this man's genesis. Two complete opposites are involved in his case: his real mother, the one who gave birth to him, was the most sensible person in the world, while his supposed mother, who took him as her own child, was the stupidest of all women. Evidence: the former sold him as soon as he was born; the latter could have bought a better one at the price, and this is the one she purchased! **150** And so for this reason he has got possession of wealth which doesn't belong to him, and has become a citizen of a city where the rule of law probably prevails more than in any other; and I suppose he's completely unable to tolerate or make use of these circumstances, but the truly barbarian and devilish part of his nature is hustling and violent, and makes it obvious that he uses what he has as if it were not his own—as indeed is the case.

151 So you see the range and character of the life led by this loathsome, shameless man. Yet some of his friends came to me, men of the jury, advising me to withdraw and drop this case. When I didn't agree, they didn't go so far as to deny

οὐ πολλὰ καὶ δεινὰ πεποίηκεν οὗτος καὶ δίκην ἥντινοῦν ἂν
δοίη δικαίως τῶν πεπραγμένων, οὐκ ἐτόλμων λέγειν, ἐπὶ
ταῦτα δ' ἀπήντων ὡς "ἥλωκεν ἤδη καὶ κατεψήφισται· τίνος
τιμήσειν αὐτῷ προσδοκᾷς τὸ δικαστήριον; οὐχ ὁρᾷς ὅτι
πλουτεῖ καὶ τριηραρχίας ἐρεῖ καὶ λειτουργίας; σκόπει δὴ μὴ 5
τούτοις αὐτὸν ἐξαιτήσηται, καὶ ἐλάττω πολὺ τῇ πόλει κατα-
152 θεὶς ἢ ὅσα σοι δίδωσι καταγελάσῃ." ἐγὼ δὲ πρῶτον μὲν οὐδὲν
ἀγεννὲς ὑμῶν καταγιγνώσκω, οὐδ' ὑπολαμβάνω τιμήσειν
οὐδενὸς ἐλάττονος τούτῳ ἢ ὅσον καταθεὶς οὑτοσὶ παύσεται
τῆς ὕβρεως· τοῦτο δ' ἐστὶ μάλιστα μὲν θάνατος, εἰ δὲ μή, 10
πάντα τὰ ὄντα ἀφελέσθαι. ἔπειθ' ὑπὲρ τῶν τούτου λειτουργιῶν καὶ τῶν τριηραρχιῶν καὶ τῶν τοιούτων λόγων ὡδὶ γιγνώ-
153 σκω. εἰ μέν ἐστιν, ὦ ἄνδρες Ἀθηναῖοι, τὸ λειτουργεῖν τοῦτο,
τὸ ἐν ὑμῖν λέγειν ἐν ἁπάσαις ταῖς ἐκκλησίαις καὶ πανταχοῦ
"ἡμεῖς οἱ λειτουργοῦντες, ἡμεῖς οἱ προεισφέροντες ὑμῖν, 15
ἡμεῖς οἱ πλούσιοί ἐσμεν", εἰ τὸ τὰ τοιαῦτα λέγειν, τοῦτ' ἔστι
λειτουργεῖν, ὁμολογῶ Μειδίαν ἁπάντων τῶν ἐν τῇ πόλει
λαμπρότατον γεγενῆσθαι· ἀποκναίει γὰρ ἀηδίᾳ δήπου καὶ
154 ἀναισθησίᾳ καθ' ἑκάστην τὴν ἐκκλησίαν ταῦτα λέγων. εἰ
μέντοι τί ποτ' ἐστὶν ἃ λειτουργεῖ τῇ ἀληθείᾳ δεῖ σκοπεῖν, ἐγὼ 20
πρὸς ὑμᾶς ἐρῶ· καὶ θεάσασθε ὡς δικαίως αὐτὸν ἐξετάσω,
πρὸς ἐμαυτὸν κρίνων.

Οὗτος, ὦ ἄνδρες Ἀθηναῖοι, γεγονὼς ἔτη περὶ πεντήκοντα
ἴσως ἢ μικρὸν ἐλάττω οὐδὲν ἐμοῦ πλείους λειτουργίας ὑμῖν
λελειτούργηκεν, ὃς †δύο† καὶ τριάκοντα ἔτη γέγονα. κἀγὼ 25
μὲν κατ' ἐκείνους τοὺς χρόνους ἐτριηράρχουν εὐθὺς ἐκ παίδων
ἐξελθών, ὅτε σύνδυο ἦμεν οἱ τριήραρχοι καὶ τἀναλώματα

6–8 ἐξαιτήσηται ... τιμήσειν Π7 7–8 ἐγὼ ... καταγιγνώσκω *AB* 1.150.16
17–26 Μειδίαν ... κατ' Π7 21–2 ὡς ... κρίνων lo. Sard. 181.1

---

1 δεινὰ καὶ πολλὰ A οὑτοσὶ AFP^{pc}_{4} 1–2 δικαίως ἂν δοίη A: ἂν δικαίως δοίη F 6 ἐξαιτήσηται AFP^{sl}_{4}: ἐξαιτήσεται $SF^{sl}_{1}YP$:]ηται Π7 6–7 καταθείσῃ $S^{ac}YP^{ac}$ 7 καταγελάσει F: [Π7P^{ac}] 8 καταγιν- A *AB*: [Π7] 9 ἢ ὅσον] ἧς ὃν S καθεὶς A οὑτοσὶ FYP: οὗτος SA 10 -τιν S 12 τῶν (ante τριηραρχιῶν) om. AF γιν- A 14 ἐν ἁπάσαις λέγειν Y 15 ἡμῖν YP^{ac} 16 τὰ om. Σ (525 Dilts) ἔστι τὸ F 18 γενέσθαι A 19 τὴν om. AFP^{pc}_{4} 20 εστι α λιτουργει Π7 ἀληθείαι μειδίας A 21 ως· και ως Π7 21–2 ἐξετάσεις ὡς πρὸς Sard. 22 κρίνω S^{ac} 23 -κοντ Π7 24 μεικρον Π7 ἐλάττον S: [Π7]: ἔλαττον Bekker 24–5 λιτουργιας υμεν λελιτουργηκεν Π7 25 δύο Π7 SAFYP: δ' Goodwin: ἕπτα (sc. *ΔΔΔΠΙΙ* pro *ΔΔΔΙΙ*) MacDowell ετηι Π7 καγωι Π7 27 ἦμεν $SAF^{sl}_{1}YP$: ἦσαν $FP^{\gamma\rho}_{4}$ τὰ ἀνα- AF

that he'd committed many serious offences justifying any penalty you can think of, but they resorted to the argument 'Suppose he's been convicted already, and the vote has gone against him: what penalty do you expect the court to award? Don't you see he's rich, and he'll talk about trierarchies and liturgies? Look out that he doesn't beg himself off in that way, and have the last laugh by paying to the city much less than the amount he's now offering to you!' But as far as I'm **152** concerned, in the first place I don't have a low opinion of you, or suppose that you'll fix his penalty at anything less than a payment which will make the man cease his insolence; that is preferably death, or else loss of all his property. Secondly, with regard to his liturgies and his trierarchies and his talk of that sort, this is my opinion. If performing liturgies, **153** men of Athens, consists simply of saying to you, at all the meetings of the Ekklesia and everywhere, 'We are the men who perform liturgies! We are the men who pay *proeisphora* for you! We are the rich men!'—if saying that sort of thing is performing liturgies, I concede that Meidias is the most distinguished man in Athens; he wears us out at every Ekklesia with this sickening and tactless talk. If, however, we need to **154** consider what his liturgies truly amount to, I will tell you. Notice how fair my examination of him will be: I shall compare him with myself.

Meidias, men of Athens, though he is perhaps fifty years old or a little less, has performed no more liturgies for you than I, who am thirty-†two†. Besides, I began service as a trierarch as soon as I came of age, in the days when we served in pairs, and all the expenses came out of our own properties,

πάντα ἐκ τῶν ἰδίων οἴκων καὶ τὰς ναῦς ἐπληρούμεθ' αὐτοί·
155 οὗτος δ', ὅτε μὲν κατὰ ταύτην τὴν ἡλικίαν ἦν ἣν ἐγὼ νῦν,
οὐδέπω λειτουργεῖν ἤρχετο, τηνικαῦτα δὲ τοῦ πράγματος
ἧπται, ὅτε πρῶτον μὲν διακοσίους καὶ χιλίους πεποιήκατε
συντελεῖς ὑμεῖς, παρ' ὧν εἰσπραττόμενοι τάλαντον ταλάντου 5
μισθοῦσι τὰς τριηραρχίας οὗτοι, εἶτα πληρώματα ἡ πόλις
παρέχει καὶ σκεύη δίδωσιν, ὥστ' αὐτῶν ἐνίοις τῇ ἀληθείᾳ τὸ
μηδὲν ἀναλῶσαι καὶ δοκεῖν λελειτουργηκέναι καὶ τῶν ἄλλων
156 λειτουργιῶν ἀτελεῖς γεγενῆσθαι περίεστιν. ἀλλὰ μὴν τί ἄλλο;
τραγῳδοῖς κεχορήγηκέν ποθ' οὗτος, ἐγὼ δὲ {αὐληταῖς} ἀν- 10
δράσιν· καὶ ὅτι τοῦτο τὸ ἀνάλωμα ἐκείνης τῆς δαπάνης πλέον
ἐστὶν πολλῷ, οὐδεὶς ἀγνοεῖ δήπου. κἀγὼ μὲν ἐθελοντὴς νῦν,
οὗτος δὲ καταστὰς ἐξ ἀντιδόσεως τότε, οὗ χάριν οὐδεμίαν
δήπου αὐτῷ δικαίως ἄν τις ἔχοι. τί ἔτι; εἱστίακα τὴν φυλὴν
ἐγὼ καὶ Παναθηναίοις κεχορήγηκα, οὗτος δ' οὐδέτερα. 15
157 ἡγεμὼν συμμορίας ὑμῖν ἐγενόμην ἐγὼ ἔτη δέκα, ἴσον Φορμίωνι καὶ Λυσιθείδῃ καὶ Καλλαίσχρῳ καὶ τοῖς πλουσιωτάτοις
εἰσφέρων, οὐκ ἀπὸ ὑπαρχούσης οὐσίας (ὑπὸ γὰρ τῶν
ἐπιτρόπων ἀπεστερήμην), ἀλλ' ἀπὸ τῆς δόξης ὧν ὁ πατήρ μοι
κατέλιπεν καὶ ὧν δίκαιον ἦν με δοκιμασθέντα κομίσασθαι. 20
ἐγὼ μὲν οὖν οὕτως ὑμῖν προσενήνεγμαι, Μειδίας δὲ πῶς;
οὐδέπω καὶ τήμερον συμμορίας ἡγεμὼν γέγονεν, οὐδὲν τῶν
πατρῴων ἀποστερηθεὶς ὑπ' οὐδενός, ἀλλὰ παρὰ τοῦ πατρὸς
πολλὴν οὐσίαν παραλαβών.
158 Τίς οὖν ἐστιν ἡ λαμπρότης, ἢ τίνες αἱ λειτουργίαι καὶ τὰ 25
σεμνὰ ἀναλώματα τούτου; ἐγὼ μὲν γὰρ οὐχ ὁρῶ, πλὴν εἰ
ταῦτά τις θεωρεῖ· οἰκίαν ᾠκοδόμηκεν Ἐλευσῖνι τοσαύτην
ὥστε πᾶσιν ἐπισκοτεῖν τοῖς ἐν τῷ τόπῳ, καὶ εἰς μυστήρια τὴν
γυναῖκ' ἄγει, κἂν ἄλλοσέ ποι βούληται, ἐπὶ τοῦ λευκοῦ ζεύγους τοῦ ἐκ Σικυῶνος, καὶ τρεῖς ἀκολούθους ἢ τέτταρας 30

1 οἴκων SFYP: ἐδαπανῶμεν A ἐπληροῦμεν AFP^{mg}_4 αὐτοί A
2 δὲ SFYP ἦν om. SYP^{ac} 4 ἧπται SAFYP: ἧρκται $S^{γρ}_3F^{γρ}_1P^{γρ}_4$
5 τάλαντον AFYP: ταλάντων S 6 -σιν A ἔπειτα A: εἶτα τὰ $S^{pc}F$ 7 με-
σκευη S^{ac} 8 μηδένα $S^{ac}YP^{ac}$ 9 ἀτελέσι FP^{mg}_4 10 -κέ AF^{pc}
-τε SFYP αὐληταῖς del. MacDowell 11 -σι AF^{pc} πλεῖον FYP
11–12 πολλῷ ante πλεῖον F 12 -τι AYP 14 αὐτῷ om. S: ante
δήπου FYP τί ἔτι; om. A ἑστίακα A 17 λυσιθίδῃ AF^{ac} 18 ἀπὸ
τῆς $Lamb^{vl}$ 21 οὖν om. F 22 ἡγεμὼν συμμορίας A 25 ἐστιν
om. F 26 -ματα τὰ A γὰρ om. A 27 -κε S ἐν ἐλευσῖνι AF
29 -κα AFYP ὅποι S^{sl} βούλεται A 30–182.1 διὰ … σοβεῖ ante τρεῖς
Σ 24.200 (361a Dilts)

and we found crews for the ships ourselves; but when Meidias **155** was as old as I am now, he was not yet beginning to perform liturgies. He has only put his hand to it at the time when, in the first place, you have made twelve hundred men contributors, from whom these fellows collect a talent, and then let contracts for the trierarchies at a talent! Besides, the state provides crews and supplies gear, so that some of them in fact end up spending nothing, and are considered to have performed a liturgy, and so get exemption from the other liturgies. Well, what is there besides? He has been a chorus- **156** producer for tragedies, and I for a men's chorus. The latter expense is much more than the former, as surely everyone knows; and I was a volunteer on this occasion, whereas he was appointed that time as a result of a challenge to exchange property; and surely he can't fairly be given credit for that. What else? I have feasted my tribe and served as chorus-producer at the Panathenaia; he has done neither. I **157** was leader of a symmory for you for ten years, contributing as much as Phormion and Lysitheides and Kallaiskhros and the richest men—not from property which I had (because my guardians had deprived me of it) but from the repute of what my father left me and I was entitled to recover when I came of age. So that's how I have conducted myself towards you: but how has Meidias? Never to this day has he been leader of a symmory, though he was certainly not deprived of his inheritance by anyone, but received substantial property from his father.

So what is his distinction? What are his liturgies, and his **158** impressive expenditures? I can't see any—unless these are the items that one considers: he has built a house at Eleusis so big that it overshadows everyone in the neighbourhood; he takes his wife to celebrations of mysteries, and anywhere else he wishes, in a carriage drawn by the pair of white horses he got from Sikyon; and he clears a way for himself through the

αὐτὸς ἔχων διὰ τῆς ἀγορᾶς σοβεῖ, κυμβία καὶ ῥυτὰ καὶ φιάλας
159 ὀνομάζων οὕτως ὥστε τοὺς παριόντας ἀκούειν. ἐγὼ δ', ὅσα
μὲν τῆς ἰδίας τρυφῆς ἕνεκα Μειδίας καὶ περιουσίας κτᾶται,
οὐκ οἶδ' ὅ τι τοὺς πολλοὺς ὑμῶν ὠφελεῖ· ἃ δ' ἐπαιρόμενος
τούτοις ὑβρίζει, ἐπὶ πολλοὺς καὶ τοὺς τυχόντας ἡμῶν ἀφικ- 5
νούμενα ὁρῶ. οὐ δὴ δεῖ τὰ τοιαῦτα ἑκάστοτε τιμᾶν οὐδὲ θαυμάζειν ὑμᾶς, οὐδὲ τὴν φιλοτιμίαν ἐκ τούτων κρίνειν, εἴ τις
οἰκοδομεῖ λαμπρῶς ἢ θεραπαίνας κέκτηται πολλὰς ἢ σκεύη
καλά, ἀλλ' ὃς ἂν ἐν τούτοις λαμπρὸς καὶ φιλότιμος ᾖ, ὧν
ἅπασι μέτεστι τοῖς πολλοῖς ὑμῶν· ὧν οὐδὲν εὑρήσετε τούτῳ 10
προσόν.
160 Ἀλλὰ νὴ Δία τριήρη ἐπέδωκεν· ταύτην γὰρ οἶδ' ὅτι
θρυλήσει, καὶ φήσει "ἐγὼ ὑμῖν τριήρη ἐπέδωκα". οὑτωσὶ δὴ
ποιήσατε. εἰ μέν, ὦ ἄνδρες Ἀθηναῖοι, φιλοτιμίας ἕνεκα
ταύτην ἐπέδωκεν, ἣν προσήκει τῶν τοιούτων ἔχειν χάριν, 15
ταύτην ἔχετε αὐτῷ καὶ ἀπόδοτε, ὑβρίζειν δὲ μὴ δῶτε· οὐδενὸς
γὰρ πράγματος οὐδ' ἔργου τοῦτο συγχωρητέον. εἰ δὲ δὴ καὶ
δειλίας καὶ ἀνανδρίας ἕνεκα δειχθήσεται τοῦτο πεποιηκώς,
μὴ παρακρουσθῆτε. πῶς οὖν εἴσεσθε; ἐγὼ καὶ τοῦτο διδάξω·
161 ἄνωθεν δέ, βραχὺς γάρ ἐσθ' ὁ λόγος, λέξω. ἐγένοντο εἰς 20
Εὔβοιαν ἐπιδόσεις παρ' ὑμῖν πρῶται· τούτων οὐκ ἦν Μειδίας,
ἀλλ' ἐγώ, καὶ συντριήραρχος ἦν μοι Φιλῖνος ὁ Νικοστράτου.
ἕτεραι δεύτεραι μετὰ ταῦτα εἰς Ὄλυνθον· οὐδὲ τούτων ἦν
Μειδίας. καίτοι τόν γε δὴ φιλότιμον πανταχοῦ προσῆκεν
ἐξετάζεσθαι. τρίται τοίνυν αὗται γεγόνασιν ἐπιδόσεις· 25
ἐνταῦθα ἐπέδωκεν. πῶς; ἐν τῇ βουλῇ γιγνομένων ἐπιδόσεων
162 παρὼν οὐκ ἐπεδίδου τότε· ἐπειδὴ δὲ πολιορκεῖσθαι τοὺς ἐν
Ταμύναις στρατιώτας ἐξηγγέλλετο, καὶ πάντας ἐξιέναι τοὺς

1 κυμβία ... φιάλας Ath. 496f, Hdn. ap. Σ **133** (471b Dilts) 20–1 ἐγένοντο ... ἐπιδόσεις Anon. Seg. (Hammer) 364.18

---

1 κυμβιτα S^{ac} ῥυτὰ καὶ κυμβία Ath. Hdn. φιάλας καὶ τὰ τοιαῦτα A
2 παρόντας FP_4^{mg} 3 Μειδίας om. A 5 τούτους Y^{pc} ὑμῶν
AFP_4^{sl} 6 δεῖ δὴ S τὰ τοι τοιαῦτα F 9 καλά $SYP_4^{\gamma\rho}$: λαμπρά AP:
πολλά $FP_4^{\gamma\rho}$ ᾖ] ἦν YP^{ac} 12 νὴ] μὴν $S^{pc}F_1^{\gamma\rho}P_4^{\gamma\rho}$ διὰ $F_1^{\gamma\rho}P_4^{\gamma\rho}$
τριήρη ἦν SFYP -κε AFYP εὖ οἶδ' AF 13 θρυλλήσει S οὕτω
ουδε S^{ac}: οὑτωσὶ δε S^{pc} 14 εἵνεκα SYP: εἵνεκεν F^{ac}: εἵνεκε F^{pc} 15 -κε SA
17 γὰρ om. Y^{ac} -δὲ A 19 παρακρουσῆτε S^{ac} 20 ἄνωθεν ... λέξω
om. A γάρ om. S^{ac} λόγος ὃν $S_3^{pc}FYP$ λέξω κἂν ἄνωθεν ἄρχεσθαι
δοκῇ FYP 20–1 ἐπιδόσεις εἰς Εὔβοιαν Seg. 23 ετεροι δευτεροι Π13
24 δὴ om. A πανταχῇ A 25 τοίνυν SYP^{ac}: νῦν AFP_4^{pc} 26 -κε A
γιν- A 27 παρ' ὧν F

Agora with an escort of three or four slaves, talking about 'cups' and 'drinking-horns' and 'chalices' loudly enough for the passers-by to hear. Well, when Meidias acquires possessions **159** for the sake of his personal luxury and advantage, I don't know what use they are to the majority of you; but when he's impelled by them to behave insolently, I can see that does affect many ordinary people among us. That surely isn't the kind of conduct you should honour and admire when it occurs; nor should you judge aspiration to honour by these criteria—whether a man builds a distinguished house or possesses a lot of maidservants or fine furniture: you should look for a man whose distinction and aspiration to honour are in things of which the majority of you all have a share. You'll find that none of this applies to Meidias.

Oh, but he donated a trireme; I know he'll drool on about **160** that and say 'I donated you a trireme!' Well then, this is what you should do, men of Athens. If it was because he aspired to honour that he donated it, show and render to him the gratitude which is appropriate to such acts, but don't permit him to be insolent; there's no action or deed in return for which that concession should be made. If on the other hand it's shown that he did it because of cowardice and unmanliness, don't be misled. So how will you know which it was? I'll explain this too; since the tale is short, I'll tell it from some way back. It was for Euboia that donations were first **161** made in Athens; Meidias was not among those, but I was, and my fellow-trierarch was Philinos son of Nikostratos. After that there was a second round of donations for Olynthos; Meidias was not among those either. Yet surely a man who really aspired to honour should have put himself forward on every occasion. So these donations are the third to have been made; this was when he donated. In what way? When donations were being made in the Boule, though he was present, he made no offer then. But when reports were **162** coming in that the soldiers at Tamynai were under siege, and

ὑπολοίπους ἱππέας, ὧν εἷς οὗτος ἦν, προεβούλευσεν ἡ βουλή, τηνικαῦτα φοβηθεὶς τὴν στρατείαν ταύτην εἰς τὴν ἐπιοῦσαν ἐκκλησίαν, πρὶν καὶ προέδρους καθίζεσθαι, παρελθὼν ἐπέδωκεν.

Τῷ δῆλον, ὥστε μηδ' ἀντειπεῖν αὐτὸν ἔχειν, ὅτι τὴν στρατείαν φεύγων, οὐ φιλοτιμίᾳ, τοῦτ' ἐποίησεν; τοῖς μετὰ **163** ταῦτα πραχθεῖσιν ὑπ' αὐτοῦ. τὸ μὲν γὰρ πρῶτον, ὡς οὐκ ἐδόκει, προϊούσης τῆς ἐκκλησίας καὶ λόγων γιγνομένων, τῆς τῶν ἱππέων βοηθείας ἤδη δεῖν, ἀλλ' ἀνεπεπτώκει τὰ τῆς ἐξόδου, οὐκ ἀνέβαινεν ἐπὶ τὴν ναῦν ἣν ἐπέδωκεν, ἀλλὰ τὸν μέτοικον ἐξέπεμψε τὸν Αἰγύπτιον, Πάμφιλον, αὐτὸς δὲ μένων ἐνθάδε τοῖς Διονυσίοις διεπράττετο ταῦτ' ἐφ' οἷς νυνὶ κρί- **164** νεται· ἐπειδὴ δὲ ὁ στρατηγὸς Φωκίων μετεπέμπετο τοὺς ἐξ Ἀργούρας ἱππέας ἐπὶ τὴν διαδοχὴν καὶ κατείληπτο σοφιζόμενος, τότε ὁ δειλὸς καὶ κατάρατος οὑτοσὶ λιπὼν τὴν τάξιν ταύτην ἐπὶ τὴν ναῦν ᾤχετο, καὶ ὧν ἱππαρχεῖν ἠξίωσε παρ' ὑμῖν ἱππέων, τούτοις οὐ συνεξῆλθεν. εἰ δ' ἐν τῇ θαλάττῃ **165** κίνδυνός τις ἦν, εἰς τὴν γῆν δῆλον ὅτι ᾤχετ' ἄν. οὐ μὴν Νικήρατός γε οὕτως ὁ τοῦ Νικίου, ὁ ἀγαπητός, ὁ ἄπαις, ὁ παντάπασιν ἀσθενὴς τῷ σώματι· οὐδ' Εὐκτήμων ὁ τοῦ Αἰσίωνος, οὐχ οὕτως· οὐδ' Εὐθύδημος ὁ τοῦ Στρατοκλέους· ἀλλ' αὐτῶν ἕκαστος ἑκὼν ἐπιδοὺς τριήρη οὐκ ἀπέδρα ταύτην τὴν στρατείαν, ἀλλὰ τὴν μὲν {ἐπίδοσιν} ἐν χάριτος μέρει καὶ δωρεᾶς παρεῖχον πλέουσαν τῇ πόλει, οὗ δὲ ὁ νόμος προσέταττεν, **166** ἐνταῦθα τοῖς σώμασιν αὐτοὶ λειτουργεῖν ἠξίουν. ἀλλ' οὐχ ὁ ἵππαρχος Μειδίας, ἀλλὰ τὴν ἐκ τῶν νόμων τάξιν λιπών, οὗ δίκην ὀφείλει τῇ πόλει δοῦναι, τοῦτ' ἐν εὐεργεσίας ἀριθμήσει μέρει. καίτοι τὴν τοιαύτην τριηραρχίαν, ὦ πρὸς θεῶν, πότερον τελωνίαν καὶ πεντηκοστὴν καὶ λιποτάξιον καὶ στρατείας ἀπόδρασιν καὶ πάντα τὰ τοιαῦτα ἁρμόττει καλεῖν, ἢ φιλοτιμίαν; οὐδένα γὰρ τρόπον ἄλλον ἐν τοῖς ἱππεῦσιν αὐτὸν

5–6 τῷ … ἐποίησεν Syrian. (Rabe 2) 27.21, Io. Sard. 101.20

2 στρατιὰν A 3 προελθὼν $F^{sl}_{1}P^{mg}_{4}$ 4 -κε AF^{pc} 5 δῆλον] δημωι S^{pc}: [S^{ac}] ὡς Syrian. Sard. 6 στρατιὰν A -σε AF^{pc} 8 προϊούσης ἤδη A γιν- A 9 -λὰ A ἀνεπεπτώκει Dindorf: ἀναπεπτώκει SAFYP 10 ἐνέβαινεν AF 12 ενταυθα δε S^{ac}: ἐνθάδε ἐν YP -τα AF 14 αργυρας S^{ac} 15 οὗτος F 18 δηλονότι F 19 γε οὐχ A οὗτος YP^{ac} παῖς AFP^{pc}_{4} 20 -δὲ A 21 -δὲ A 23 στρατιὰν A ἐπίδοσιν del. Bekker 24 ὁ om. S^{pc}: [S^{ac}] 27 -το A εὐεργεσίαις SP^{ac} 28 τῶν θεῶν AF 29 τελωνείαν F λιποτάξιον Cobet: λιποταξίαν SAFYP 30 στρατιᾶς A 31 αὐτὸν $SFYP^{ac}$

the Boule resolved to propose that all the remaining cavalry, of whom he was one, should go out, then, being afraid of this expedition, he came forward at the ensuing Ekklesia, before the proedroi even took their seats, and made his donation.

What is it that shows, so clearly that he can't even deny it, that he did this to evade the expedition, not from aspiration to honour? His subsequent actions. First, when it did not **163** seem, as the meeting proceeded and speeches were made, that there was now any need for the cavalry's support, and the matter of the expedition had receded, he didn't proceed to embark on the ship which he donated, but sent out the Egyptian metic, Pamphilos, while he himself stayed in Athens and committed at the Dionysia the acts for which he's now on trial. But when the general, Phokion, was sending for **164** the cavalry from Argoura to take over, and he'd been caught out in his ruse, that was when this damned coward, abandoning that post, took to his ship; he thought fit to be a commander of the cavalry at home, but he wouldn't go on an expedition with them. If there'd been some danger at sea, of course he'd have made for the land. Not so Nikeratos son of **165** Nikias—though he has no brother and no son, and is in very poor health; not so Euktemon son of Aision, nor Euthydemos son of Stratokles. They each voluntarily donated a trireme, but they didn't evade this campaign. They put their ships in commission for Athens by way of a free gift, but they also thought it was their duty to perform personal service where that was legally required. But that's not what Meidias the **166** hipparch did. After deserting the post assigned by law, he'll count as a piece of generosity an act for which he ought to pay the city a penalty. A trierarchy like this—I ask you!—is it suitable to call it a tax-purchase, and a two-per-cent exemption, and desertion, and evasion of military service, and every name of that sort—or 'aspiring to honour'? Because, since he couldn't make himself exempt from service in the

ἀτελῆ ποιῆσαι στρατείας δυνάμενος ταύτην εὕρηκε Μειδίας
167 καινὴν ἱππικῆς τινα πεντηκοστήν. καὶ γὰρ αὖ τοῦτο· τῶν
ἄλλων ἁπάντων τῶν ἐπιδόντων τριηράρχων παραπεμπόντων
ὑμᾶς ὅτε δεῦρ' ἀπεπλεῖτε ἐκ Στύρων, μόνος οὗτος οὐ παρέ-
πεμπεν, ἀλλ' ἀμελήσας ὑμῶν χάρακας καὶ βοσκήματα καὶ 5
θυρώματα ὡς αὑτὸν καὶ ξύλα εἰς τὰ ἔργα τὰ ἀργύρεια
ἐκόμιζεν, καὶ χρηματισμός, οὐ λειτουργία γέγονεν ἡ τριηρ-
αρχία τῷ καταπτύστῳ τούτῳ. ἀλλὰ μὴν ὡς ἀληθῆ λέγω,
σύνιστε μὲν τὰ πολλὰ τούτων, ὅμως δὲ καὶ μάρτυρας ὑμῖν
καλῶ. 10

168 ΜΑΡΤΥΡΕΣ

{Κλέων Σουνιεύς, Ἀριστοκλῆς Παιανιεύς, Πάμφιλος,
Νικήρατος Ἀχερδούσιος, Εὐκτήμων Σφήττιος, καθ' ὃν
καιρὸν ἐκ Στύρων ἀπεπλέομεν δεῦρο τῷ στόλῳ παντί,
ἐτύχομεν τριηραρχοῦντες καὶ αὐτοὶ καὶ Μειδίας ὁ νῦν 15
κρινόμενος ὑπὸ Δημοσθένους, ᾧ μαρτυροῦμεν. παντὸς
δὲ τοῦ στόλου πλεόντων ἐν τάξει, καὶ τῶν τριηράρχων
ἐχόντων παράγγελμα μὴ χωρίζεσθαι ἕως ἂν δεῦρο
καταπλεύσωμεν, Μειδίας {δ'} ὑπολειφθεὶς τοῦ στόλου,
καὶ γεμίσας τὴν ναῦν ξύλων καὶ χαράκων καὶ βοσκημά- 20
των καὶ ἄλλων τινῶν, κατέπλευσεν εἰς Πειραιᾶ μόνος
μεθ' ἡμέρας δύο, καὶ οὐ συγκατέστησε τὸν στόλον μετὰ
τῶν ἄλλων τριηράρχων.}

169 Εἰ τοίνυν ὡς ἀληθῶς, ἄνδρες Ἀθηναῖοι, ἅπερ φήσει καὶ
καταλαζονεύσεται πρὸς ὑμᾶς αὐτίκα δὴ μάλα, τοιαῦτ' ἦν 25
αὐτῷ τὰ λελειτουργημένα καὶ πεπραγμένα, καὶ μὴ τοιαῦτα
οἷα ἐγὼ δεικνύω, οὐδ' οὕτω δήπου τό γε δοῦναι δίκην

7–8 *καὶ ... τούτῳ* Syrian. (Rabe 2) 27.24 19–20 *Μειδίας ... ξύλων AB* 1.177.17 19–21 *Μειδίας ... τινῶν AB* 1.131.9

---

1 *δυνάμενος στρατιᾶς* A 2 *κενην ιππικην* S *αὖ τοῦτο* AP$^{pc}_{4}$: *αὐτοῦ τὸ* SYP^{ac}: *αὖ τότε* F$^{pc}_{1}$: [F^{ac}] 4 *ἐπεπλεῖτε* Y: [P^{ac}] *σκυθῶν* S^{pc} 6 *αὑτὸν* P: *αυτὸν* SA: *αὐτὸν* FY *καὶ ξύλα* om. A *τὰ* (post *ἔργα*) om. SYP^{ac} *ἀργύρια* SAYP 7 *-ζε* AF^{pc} *οὐχὶ* A 8 *τουτῳί* Syrian. 9 *μὲν τὰ* SAFP$^{sl}_{4}$: *μετὰ* YP: *μοι τὰ* Vg^{pc}Ald^{vl} *τούτων ὑμεῖς* A 12–23 testimonium om. A 12 *σουννιεύς* F 13 *Νικήρατος* del. Boeckh 14 *σκυθῶν* S^{mg} 15 *καὶ* (post *αὐτοὶ*) F: om. SYP 17 *δὲ τοῦ* FP$^{pc}_{4}$: *δεκάτου* S^{ac}: *δε καὶ τοῦ* S^{pc}: *δεκατοῦ* Y: [P^{ac}] 19 *δ'* om. F^{pc} *ὑπολιφθεις* S: *ἀπολειφθεὶς AB* 1.131.9 21 *κατεπλευσαι* S^{ac}: *κατέπλευσε* S^{pc} 24 *ὦ ἄνδρες Ἀθηναῖοι* A^{cp}F *οἷάπερ* Reiske *φήσει* SFP$^{sl}_{4}$: *φησὶ* AYP 25 *-νεύσεται* FP$^{sl}_{4}$: *-νεύεται* SAYP 25–6 *τοιαύτην αὐτῷ καὶ τὰ* YP^{ac}

cavalry in any other way, this is a new sort of two-per-cent exemption from riding that Meidias has invented. There's **167** this point too: when all the other trierarchs who donated triremes escorted your army on the voyage home from Styra, he was the only one who didn't join the escort. He ignored you, and spent the time transporting vine-stakes, cattle, and doors to his house, and logs to the silver-mines; it was profit-making, not public service, this abominable man's trierarchy! For the truth of what I say, you already know much of it as well as I do, but still I will also call witnesses for you.

WITNESSES **168**

{We, Kleon of Sounion, Aristokles of Paiania, Pamphilos, Nikeratos of Akherdous, and Euktemon of Sphettos, were trierarchs at the time when we were on the voyage home from Styra with the whole armada, and so was Meidias, who is now prosecuted by Demosthenes, for whom we testify. While the whole armada were sailing in order, and the trierarchs had instructions not to break away until we put into port here, Meidias fell behind the armada, loaded his ship with logs, vine-stakes, cattle, and other things, and put into Peiraieus by himself two days later, and did not join with the other trierarchs in bringing the armada back.}

If it were indeed true, men of Athens, that his public **169** services and conduct had been such as he'll tell you and brag to you in a few minutes from now, and not such as I'm showing you they were, even so it surely wouldn't be right for

ὧν ὕβρικεν ἐκφυγεῖν ταῖς λειτουργίαις δίκαιος ἂν ἦν. ἐγὼ
γὰρ οἶδ' ὅτι πολλοὶ πολλὰ κἀγαθὰ ὑμᾶς εἰσιν εἰργασμένοι, οὐ
κατὰ τὰς Μειδίου λειτουργίας, οἱ μὲν ναυμαχίας νενικηκότες,
οἱ δὲ πόλεις εἰληφότες, οἱ δὲ πολλὰ καὶ καλὰ ὑπὲρ τῆς πόλεως
170 στήσαντες τρόπαια· ἀλλ' ὅμως οὐδενὶ πώποτε τούτων δεδώ- 5
κατε τὴν δωρεὰν ταύτην οὐδ' ἂν δοίητε, ἐξεῖναι τοὺς ἰδίους
ἐχθροὺς ὑβρίζειν αὐτῶν ἑκάστῳ, ὅπου ἂν βούληται καὶ ὃν ἂν
δύνηται τρόπον. οὐδὲ γὰρ Ἁρμοδίῳ καὶ Ἀριστογείτονι
(τούτοις γὰρ δὴ μέγισται δέδονται δωρεαὶ παρ' ὑμῶν καὶ
ὑπὲρ μεγίστων), οὐδ' ἂν ἠνέσχεσθε, εἰ προσέγραψέ τις ἐν τῇ 10
στήλῃ "ἐξεῖναι δὲ καὶ ὑβρίζειν αὐτοῖς ὃν ἂν βούλωνται"· ὑπὲρ
γὰρ αὐτοῦ τούτου τὰς ἄλλας ἔλαβον δωρεάς, ὅτι τοὺς ὑβρίζοντας ἔπαυσαν.

171 Ὅτι τοίνυν καὶ κεκόμισται χάριν, ὦ ἄνδρες Ἀθηναῖοι, παρ'
ὑμῶν οὐ μόνον ὧν αὐτὸς λελειτούργηκεν λειτουργιῶν ἀξίαν 15
(μικρὰ γὰρ αὕτη γέ τις ἦν) ἀλλὰ καὶ τῶν μεγίστων, καὶ τοῦτο
βούλομαι δεῖξαι, ἵνα μηδ' ὀφείλειν οἴησθέ τι τῷ καταπτύστῳ
τούτῳ. ὑμεῖς γάρ, ὦ ἄνδρες Ἀθηναῖοι, ἐχειροτονήσατε τοῦτον
τῆς Παράλου ταμίαν, ὄντα τοιοῦτον οἷός ἐστιν, καὶ πάλιν
ἵππαρχον, ὀχεῖσθαι διὰ τῆς ἀγορᾶς ταῖς πομπαῖς οὐ δυνά- 20
μενον, καὶ μυστηρίων ἐπιμελητὴν καὶ ἱεροποιόν ποτε καὶ
172 βοώνην καὶ τὰ τοιαῦτα δή. εἶτα πρὸς τῶν θεῶν τὸ τὴν τῆς
φύσεως κακίαν καὶ ἀνανδρίαν καὶ πονηρίαν ταῖς παρ' ὑμῶν
ἀρχαῖς καὶ τιμαῖς καὶ χειροτονίαις ἐπανορθοῦσθαι μικρὰν
ὑπολαμβάνετ' εἶναι δωρεὰν καὶ χάριν; καὶ μὴν εἴ τις αὐτοῦ 25
ταῦτ' ἀφέλοιτο "ἱππάρχηκα, τῆς Παράλου ταμίας γέγονα",
173 τίνος ἄξιός ἐστιν οὗτος; ἀλλὰ μὴν κἀκεῖνό γ' ἐπίστασθε, ὅτι
τῆς μὲν Παράλου ταμιεύσας Κυζικηνῶν ἥρπασε πλεῖον ἢ
πέντε τάλαντα, ὑπὲρ ὧν ἵνα μὴ δῷ δίκην, πάντα τρόπον περι-
ωθῶν καὶ ἐλαύνων τοὺς ἀνθρώπους καὶ τὰ σύμβολα συγχέων 30

4 οἵ ... εἰληφότες *AB* 1.155.7 17–18 ἵνα ... τούτῳ *AB* 1.160.15
27–190.14 ἀλλὰ ... ἱππεῖς Greg. Cor. (Walz 7) 1200.9

1 ὧν ἐμὲ FP$_4^{mg}$ ἐκφεύγειν F 2 εἰσιν ὑμᾶς A 4 ὑπὲρ τῆς πόλεως SFYP: τῇ πόλει A 5 ὅμως ὑμεῖς AP$_4^{pc}$ τούτων πώποτε F ἐδώ-
κατε AF 7 ὅπου ἐὰν SYP^{ac}: ὁπόταν A: ὁπότ' ἂν FP$_2^{pc}$: ὅπως ἂν P$_4^{\gamma\rho}$
9 δὴ om. S$_3^{\gamma\rho}$ δέδονται ante μέγισται S$_3^{\gamma\rho}$: post δωρεαὶ A 10 μεγίστων
οὐ δέδοται τοῦτο S$_3^{\gamma\rho}$ 14 καὶ om. S 15 -κε AF^{pc} 16 γὰρ ἂν
Spalding καὶ (ante τοῦτο) om. A 17 οἴεσθέ Y 18 τοῦτον
ἐχειροτονήσατε A 20 ὀρχεῖσθαι Ald 22 δή om. A 25 -τε AF
26 -τα SFYP 27 ἄξιός ἐστιν A: ἄλλου ἔστ' ἄξιος SYP: ἔστ' ἄλλου ἄξιος F
γε AF 29–30 περιάγων S^{mg}F$_1^{mg}$P$_4^{mg}$

him to use his public services as a means to escape being punished for his insolence. I can think of many men who have served Athens well, not in the manner of Meidias' services—some winning battles, some capturing cities, some setting up many fine trophies for Athens. Nevertheless you've **170** never given to any of them, nor would you give, the reward of permitting each of them to treat his personal enemies with insolence, wherever he wishes and in whatever way he can. You didn't grant that even to Harmodios and Aristogeiton (to name men to whom you have given outstanding rewards for outstanding services), nor would you have tolerated it if someone had added to the inscription on the stone 'They are also to be permitted to treat with insolence whomever they wish'; it was precisely because they stopped men being insolent that they received the other rewards.

He has in fact received from you, men of Athens, a reward **171** appropriate not merely to the services which he has performed (a small one that was!) but to the greatest services—as I should like to explain, so that you won't think you owe anything to this abominable man. Your votes, men of Athens, made him treasurer of the *Paralos* (a man like that!), and also a hipparch (though he wasn't capable of riding through the Agora in processions), an administrator of the Mysteries, a performer of some ritual, a purchaser of cattle for sacrifice, and so on. Well then, I ask you! Do you consider it a small **172** reward and favour that his natural badness and unmanliness and wickedness is set right by the offices and honours and votes which you give him? Indeed, if he were deprived of the boasts 'I have been a hipparch! I have been treasurer of the *Paralos*!', what worth does the man have? Besides which, as **173** you know, after becoming treasurer of the *Paralos*, he seized more than five talents from some Kyzikenes, and to avoid conviction for it he pushed the fellows around and harassed them in every way, and by throwing the treaty into confusion

τὴν μὲν πόλιν ἐχθρὰν τῇ πόλει πεποίηκεν, τὰ χρήματα δ' αὐτὸς ἔχει. ἵππαρχος δὲ χειροτονηθεὶς λελύμανται τὸ ἱππικὸν ὑμῶν, τοιούτους θεὶς νόμους οὓς πάλιν αὐτὸς ἔξαρνος ἦν μὴ **174** τεθεικέναι. καὶ τῆς μὲν Παράλου ταμιεύων τότε, ὅτε τὴν ἐπὶ Θηβαίους ἔξοδον εἰς Εὔβοιαν ἐποιεῖσθ' ὑμεῖς, δώδεκα τῆς 5 πόλεως τάλαντα ἀνάλισκειν ταχθείς, ἀξιούντων ὑμῶν πλεῖν καὶ παραπέμπειν τοὺς στρατιώτας οὐκ ἐβοήθησεν, ἀλλ' ἤδη τῶν σπονδῶν γεγονυιῶν, ἃς Διοκλῆς ἐσπείσατο Θηβαίοις, ἧκεν. καὶ τότε ἡττᾶτο πλέων τῶν ἰδιωτικῶν τριήρων μιᾶς· οὕτως εὖ τὴν ἱερὰν τριήρη παρεσκευάκει. ἱππαρχῶν τοίνυν, τί 10 οἴεσθε τἄλλα; ἀλλ' ἵππον, ἵππον οὐκ ἐτόλμησεν ὁ λαμπρὸς καὶ πλούσιος οὗτος πρίασθαι, ἀλλ' ἐπ' ἀλλοτρίου τὰς πομπὰς ἡγεῖτο, τοῦ Φιλομήλου τοῦ Παιανιέως ἵππου· καὶ ταῦτα πάντες ἴσασιν οἱ ἱππεῖς. καὶ ὅτι ταῦτ' ἀληθῆ λέγω, κάλει μοι καὶ τούτων τοὺς μάρτυρας. 15

ΜΑΡΤΥΡΕΣ

175 Βούλομαι τοίνυν ὑμῖν, ὦ ἄνδρες Ἀθηναῖοι, καὶ ὅσων ἤδη καταχειροτονήσαντος τοῦ δήμου περὶ τὴν ἑορτὴν ἀδικεῖν ὑμεῖς κατεγνώκατε εἰπεῖν, καὶ δεῖξαι τί πεποιηκότες αὐτῶν ἔνιοι τίνος ὀργῆς τετυχήκασι παρ' ὑμῶν, ἵνα ταῦτα πρὸς τὰ 20 τούτῳ πεπραγμένα ἀντιθῆτε. πρῶτον μὲν τοίνυν, ἵνα πρώτης τῆς τελευταίας γεγονυίας μνησθῶ καταγνώσεως, περὶ τὰ μυστήρια ἀδικεῖν Εὐάνδρου κατεχειροτόνησεν ὁ δῆμος τοῦ Θεσπιέως, προβαλομένου αὐτὸν Μενίππου, Καρός τινος ἀνθρώπου. ἔστι δ' ὁ αὐτὸς νόμος τῷδε τῷ περὶ τῶν Διονυσίων 25 **176** ὁ περὶ τῶν μυστηρίων, κἀκεῖνος ὕστερος τοῦδε ἐτέθη. τί οὖν ποιήσαντος, ὦ ἄνδρες Ἀθηναῖοι, κατεχειροτονήσατε τοῦ Εὐάνδρου; τοῦτ' ἀκούσατε. ὅτι δίκην ἐμπορικὴν καταδικασάμενος τοῦ Μενίππου, οὐκ ἔχων πρότερον λαβεῖν αὐτόν, ὡς ἔφη, τοῖς μυστηρίοις ἐπιδημοῦντος ἐπελάβετο, κατεχειροτόνησατε 30

4–5 *καὶ … ὑμεῖς* *AB* 1.175.31 11–12 *ἀλλ' … πρίασθαι* Hermog. 425.6

---

1 *-κε* AF^{pc} 2 *αὐτὰ* Greg. 3 *ἡμῶν* Greg. *οὓς* om. S^{ac} 4 *τεθηκέναι* Goodwin *ὅτε* om. Greg. 5 *Θηβαίους ἐποιεῖτο* Greg. *ἐποιεῖσθ' ὑμεῖς* om. Greg. *-σθε* AF 9 *-κε* SF^{pc}YP *πλέον* Greg. 10 *ἱππάρχω* YP^{ac} 11–12 *μὲν οὐκ ἐτόλμησεν ὁ κατάρατος οὑτοσὶ* Hermog. 13 *ἡγεῖτο*] *ἐποιεῖτο* S$_3^{γρ}$ *ταῦτα ἐξιόντων* SYP^{ac} 14 *πάντα* F *συνιππεῖς* A Greg. *καὶ* SYP: *ἀλλὰ μὴν* AF 15 *τοὺς* om. SFYP 19 *αὐτῶν* om. A 20 *ἵν' αὐτὰ* AF: *ἵνα αὐτὰ* P$_2^{pc}$ 21 *τιθῆτε* YP^{ac} *πρώτης* SFYP: *πρῶτον* A 22 *γεγονυιης* S 24 *θεσπειέως* A^{ac} *προβαλλομένου* AYP^{ac} *με ιππου* P^{ac} 25 *δὲ* AF 28 *-το* A

he has created hostility between the two cities, while he's kept the money for himself. After being elected hipparch, he's done harm to your cavalry by the rules he made, and then turned round and denied having made them. He was **174** treasurer of the *Paralos* at the time when you were making your expedition to oppose the Thebans in Euboia, and was appointed to spend twelve talents of public money; but when you requested him to set sail and escort the soldiers, he failed to give his support, but arrived after the treaty which Diokles concluded with the Thebans had already been made. And then he was overtaken on the voyage by one of the privately maintained triremes: that's how well he'd equipped the sacred ship! And then when he was a hipparch—well, can you imagine the rest? But as for a horse—a horse!—this distinguished and wealthy man never went to the length of buying one, but led the processions on someone else's, the horse belonging to Philomelos of Paiania, as all the cavalrymen well know. To show that I'm telling the truth, please call the witnesses to these statements also.

WITNESSES

Now I should also like to mention to you, men of Athens, **175** the previous cases of persons whom, after an adverse vote by the people, you have convicted of offences concerning the festival, and to explain what some of them have done and how severely you have expressed your anger at them, so that you may compare what they did with the actions of Meidias. First, then (to mention first the most recent conviction), the people voted that Euandros the Thespian was guilty of an offence concerning the Mysteries, when a *probole* was brought against him by a Karian named Menippos. The law about the Mysteries is the same as this one about the Dionysia, and was enacted later than it. Well, what had Euandros done, **176** men of Athens, that you voted against him? Let me tell you. Because, after winning a mercantile case against Menippos and not being able to find him earlier (so he said), he laid hold of him when he was in Attika at the Mysteries, you

μὲν διὰ ταῦτα, καὶ οὐδ᾿ ὁτιοῦν ἄλλο προσῆν· εἰσελθόντα δ᾿ εἰς τὸ δικαστήριον ἠβούλεσθε μὲν θανάτῳ κολάσαι, τοῦ δὲ προβαλομένου πεισθέντος τὴν δίκην τε πᾶσαν ἀφεῖναι ἠναγκάσατε αὐτόν, ἣν ᾐρήκει πρότερον (ἦν δὲ δυοῖν αὕτη ταλάντοιν), καὶ προσετιμήσατε τὰς βλάβας, ἃς ἐπὶ τῇ καταχειροτονίᾳ **177** μένων ἐλογίζετο αὑτῷ γεγενῆσθαι πρὸς ὑμᾶς ἅνθρωπος. εἷς μὲν οὗτος ἐξ ἰδίου πράγματος, οὐδεμιᾶς ὕβρεως προσούσης, ὑπὲρ αὐτοῦ τοῦ παραβῆναι τὸν νόμον τοσαύτην ἔδωκε δίκην. εἰκότως· τοῦτο γάρ ἐσθ᾿ ὃ φυλάττειν ὑμᾶς δεῖ, τοὺς νόμους, τὸν ὅρκον. ταῦτ᾿ ἔχεθ᾿ ὑμεῖς οἱ δικάζοντες ἀεὶ παρὰ τῶν ἄλλων ὡσπερεὶ παρακαταθήκην, ἣν ἅπασιν, ὅσοι μετὰ τοῦ **178** δικαίου πρὸς ὑμᾶς ἔρχονται, σῶν ὑπάρχειν δεῖ. ἕτερος ἀδικεῖν ποτ᾿ ἔδοξεν ὑμῖν περὶ τὰ Διονύσια, καὶ κατεχειροτονήσατε αὐτοῦ παρεδρεύοντος ἄρχοντι τῷ υἱεῖ, ὅτι θέαν τινὸς καταλαμβάνοντος ἥψατο, ἐξείργων ἐκ τοῦ θεάτρου· ἦν δ᾿ **179** οὗτος ὁ τοῦ βελτίστου πατὴρ Χαρικλείδου τοῦ ἄρξαντος. καὶ μέγα γ᾿ ὑμῖν τοῦτ᾿ ἐδόκει καὶ δίκαιον ἔχειν ὁ προβαλόμενος λέγειν· "εἰ κατελάμβανον, ἄνθρωπε, θέαν, εἰ μὴ τοῖς κηρύγμασιν, ὡς σύ με φῄς, ἐπειθόμην, τίνος ἐκ τῶν νόμων εἶ κύριος, καὶ ὁ ἄρχων αὐτός; τοῖς ὑπηρέταις ἐξείργειν εἰπεῖν, οὐκ αὐτὸς τύπτειν. οὐδ᾿ οὕτω πείθομαι· ἐπιβολὴν ἐπιβάλλειν, πάντα μᾶλλον πλὴν αὐτὸς ἅψασθαι τῇ χειρί. πολλὰ γὰρ πρὸ τοῦ μὴ τὸ σῶμα ἕκαστον ὑβρίζεσθαι πεποιήκασιν οἱ νόμοι." ταῦτ᾿ ἔλεγεν μὲν ἐκεῖνος, κατεχειροτονήσατε δ᾿ ὑμεῖς· οὐ μὴν εἰσῆλθεν εἰς τὸ δικαστήριον οὗτος, ἀλλ᾿ ἐτελεύτησεν πρότερον. **180** ἑτέρου τοίνυν ὅ τε δῆμος ἅπας κατεχειροτόνησεν ἀδικεῖν περὶ τὴν ἑορτήν, καὶ ὑμεῖς εἰσελθόντα ἀπεκτείνατε τοῦτον, Κτησικλέα λέγω. διὰ τί δή; ὅτι σκῦτος ἔχων ἐπόμπευεν, καὶ τούτῳ μεθύων ἐπάταξέν τινα ἐχθρὸν ὑπάρχοντα ἑαυτῷ· ἐδόκει γὰρ ὕβρει καὶ οὐκ οἴνῳ τύπτειν, ἀλλὰ τὴν ἐπὶ τῆς πομπῆς

1 δὲ AF   2 ἐβουλεύσασθε A: ἐβούλεσθε FP_2^{pc} κολάσαι SAYP: ζημιῶσαι $FP_4^{\gamma\rho}$ 3 τε om. $S^{pc}FP_4^{pc}$ 4 εἰρήκει AYP   5 χειροτονίᾳ S^{ac} 6 αὐτῷ Taylor: αυτῷ SA: αὐτῷ FYP   ἅνθρωπος Bekker: άνθρωπος AF: ἄνθρωπος SYP   9 ἐστιν A   11 ωσπερ F^{ac} 12 σῶν A: σώαν S_3^{pc}YP: σώιαν F: [S^{ac}]   13 -τε A   15 καταλαβόντος $AP_4^{\gamma\rho}$ 17 γε A   καὶ om. A   προβαλλόμενος SP   18 ἄνθρωπε, θέαν om. $AP_4^{\gamma\rho}$: ἄνθρωπε εαν S^{ac} 19 μεφῆς S: μ᾿ ἔφης Y: φῄς $AP_4^{\gamma\rho}$ 21 ἐπιβολὴν ζημίαν P^{ac} 22 μᾶλλον om. A   πλὴν] ἢ $P_2^{\gamma\rho}$ πρὸ τοῦ SAYP: πρὸς τὸ F   24 -γε AFYP   ἐχειροτονήσατε S   δὲ AF   25 -σε AFYP   27 τοῦτον om. AF   28 λέγω. διὰ τί δή $S_3^{\gamma\rho}$AFYP: om. S   ἀπεκτείνατε τοῦτον (post δή) $S_3^{\gamma\rho}$A: τοῦτον ἀπεκτείνατε F: om. SYP   -ευε AF^{pc} 29 -ξέ AF   αυτῷ A: αὑτῷ F

voted against him—just for that, when there was absolutely no other reason in addition. And when he came into court for trial, you were ready to impose the death penalty, and, when the prosecutor was persuaded out of that, you required Euandros to forfeit the whole of the award made to him in the previous case, amounting to two talents, and awarded damages against him in addition, the amount which the other man reckoned he had incurred in connection with your court while waiting here in consequence of the vote. That's one **177** man who, as a result of a private matter, not involving any insolence, paid so severe a penalty simply for transgressing the law. Quite properly: for that is what you have to guard —the laws and your oath. You, the jurors in office, have these as a trust from the rest of us, which must be preserved for all who come to you with a just cause. Another man was **178** once considered by you to be guilty of an offence concerning the Dionysia, and you voted against him, when he was assessor to his son, who was arkhon, because he touched someone who was taking a place as a spectator, when excluding him from the theatre. This was the father of that excellent man Kharikleides, the one who was arkhon. And the prosecutor **179** seemed to you to have a strong case to put: 'For goodness' sake, man, if I was taking a place as a spectator, if I was disobeying the announcements, as you allege, what do you, and the arkhon himself, have legal authority to do? To tell the attendants to exclude me, not to strike me yourself. Suppose I don't obey even so: you can impose a fine, anything rather than touch me with your own hand. The laws have provided many safeguards against insolence to the person of an individual.' That's what he said, and you voted for condemnation; however, the man didn't come to trial, but died before that. Then there was another man against whom the **180** whole people voted that he was guilty of an offence concerning the festival, and when he came to trial you put him to death: I mean Ktesikles. Why was this? Because he had a whip in the procession, and with this, being drunk, he struck a man who was on bad terms with him. It was thought that it was insolence, not wine, which prompted him to strike, and

καὶ τοῦ μεθύειν πρόφασιν λαβὼν ἀδικεῖν, ὡς δούλοις χρώ-
181 μενος τοῖς ἐλευθέροις. ἁπάντων τοίνυν, ὦ ἄνδρες Ἀθηναῖοι,
τούτων, ὧν ὁ μὲν ὧν εἷλεν ἀποστάς, ὁ δὲ καὶ θανάτῳ ζημιω-
θεὶς φαίνεται, πολλῷ δεινότερ' εὖ οἶδ' ὅτι πάντες ἂν εἶναι
φήσειαν τὰ Μειδίᾳ πεπραγμένα. οὔτε γὰρ πομπεύων οὔτε 5
δίκην ᾐρηκὼς οὔτε παρεδρεύων οὔτ' ἄλλην σκῆψιν ἔχων
οὐδεμίαν πλὴν ὕβριν, τοιαῦτα πεποίηκεν οἷα οὐδεὶς ἐκείνων.
182 Καὶ τούτους μὲν ἐάσω· ἀλλὰ Πύρρον, ὦ ἄνδρες Ἀθηναῖοι,
τὸν Ἐτεοβουτάδην, ἐνδειχθέντα δικάζειν ὀφείλοντα τῷ
δημοσίῳ, θανάτῳ ζημιῶσαί τινες ὑμῶν ᾤοντο χρῆναι, καὶ 10
τέθνηκεν ἁλοὺς παρ' ὑμῖν· καίτοι τοῦτο τὸ λῆμμα δι' ἔνδειαν,
οὐ δι' ὕβριν λαμβάνειν ἐπεχείρησεν ἐκεῖνος. καὶ πολλοὺς ἂν
ἑτέρους ἔχοιμι λέγειν, ὧν οἱ μὲν τεθνᾶσιν, οἱ δ' ἠτιμωμένοι
διὰ πολλῷ τούτων εἰσὶν ἐλάττω πράγματα· ὑμεῖς δ', ὦ ἄνδρες
Ἀθηναῖοι, Σμίκρῳ δέκα ταλάντων ἐτιμήσατε καὶ Σκίτωνι 15
τοσούτων ἑτέρων, δόξαντι παράνομα γράφειν, καὶ οὔτε
παιδία οὔτε φίλους οὔτε συγγενεῖς οὐθ' ὁντινοῦν ἠλεήσατε
183 τῶν παρόντων ἐκείνοις. μὴ τοίνυν, ἐὰν μὲν εἴπῃ τις παράνομα,
οὕτως ὀργιζόμενοι φαίνεσθε, ἐὰν δὲ ποιῇ, μὴ λέγῃ, πράως
διάκεισθε. οὐδὲν γὰρ ῥῆμα οὐδ' ὄνομα οὕτως ἐστὶ τοῖς πολ- 20
λοῖς ὑμῶν χαλεπόν, ὡς ὅσα ὑβρίζων τις τὸν ἐντυχόντα ὑμῶν
διαπράττεται. μὴ τοίνυν αὐτοὶ καθ' ὑμῶν αὐτῶν δεῖγμα
τοιοῦτον ἐξενέγκητε, ἄνδρες Ἀθηναῖοι, ὡς ἄρα ὑμεῖς, ἐὰν μὲν
τῶν μετρίων τινα καὶ δημοτικῶν λάβητε ὁτιοῦν ἀδικοῦντα,
οὔτ' ἐλεήσετε οὔτ' ἀφήσετε ἀλλ' ἀποκτενεῖτε ἢ ἀτιμώσετε, 25
ἐὰν δὲ πλούσιος ὤν τις ὑβρίζῃ, συγγνώμην ἕξετε. μὴ δῆτα, οὐ
γὰρ δίκαιον· ἀλλ' ἐπὶ πάντων ὁμοίως ὀργιζόμενοι φαίνεσθε.
184 Ἃ τοίνυν οὐδενὸς τῶν εἰρημένων ἧττον ἀναγκαῖον εἶναι
νομίζω πρὸς ὑμᾶς εἰπεῖν, ταῦτ' εἰπὼν ἔτι καὶ βραχέα περὶ
τούτων διαλεχθεὶς καταβήσομαι. ἔστιν, ὦ ἄνδρες Ἀθηναῖοι, 30
μεγάλη τοῖς ἀδικοῦσιν ἅπασι μερὶς καὶ πλεονεξία ἡ τῶν

1 λαμβάνων A 3 ὧν (post μὲν) AP$_2^{\gamma\rho}$: ὃν P$_4^{\gamma\rho}$: om. SFYP 4 -ρα AF εὖ om. F ἅπαντες A 5 φήσετε A 6 -τε A 6–7 οὐδεμίαν ἔχων SYP 7 οὐδεεὶς A: οὐδὲ εἷς CdKOd 8 μὲν δὴ F ἐῶμεν FP$_4^{\gamma\rho}$ 10 θάνατον A 11 αλουσα S^{ac} 12 ἂν om. A 13 δὲ A 14 ὑμεῖς δ' SYP: ὑμεῖς δὲ F: ἀλλ' ὑμεῖς A: ἔτι τοίνυν ὑμεῖς S$_3^{\gamma\rho}$: ὑμεῖς U 15 Σμίκρῳ Bekker: σμικρω S^{ac}: σμικρῷ AYP^{ac}: σμίκρωνι S^{pc}FP$_4^{pc}$ 17 -τε A ὁντινοῦν BPrX: ὅντιν' οὖν SFYP: ἄλλον οὐδένα A 18 ἂν AFP$_4^{pc}$ 19 ἂν AFP$_4^{pc}$ δ' ἃ Y: [P^{ac}] 20 οὐδὲν AFYP: οὐδὲ S οὕτω τοῖς πολλοῖς ἐστὶν A 21 ἐντυγχάνοντα A 23 ὦ ἄνδρες F ἂν SFYP 24 δημωτικῶν P 25 -τε (ante ἀφήσετε) A ἀποκτενεῖτε AFP$_4^{mg}$: ἀποκτείνετε SYP 26 ἂν F τις om. A 27 ὁμοίως om. A 29–30 βραχέα post τούτων A: om. F

that he made the procession and his intoxication the excuse for his offence, treating free men as slaves. Now all these, men **181** of Athens, one of whom, as you see, gave up the award which he obtained in court while another actually incurred the death penalty, were far less shocking, I'm sure everyone would say, than what Meidias did. He was not in a procession; he had not won a court case; he was not an arkhon's assessor; he had no other excuse, except insolence, for committing acts worse than any of them.

Apart from these cases, there was Pyrrhos, men of Athens, **182** one of the Eteoboutadai, who was indicted for acting as a juror while in debt to the treasury. Some of you thought it right to impose the death penalty, and he died after conviction in your court. Yet it was neediness, not insolence, which prompted him to try to get that payment. And I could mention many more, some of whom have died and others have been disfranchised for offences far less than these. You also, men of Athens, imposed a fine of ten talents on Smikros and the same amount on Skiton, after finding them guilty of proposing an illegal decree; you were not moved to mercy by children or friends or relatives or anyone else who supported them. So do not, while showing such severity towards some- **183** one who proposes something illegal, be leniently disposed to one who does not just say something illegal but does it. No utterance or word gives as much trouble to most of you as the behaviour of an insolent man towards any one of you who meets him. So do not provide such evidence against yourselves, men of Athens, to show that, if you find one of the ordinary people guilty of any offence, you will not pity him or let him off but will execute or disfranchise him, and yet if a rich man behaves insolently you will pardon him. Do not do it, for it is not right; but show equal severity in every case.

Now, I shall speak of some other matters which I consider **184** it no less necessary to tell you about than those already mentioned, and after saying a few words about them I shall stand down. One thing which contributes greatly to the advantage of all offenders, men of Athens, is your habitual lenience. Let

ὑμετέρων τρόπων πραότης. ὅτι δὴ ταύτης οὐδ' ὁτιοῦν ὑμῖν
μεταδοῦναι τούτῳ προσήκει, ταῦτ' ἀκούσατέ μου. ἐγὼ
νομίζω πάντας ἀνθρώπους ἐράνους φέρειν παρὰ τὸν βίον
αὑτοῖς, οὐχὶ τούσδε μόνους οὓς συλλέγουσί τινες καὶ ὧν
185 πληρωταὶ γίγνονται, ἀλλὰ καὶ ἄλλους. οἷον ἔστι μέτριος καὶ 5
φιλάνθρωπός τις ἡμῶν καὶ πολλοὺς ἐλεῶν· τούτῳ ταὐτὸ δίκαιον ὑπάρχειν παρὰ πάντων, ἄν ποτε εἰς χρείαν καὶ ἀγῶνα
ἀφίκηται. ἄλλος οὑτοσί τις ἀναιδὴς καὶ πολλοὺς ὑβρίζων, καὶ
τοὺς μὲν πτωχούς, τοὺς δὲ καθάρματα, τοὺς δ' οὐδὲν ὑπολαμ-
βάνων εἶναι· τούτῳ τὰς αὐτὰς δίκαιον ὑπάρχειν φοράς, ἅσπερ 10
αὐτὸς εἰσενήνοχεν τοῖς ἄλλοις. ἂν τοίνυν ὑμῖν ἐπίῃ σκοπεῖν,
τούτου πληρωτὴν εὑρήσετε Μειδίαν ὄντα τοῦ ἐράνου καὶ οὐκ
ἐκείνου.

186 Οἶδα τοίνυν ὅτι καὶ τὰ παιδία ἔχων ὀδυρεῖται, καὶ πολλοὺς
λόγους καὶ ταπεινοὺς ἐρεῖ, δακρύων καὶ ὡς ἐλεεινότατον 15
ποιῶν ἑαυτόν. ἔστι δ', ὅσῳ περ ἂν αὐτὸν νῦν ταπεινότερον ποιῇ,
τοσούτῳ μᾶλλον ἄξιον μισεῖν αὐτόν, ὦ ἄνδρες Ἀθηναῖοι. διὰ
τί; ὅτι εἰ μὲν μηδαμῶς δυνηθεὶς ταπεινὸς γενέσθαι οὕτως
ἀσελγὴς καὶ βίαιος ἦν ἐπὶ τοῦ παρεληλυθότος βίου, τῇ φύσει
καὶ τῇ τύχῃ, δι' ἣν τοιοῦτος ἐγένετο, ἄξιον ἦν ἄν τι τῆς ὀργῆς 20
ἀνεῖναι· εἰ δ' ἐπιστάμενος μέτριον παρέχειν ἑαυτόν, ὅταν
βούληται, τὸν ἐναντίον ᾗ τοῦτον τὸν τρόπον εἵλετο ζῆν,
εὔδηλον δήπου τοῦθ', ὅτι, καὶ νῦν ἐὰν διακρούσηται, πάλιν
187 αὐτὸς ἐκεῖνος ὃν ὑμεῖς ἴστε γενήσεται. οὐ δὴ δεῖ προσέχειν,
οὐδὲ τὸν παρόντα καιρόν, ὃν οὗτος ἐξεπίτηδες πλάττεται, 25
κυριώτερον οὐδὲ πιστότερον τοῦ παντός, ὃν αὐτοὶ σύνιστε,
χρόνου ποιήσασθαι. ἐμοὶ παιδία οὐκ ἔστιν, οὐδ' ἂν ἔχοιμι
ταῦτα παραστησάμενος κλαίειν καὶ δακρύειν ἐφ' οἷς
ὑβρίσθην· διὰ τοῦτ' ἄρα τοῦ πεποιηκότος ὁ πεπονθὼς ἔλαττον

1 δὴ SFY: δὲ AP$^{pc}_{2}$: ἂν F$^{sl}_{1}$: [P^{ac}] 2 τούτῳ προσήκει SYP: μειδίᾳ προσήκει A: προσήκει μειδίᾳ F ταῦτ' SYP: τοῦτ' AF 3 εἰσφέρειν AP$^{pc}_{2}$ παρὰ πάντα A 4 τούσδε μόνους S^{pc}AFP$^{pc}_{2}$: τοὺς δεομένους S^{ac}YP^{ac} 5 γιν- A 5–6 τις ante μέτριος AF 6 ὑμῶν A 7 ὑπάρχειν εἰκὸς A: εἰκὸς ὑπάρχειν S$^{sl}_{3}$F$^{mg}_{1}$P$^{\gamma\rho}_{4}$ 9 τοὺς μὲν καθάρματα τοὺς δὲ πτωχοὺς F οὐδὲν] οὐδὲ ἀνθρώπους Markland 10 εἶναι om. AF δίκαιόν ἐστι φορὰς ὑπάρχειν A 11 -χε AF ὑμῖν ὀρθῶς AFYP σκοπεῖν ὀρθῶς S$^{sl}_{2}$ 12 καὶ om. AF 14 καὶ (ante τὰ) AFP$^{sl}_{4}$: om. SYP 16 δὲ A ἂν S$^{sl}_{3}$AFP$^{sl}_{4}$: om. SYP νῦν om. A 19 ασελγις S^{ac} 20 ἀντὶ RLp 21 δὲ A αυτον A: αὑτὸν CdKR 21–2 ὅταν βούληται om. A 22 τοὐναντίον A 23 ἐὰν A: ἂν S$^{sl}_{2}$FYP: om. S διακρούσητε SYP^{ac} 24 αὑτὸς Cobet: αυτὸς S: ὁ αὐτὸς A: αὐτὸς FYP: αὐτὸς post ἐκεῖνος Σ (633 Dilts) δεῖ δὴ SYP 25 πλάτττεται Y 26 οὐδὲ πιστότερον om. A 29 ταῦτ' A

me explain why it's not appropriate for you to extend any of it whatever to this man. I consider that all men contribute to loans throughout their life for their own benefit—not just these loans which people collect and which have contributors, but others too. For example, one of us is well-behaved, **185** humane, and sympathetic to many: it's right for him to get the same treatment from everyone, whenever he gets into a difficulty or dispute. Suppose someone else is shameless and insolent to many, regarding some men as paupers, some as rubbish, and some as negligible: it's right for him to get the same payments as he's given to the others. So, if you will look into the matter, you'll find that Meidias is a contributor to the latter kind of loan, not the former.

Now, I know that he'll have his children here too, and he'll **186** lament and make a long humble speech, weeping and making himself as pitiable as possible. But in fact, the more humble he makes himself now, the more appropriate it is to hate him, men of Athens. Why? Because, if this bullying and violence during his past life had resulted from inability to be humble, it would be appropriate to abate your anger to some extent because of the accident of nature which made him a man of that sort; but if, though knowing how to behave well when he wishes, he chose to live in the opposite manner, surely it's obvious that, if he gets away with it even now, he'll once again become the same old Meidias that you all know. You must pay no attention, and not attach greater import- **187** ance or credence to the present exigency, which he is purposely fabricating for himself, than to all the rest of the time, of which you have personal knowledge. I have no children; I can't bring them forward and then cry and weep about the insults which I suffered. Does that mean you'll give the per-

188 ἔξω παρ’ ὑμῖν; μὴ δῆτα· ἀλλ’ ὅταν οὗτος ἔχων τὰ παιδία
τούτοις ἀξιοῖ δοῦναι τὴν ψῆφον ὑμᾶς, τότε ὑμεῖς τοὺς νόμους
ἔχοντά με πλησίον ἡγεῖσθε παρεστάναι {καὶ τὸν ὅρκον ὃν
ὀμωμόκατε} τούτοις ἀξιοῦντα καὶ ἀντιβολοῦντα ἕκαστον
ὑμῶν ψηφίσασθαι. οἷς ὑμεῖς κατὰ πολλὰ δικαιότερον προσ- 5
θοῖσθ’ ἂν ἢ τούτῳ· καὶ γὰρ ὀμωμόκατε, ὦ ἄνδρες Ἀθηναῖοι,
τοῖς νόμοις πείσεσθαι, καὶ τῶν ἴσων μέτεστιν ὑμῖν διὰ τοὺς
νόμους, καὶ πάνθ’ ὅσα ἐστὶν ἀγαθὰ ὑμῖν διὰ τοὺς νόμους
ἐστίν, οὐ διὰ Μειδίαν οὐδὲ διὰ τοὺς Μειδίου παῖδας.

189 Καὶ “ῥήτωρ ἐστὶν οὗτος” ἴσως ἐμὲ φήσει λέγων. ἐγὼ δ’, εἰ 10
μὲν ὁ συμβουλεύων ὅ τι ἂν συμφέρειν ὑμῖν ἡγῆται, καὶ τοῦτ’
ἄχρι τοῦ μηδὲν ὑμῖν ἐνοχλεῖν μηδὲ βιάζεσθαι, ῥήτωρ ἐστίν,
οὔτε φύγοιμ’ ἂν οὔτ’ ἀπαρνοῦμαι τοῦτο τοὔνομα. εἰ μέντοι
ῥήτωρ ἐστὶν οἵους ἐνίους τῶν λεγόντων ἐγὼ καὶ ὑμεῖς δὲ
ὁρᾶτε, ἀναιδεῖς καὶ ἐξ ὑμῶν πεπλουτηκότας, οὐκ ἂν εἴην 15
οὗτος ἐγώ· εἴληφα μὲν γὰρ οὐδ’ ὁτιοῦν παρ’ ὑμῶν, τὰ δὲ ὄντα
εἰς ὑμᾶς πλὴν πάνυ μικρῶν ἅπαντ’ ἀνήλωκα. καίτοι καὶ εἰ
τούτων ἦν πονηρότατος, κατὰ τοὺς νόμους ἔδει παρ’ ἐμοῦ
190 δίκην λαμβάνειν, οὐκ ἐφ’ οἷς ἐλειτούργουν ὑβρίζειν. ἔτι τοίνυν
οὐδὲ εἷς ἐστιν ὅστις ἐμοὶ τῶν λεγόντων συναγωνίζεται. καὶ 20
οὐδενὶ μέμφομαι· οὐδὲ γὰρ αὐτὸς οὐδενὸς ἕνεκα τούτων οὐδὲν
ἐν ὑμῖν πώποτ’ εἶπον, ἀλλ’ ἁπλῶς κατ’ ἐμαυτὸν ἔγνων καὶ
λέγειν καὶ πράττειν ὅ τι ἂν συμφέρειν ὑμῖν ἡγῶμαι. ἀλλὰ
τούτῳ πάντας αὐτίκα δὴ μάλα συνεξεταζομένους τοὺς
ῥήτορας ὄψεσθ’ ἐφεξῆς. καίτοι πῶς ἐστι δίκαιον, τοὔνομα μὲν 25
τοῦτο ὡς ὄνειδος προφέρειν, διὰ τούτων δ’ αὐτὸν τῶν ἀνδρῶν
ἀξιοῦν σωθῆναι;

191 Τάχα τοίνυν ἴσως καὶ τὰ τοιαῦτα ἐρεῖ, ὡς ἐσκεμμένα καὶ

1–3 ἀλλ’ … παρεστάναι schol. Hermog. (Walz 4) 419.17 1–4 ἀλλ’ … ὀμωμόκατε schol. Hermog. (Walz 7) 335.19 2–5 ὑμεῖς … ὑμῶν *AB* 1.126.5 20–1 καὶ … μέμφομαι *AB* 1.156.7

---

1 οὗτος om. schol. Hermog. (Walz 4) 2 τούτοις … ὑμᾶς] κλαίῃ schol. Hermog. (Walz 4) ὑμᾶς ante ἀξιοῖ schol. Hermog. (Walz 7) 3 ἡγεῖσθαι Y 3–4 καὶ … ὀμωμόκατε del. Dobree 4 ἀντικαλοῦντα A^{ac} 5–6 πρόσθοισθ’ AFYP 7 πείσεσθαι Herwerden: πείθεσθαι SAFYP μέτεστιν bis A^{ac} 11 ἡγῶμαι AF$_1^{\gamma\rho}$P$_4^{\gamma\rho}$ -το SYP 13 τοῦτο] τοῦ F^{ac}: om. F^{pc} 15 ἐξ AFYP: ὑφ’ S: ἀφ’ Cobet 16 δέοντα A^{ac} 17 μικρὸν A εἰ καὶ A 20 οὐδείς AF 21 οὔτε A αὐτὸ S^{ac} τούτων om. P$_4^{\gamma\rho}$ 21–2 οὐδὲν ὑμῖν S^{ac}P$_4^{\gamma\rho}$: ἐν οὐδενὶ ὑμῶν F$_1^{\gamma\rho}$YP 22 ἁπλῶς AFP$_4^{pc}$: ὅπως SYP^{ac} 23 ὅταν συμφέρει S^{ac}: ὅτι συμφέρειν A 24 ἐξεταζομένους A 25 -σθε AF 26 προφέρειν ἐμοὶ FYP αὐτὸν S 28 τὰ om. YP^{ac}

petrator preference over me, the victim? No; when Meidias **188** with his children requests you to give your votes to them, you must imagine me standing alongside with the laws, requesting and entreating each of you to vote for *them*. Many considerations make it right for you to side with the laws rather than with Meidias: you are under oath, men of Athens, to obey the laws; your equality of rights is due to the laws; all the good things you have are due to the laws—not to Meidias, nor to Meidias' children!

'The man is an orator': perhaps he'll say that too of me in **189** his speech. In my view, if a man who speaks in favour of whatever he considers beneficial to Athens, and does so without annoying or browbeating you, is 'an orator', I can't avoid that name and don't repudiate it. But if 'an orator' is a man like some of the speakers I see, and you do too—men without shame, who have got rich at your expense—that can't be what I am. I haven't taken a penny from you; I've spent all my money on you, except a very small amount. Yet even if I were the wickedest of them all, legal punishment is what I should undergo, not insolence for the liturgies I performed. Besides, not a single one of the speakers is supporting **190** me in this trial. Not that I blame anyone for that; I've never spoken for any of them in court either. I've simply kept to my own policy of saying and doing whatever I consider beneficial to Athens. But presently you'll see all the orators taking the side of Meidias, one after another. Yet how can it be right, while applying this name to me as a reproach, to use these very men for requesting acquittal himself?

Then perhaps he'll also say something like this, that the **191**

παρεσκευασμένα πάντα λέγω νῦν. ἐγὼ δ' ἐσκέφθαι μέν, ὦ
ἄνδρες Ἀθηναῖοι, φημὶ καὶ οὐκ ἂν ἀρνηθείην, καὶ μεμελετηκέναι γ' ὡς ἐνῆν μάλιστα ἐμοί (καὶ γὰρ ἂν ἄθλιος ἦν, εἰ τοιαῦτα
παθὼν καὶ πάσχων ἠμέλουν ὧν περὶ τούτων ἐρεῖν ἤμελλον
192 πρὸς ὑμᾶς), γεγραφέναι μέντοι μοι τὸν λόγον Μειδίαν. ὁ γὰρ 5
τὰ ἔργα παρεσχηκὼς περὶ ὧν εἰσιν οἱ λόγοι δικαιότατ' ἂν
ταύτην ἔχοι τὴν αἰτίαν, οὐχ ὁ ἐσκεμμένος οὐδ' ὁ μεριμνήσας
τὰ δίκαια λέγειν νῦν. ἐγὼ μὲν οὖν τοῦτο ποιῶ, ὦ ἄνδρες
Ἀθηναῖοι, καὶ αὐτὸς ὁμολογῶ. Μειδίαν μέντοι μηδὲν ἐσκέφθαι πώποτε ἐν παντὶ τῷ βίῳ δίκαιον εἰκός ἐστιν· εἰ γὰρ καὶ 10
κατὰ μικρὸν ἐπῄει τὰ τοιαῦτ' αὐτῷ σκοπεῖν, οὐκ ἂν τοσοῦτον
διημάρτανε τοῦ πράγματος.

193 Οἶμαι τοίνυν αὐτὸν οὐδὲ τοῦ δήμου κατηγορεῖν ὀκνήσειν
οὐδὲ τῆς ἐκκλησίας, ἀλλ' ἅπερ τότε ἐτόλμα λέγειν ὅτ' ἦν ἡ
προβολή, ταῦτα καὶ νῦν ἐρεῖ, ὡς ὅσοι δέον ἐξιέναι κατέμενον, 15
καὶ ὅσοι τὰ φρούρια ἦσαν ἔρημα λελοιπότες, ἐξεκλησίασαν,
καὶ χορευταὶ καὶ ξένοι καὶ τοιοῦτοί τινες ἦσαν οἱ κατεχει-
194 ροτόνησαν αὐτοῦ. εἰς γὰρ τοῦτο θράσους καὶ ἀναιδείας τότε
ἀφίκετο, ὦ ἄνδρες δικασταί, ὡς ἴσασιν ὅσοι παρῆσαν ὑμῶν,
ὥστε κακῶς λέγων καὶ ἀπειλῶν καὶ βλέπων εἰς τὸν ἀεὶ θορυ- 20
βοῦντα τόπον τῆς ἐκκλησίας καταπλήξειν ᾤετο τὸν δῆμον
ἅπαντα· ᾗ καὶ γέλοια εἶναι τὰ νῦν, οἶμαι, δάκρυα εἰκότως ἂν
195 αὐτοῦ δοκοίη. τί λέγεις, ὦ μιαρὰ κεφαλή; σὺ τὰ σαυτοῦ
παιδία ἀξιώσεις ἐλεεῖν ἢ σὲ τούσδε, ἢ σπουδάζειν εἰς τὰ σά,
τοὺς ὑπὸ σοῦ δημοσίᾳ προπεπηλακισμένους; σὺ μόνος τῶν 25
ὄντων ἀνθρώπων ἐπὶ μὲν τοῦ βίου τοσαύτης ὑπερηφανίας
μεστὸς ὢν πάντων ἀνθρώπων ἔσει φανερώτατος, ὥστε καὶ
πρὸς οὓς μηδέν ἐστί σοι πρᾶγμα, λυπεῖσθαι τὴν σὴν θρα-

1–4 *ἐγὼ ... ἤμελλον* Plu. *Mor.* 6d 15 *ὡς ... κατέμενον* *AB* 1.132.30
23–5 *τί ... προπεπηλακισμένους* Aristid. *Rh.* (Schmid) 45.19

---

1 δὲ A 2 ἄνδρες om. Plu. ἂν ἀρνηθείην $S_3^{\gamma\rho}$AFP_4^{mg} Plu.: ἀπαρνηθείην SYP καὶ με-] καταμε- Plu. 3 γε SFYP 4 ἔμελλον P_2^{pc} Plu. 6 δικ-αιοτα S^{ac}: δικαιότατα S_3^{pc} 8 ποιῶ SAY^{pc}P: ποιῶν F: ποιεῖν $Lamb^{vl}$: om. Y^{ac} 10 πώποτε AF: ποτὲ SYP ἅπαντι FP_4^{pc} 11 ἐπίῃ A: ἐπῆν F -τα AF -το A 12 διημάρτανε $S_3^{\gamma\rho}$FYP: διημαρτάνει S: διήμαρτε A 14 ὅτε A 15 ἐρεῖν A 16 ἔρημα om. A ἐξεκλη- S^{pc}: ἐκκλη- A$P_4^{\gamma\rho}$: ἐξεκκλη- FYP: [S^{ac}] -σίασαν SFY^{pc}P: -σίαζον A$F_1^{sl}$$P_4^{\gamma\rho}$: [$Y^{ac}$] 18 καὶ bis Y 19 ἀφίκτο A 20 βλέπων οὗτος AFP_4^{mg} θορυβοῦν Y^{ac} 22 ᾗ SF: ἡ A: ἢ YP γελοῖα YP_4^{pc} οἶμαι om. A 24 ἀξιώσεις ... σά] σώζειν ἢ ἐλεεῖν σε τούτους Aristid. εἰς om. S^{ac} 25 προπηλακ- S^{ac} 26 καὶ ὑπεροψίας (post ὑπερηφανίας) AF: καὶ τοσαύτης ὑπεροψίας YP: om. S 27 μεστὸς AFP_4^{mg}: πλήρης SYP πάντων ἀνθρώπων ante μεστὸς F: del. Blass ἔσῃ AFYP

whole of this speech that I'm making has been thought out and prepared. That I've thought it out, men of Athens, I admit, I can't deny it—and indeed that I've rehearsed it, as far as was possible for me. I should in fact have been foolish, when I have been and am subjected to such treatment, if I hadn't taken trouble over what I intended to say to you about it. However, I say it's Meidias who has written the speech for me. For this responsibility should be attributed to **192** the man who has performed the deeds which are the subject of speeches, not the man who has taken thought or who has fretted to say the right things today. That's what I've been doing, men of Athens; I grant it myself. But Meidias has probably never thought out anything that's right in his whole life; if it had occurred to him to give even a little thought to such matters, he wouldn't be making such a mess of the business.

Then I think he won't hesitate even to accuse the people **193** and the Ekklesia, and he'll repeat today what he was bold enough to say at the time of the *probole*—that the meeting comprised all those who had stayed in Athens when they should have gone on the campaign, and all those who had left the forts unmanned, and that it was choristers and aliens and men of that sort who voted him down. He reached such **194** a peak of audacity and shamelessness that day, men of the jury, as those of you who were present know, that, by abuse and threats and staring at any part of the Ekklesia which was making a hubbub, he thought he would terrorize the whole people—which, I think, may well make his tears today seem ridiculous. What do you mean by it, you scoundrel? Are you **195** going to make a request for mercy for your children or yourself, or for support for your case, to *these* men, the men whom you've publicly abused? Are you going to be the only person in the world who, though proved to be during your life so full of arrogance towards everybody that even people who have nothing to do with you are upset at the sight of your audaci-

σύτητα καὶ φωνὴν καὶ τὸ σὸν σχῆμα καὶ τοὺς σοὺς ἀκολούθους καὶ πλοῦτον καὶ ὕβριν θεωροῦντας, ἐν δὲ τῷ κρίνεσθαι
196 παραχρῆμα ἐλεηθήσει; μεγάλην μένταν ἀρχήν, μᾶλλον δὲ
τέχνην, εἴης εὑρηκώς, εἰ δύο τἀναντιώτατα ἑαυτοῖς ἐν οὕτω
βραχεῖ χρόνῳ περὶ σεαυτὸν δύναιο ποιεῖσθαι, φθόνον ἐξ ὧν 5
ζῇς, καὶ ἐφ' οἷς ἐξαπατᾷς ἔλεον. οὐκ ἔστιν οὐδαμόθεν σοι
προσήκων ἔλεος οὐδὲ καθ' ἕν, ἀλλὰ τοὐναντίον μῖσος καὶ
φθόνος καὶ ὀργή· τούτων γὰρ ἄξια ποιεῖς.

Ἀλλ' ἐπ' ἐκεῖνο ἐπάνειμι, ὅτι τοῦ δήμου κατηγορήσει καὶ
197 τῆς ἐκκλησίας. ὅταν οὖν τοῦτο ποιῇ, ἐνθυμεῖσθε παρ' ὑμῖν 10
αὐτοῖς, ἄνδρες δικασταί, ὅτι οὗτος τῶν μεθ' ἑαυτοῦ στρατευ-
σαμένων ἱππέων, ὅτ' εἰς Ὄλυνθον διέβησαν, ἐλθὼν πρὸς ὑμᾶς
εἰς τὴν ἐκκλησίαν κατηγόρει. πάλιν νῦν μείνας πρὸς τοὺς
ἐξεληλυθότας τοῦ δήμου κατηγορήσει. πότερον οὖν ὑμεῖς, ἐάν
τε μένητε ἐάν τε ἐξίητε, ὁμολογήσετ' εἶναι τοιοῦτοι οἵους 15
Μειδίας ὑμᾶς ἀποφαίνει, ἢ τοὐναντίον τοῦτον ἀεὶ καὶ
πανταχοῦ θεοῖς ἐχθρὸν καὶ βδελυρόν; ἐγὼ μὲν οἶμαι τοῦτον
τοιοῦτον· ὃν γὰρ οὐχ ἱππεῖς, οὐ συνάρχοντες, οὐ φίλοι δύ-
198 νανται φέρειν, τί τοῦτον εἴπῃ τις; ἐμοὶ μὲν νὴ τὸν Δία καὶ τὸν
Ἀπόλλω καὶ τὴν Ἀθηνᾶν (εἰρήσεται γάρ, εἴτ' ἄμεινον εἴτε 20
μή), ὅθ' οὗτος ὡς ἀπήλλαγμαι περιιὼν ἐλογοποίει, ἔνδηλοί
τινες ἦσαν ἀχθόμενοι τῶν πάνυ τούτῳ λαλούντων ἡδέως. καὶ
νὴ Δία αὐτοῖς πολλὴ συγγνώμη· οὐ γάρ ἐστι φορητὸς ἅνθρωπος, ἀλλὰ καὶ πλουτεῖ μόνος καὶ λέγειν δύναται μόνος, καὶ
πάντες εἰσὶ τούτῳ καθάρματα καὶ πτωχοὶ καὶ οὐδ' ἄνθρωποι. 25
199 τὸν οὖν ἐπὶ ταύτης τῆς ὑπερηφανίας ὄντα, νῦν ἂν ἀποφύγῃ, τί
ποιήσειν οἴεσθε; ἐξ ὅτου δὲ τοῦτ' ἂν εἰδείητε, ἐγὼ φράσω· εἰ
τοῖς μετὰ τὴν καταχειροτονίαν τεκμηρίοις θεωρήσετε. τίς γάρ
ἐστιν ὅστις καταχειροτονηθὲν αὐτοῦ, καὶ ταῦτ' ἀσεβεῖν περὶ
τὴν ἑορτήν, εἰ καὶ μηδεὶς ἄλλος ἐπῆν ἀγὼν ἔτι μηδὲ κίνδυνος, 30

1 τὸ σὸν AFP^{sl}_{2}: τὸ SYP: del. Weil σοὺς om. Fu 2 κρίνεσθαί σε A 3 ἐλεηθήσῃ AFYP 4 εἴης ἂν AFYP τὰ ἐν- SFYP ἐν om. YP^{ac} οὕτω SFYP: τῷ A 5 σαυτὸν F ποιήσασθαι F ἐξ ὧν SFYP: ἐφ' οἷς A 11 ὦ ἄνδρες Ἀθηναῖοι A^{cp} 12 -τε AF 13 νῦν πάλιν AF 14–15 ἄν ... ἄν AF 15 -τε AF 16 καὶ FP^{sl}_{4}: om. SAYP 19 εἴποι P^{pc}_{2}: ἂν εἴποι F: [P^{ac}] μὲν om. AF 21 ὅτε A ἀπήλαγμαι S περιιὼν S^{pc}FP^{pc}_{3}: περιὼν $S^{ac}A^{pc}P^{ac}$: περὶ ὧν A^{ac}Y 22 τούτων YP^{ac} 23–4 ἅνθρωπος Bekker: ἀνθρωπος SA: ἄνθρωπος FYP 24 καὶ λέγειν δύναται μόνος om. P^{ac} 25 -δὲ AF 26 τὸν οὖν] τουτονὶ τοίνυν τὸν $S^{\gamma\rho}_{3}$ ἐὰν A 27 ταῦτ' P: ταῦτα A 28 καταχειροτονίαν S^{sl}_{2}AFYP: χειροτονίαν S τεκμη-ρίοις om. A 29 ὅστις ἂν S^{sl}_{3} καταχειροτονηθὲν S^{ac}AP^{pc}_{4}: καταχειροτον-ηθέντος S^{pc}_{3}FYP^{ac} -τα AF 30 κίνδυνος μηδὲ ἀγὼν AF

ous talking, your posturing, your lackeys and wealth and insolence—when it comes to your trial, are you going to obtain mercy, just like that? You would certainly have **196** acquired a great power, or rather skill, if you were able to secure for yourself two exactly opposite things in such a short time—resentment because of your lifestyle, and mercy for your lies! No, mercy is not appropriate for you, from any source or in any respect, but, on the contrary, loathing, resentment, and anger; those are what your acts deserve.

But I'll go back to the other point: that he's going to accuse the people and the Ekklesia. When he does that, bear **197** in mind, men of the jury, that this is the man who came to you and accused before the Ekklesia his own fellow-cavalrymen, at the time when they crossed over to Olynthos. Now, on the other hand, he, who remained here, is going to accuse the people, in the presence of those who went out on the campaign. Which, then, is your opinion—will you, whether you stay at home or go out, admit that you are the sort of men Meidias declares you to be? Or do you consider, on the contrary, that it's he who, always and everywhere, is a loathsome devil? The latter, I think; for what name is one to give to a man whom neither cavalry nor colleagues nor friends can bear? To me—I affirm it by Zeus and Apollo and Ath- **198** ena; for it shall be said, whether for good or ill—it was plain, at the time when Meidias was going around alleging that I had dropped the case, that some of those who are perfectly happy to talk to him were sorry to hear it. And indeed they had every excuse, for the fellow's unbearable: oh yes, he's the only man who's rich, he's the only man who can speak, to him all the others are rubbish and paupers and not human beings at all! So what do you think a man as arrogant as that **199** will do if he's acquitted today? I'll tell you how you can find out: by considering the question on the evidence of what happened after the adverse vote. For who on earth, when there'd been a vote against him, and that for impiety concerning the festival, even if no other trial or jeopardy still

οὐκ ἂν ἐπ' αὐτῷ τούτῳ κατέδυ καὶ μέτριον παρέσχεν ἑαυτὸν
τόν γε δὴ μέχρι τῆς κρίσεως χρόνον, εἰ καὶ μὴ πάντα; οὐδεὶς
200 ὅστις οὐκ ἄν. ἀλλ' οὐ Μειδίας, ἀλλ' ἀπὸ ταύτης τῆς ἡμέρας
λέγει, λοιδορεῖται, βοᾷ. χειροτονεῖταί τις· Μειδίας Ἀναγυ-
ράσιος προβέβληται. Πλουτάρχου προξενεῖ, τἀπόρρητ' οἶδεν, 5
ἡ πόλις αὐτὸν οὐ χωρεῖ. καὶ ταῦτα πάντα ποιεῖ δῆλον ὅτι
οὐδὲν ἄλλο ἐνδεικνύμενος ἢ ὅτι "ἐγὼ οὐδὲν πέπονθα ὑπὸ τῆς
καταχειροτονίας, οὐδὲ δέδοικα οὐδὲ φοβοῦμαι τὸν μέλλοντα
201 ἀγῶνα." ὃς οὖν, ὦ ἄνδρες Ἀθηναῖοι, τὸ μὲν ὑμᾶς δεδιέναι
δοκεῖν αἰσχρὸν ἡγεῖται, τὸ δὲ μηδὲν φροντίζειν ὑμῶν νεα- 10
νικόν, τοῦτον οὐκ ἀπολωλέναι δεκάκις προσήκει; οὐδὲ γὰρ
ἕξειν ὑμᾶς ὅ τι χρήσεσθε αὐτῷ νομίζει. πλούσιος, θρασύς,
μέγα φρονῶν, μέγα φθεγγόμενος, βίαιος, ἀναιδής—ποῦ
ληφθήσεται, νῦν ἐὰν διακρούσηται;

202 Ἀλλ' ἔγωγε, εἰ μηδενὸς ἕνεκα τῶν ἄλλων, τῶν γε δημηγορ- 15
ιῶν ὧν ἑκάστοτε δημηγορεῖ, καὶ ἐν οἷς καιροῖς, τὴν μεγίστην
ἂν αὐτὸν δικαίως οἶμαι δίκην δοῦναι. ἴστε γὰρ δήπου τοῦθ',
ὅτι, ἂν μέν τι τῶν δεόντων ἀπαγγελθῇ τῇ πόλει καὶ τοιοῦτον
οἷον εὐφρᾶναι πάντας, οὐδαμοῦ πώποτε Μειδίας τῶν συνη-
203 δομένων οὐδὲ τῶν συγχαιρόντων ἐξητάσθη τῷ δήμῳ, ἂν δέ τι 20
φλαῦρον, ὃ μηδεὶς ἂν βούλοιτο τῶν ἄλλων, πρῶτος ἀνέστηκεν
εὐθέως καὶ δημηγορεῖ, ἐπεμβαίνων τῷ καιρῷ καὶ τῆς σιωπῆς
ἀπολαύων ἣν ἐπὶ τῷ περὶ τῶν συμβεβηκότων ἄχθεσθαι ποιεῖσθε ὑμεῖς, "τοιοῦτοι γάρ ἐστε, ὦ ἄνδρες Ἀθηναῖοι· οὐ γὰρ
ἐξέρχεσθε, οὐδ' οἴεσθε δεῖν χρήματα εἰσφέρειν. εἶτα θαυμά- 25
ζετε εἰ κακῶς τὰ πράγματα ὑμῖν ἔχει; ἐμὲ οἴεσθε ὑμῖν εἰσοίσειν, ὑμεῖς δὲ νεμεῖσθε; ἐμὲ οἴεσθε τριηραρχήσειν, ὑμεῖς δ'

3–6 ἀλλ' οὐ … χωρεῖ Plu. *Mor.* 1010f 4–6 Μειδίας … χωρεῖ Aristid. *Rh.* (Schmid) 13.25, 46.3 5 Πλουτάρχου προξενεῖ *AB* 1.163.21 12–13 πλούσιος … ἀναιδής Aristid. *Rh.* (Schmid) 46.1

---

3 τῆς ἡμέρας ταύτης SYP 4 τι Plu. Ἀναγυρράσιος Plu. 5 Πλουτάρχου προξενεῖ SAFYP *AB*: Πλουτάρχῳ προξενεῖ Plu.: om. Aristid. τὰ ἀπ- AFYP Plu. Aristid. 6 δηλονότι FP 7 οὐδὲν πέπονθα AF Σ (Patm.): πέπονθα οὐδὲν SYP 8 οὐ δέδοικα AY^{ac} 9 δεδειέναι S^{ac} 11 προσήκει. ἐγὼ μὲν ἡγοῦμαι AP^{sl}_4 12 χρήσεσθαι F 12–13 θρασὺς καὶ μεγαλόφρων Σ (672 Dilts) 14 ἂν F 16 ὧν] τῶν S^{ac} -τοτε δημ- YP^{ac}: -τοτ' ἐδημ- SFP^{pc}_4: -τοτε ἐδημ- A οἷς om. $Lamb^{vl}$: τοῖς Lamb 17 οἶμαι δικαίως A: δικαίως οἴομαι F^{ac} διδόναι A 18 ἐὰν SYP ἀγγελθῇ A 20 οὐδὲ τῶν συγχαιρόντων SFYP: ἢ τῶν συνευφραινομένων A ἐὰν A 20–1 τι λυπηρὸν ἢ φλαῦρον AP^{mg}_4 21 οὗ μηδείς γ' A ἀνέστησεν F: ἀνέστηκεν κεν Y 24 οὐ AFP^{pc}_4: οὐδὲ SYP^{ac} 25 οὐδ' SYP: οὐ γὰρ AFP^{sl}_4 χρήματα δεῖν S^{ac} 26 ὑμῖν τὰ πράγματ' F ἐμὲ δὲ A 27 ὑμᾶς δὲ νεμεῖσθαι Σ **2** (12 Dilts) τριηραρχεῖν A δὲ A

impended, would not have, on that ground alone, kept a low profile and conducted himself modestly, at least for the period until the trial, if not for ever? Anyone would have **200** done that. But not Meidias; from that day on he speaks, he reviles, he shouts. An election is being held: Meidias of Anagyrous has been proposed! He's the representative of Ploutarkhos, he knows all the confidential information, the city isn't big enough to hold him! And he does all these things, obviously, just to demonstrate that 'The adverse vote hasn't done me any harm! I'm not frightened or afraid of the coming trial!' Surely, men of Athens, a man who considers it shame- **201** ful to seem to be frightened of you, but macho not to care about you, deserves death ten times over. He thinks you won't know how to deal with him. Rich, bold, with a big head and a big voice, violent, shameless—where will he ever be convicted if he gets off this time?

In my opinion, quite apart from anything else, he deserves **202** the severest punishment for the public speeches which he keeps making, and for the occasions when he makes them. As you know, whenever a report of any success reaches the city, the sort of thing to cheer everyone up, Meidias has never joined in anywhere with those sharing the people's pleasure and rejoicing. But if there's some bad news, something that no **203** one else would wish for, he's the first to be on his feet straightaway and make a speech, seizing the opportunity and taking advantage of the hush which falls upon you in grief at the circumstances—'That's the sort you are, men of Athens: you don't go out on campaigns, and you don't think you need contribute money. Then are you surprised that your affairs are in a bad way? Do you think I shall contribute money for you, while you're going to share it out among yourselves? Do you think I shall serve as a trierarch, while you're not going

204 οὐκ ἐμβήσεσθε;”—τοιαῦτα ὑβρίζων καὶ τὴν ἀπὸ τῆς ψυχῆς πικρίαν καὶ κακόνοιαν, ἣν κατὰ τῶν πολλῶν ὑμῶν ἔχων ἀφανῆ παρ' ἑαυτῷ περιέρχεται, φανερὰν ἐπὶ τοῦ καιροῦ καθιστάς. δεῖ τοίνυν, ὦ ἄνδρες Ἀθηναῖοι, καὶ ὑμᾶς οὕτω νῦν, ὅταν ἐξαπατῶν καὶ φενακίζων ὀδύρηται καὶ κλαίῃ καὶ δέηται, 5 τοῦθ' ὑποβάλλειν αὐτῷ· “τοιοῦτος γὰρ εἶ, Μειδία· ὑβριστὴς γὰρ εἶ, καὶ οὐκ ἐθέλεις ἔχειν παρὰ σεαυτῷ τὼ χεῖρε. εἶτα θαυμάζεις εἰ κακὸς κακῶς ἀπολεῖ; ἀλλὰ νομίζεις ἡμᾶς μὲν ἀνέξεσθαί σου, αὐτὸς δὲ τυπτήσειν; καὶ ἡμᾶς μὲν ἀποψηφιεῖσθαί σου, σὲ δὲ οὐ παύσεσθαι;” 10

205 Καὶ βοηθήσουσιν οἱ λέγοντες ὑπὲρ αὐτοῦ, οὐχ οὕτω τούτῳ χαρίζεσθαι μὰ τοὺς θεοὺς βουλόμενοι, ὡς ἐπηρεάζειν ἐμοὶ διὰ τὴν ἰδίαν ἔχθραν, ἣν οὗτος αὑτῷ πρὸς ἐμέ, ἄν τ' ἐγὼ φῶ ἄν τε μὴ φῶ, φησὶν εἶναι καὶ βιάζεται, οὐκ ὀρθῶς. ἀλλὰ κινδυνεύει τὸ λίαν εὐτυχεῖν ἐνίοτε ἐπαχθεῖς ποιεῖν· ὅπου γὰρ ἐγὼ μὲν 15 οὐδὲ πεπονθὼς κακῶς ἐχθρὸν εἶναί μοι τοῦτον ὁμολογῶ, οὗτος δ' οὐδ' ἀφιέντα ἀφίησιν, ἀλλὰ καὶ ἐπὶ τοῖς ἀλλοτρίοις ἀγῶσιν ἀπαντᾷ καὶ νῦν ἀναβήσεται μηδὲ τῆς κοινῆς τῶν νόμων ἐπικουρίας ἀξιῶν ἐμοὶ μετεῖναι, πῶς οὐχ οὗτος ἐπαχθής ἐστιν ἤδη καὶ μείζων ἢ καθ' ὅσον ἡμῶν ἑκάστῳ συμφέρει; 20

206 ἔτι τοίνυν παρῆν, ὦ ἄνδρες Ἀθηναῖοι, καὶ ἐκάθητο Εὔβουλος ἐν τῷ θεάτρῳ, ὅτε ὁ δῆμος κατεχειροτόνησε Μειδίου, καὶ καλούμενος ὀνομαστὶ καὶ ἀντιβολοῦντος τούτου καὶ λιπαροῦντος, ὡς ὑμεῖς ἴστε, οὐκ ἀνέστη. καὶ μὴν εἰ μὲν μηδὲν ἠδικηκότος ἡγεῖτο τὴν προβολὴν γεγενῆσθαι, τότ' ἔδει τόν γε 25 φίλον δήπου συνειπεῖν καὶ βοηθῆσαι· εἰ δὲ καταγνοὺς ἀδικεῖν τότε, διὰ ταῦτ' οὐχ ὑπήκουσεν, νῦν δέ, ὅτι προσκέκρουκεν ἐμοί, διὰ ταῦτα τοῦτον ἐξαιτήσεται, ὑμῖν οὐχὶ καλῶς ἔχει

207 χαρίσασθαι. μὴ γὰρ ἔστω μηδεὶς ἐν δημοκρατίᾳ τηλικοῦτος,

8–10 ἀλλὰ ... παύσεσθαι *AB* 1.122.27 14–15 ἀλλὰ ... ποιεῖν *AB* 1.151.14

---

1 ἐμβήσεσθαι Xe 2 ἔχει F$^{pc}_{1}$: [F^{ac}] 4 ὅταν οὗτος F 6 τοῦθ' S: τοιαῦθ' A: ταῦθ' FYP 7 καὶ AFP$^{pc}_{3}$: om. SYP^{ac} σαυτῷ AF τῶι S 8 κακὸς om. Y^{ac} ἀπολῇ AF^{sl}YP μὲν om. *AB* 9 σου om. AF 10 σὲ SFYP: σὺ A *AB* οὐ SFYP *AB*: οὐδὲ A 11 βοηθήσουσιν S$^{sl}_{4}$AFYP: βοηθοῦσιν S οὕτως F 12 χαρίζεσθαι AFP$^{sl}_{3}$: χαρίσασθαι SF$^{sl}_{1}$YP 13 οὗτος] εὔβουλος S^{sl}P^{mg} αὑτῷ P$^{pc}_{4}$: αυτῷ SAP^{ac}: αὐτῷ FY ἐάν ... ἐάν A τε SYP φῶι S 14 βιάζεται τοῦτο FP$^{sl}_{3}$ κινδυνεύει καὶ *AB* 16 οὐδὲ SAFP$^{mg}_{4}$: ὁ YP: οὐ P$^{γρ}_{2}$ 17 δὲ SFYP ἀφίησιν S$^{sl}_{4}$AFP$^{sl}_{4}$: ἀνίησιν SYP 19–20 ἀπεχθής A 20 μεῖζον SAYP 21 ὦ om. F καθῆτο F: ἐκαθεῖτο P$^{γρ}_{4}$ σύμβουλος S^{ac} 23 καὶ (post ὀνομαστὶ) om. AFP$^{pc}_{4}$ 27 ταῦθ' S: τοῦτο A -σε AF^{pc} 28 ταῦτα S^{pc}FYP: ταυτας S^{ac}: τοῦτο A ἔχει αὐτῷ Lamb^{vl}

to man the ships?'—delivering that sort of insult, and making manifest on this occasion the heartfelt hostility and ill-will towards you ordinary people which he normally conceals within himself. So now you must do the same, men of Athens: when he tries to cheat and trick you by wailing and weeping and entreating, interrupt him with this retort—'That's the sort you are, Meidias: you're insolent, you aren't prepared to keep your hands to yourself. Then are you surprised that, being a bad man, you're going to die a bad death? Do you suppose that we shall put up with you and you yourself can go on hitting? And that we shall acquit you and you needn't stop doing it?' **204**

His supporting speakers will also come to his aid—not so much wanting to do him any favour, I can tell you that, as to create difficulties for me, because of the personal hostility which that man ⟨Euboulos⟩, whether I agree or disagree, says exists between himself and me, and insists that it is so—quite wrongly. But perhaps extreme success does sometimes make people oppressive; for when I, even though badly treated, don't admit that there is a quarrel between us, while he won't release me from it though I release him, but confronts me at the trials of other men and in this case is going to come up and request that I should be denied even the legal protection which is available to everyone—surely that amounts to oppression, and greater power than is good for us individuals. Besides, men of Athens, Euboulos was present and seated in the theatre on the occasion when the people voted against Meidias, and, though he was called on by name and Meidias begged and implored him, as you know, he didn't stand up. Now, if he thought that Meidias was innocent when the *probole* was brought against him, he surely ought, being his friend, to have spoken for him and supported him at that time. But if the reason why he didn't respond was that he'd decided he was guilty then, and yet now, just because he's had a disagreement with me, he's going to ask for Meidias' acquittal, it's not proper for you to grant it. Let no one in a democracy be so mighty that his **205** **206** **207**

ὥστε συνειπὼν τὸν μὲν ὑβρίσθαι, τὸν δὲ μὴ δοῦναι δίκην ποιῆσαι. ἀλλ' εἰ κακῶς ἐμὲ βούλει ποιεῖν, Εὔβουλε, ὡς ἔγωγε μὰ τοὺς θεοὺς οὐκ οἶδα ἀνθ' ὅτου, δύνασαι μὲν καὶ πολιτεύει, κατὰ τοὺς νόμους δ' ἥντινα βούλει παρ' ἐμοῦ δίκην λάμβανε, ὧν δ' ἐγὼ παρὰ τοὺς νόμους ὑβρίσθην, μή μ' ἀφαιροῦ τὴν τιμωρίαν. εἰ δ' ἀπορεῖς ἐκείνως με κακῶς ποιῆσαι, εἴη ἂν καὶ τοῦτο σημεῖον τῆς ἐμῆς ἐπιεικείας, εἰ τοὺς ἄλλους ῥᾳδίως κρίνων ἐμὲ μηδὲν ἔχεις ἐφ' ὅτῳ τοῦτο ποιήσεις.

208 Πέπυσμαι τοίνυν καὶ Φιλιππίδην καὶ Μνησαρχίδην καὶ Διότιμον τὸν Εὐωνυμέα καὶ τοιούτους τινὰς πλουσίους καὶ τριηράρχους ἐξαιτήσεσθαι τοῦτον καὶ λιπαρήσειν παρ' ὑμῶν, αὑτοῖς ἀξιοῦντας δοθῆναι τὴν χάριν ταύτην. περὶ ὧν οὐδὲν ἂν εἴποιμι πρὸς ὑμᾶς φλαῦρον ἐγώ (καὶ γὰρ ἂν μαινοίμην), ἀλλ' ἃ θεωρεῖν ὑμᾶς, ὅταν οὗτοι δέωνται, δεῖ καὶ λογίζεσθαι, ταῦτ' **209** ἐρῶ. ἐνθυμεῖσθ', ὦ ἄνδρες δικασταί, εἰ γένοιντο (ὃ μὴ γένοιτ', οὐδ' ἔσται) οὗτοι κύριοι τῆς πολιτείας μετὰ Μειδίου καὶ τῶν ὁμοίων τούτῳ, καί τις ὑμῶν τῶν πολλῶν καὶ δημοτικῶν ἀνθρώπων ἁμαρτὼν εἴς τινα τούτων, μὴ τοιαῦθ' οἷα Μειδίας εἰς ἐμέ, ἀλλ' ὁτιοῦν ἄλλο, εἰς δικαστήριον εἰσίοι πεπληρωμένον ἐκ τούτων, τίνος συγγνώμης ἢ τίνος ἐλέου τυχεῖν ἂν οἴεσθε; ταχύ γ' ἂν χαρίσαιντο (οὐ γάρ;) ἢ δεηθέντι τῳ τῶν πολλῶν πρόσσχοιεν. ἀλλ' οὐκ ἂν εὐθέως εἴποιεν "τὸν δὲ βάσκανον, τὸν δὲ ὄλεθρον· τοῦτον δὲ ὑβρίζειν, ἀναπνεῖν δέ· ὃν εἴ τις ἐᾷ **210** ζῆν, ἀγαπᾶν δεῖ"; μὴ τοίνυν, ὦ ἄνδρες Ἀθηναῖοι, τούτοις τοῖς οὕτω χρησαμένοις ἂν ὑμῖν ἄλλως πως ἔχετε ὑμεῖς, μηδὲ τὸν πλοῦτον μηδὲ τὴν δόξαν τὴν τούτων θαυμάζετε, ἀλλ' ὑμᾶς αὐτούς. πολλὰ τούτοις ἀγαθά ἐστιν, ἃ τούτους οὐδεὶς κωλύει κεκτῆσθαι· μὴ τοίνυν μηδ' οὗτοι τὴν ἄδειαν, ἣν ἡμῖν κοινὴν οὐσίαν οἱ νόμοι παρέχουσιν, κωλυόντων κεκτῆσθαι.

3 πολιτεύει Dindorf: πολιτεύῃ SAFYP 4 δὲ A 9 πέπυσμαι $S^{pc}FP^{sl}$: πέπεισμαι S^{ac}AYP 10 δότιμον A 11 τοῦτον A: om. SFYP ὑμῶν αὐτὸν SFYP 12 αὑτοῖς Felicianus: αυτοῖς SA: αὐτοῖς FYP 13 μενοίμην S^{ac} 14 αὐτοὶ Y^{ac} 15 -σθε AF ὦ ἄνδρες ἀθηναῖοι S^{pc} 16 ἔσταί ποτε A νῦν οὗτοι FP^{sl}_4 17–18 ἀνθρώπων om. F 18 -τα A 20 ἐλέου AFYP: λόγου S ἂν om. S 21 ἢ A: οὐ S^{pc}_3: δὴ FP^{sl}_4: om. YP: [S^{ac}] 22 πρόσσχοιεν MacDowell (προσσχοῖεν Bekker): προσχοῖεν S: προσέχοιεν A: πρόσχοιεν FYP οὐκ ἂν εὐθέως εἴποιεν SYP: οὐ AF 22–3 τόνδε ... τόνδε A 23 δὲ (post τοῦτον) om. XPrFu δέ· ὃν AYP: δέον SFP^{mg}_2 24 ζηιν S ἔδει AFP^{pc}_4 ὦ ἄνδρες δικασταί A^{cp} 25 ὑμῖν] ὑμεῖς S^{ac} ὑμεῖς ἔχετε A 26 τὴν (post δόξαν) AFP^{pc}_4: om. SYP^{ac} 27 τούτους] τούτοις A οὐδε εἰς A: οὐδὲ εἷς CdTWd 28–9 κοινὴν ἡμῖν A 29 οὖσαν $S^{\gamma\rho}_3P^{\gamma\rho}_4$ -σι AF

advocacy causes the one man to continue bearing the burden of insolence and the other to go unpunished! But if you wish to attack me, Euboulos—and I swear I don't know what you wish to requite—you have influence and are active in the state: exact from me what legal penalty you wish, but don't deprive me of requital for the illegal insolence I suffered. If you have no means of attacking me in that way, that may be evidence of my innocence—if, though you find it easy to put other men on trial, you have no grounds for doing that to me.

I've also been told that Philippides, Mnesarkhides, Dioti- **208** mos of Euonymon, and some men of that sort, rich trierarchs, will beg him off and entreat you, asking for this to be done as a favour to them. I don't want to say anything disparaging about them to you (I should be mad to do so), but I will say what you ought to observe and bear in mind when they make their request. Just consider, men of the jury: if it should **209** happen (as I hope and am sure it will not) that these men got control of the state, along with Meidias and those like him, and then one of you, the general public, committed an offence against one of them—not the sort of thing Meidias did to me, but anything else—and came before a court manned by these men, what pardon or what mercy do you think he would get? They'd soon do him a favour, wouldn't they, or pay careful attention to the request of an ordinary member of the public! Wouldn't they, rather, say at once 'The menace! The pest! That he should be guilty of insolence, and still breathe! He ought to be satisfied if he's allowed to live'? So, **210** men of Athens, since they would treat you in that way, your attitude to them should not be any different. Show respect not for these men's wealth or reputation, but for yourselves. They have many assets, which no one prevents them from possessing; so let not even them prevent the possession of the

211 οὐδὲν δεινὸν οὐδ' ἐλεεινὸν Μειδίας πείσεται, ἂν ἴσα κτήσηται
τοῖς πολλοῖς ἡμῶν, οὓς νῦν ὑβρίζει καὶ πτωχοὺς ἀποκαλεῖ, ἃ
δὲ νῦν περιόντ' αὐτὸν ὑβρίζειν ἐπαίρει, περιαιρεθῇ. οὐδ' οὗτοι
δήπου ταῦθ' ὑμῶν εἰσι δίκαιοι δεῖσθαι, "μὴ κατὰ τοὺς νόμους
δικάσητε, ἄνδρες δικασταί· μὴ βοηθήσητε τῷ πεπονθότι 5
δεινά, μηδ' εὐορκεῖτε· ἡμῖν δότε τὴν χάριν ταύτην." ταῦτα
γάρ, ἄν τι δέωνται περὶ τούτου, δεήσονται, κἂν μὴ ταῦτα
212 λέγωσι τὰ ῥήματα. ἀλλ' εἴπερ εἰσὶ φίλοι καὶ δεινὸν εἰ μὴ
πλουτήσει Μειδίας ἡγοῦνται, εἰσὶ μὲν εἰς τὰ μάλιστ' αὐτοὶ
πλούσιοι, καὶ καλῶς ποιοῦσι, χρήματα δ' αὐτῷ παρ' ἑαυτῶν 10
δόντων, ἵν' ὑμεῖς μὲν ἐφ' οἷς εἰσήλθετε ὀμωμοκότες δικαίως
ψηφίσησθε, οὗτοι δὲ παρ' αὐτῶν τὰς χάριτας, μὴ μετὰ τῆς
ὑμετέρας αἰσχύνης, ποιῶνται. εἰ δ' οὗτοι χρήματ' ἔχοντες μὴ
προοῖντ' ἄν, πῶς ὑμῖν καλὸν τὸν ὅρκον προέσθαι;

213 Πλούσιοι πολλοὶ συνεστηκότες, ὦ ἄνδρες Ἀθηναῖοι, τὸ 15
δοκεῖν τινες εἶναι δι' εὐπορίαν προσειληφότες, ὑμῶν παρίασι
δεησόμενοι. τούτων μηδενί με, ὦ ἄνδρες Ἀθηναῖοι, προῆσθε,
ἀλλ' ὥσπερ ἕκαστος τούτων ὑπὲρ τῶν ἰδίᾳ συμφερόντων καὶ
ὑπὲρ τούτου σπουδάσεται, οὕτως ὑμεῖς ὑπὲρ ὑμῶν αὐτῶν καὶ
τῶν νόμων καὶ ἐμοῦ τοῦ ἐφ' ὑμᾶς καταπεφευγότος σπουδά- 20
214 σατε, καὶ τηρήσατε τὴν γνώμην ταύτην ἐφ' ἧς νῦν ἐστε. καὶ
γὰρ εἰ μέν, ὦ ἄνδρες Ἀθηναῖοι, τότε, ὅτ' ἦν ἡ προβολή, τὰ
πεπραγμένα ὁ δῆμος ἀκούσας ἀπεχειροτόνησε Μειδίου, οὐκ
ἂν ὁμοίως ἦν δεινόν· καὶ γὰρ ⟨ὕβριν⟩ μὴ γεγενῆσθαι, καὶ μὴ
περὶ τὴν ἑορτὴν ἀδικήματα ταῦτ' εἶναι, καὶ πολλὰ ἂν εἶχέν τις 25
215 αὐτὸν παραμυθήσασθαι. νῦν δὲ τοῦτο καὶ πάντων ἄν μοι
δεινότατον συμβαίη, εἰ παρ' αὐτὰ μὲν τἀδικήματα οὕτως
ὀργίλως καὶ πικρῶς καὶ χαλεπῶς ἅπαντες ἔχοντες ἐφαίνεσθε,
ὥστε Νεοπτολέμου καὶ Μνησαρχίδου καὶ Φιλιππίδου καί

1–2 *οὐδὲν … ἡμῶν* Aristid. *Rh.* (Schmid) 9.3 15–16 *πλούσιοι … προσειληφότες* Aristid. *Rh.* (Schmid) 9.5 21–4 *καὶ γὰρ … δεινόν* *AB* 1.122.31

---

1 *-δὲ* A *ἐλεινὸν* Aristid. *πείσεται μειδίας* A *ἐὰν* A Aristid. 2 *ἡμῶν* SYP: *ὑμῶν* AFP_4^{sl}: *ὑμῖν* Aristid. 3 *περιόντ'* S: *περιόντα* A^{ac}: *περιιόντα* A^{pc}: *περιιόντ'* FYP 5 *δικάσητε* S_3^{sl}AFYP: *δικάσηται* S *ὦ ἄνδρες δικασταί* A^{cp}F *βοηθήσετε* A 6 *μηδ'* SFYP: *μὴ* A *εὐορκῆτε* F 7 *ταὐτὰ* FP_4^{mg} 8 *-ωσιν* S 9 *πλουτήσῃ* F *εἰς*] *εἰσι* $S_3^{γρ}$A *-τα* $S_3^{γρ}$AF 10 *δὲ* A *ἑαυτῶν*] *αὑτῷ* F^{ac}: *αὑτῶν* F^{pc} 11 *-να* A *εἰσεληλύθατε* F$P_4^{γρ}$ 13 *-τα* AF 14 *προοῖντ'* MacDowell: *πρόοιντ'* SAFYP 15 *ὦ ἄνδρες Ἀθηναῖοι* om. Aristid. 16 *προειληφότες* Aristid. 17 *με* SFYP: *μὲν* A *προῆσθε* Bekker: *πρόησθε* SAFYP 18 *ἕκαστοι* S^{ac} *ἰδίων* FP_4^{sl} 19 *αὐτῶν* om. Y^{ac} 22 *ὅτε* A 24 *ὕβριν* add. Reiske 25 *-χέ* AF 27 *μὲν* FP_4^{sl}: om. SAYP 28 *ἔχοντες πάντες* A *φαινεσθε* S

legal protection which is our common property. Meidias **211** won't be suffering anything terrible or pitiable if he possesses equality with most of us (whom he now insults and calls paupers) but is stripped of the superfluous property which at present induces him to be insolent. Surely not even these men can fairly make these requests of you: 'Don't give a verdict in accordance with the laws, men of the jury; don't support the victim of maltreatment or keep your oath; do us this favour!' For, if they make any plea about Meidias, that's what it will be, even if those aren't the words they use. But if they are **212** friends, and consider it a dreadful thing if Meidias is not going to be rich, they themselves have riches in plenty, and no harm in that: let them give him money from their own pockets, so that, while you give a just verdict in the case for which you've come into court under oath, they may do their favours at their own expense without disgrace to you. But if they have money and are not prepared to sacrifice it, how can it be right for you to sacrifice your oath?

A large number of rich men have gathered here, men of **213** Athens; their affluence has given them a reputation for being important people, and they will come forward to address a plea to you. Don't sacrifice me to any of them, men of Athens! Just as each of them will exert himself in defence of his personal interests and in defence of Meidias, so you must exert yourselves in defence of yourselves, the laws, and me, who have taken refuge with you, and you must keep to this purpose which you have at present. If, men of Athens, at the **214** time when the *probole* was held, the people, after hearing the facts, had voted to acquit Meidias, that wouldn't have been so serious; one could have consoled oneself by saying that insolence had not been committed, that these acts were not offences concerning the festival, and so on. But as it is, it **215** would really be the most dreadful thing that could happen to me, if, whereas at the actual time of the offences it was obvious that you all felt so angry and bitter and severe that, when Neoptolemos and Mnesarkhides and Philippides and

τινος τῶν σφόδρα τούτων πλουσίων δεομένων καὶ ἐμοῦ καὶ
ὑμῶν ἐβοᾶτε μὴ ἀφεῖναι, καὶ προσελθόντος μοι Βλεπαίου τοῦ
τραπεζίτου τηλικοῦτ' ἀνεκράγετε ὡς, τοῦτ' ἐκεῖνο, χρήματά
216 μου ληψομένου, ὥστε με, ὦ ἄνδρες Ἀθηναῖοι, φοβηθέντα τὸν
ὑμέτερον θόρυβον θοἰμάτιον προέσθαι καὶ μικροῦ γυμνὸν ἐν 5
τῷ χιτωνίσκῳ γενέσθαι φεύγοντα ἐκεῖνον ἕλκοντά με, καὶ
μετὰ ταῦτα ἀπαντῶντες "ὅπως ἐπέξει τῷ μιαρῷ καὶ μὴ
διαλύσει· θεάσονταί σε τί ποιήσεις Ἀθηναῖοι" τοιαῦτα λέγοντες, ἐπειδὴ δὲ κεχειροτόνηται μὲν ὕβρις τὸ πρᾶγμα εἶναι,
ἐν ἱερῷ δ' οἱ ταῦτα κρίνοντες καθεζόμενοι διέγνωσαν, διέ- 10
μεινα δὲ ἐγὼ καὶ οὐ προύδωκα οὔθ' ὑμᾶς οὔτ' ἐμαυτόν, τηνι-
217 καῦτ' ἀποψηφιεῖσθε ὑμεῖς. μηδαμῶς· πάντα γὰρ τὰ αἴσχιστα
ἔνεστιν ἐν τῷ πράγματι, εἰμὶ δ' οὐ τούτων ὑμῖν ἄξιος (πῶς
γάρ, ὦ ἄνδρες Ἀθηναῖοι;) κρίνων ἄνθρωπον καὶ δοκοῦντα καὶ
ὄντα βίαιον καὶ ὑβριστήν, ἡμαρτηκότα ἀσελγῶς ἐν πανηγύ- 15
ρει, μάρτυρας τῆς ὕβρεως τῆς ἑαυτοῦ πεποιημένον οὐ μόνον
ὑμᾶς ἀλλὰ καὶ τοὺς ἐπιδημήσαντας ἅπαντας τῶν Ἑλλήνων.
ἤκουσεν ὁ δῆμος τὰ πεπραγμένα τούτῳ. τί οὖν; ὑμῖν κατα-
218 χειροτονήσας παρέδωκεν. οὐ τοίνυν οἷόν τε ἀφανῆ τὴν γνῶσιν ὑμῶν γενέσθαι οὐδὲ λαθεῖν, οὐδ' ἀνεξέταστον εἶναι τί ποθ' 20
ὡς ὑμᾶς τοῦ πράγματος ἐλθόντος ἔγνωτε· ἀλλ' ἐὰν μὲν
κολάσητε, δόξετε σώφρονες εἶναι καὶ καλοὶ κἀγαθοὶ καὶ
μισοπόνηροι, ἂν δ' ἀφῆτε, ἄλλου τινὸς ἡττῆσθαι. οὐ γὰρ
ἐκ πολιτικῆς αἰτίας, οὐδ' ὥσπερ Ἀριστοφῶν ἀποδοὺς τοὺς
στεφάνους ἔλυσε τὴν προβολήν, ἀλλ' ἐξ ὕβρεως, καὶ ἐκ τοῦ 25
μηδὲν ἂν ὧν πεποίηκεν ἀναλῦσαι δύνασθαι κρίνεται. πότερ'
οὖν τούτου γενομένου κρεῖττον αὖθις ἢ νυνὶ κολάσαι; ἐγὼ μὲν
οἶμαι νῦν· κοινὴ γὰρ ἡ κρίσις, καὶ τἀδικήματα πάντα ἐφ' οἷς
νῦν κρίνεται κοινά.

219 Ἔτι δὲ οὐκ ἐμὲ ἔτυπτεν, ἄνδρες Ἀθηναῖοι, μόνον οὗτος 30

1 τινος SF$^{sl}_{1}$YP: τῶν ἄλλων A: τινων FP$^{sl}_{4}$ πλουσίων τούτων πλουσίων S^{ac}: πλουσίων τούτων YP δεομένου S^{pc}: [S^{ac}] 3 εκεινου S 4 ληψομένου S^{ac} ὦ ἄνδρες Ἀθηναῖοι om. A 5 τὸ ἱμ- A 6 ἐκεῖνον δ' SAYP 7 επεξηι A: ἐπεξῇ Y μὴ om. A 8 διαλύσῃ AP: [S^{ac}] 9 ἐπεὶ F^{pc} δὲ FP$^{pc}_{4}$: om. SAYP^{ac} 10 δὲ SYP κρίναντες A 11 κἀγὼ καὶ AF 12 -τα A ὑμεῖς; AY 13 δὲ A 13–14 πῶς γὰρ ὅστις ὦ ἄνδρες ἀθηναῖοι κρίνω S$^{\gamma\rho}_{3}$F$^{mg}_{1}$P$^{\gamma\rho}_{4}$ 16 αὐτοῦ F 17 ἐπιδημοῦντας A ἅπαντα S^{ac}Y^{ac}: πάντας A 22 καὶ ἀγαθοὶ F 23 ἡττῆσθαι Lamb^{vl}: ἥττησθε SYP: ἡττᾶσθε AFP$^{sl}_{4}$: ἡττᾶσθαι Cd^{pc}Pr 25 προσβολήν S^{ac} καὶ F$^{pc}_{1}$: om. SAF^{ac}YP 26 πότερον AF 27 κρεῖττον ἦν SFYP 29 νῦν SF$^{\gamma\rho}_{1}$YP: om. AF 30 ὦ ἄνδρες ἀθηναῖοι A^{cp}F

those very rich men one and all were making their pleas both to me and to you, you kept calling out 'Don't let him off!', and when Blepaios the banker came up to me you let out such a shout, thinking 'Here we go again! He's going to accept money!', that in my fright at the noise you were mak- **216** ing, men of Athens, I let my cloak fall off and was left almost nude in my tunic as he was trying to catch hold of me and I was trying to get away from him; and later, when you met me, you kept saying 'Be sure to proceed against the scoundrel, and not come to terms with him! People in Athens will be watching to see what you will do!', all that sort of thing—yet after the act has been voted to be one of insolence, and the judges who gave the verdict were sitting in a sacred place, and I persisted and didn't betray either you or myself, in the end you're going to acquit him! Never; for to **217** do so entails the greatest disgrace, and I don't deserve that treatment from you, surely, men of Athens, when I'm putting on trial a man who is reputed to be, and indeed is, violent and insolent, who has behaved as a bully at a national festival, and who has displayed his insolence not merely before you, but before all the Greeks who were in Athens. His conduct was reported to the people; and what did they do? They voted against him and passed him on to you. So it isn't **218** possible for your verdict to be unobtrusive or unnoticed, nor that the question should not be asked what you decided when the case came before you. If you punish him, you'll be thought prudent and honourable opponents of crime; if you acquit him, you'll be thought to have yielded to some other consideration. For the charge is not a political one, nor is it like the case in which Aristophon disposed of the prosecution by delivering the crowns; Meidias' trial arises from insolence, and from the impossibility of undoing any of what he has done. That being so, is it better to punish him on some future occasion, or now? I think now; for this is a public trial, and all the offences for which he is now on trial are public ones.

Besides, men of Athens, it was not only I that he was **219**

οὐδ' ὕβριζε τῇ διανοίᾳ τότε ποιῶν οἷα ἐποίει, ἀλλὰ πάντας
ὅσους περ ἂν οἴηταί τις ἧττον ἐμοῦ δύνασθαι δίκην ὑπὲρ
αὐτῶν λαβεῖν. εἰ δὲ μὴ πάντες ἐπαίεσθε μηδὲ πάντες ἐπηρ-
εάζεσθε χορηγοῦντες, ἴστε δήπου τοῦθ', ὅτι οὐδὲ ἐχορηγεῖθ'
ἅμα πάντες, οὐδὲ δύναιτ' ἄν ποθ' ὑμᾶς οὐδεὶς ἅπαντας μιᾷ 5
220 χειρὶ προπηλακίσαι. ἀλλ' ὅταν εἷς ὁ παθὼν μὴ λάβῃ δίκην,
τόθ' ἕκαστον αὐτὸν χρὴ προσδοκᾶν τὸν πρῶτον μετὰ ταῦτα
ἀδικησόμενον γενήσεσθαι, καὶ μὴ παρορᾶν τὰ τοιαῦτα μηδ'
ἐφ' ἑαυτὸν ἐλθεῖν περιμένειν, ἀλλ' ὡς ἐκ πλείστου φυλάττεσθαι. μισεῖ Μειδίας ἴσως ἐμέ, ὑμῶν δέ γε ἕκαστον ἄλλος 10
τις· ἆρ' οὖν συγχωρήσαιτ' ἂν τοῦτον, ὅστις ἐστὶν ἕκαστος ὁ
μισῶν, κύριον γίγνεσθαι τοῦ ταῦθ' ἅπερ οὗτος ἐμὲ ὑμῶν
ἕκαστον ποιῆσαι; ἐγὼ μὲν οὐκ οἶμαι. μὴ τοίνυν μηδ' ἐμέ, ὦ
221 ἄνδρες Ἀθηναῖοι, προῆσθε τούτῳ. ὁρᾶτε δέ· αὐτίκα δὴ μάλα,
ἐπειδὰν ἀναστῇ τὸ δικαστήριον, εἷς ἕκαστος ὑμῶν, ὁ μὲν 15
θᾶττον ἴσως, ὁ δὲ σχολαίτερον, οἴκαδε ἄπεισιν οὐδὲν
φροντίζων οὐδὲ μεταστρεφόμενος οὐδὲ φοβούμενος, οὔτ' εἰ
φίλος οὔτ' εἰ μὴ φίλος αὐτῷ συντεύξεταί τις, οὐδέ γε εἰ μέγας
ἢ μικρός, οὐδ' εἰ ἰσχυρὸς ἢ ἀσθενής, οὐδὲ τῶν τοιούτων
οὐδέν. τί δήποτε; ὅτι τῇ ψυχῇ τοῦτ' οἶδε καὶ θαρρεῖ καὶ 20
πεπίστευκε τῇ πολιτείᾳ, μηδένα ἕλξειν μηδ' ὑβριεῖν μηδὲ
222 τυπτήσειν. εἶτ' ἐφ' ἧς ἀδείας αὐτοὶ πορεύεσθε, ταύτην οὐ
βεβαιώσαντες ἐμοὶ βαδιεῖσθε; καὶ τίνι χρή με λογισμῷ
περιεῖναι ταῦτα παθόντα, εἰ περιόψεσθέ με νῦν ὑμεῖς; "θάρρει
νὴ Δία" φήσειέ τις ἄν, "οὐ γὰρ ἔτ' οὐδὲν ὑβρισθήσει." ἐὰν δέ, 25
τότε ὀργιεῖσθε, νῦν ἀφέντες; μηδαμῶς, ὦ ἄνδρες δικασταί, μὴ
προδῶτε μήτ' ἐμὲ μήθ' ὑμᾶς αὐτοὺς μήτε τοὺς νόμους.

8–10 *μὴ ... φυλάττεσθαι AB* 1.138.19 10 *μισεῖ ... ἐμέ* Aristid. *Rh.* (Schmid) 9.6

---

1 -ζεν A οἷα SFYP: ἃ A ἀλλ' ἅπαντας F 4 οὐδὲ ἐχορ- SA: οὐδὲ χορ- FP: οὐδ' ἐχορ- Y -τε A 5 ποτε A: οὐδέποθ' FP$_4^{\gamma\rho}$ 5–6 οὐδεὶς post χειρὶ SFYP ἅμα πάντας AF: ἅμα ἅπαντας P$_4^{\gamma\rho}$ 7 τότε A: τοῦθ' Y αυτὸν SA: αὑτὸν BcLp^{pc} 8 ἀδικηθησόμενον AF 11 συγχωρήσετ' A τουτω F$_1^{sl}$ ἕκαστον Mp 12 γίνεσθαι A: γενέσθαι F -τα A 12–13 ἕκαστον ὑμῶν A 13 οἴομαι A -δὲ SYP 14 προῆσθε V^{pc}: πρόησθε SAYP: πρόεσθε F δέ om. SFYP 16 σχολερον S^{ac} οὐδέν γε AFP$_4^{mg}$ 18 οὔτ' εἰ μὴ φίλος om. F εἰ (ante μέγας) FYP: ἢ SAP$_4^{sl}$ 19 οὐδ' εἰ (ante ἰσχυρὸς) Reiske: οὐδ' A: ἢ SFY: ἢ ἡ P 20 τοῦτο οἶδεν A 21 μηδένα αὐτὸν AFYP ἐλπίζειν P μηδὲ ὑβρίσειν A 22 -τα A ἧς ἀδείας Reiske: ἣν ἄδειαν SAFYP ταύτην SAFP$_4^{mg}$: om. YP 23 με χρὴ S^{mg} 24 περιεῖναι om. S$_3^{\gamma\rho}$A παθόντα ζῆν S$_3^{\gamma\rho}$AF: παθόντα καὶ ζῆν Pr εἴπερ ὄψεσθέ A ὑμεῖς om. F 25 -τι A ὑβρισθήσῃ AFYP 26 ὦ ἄνδρες ἀθηναῖοι A^{cp}F 26–7 μὴ προδῶτε ante ὦ A

hitting and insulting, in his intention, at the time when he did what he did, but all who may be considered less capable than I am of obtaining redress for themselves. If you were not all struck, and not all obstructed in service as chorus-producers, you know of course that neither were you all chorus-producers at once, and no one could ever assail you all with a single hand. But if one victim doesn't obtain re- **220** dress, then each person must expect that he himself will be the next one to be attacked, and must not neglect such events or wait for them to come his way, but must guard against them as far ahead as possible. Meidias hates me, maybe; but each of you too is hated by someone else: would you agree, then, that each man who hates someone should be authorized to do to each of you what Meidias did to me? I think not. Well then, don't sacrifice me to him either, men of Athens. Consider: in a moment, when the court rises, each one of **221** you, one perhaps more quickly and another more slowly, will go away home, without worrying or turning aside or being afraid, not wondering whether someone friendly or whether someone unfriendly will meet him on the way, nor if he will be big or small, or if he will be strong or weak, or anything of that sort. Why so? Because in his heart he knows and is sure and has put his trust in the constitution that no one will take hold of him or be insolent to him or hit him. Then won't you **222** confirm for me, before you go, the security in which you yourselves go on your way? What reason is there to expect me to survive, after suffering this treatment, if you overlook it now? 'Cheer up!' someone might say, 'you won't suffer any insolence in future.' And if I do, will you show severity then, after letting it go this time? Don't do it, men of the jury; don't betray me or yourselves or the laws.

223 Καὶ γὰρ αὐτὸ τοῦτο εἰ ᾽θέλοιτε σκοπεῖν καὶ ζητεῖν, τῷ ποτ᾽ εἰσὶν ὑμῶν οἱ ἀεὶ δικάζοντες ἰσχυροὶ καὶ κύριοι τῶν ἐν τῇ πόλει πάντων, ἐάν τε διακοσίους ἐάν τε χιλίους ἐάν τε ὁποσουσοῦν ἡ πόλις καθίσῃ, οὔτε τῷ μεθ᾽ ὅπλων εἶναι συντεταγμένοι μόνοι τῶν ἄλλων πολιτῶν, εὕροιτ᾽ ἄν, οὔτε τῷ τὰ σώματ᾽ ἄριστα ἔχειν καὶ μάλιστα ἰσχύειν {τοὺς δικάζοντας}, οὔτε τῷ τὴν ἡλικίαν εἶναι νεώτατοι, οὔτε τῶν τοιού-
224 των οὐδενί, ἀλλὰ τῷ τοῖς νόμοις ἰσχύειν. ἡ δὲ τῶν νόμων ἰσχὺς τίς ἐστιν; ἆρα ἐάν τις ὑμῶν ἀδικούμενος ἀνακράγῃ, προσδραμοῦνται καὶ παρέσονται βοηθοῦντες; οὔ· γράμματα γὰρ γεγραμμένα ἐστίν, καὶ οὐχὶ δύναιτ᾽ ἂν τοῦτο ποιῆσαι. τίς οὖν ἡ δύναμις αὐτῶν ἐστιν; ὑμεῖς, ἐὰν βεβαιῶτε αὐτοὺς καὶ παρέχητε κυρίους ἀεὶ τῷ δεομένῳ. οὐκοῦν οἵ νόμοι τε ὑμῖν
225 εἰσιν ἰσχυροὶ καὶ ὑμεῖς τοῖς νόμοις. δεῖ τοίνυν τούτοις βοηθεῖν ὁμοίως ὥσπερ ἂν αὑτῷ τις ἀδικουμένῳ, καὶ τὰ τῶν νόμων ἀδικήματα κοινὰ νομίζειν, ἐφ᾽ ὅτου περ ἂν λαμβάνηται, καὶ μήτε λειτουργίας μήτ᾽ ἔλεον μήτ᾽ ἄνδρα μηδένα μήτε τέχνην μηδεμίαν μήτ᾽ ἄλλο μηδὲν εὑρῆσθαι, δι᾽ ὅτου παραβάς τις τοὺς νόμους οὐ δώσει δίκην.

226 Ὑμῶν οἱ θεώμενοι τοῖς Διονυσίοις εἰσιόντα εἰς τὸ θέατρον τοῦτον ἐσυρίττετε καὶ ἐκλώζετε, καὶ πάντα ἃ μίσους ἐστὶ σημεῖα ἐποιεῖτε, οὐδὲν ἀκηκοότες πω περὶ αὐτοῦ παρ᾽ ἐμοῦ. εἶτα, πρὶν μὲν ἐλεγχθῆναι τὸ πρᾶγμα, ὠργίζεσθε, προυκαλεῖσθ᾽ ἐπὶ τιμωρίαν τὸν παθόντα, ἐκροτεῖτε ὅτε προυβαλόμην
227 αὐτὸν ἐν τῷ δήμῳ· ἐπειδὴ δὲ ἐξελήλεγκται, καὶ προκατέγνωκεν ὁ δῆμος τούτου εἰς ἱερὸν καθεζόμενος, καὶ τἆλλα προσεξήτασται τὰ πεπραγμένα τῷ μιαρῷ τούτῳ, καὶ δικάσοντες εἰλήχατε, καὶ πάντ᾽ ἔστιν ἐν ὑμῖν μιᾷ ψήφῳ διαπράξασθαι, νῦν ὀκνήσετε ἐμοὶ βοηθῆσαι, τῷ δήμῳ χαρίσασθαι, τοὺς ἄλλους σωφρονίσαι, μετὰ πολλῆς ἀσφαλείας αὐτοὶ τὸ λοιπὸν διάγειν, παράδειγμα ποιήσαντες τοῦτον τοῖς ἄλλοις;

1 τίνι τῳ F 2 -τε A 3 πολιτείᾳ A ἐάν ... ἐάν ... ἄν A: ἄν ... ἄν ... ἄν $\mathrm{FP}^{\mathrm{pc}}_{4}$ 4 ὁποσουσοῦν $\mathrm{Cd}^{\mathrm{pc}}\mathrm{Od}$: ὁπόσους οὖν A: ὁπόσους ἂν SFYP 6 -τα AF 6–7 τοὺς δικάζοντας del. Bekker 8 τῷ τοῖς νόμοις ἰσχύειν AFP: τῇ τῶν νόμων ἰσχύι $\mathrm{SYP}^{\mathrm{mg}}_{4}$: τῷ τοὺς νόμους ἰσχύειν $\mathrm{Lamb}^{\mathrm{vl}}$ 11 -τί $\mathrm{AF}^{\mathrm{pc}}$ δύναιτ᾽ $\mathrm{S}^{\mathrm{pc}}_{1}\mathrm{FYP}$: δυναταιτ᾽ S^{ac}: δύναιντ᾽ A 12 αὐτῶν ἡ δύναμίς AF 13 κυρίως A 15 αυτῷ S: αὐτῷ FY 17 μήτε ἔλεον μήτε AF 18 μήτ᾽ ἄλλο μηδὲν A: om. SFYP 21 καὶ ἐκλώζετε $\mathrm{Lamb}^{\mathrm{vl}}$: καὶ ἐκεκράγετε $\mathrm{SF}^{\gamma\rho}_{1}\mathrm{YP}$: om. AF καὶ πάντα $\mathrm{SF}^{\gamma\rho}_{1}\mathrm{YP}$: ὥστ᾽ A: ὥστε F ἐστὶ om. $\mathrm{F}^{\gamma\rho}_{1}$ 22 σημεῖα $\mathrm{SF}^{\gamma\rho}_{1}\mathrm{YP}$: σημεῖα ταῦτα A: ταῦτ᾽ F 24 -σθε AF 25 ἐξελήλεκται S^{ac} 30 αὐτοὶ om. A 31 τοῖς ἄλλοις $\mathrm{S}^{\mathrm{pc}}_{3}\mathrm{AFYP}$: τοῖ S^{ac}

For in fact, if you cared to consider and investigate the **223** question what it is that gives power and control over everything in the state to those of you who are jurors at any time, whether the state convenes two hundred or a thousand or any other number, you'd find that the reason is not that you alone of the citizens are mobilized and armed, nor that you are physically the best and strongest, nor that you are youngest in age, nor anything of that sort, but that your power is derived from the laws. And what is the power of the laws? Is **224** it that, if any of you is attacked and gives a shout, they'll come running to your aid? No; they're written documents, and they couldn't do that. What is their strength then? *You* are, if you guarantee them and make them effective on each occasion for anyone who asks. So the laws get their power from you, and you from the laws. You must therefore stand **225** up for them in the same way as anyone would stand up for himself if attacked. You must take the view that offences against the laws concern everyone, no matter in whose case they are detected, and that neither liturgies nor mercy nor any individual man nor any skill nor anything else has ever been discovered which will enable anyone to transgress the laws without punishment.

Those of you who were in the audience at the Dionysia **226** hissed and booed Meidias when he entered the theatre, and did everything that showed loathing of him, even though you hadn't yet heard anything about him from me. So, before the case was proved, you showed anger, you called on the victim to take revenge, you applauded when I brought a *probole* against him in the Ekklesia; yet now that it has been demon- **227** strated, and the Ekklesia sitting in a sacred precinct has given a preliminary condemnation of him, and the rest of the scoundrel's activities have been examined in addition, and you have been appointed by lot to judge him, and it's in your power to deal with it all by a single vote, will you now hesitate to support me, to gratify the Ekklesia, to teach everyone else a lesson, and to secure a safe life for yourselves in future by making him an example to everyone?

Πάντων οὖν ἕνεκα τῶν εἰρημένων, καὶ μάλιστα τοῦ θεοῦ χάριν περὶ οὗ τὴν ἑορτὴν ἀσεβῶν οὗτος ἥλωκεν, τὴν ὁσίαν καὶ δικαίαν θέμενοι ψῆφον τιμωρήσασθε τοῦτον.

1 εἵνεκα SYP 3 τὴν δικαίαν F Subscriptum *ΚΑΤΑ ΜΕΙΔΙΟΥ XXIII ΔΙΩΡΘΩΤΑΙ ΜΕΤΡΙΩΣ* in S: *ΚΑΤΑ ΜΕΙΔΙΟΥ ΠΕΡΙ ΤΟΥ ΚΟΝΔΥΛΟΥ XXIII* in FY: *ΔΗΜΟΣΘΕΝΟΥΣ ΚΑΤΑ ΜΕΙΔΙΟΥ* in A: om. P

So for all the reasons I have given, and especially for the sake of the god whose festival he has been caught treating impiously, cast the vote which is holy and just, and punish him.

COMMENTARY

1–8. *Introduction. Demosthenes reminds the jurors that the case results from the people's own vote against Meidias. He asks for a favourable hearing, both for his own sake and in the public interest.*

1. ἀσέλγειαν: 'aggressiveness', 'bullying'. Unlike ὕβρις, αἴκεια, and ἀσέβεια, this is not the name of an offence in law. It may have been a more colloquial term. It is not used in serious poetry; its earliest occurrences are in Old Comedy (Ar. *Wasps* 61, Eupolis 244, 320 Kock = 261, 345 Kassel and Austin, Pl. com. 210). In the fourth century it became a standard part of oratorical vocabulary, and *Meidias* marks its zenith; there are more instances of it here than in any other work. It is often linked with ὕβρις, as here (e.g. 23.56, 24.143, 54.13), and the two words may appear to be simply synonyms. Often ἀσέλγεια, like ὕβρις, refers to a self-indulgent and extravagant way of life (e.g. 36.45, 59.30, Isok. 15.305, Ais. 3.170) and sexual misconduct (e.g. 59.114, Isai. 3.13, Ais. 1.137). But its commonest use is for aggressive, pushful behaviour, such as hitting or manhandling or thrusting one's way into a house. In his political speeches D. uses it to refer to Philip's aggression in the 340s (4.9, 9.35, 10.2, 19.342).

ὕβριν: see pp. 18–23. No particular distinction between ἀσέλγεια and ὕβρις is intended here. One of D.'s commonest devices for emphasis is pleonasm, the use of two words which are synonyms or almost synonyms; cf. Dion. Hal. *Dem.* 58 and Ronnet *Style* 71–3. Separation of the two by a vocative, creating an interval between them, adds further emphasis; cf. Ronnet *Style* 48–9. For the general pattern of the sentence cf. **88** τὴν ἀσέλγειαν καὶ τὴν ὠμότητα, ᾗ καθ' ἁπάντων χρῆται τῶν ἐντυγχανόντων, 10.2 ἡ μὲν οὖν ἀσέλγεια καὶ πλεονεξία, ᾗ πρὸς ἅπαντας ἀνθρώπους Φίλιππος χρῆται. In all these passages 'all' is a sweeping exaggeration.

οὐδένα ... ἀγνοεῖν: a commonplace, e.g. **156**, 18.81, 39.22. Likewise D. may say 'Everyone knows ...', e.g. **71**, **137**, or 'You are all witnesses ...', e.g. **18**. If an orator can make the jury think that they already know something, that may enable him to get away without providing adequate evidence of it.

ὅπερ ἂν καὶ ὑμῶν ..., τοῦτο καὶ αὐτὸς ...: the parallelism is emphatic; D. wants the jurors to think that he is a man like themselves, whereas Meidias is a different sort of person. The rhetorical danger in such parallel clauses is that the hearers may guess the last word before it arrives; D. therefore varies his vocabulary at that point (πρᾶξαι ... ἐποίησα).

προυβαλόμην: the verb of the legal procedure *probole*, for which see pp. 13–16. Here it takes accusative and present infinitive like other verbs of prosecution, e.g. Ar. *Wasps* 894–6 *ἐγράψατο Κύων Κυδαθηναιεὺς Λάβητ' Αἰξωνέα ἀδικεῖν*, Lys. 10.1 *Λυσίθεος Θεόμνηστον εἰσήγγελλε τὰ ὅπλα ἀποβεβληκότα δημηγορεῖν*.

τοῖς Διονυσίοις: the city Dionysia in the spring of 348.

παρὰ: 'throughout' a period of time; cf. LSJ *παρά* C.I.10.d. So also in **14**, **101**, **184**.

2. **ἐπειδὴ** governs everything down to *κατεχειροτόνησεν αὐτοῦ*. This clause tells what happened at the hearing of the *probole* in the Ekklesia, the main clause what happened afterwards.

ὁ δῆμος ἅπας: the Ekklesia. At this point D. wishes to establish firmly in the jury's minds the belief that the whole people of Athens condemned Meidias. Not until much later (**193**) do we find that Meidias disputed this on the ground that many citizens were away from Athens on campaign at that time.

ἐσπούδασεν: regularly used of showing support for a litigant; so also in **4**, **195**, **213**.

ἐφ' οἷς: equivalent to *ἐπὶ τούτοις ἅ*, in which *ἅ* would be an internal accusative with *ἠδικημένῳ*: 'at the wrongs which the people realized I had been done'.

πάντα ποιοῦντος: arguments and entreaties; *ποιοῦντος* does not necessarily refer to any activity other than speaking.

οὐδ' ἀπέβλεψεν εἰς τὰς οὐσίας: 'was not distracted by their wealth'. This implies that even in democratic Athens people were generally more reluctant to condemn a rich man than a poor one. In **98** D. rejects wealth as an excuse.

τὰς ὑποσχέσεις: promises that Meidias would behave better in future, or that he would perform liturgies or other services to Athens.

προσιόντες: personal approaches to D. in the street or elsewhere.

ἐπεξελθεῖν: to proceed against him legally. After the hearing in the Ekklesia D. was free either to proceed or not; see p. 13–14.

μὲν: often inserted in expressions meaning 'I think'; without actually adding a *δέ*-clause the speaker implies (with real or assumed modesty) that another opinion is possible.

ὦ ἄνδρες Ἀθηναῖοι, νὴ τοὺς θεούς: juxtaposition of a vocative and an oath is not common, perhaps because it makes a rather long interruption in the sentence, but it occurs in 4.49, 24.121, 42.6. Here the assertiveness of the oath counteracts the tentativeness of *ὡς μὲν ἐμοὶ δοκεῖ*.

ὧν: equivalent to *τούτων ἅ*, in which *ἅ* would be an accusative limiting *θρασὺν* ...: 'wishing to punish him for the acts in which they had observed him being audacious ...'.

ἐπὶ τῶν ἄλλων: masculine (cf. **1** *πρὸς ἅπαντας*), 'in the cases of the other people', contrasted with *ἐμέ*. D. uses *ἐπί* with the genitive not only in its normal local sense of 'on' (e.g. **133** *ἐπ' ἀστράβης*) but also to refer to a wide range of occasions and circumstances (cf. LSJ *ἐπί* A.III.3). These include:

(*a*) The period of time within which events occur, e.g. **186** *ἐπὶ τοῦ παρεληλυθότος βίου*, 'during his past life'.

(*b*) The particular occasion when an event occurs, e.g. **72** *ἐπὶ τῆς ἀληθείας καὶ τοῦ πράγματος*, 'at the moment of the actual deed'; **204** *ἐπὶ τοῦ καιροῦ*, 'on this occasion'.

(*c*) Particular instances, e.g. **38** *οὐκ ἐπὶ τούτου μόνον, ἀλλ' ἐπὶ πάντων*, 'not only in this case but in all cases'; **54** *ἐφ' ἑκάστης μαντείας*, 'in the case of each oracle'.

(*d*) A person's external circumstances, e.g. **138** *ἐπ' ἐξουσίας καὶ πλούτου*, 'in circumstances of freedom and wealth'; **222** *ἐφ' ἧς ἀδείας αὐτοὶ πορεύεσθε*, 'the security in which you yourselves go on your way'.

(*e*) A person's internal state of mind, e.g. **199** *ἐπὶ ταύτης τῆς ὑπερηφανίας ὄντα*, 'being so arrogant'; **213** *τὴν γνώμην ταύτην ἐφ' ἧς νῦν ἐστε*, 'your present purpose'.

οὐδὲ καθεκτὸν ἔτι: 'out of control'. *καθεκτός* (from *κατέχω*) occurs only here in Attic authors. Vince's translation 'who could no longer be tolerated' is wrong, resulting from confusion with *ἀνεκτός*.

3. φυλαχθῆναι: D. means that by proceeding with the case he has preserved for the Athenians the opportunity to punish Meidias, which they would otherwise have lost; cf. **40** *τὴν ὑπὲρ ὑμῶν τιμωρίαν δικαίως φυλάξας*.

τις εἰσάγει: this verb is regularly used of the magistrate who arranges and presides over the trial of a case. Cases of *probole* were brought into court by the thesmothetai; cf. **32** *τῶν θεσμοθετῶν τούτων* (where *τούτων* implies that D. is referring to someone present in court) and *AP* 59.2, where a list of functions of the thesmothetai includes the words *τὰς προβολὰς ἁπάσας εἰσάγουσιν οὗτοι*. That is a general statement in the plural, but here in **3** D. uses the singular for the particular case. This is important evidence for a point not otherwise clearly attested (and generally fudged in modern accounts of Athenian legal procedure): when a board of magistrates had responsibility for a type of legal action, each individual case was administered and presided over by one member of the board, not by the whole board collectively. It is possible that the rest of the board also attended the trial, but **32** *τῶν θεσμοθετῶν τούτων* does not prove it, since D. could use the demonstrative adjective even if only one of the thesmothetai was present. But why does he call the president of the court *τις*? Westermann *De litis instrumentis* 19 argues that the man's identity could not have been doubtful, and he therefore proposes deletion of *τις*, leaving the subject to be understood as 'the appropriate officer' (as in **8** *ἀναγνώσεται*, etc.); he explains *τις* as originating from a note *τίς;* written by some reader who wondered which magistrate presided. That may be correct; yet it is possible to keep *τις* with the sense 'one of the appropriate officers (the thesmothetai)', and D. may have wanted to avoid the hiatus *-δὴ εἰσ-*.

πάρειμι, ὡς ὁρᾶτε: a slight touch of wry humour, the ulterior purpose of which is to encourage the jury to give D. credit for persevering with the case.

ἐξόν μοι λαβεῖν: on the question whether D. did in fact accept payment to drop the prosecution see p. 23–8.

ὥστε: 'on condition that', as in 6.11 *ἐξὸν αὐτοῖς τῶν λοιπῶν ἄρχειν Ἑλλήνων ὥστ' αὐτοὺς ὑπακούειν βασιλεῖ.*

ὑπομείνας: 'standing up to', 'resisting'. The context makes the sense clear, though the word would otherwise be ambiguous; it is used of acceding to a request or bribe in 37.14, Isok. 4.94.

4. ἃ δ' ...: the verdict and penalty. This clause is contrasted with *ὅσα μὲν ...*, but only loosely attached to the main clause which follows. That is not unusual, and it is unnecessary to emend; Spalding proposes *ἃ δ' ἐν ὑμῖν, {μετὰ} ταῦτά ἐστιν ὑπόλοιπα· ὅσῳ γὰρ ...*

ἠνώχληκεν καὶ παρήγγελκεν: Ar. *Wasps* 552–8 provides a sarcastic description of guilty defendants who accost jurors on their way into court, to ask them to be merciful. For the words used cf. 19.1 *παραγγελία ... ἐνοχλοῦντας*, 19.283 *παραγγελία*, *Pr.* 55.2 *ἐνοχλεῖν καὶ παραγγέλλειν*. Those passages favour the reading *παρ-* here, even though *περι-* is probably the reading transmitted from antiquity. *παρ-* in F may be due to accident or conjecture, and has been copied from F (or from a later ms. derived from F) by a corrector of P.

ἑώρων: if it is true that what we have is a draft of the speech written before the trial and not revised later (cf. pp. 24–8), this sentence is not a simple record of fact but a piece of speculation. D. could have omitted it in delivery if it had turned out not to be true.

πρὸ τῶν δικαστηρίων: in earlier times buildings in various parts of Athens were used as courts, but by the time of D., as this phrase indicates, all or most of the courts were together, and the jurors assembled in an area in front of them each morning for allotment to individual courts; cf. *AP* 63. This area was presumably in or beside the Agora, but its exact location is uncertain; cf. Thompson and Wycherley *Agora* 52–72, R. E. Wycherley *The Stones of Athens* (Princeton 1978) 53–60, Rhodes *Comm. on AP* 700–1.

ἐσπουδάσατε αὐτοί: D. assumes that the present jurors all attended the meeting of the Ekklesia two years ago and supported him then. *πρότερον* is added in FYP, but not in S or A, and is probably an intrusive gloss.

ἵνα: sarcastic, presenting a consequence as if it were an intention.

ὀμωμοκὼς: 'when under oath', intransitive (*ἄλλο τι* goes with *ψηφιεῖται*), referring to the oath which all jurors took at the beginning of the year. It included an undertaking to give a verdict in accordance with the laws (*κατὰ τοὺς νόμους*) or, on matters not covered by laws, to take the justest view (*περὶ ὧν ἂν νόμοι μὴ ὦσι, γνώμῃ τῇ δικαιοτάτῃ κρινεῖν*); cf. **42**, **177**, **188**, 20.118, 23.96, etc. A full text of the oath is given in 24.149–51, but the genuineness of much of that text is disputed; cf. Bonner and Smith *Justice* 2.152–6.

5. παρανόμων: D. speaks as if he were giving random examples of less important accusations, but in fact he selects two which were regarded as serious public offences; thus he contrives to convey the impression that the charge which he is bringing is a very serious one indeed. *παράνομα* ('illegalities') means proposing a decree which is in conflict with an existing law. For this offence a prosecution could be brought by the procedure of *graphe*; the penalty was usually a fine, to which disfranchisement was added on a third conviction for it. In the fourth century this kind of prosecution became a common method of attacking prominent politicians. For a short account of it see MacDowell *Law* 50–2; for full discussion see H. J. Wolff *'Normenkontrolle' und Gesetzesbegriff in der attischen Demokratie* (Sitzungsberichte der Heidelberger Akad. 1970) and M. H. Hansen *The Sovereignty of the People's Court in Athens in the Fourth Century B.C. and the Public Action against Unconstitutional Proposals* (Odense 1974).

παραπρεσβείας: 'misconduct as an ambassador'. An ambassador, like all Athenian officials, was subject to an examination (*euthyna*) of his conduct in office, which might lead to a trial if any serious accusation was made against him. The existence of the words *παραπρεσβεία* and *παραπρεσβεύω* (whereas no corresponding words exist for the misconduct of other officials) probably does not mean that ambassadorial misconduct was subject to any different legal procedure, but merely that it (or accusations of it) was common. Cf. Lipsius *Recht* 104–5.

δεῖσθαι: the normal verb for an orator's request, in the introductory part of his speech, that the jury should regard his case with special favour; cf. 18.6, 27.3, 32.3, 40.4, etc.

ἐπειδὴ governs everything down to *εἰσέρχομαι*. Within that, there are participle-phrases in the genitive absolute and in the nominative, each linked by *καί*. The purpose of using this periodic structure is to convey the impression that D. has many justifications for his request for favour.

κριτὰς: the judges of the contest of men's choruses. For Meidias' corruption of the judges cf. **17**. One may think D. rather too ready to assume that, when his chorus did not win, it must have been the judges' fault.

τῆς φυλῆς: the Pandionis tribe; cf. **13**.

ἀφαιρεθείσης τὸν τρίποδα: 'deprived of the tripod'. *ἀφαιρέω* can take two accusatives in the active or middle (e.g. **207** *μή μ' ἀφαιροῦ τὴν τιμωρίαν*) and therefore can retain one in the passive; so also in **66**, **100**. A tripod was the prize for the contest of men's choruses; cf. Lys. 21.2, schol. Ais. 1.10.

6. οἷα: internal accusative with the passive of *ὑβρίζω*, as in **21**, **126**, etc.

ὑπὲρ: 'concerning'; cf. LSJ *ὑπέρ* A.III.

συνοργισθεὶς: 'angry in sympathy with me'; cf. **100**.

καταχειροτονίαν ... εἰσέρχομαι: one might have expected something like *τοῦ δήμου καταχειροτονήσαντος, ταύτην τὴν δίκην εἰσέρχομαι*. (A prosecutor coming to court for a trial may be said to be entering the case; cf. 18.105 *εἰσῆλθον τὴν γραφήν*, 28.17 *τὰς δίκας ταύτας ἔμελλον εἰσιέναι*.) But D. wishes to convey the impression that his case is identical with what the people wanted. He therefore contrives a form of words which makes the people's decision the object of the verb of which he himself is the subject. 'I am entering the condemnation' sounds strange, but the intervention of other words between this noun and verb makes the strangeness less obtrusive.

εἰ οἷόν τε τοῦτ' εἰπεῖν: 'if I may say so', introducing a paradoxical or unexpected remark; so also in 15.15, 16.18.

ἐγὼ νῦν φεύγω: 'I am a defendant in this case', a person who will suffer if he loses the case. D. is trying to secure for himself the sympathy which a jury tends to show towards a defendant.

συμφορά: regularly used of legal penalties; cf. **58**, **96**, **99**, **100**. D. is claiming that dishonour, suffered by a victim of *hybris* unless he avenges it, is equivalent to a penalty. 'Raisonnement très spécieux', comments Humbert.

7. δέομαι ... λέγοντος: commonplace phrases in a plea for favour. Cf. for example 40.4 *δέομαι οὖν ἁπάντων ὑμῶν, ὦ ἄνδρες δικασταί, μετ' εὐνοίας τέ μου ἀκοῦσαι οὕτως ὅπως ἂν δύνωμαι λέγοντος*, 45.1 *δέομαι δὲ πάντων ὑμῶν καὶ ἱκετεύω καὶ ἀντιβολῶ πρῶτον μὲν εὐνοϊκῶς ἀκοῦσαί μου*, 58.3 *δέομαι οὖν ὑμῶν ἁπάντων, ὦ ἄνδρες Ἀθηναῖοι, καὶ ἱκετεύω μετ' εὐνοίας ἀκοῦσαί μου*.

εἰς: it makes no difference to the sense whether *εἰς* is inserted or omitted with the verb *ὑβρίζω*. Its repeated insertion here perhaps serves to make the list of victims sound longer and more substantial.

καὶ εἰς τοὺς νόμους: omitted both by F and in the quotation of these words in hyp. v 2 in Π12 (see p. 430), but not omitted in Π3. Omission by haplography between other *καὶ εἰς* ... phrases was an easy error, which may have been made by two scribes independently. It is less likely that the phrase is a conjectural insertion, since nothing in the rest of the sentence suggests to a reader a reference to the laws at this point; but D. himself already has in mind his argument that the laws are victims when *hybris* is committed against an official (**32**).

βοηθῆσαι: the commonest use of this verb (and of the noun *βοήθεια*) in this speech is to refer to the jury's verdict in favour of the victim of an attack who is prosecuting his attacker, as here. But D. uses it also to refer to the victim assisting himself by prosecuting (**106**, **141**), or to a supporting speaker on either side at the trial (**127**, **132**, **205–6**).

ἀγωνιεῖται: passive. There seem to be no other instances of the future passive of this verb in the classical period to show whether it was normally *ἀγωνιοῦμαι* or *ἀγωνισθήσομαι*.

8. οὖν ... ἄρα: virtually synonyms here, combined to make a rather emphatic 'therefore'.

καὶ emphasizes the condition as a whole: 'if any of you did suppose ...'. The conjecture *κατὰ* (Dobree *Adversaria* 1.457) is not needed.

τῶν ἰδίων: *τῶν* is omitted by A, but should probably be retained. It is used because D., obviously, does also have personal reasons for prosecuting.

προσέχων: sc. *τὸν νοῦν*, as in **24**, **27**, **187**.

8–12. *The legal procedure of* probole *for offences at festivals.*

This is the part of the speech which the rhetoricians call *προκατασκευή*, explanation of some topic (often, as here, the legal procedure) before the narrative begins. In the present case, as Σ (37 Dilts) says, the account of the public procedure for dealing with an offence concerning a festival reinforces the point made by D. in his introduction, that Meidias' offence is not of merely private significance.

ἀναγνώσεται: the unexpressed subject is the clerk of the court, to whom *λέγε* is addressed.

The text of the law is omitted in A, like all the other documents included in the oration (see pp. 48–9), but its genuineness need not be doubted. Nothing in it conflicts with what is otherwise known about procedure in the Ekklesia at this period, and several details are clearly not derived from D.'s own words in **9**. A few verbal difficulties are discussed in the notes below. D. says later (**147**) that this law did not yet exist in the time of Alkibiades, and the fact that he thinks it necessary to say this suggests that it was not recently enacted; so we may assign it to either the late fifth or the early fourth century. For its relationship to other laws about *probole* see p. 14.

τοὺς πρυτάνεις: the Boule consisted of fifty citizens from each of the ten tribes (*phylai*); each fifty had charge of affairs for one-tenth of the year and were called prytaneis for that period. Their duties included the calling of meetings of the Boule and the Ekklesia. For a detailed account see Rhodes *Boule* 16–25.

ποιεῖν: jussive infinitive, as often in laws.

ἐν Διονύσου: 'in Dionysos' ⟨precinct⟩'. In this idiom *ἐν* appears to govern a genitive because a dative has been suppressed, just as in English 'in St. Paul's ⟨cathedral⟩'; cf. Isai. 5.41 *τοῦτο μὲν ἐν Διονύσου ... τοῦτο δ' ἐν Πυθίου*. The precinct of Dionysos is the theatre, beside the Akropolis. It was not used for ordinary meetings of the Ekklesia in this period (though it was so used in the Hellenistic age), but the special meeting there after the Dionysia is mentioned also in the law of Euegoros (**10**) and in **206**, Ais. 2.61, 3.52.

τῶν Πανδίων: a certain emendation. Presumably *τῶν* was misread as *ἐν* (a word which occurs shortly before and after) and *Πανδίων* was then altered in an attempt to supply a dative. Two inscriptions confirm the existence of a festival called *τὰ Πάνδια* (*IG* 2[2] 1140.5, 1172.9), but there is no other contemporary reference to it except that in **9**. The name shows that it was a general festival of Zeus, as *τὰ Παναθήναια* of Athena; cf. Pol. 1.37, 6.163. We should reject on etymological grounds two different explanations of the word which are offered by Photios in his *Lexicon*: a festival of Pandia, daughter of the moon (cf. Σ here (39 Dilts) and *Hom. hymn* 32.15), which would rather be called *τὰ Πανδίαια*, and a festival of the Athenian hero Pandion, which would rather be called *τὰ Πανδιόνια*. However, it is possible that the festival did somehow become associated with Pandion; for one of the inscriptions is a record that the Pandionis tribe honoured the priest of Pandion (actually D.'s uncle Demon) at its meeting on the day after the Pandia in 386/5 (*IG* 2[2] 1140). Cf. L. Deubner *Attische Feste* (1932) 176–7, H. W. Parke *Festivals of the Athenians* (1977) 136.

What was the date of the Pandia? Our passage implies that the Pandia followed close on the heels of the Dionysia, probably with no intervening non-festival day on which the Ekklesia could conveniently meet to consider business arising from the Dionysia. This is confirmed by two passages of Aiskhines referring to a decree proposed by D. to hold two meetings of the Ekklesia on 18 and 19 Elaphebolion 346: one passage describes these dates as *μετὰ τὰ Διονύσια τὰ ἐν ἄστει καὶ τὴν ἐν Διονύσου ἐκκλησίαν* (2.61), and the other as *εὐθὺς μετὰ τὰ Διονύσια τὰ ἐν ἄστει* (3.68). To justify *εὐθύς*, and also to make the Dionysia, which began on 10 Elaphebolion (**10** n.), long enough to accommodate all the contests held at it, we must assume that the various events were on consecutive days. Thus 15 Elaphebolion was the last day of the Dionysia, 16 Elaphebolion was the day of the Pandia, and 17 Elaphebolion was the day of the meeting of the Ekklesia *ἐν Διονύσου*. (There have been many discussions of this problem. See especially Pickard-Cambridge *Festivals*[2] 63–6, Mikalson *Calendar* 137; but neither of these, as it seems to me, draws the right conclusion from Aiskhines.)

χρηματίζειν: another jussive infinitive. In the fourth century business in the Ekklesia was not introduced by the prytaneis but by the proedroi; cf. **9**. But that does not prove that the document before us is a forgery, composed by someone ignorant of the historical facts; for the law may have been made before the proedroi were instituted, and not subsequently amended to take account of them. (When laws were inscribed on stone, we cannot assume that they were constantly amended to bring their details up to date.) Alternatively the subject of the infinitive may be 'the appropriate officers', unexpressed, as with **52** *θύειν καὶ εὔχεσθαι*, 24.20 *ἐπιχειροτονίαν ποιεῖν*, 24.45 *χρηματίζειν*, etc. Cf. Drerup *Jb. Cl. Ph.* Supp. 24 (1898) 303.

ἱερῶν: glossed by D. in **9** as 'what the arkhon has organized'. The Ekklesia would either praise or criticize the performance of the festival ceremonies; e.g. in 282 B.C. *Εὔθιος ἄρχων γενόμενος … τῆς πομπῆς τῷ Διονύσῳ ἐπεμελήθη φιλοτίμως, … ὁ δῆμος ἐπῄνεσεν καὶ ἐστεφάνωσεν ἐν τῇ ἐκκλησίᾳ τῇ ἐν Διονύσου* (*Hesperia* 7 (1938) 100–2 no. 18).

παραδιδότωσαν: probably the prytaneis brought to the Ekklesia and read out, or gave to the proedroi to read out, a written statement of any *probole* which the accuser had submitted to them. The imperative ending *-τωσαν*, common in later Greek, has been considered a reason for rejecting the document as spurious. However, this alternative to *-ντων* was already in use before the end of the fifth century: it is guaranteed by the metre in E. *Ion* 1131, *IT* 1480, and is the normal form in Thucydides (according to the mss., though in some editions, including the Oxford Classical Text, it is silently altered to *-ντων*). Cf. O. Lautensach *Glotta* 9 (1918) 80–2, B. Rosenkranz *Indog. Forsch.* 48 (1930) 153.

τοῖς Διονυσίοις: *τὰ Διονύσια* without qualification even in official documents (cf. *IG* I[3] 61.25, Th. 5.23.4) means the principal festival of Dionysos, often called *τὰ ἐν ἄστει Διονύσια* (as in **10**). The words should not be taken to include other festivals of Dionysos (as they are by Drerup *Jb. Cl. Ph.* Supp. 24 (1898) 302).

ὅσαι ἂν μὴ ἐκτετισμέναι ὦσιν: 'all that have not been paid for'. For this sense of *ἐκτίνω*, paying a fine or damages in compensation for an offence, cf. **43**. The text as we have it does not explain how a *probole* could be 'paid for'; that must have been explained in a further sentence of the law which is omitted here as irrelevant to D.'s case. The best explanation is that given by Boeckh *Abh. Berlin 1818–19* 67 and Lipsius *Recht* 215: a *probole* (like other business for the Ekklesia) must have been considered first by the Boule; and if the Boule agreed that the alleged offence was one for which a small fine was sufficient penalty, and the defendant paid the fine at once, that ended the matter and the *probole* did not need to be presented to the Ekklesia. It is unlikely that the reference is to an arrangement made privately between the prosecutor and the defendant, because the offence was a public one: any penalty should be paid to the state and approved by some state authority.

9. λέγων: 'ordering'; cf. LSJ λέγω III.5.

ἐν Διονύσου μετὰ τὰ Πάνδια: 8 n.

οἱ πρόεδροι: presidents of the Ekklesia. In the fifth century the prytaneis presided over the Ekklesia and the Boule, but in the fourth century this was the function of nine proedroi. Whenever the Ekklesia or the Boule met, the foreman (ἐπιστάτης) of the prytaneis picked by lot one proedros from each of the ten tribes except the prytaneis' own, and one out of these nine men to be their foreman; he then handed them the agenda, and they conducted the rest of the meeting (*AP* 44.2–3). The exact date of this innovation is unknown, but it was certainly between 403 and 379. Evidence that the prytaneis sat on their traditional benches in the 390s (Lys. 13.37, Ar. *Ekkl.* 87) does not help, since their position may have remained unchanged when a bench for the proedroi was placed in front of them; but the references to the proedroi in the law quoted in 24.20–3, which may be associated with the revision of laws undertaken in 403/2, makes it likely that they were instituted in that year. Cf. Lewis *BSA* 49 (1954) 31–4, Rhodes *Boule* 25–7 and *Comm. on AP* 534, MacDowell *JHS* 95 (1975) 68.

ὁ ἄρχων: the chief of 'the nine arkhons', the one later (not in D.'s time) called ἄρχων ἐπώνυμος. He was the official responsible for the organization of the Dionysia (*AP* 56.3–5).

χρηματίζειν: κελεύει, added in AFY^{pc}P, is probably a scholiast's explanation of the infinitive; in fact the infinitive can be taken with λέγων.

καλῶς: the merit which D. claims for the law is simply that it does specify a procedure for bringing festival offenders to trial, instead of allowing them impunity. He is not praising any particular feature of the procedure.

τινες: a rhetorical plural, meaning Meidias.

ἀγὼν μηδὲ κίνδυνος: in a legal sense: 'trial or risk of penalty'.

10. τοίνυν: commonly used by D. and other orators to mark a transition to another topic; so also in **23**, **29**, **36**, etc.

καὶ τὸν ἑξῆς νόμον ἀναγνῶναι τούτῳ: D. will not read out the law himself; the clerk will read it, as *λέγε* shows. Cf. 38.4 *βούλομαι δ' ὅμως καὶ τὸν νόμον ὑμῖν αὐτὸν ἀναγνῶναι. λέγε τὸν νόμον*, 19.297, 24.27, etc.; and in general, for the use of an active verb where a middle might be expected, cf. K. J. Dover *JHS* 80 (1960) 73–4. The emphasis given to 'next' by the insertion of *καί* and *τούτῳ* implies that the law stood next in the official record of laws; 'next' does not just mean 'which I am going to read next'. The deictic form *τουτονί* (AF, and added as a correction in SP) is less suitable here, whether taken with *νόμον* or as a subject for *ἀναγνῶναι*, since neither the document nor the clerk needs to be pointed out to the jury. (It is appropriate when D. addresses the clerk and points out which document he is to read, as in 18.28, 20.115, 24.32.)

εὐλάβεια: care taken to prevent interference with festivals.

The law of Euegoros concerns four festivals, which are listed in the order in which they occurred during the year: the Dionysia in Peiraieus, the Lenaia, the city Dionysia, and the Thargelia. Their common feature was that they all included contests of dithyrambs or plays or both. (Other dramatic festivals, namely the Dionysia held at Eleusis, Aixone, and elsewhere, are omitted, presumably as being of merely local importance.) The law is supplementary to the law quoted in **8** and, we must assume, to laws similar to that in **8** referring to the Dionysia in Peiraieus, the Lenaia, and the Thargelia; cf. p. 14. Whereas the law quoted in **8** is about the use of *probole* for acts concerning a festival, the law of Euegoros extends its use to certain acts which have nothing to do with festivals except that they are committed while a festival is in progress. Its purposes are to prevent interruption of the festival performances and to encourage full attendance at them. (*a*) Sometimes a creditor, having difficulty in tracking down his debtor, may have thought that a performance in the theatre was a good opportunity to catch him; but a creditor scanning the rows of the audience and pouncing on his debtor, while a play or dithyramb was going on, would cause a disturbance to the whole audience. (*b*) Sometimes a creditor, claiming the right to seize property from a defaulting debtor's house, may have thought that he could seize it more conveniently if he chose a day when the debtor was likely to be away from home attending a festival; the law enables a debtor to attend a festival without fear that his possessions may be carried off in his absence.

There is no precise evidence for the date of the law of Euegoros. But it must be subsequent to the law quoted in **8**, and his gravestone (see next note) dates from the middle of the fourth century; so we may assign the law roughly to the first half of the fourth century. There is no reason to doubt the genuineness of the text; cf.

Foucart *Rev. Phil.* 1 (1877) 168–81, Drerup *Jb. Cl. Ph.* Supp. 24 (1898) 300–3.

Εὐήγορος: unknown, except that his gravestone (*IG* 2^2 7045) calls him Euegoros son of Philoinos of Paiania and his wife Isthmonike daughter of Lysis of Aixone; her father is the title-character of Plato's *Lysis*. The identification may be regarded as certain because the name is rare. Cf. Davies *Families* 360.

ἐν Πειραιεῖ: the fact that the Dionysia in Peiraieus are included in this law shows that they were regarded as a festival of national, not merely local, importance. (That is confirmed by the fact that the demarch of Peiraieus, whose main function was to organize this festival, was appointed (by lot) by the Athenian state, not by the deme of Peiraieus; cf. Foucart *Rev. Phil.* 1 (1877) 171–4, D. Whitehead *The Demes of Attica* (1986) 394–6.) The festival was held in Poseideon (approximately December), as is shown by three inscriptions: in *IG* 2^2 456b.33, a decree passed on the last day of Maimakterion, the Dionysia in Peiraieus are mentioned as an imminent festival, to which some honoured visitors are invited; in *IG* 2^2 1496.144 this festival comes chronologically between the Theseia (in Pyanopsion) and the Lenaia (in Gamelion); and in *IG* 2^2 1672.106 money for a sacrifice at the festival is paid out (presumably in advance) in the fourth prytany of 329/8 (which began on 19 Pyanopsion and ended towards the end of Maimakterion). Our text shows that the festival included a procession (mentioned also in *IG* 2^2 380, which shows that it celebrated Zeus Soter as well as Dionysos), comedies, and tragedies (mentioned also in *IG* 2^2 1214.29), but the number of plays is unknown. Remains of the theatre have been found on the north-west side of the hill of Mounykhia. The festival and theatre existed before the end of the fifth century, when the theatre was sometimes used for public meetings and other purposes (Th. 8.93.1, Lys. 13.32, X. *Hell.* 2.4.32–3) and Euripides entered the dramatic contests (Ael. *Var. Hist.* 2.13). For a collection of evidence about the festival see Pickard-Cambridge *Festivals*^{2} 44–7; for the theatre see E. Fiechter *Antike griechische Theaterbauten* 9 (1950) 35–41.

ἐπὶ Ληναίῳ: the same phrase as in Ar. *Akh.* 504, Pl. *Prot.* 327d, *IG* 2^2 1496.74, 1496.105. If interpreted literally, it means that the Lenaia were celebrated at the precinct called Lenaion (the location of which is doubtful: see R. E. Wycherley *Hesp.* 34 (1965) 72–6, Pickard-Cambridge *Festivals*^{2} 37–9). That interpretation is maintained by C. F. Russo *Aristofane, autore di teatro* (Florence 1962) 1–21; but it seems unlikely that plays would have been performed in an open space or makeshift theatre when the theatre of Dionysos was available. So it is better to assume that it was only in early times that the whole festival was held at the Lenaion, and that *ἐπὶ Ληναίῳ* remained as a fossilized phrase after the plays were transferred to the theatre of Dionysos at some time in the fifth century.

The Lenaia were held in Gamelion (approximately January). The exact dates are unknown, but 12–15 Gamelion are the likeliest, because no meetings of the Ekklesia or the Boule are known to have been held within that period; cf. Mikalson *Calendar* 102–3, 110. Two tragedians each presented two tragedies and five comedians each presented one comedy. For an account of the festival see Pickard-Cambridge *Festivals*^{2} 25–42.

ἡ πομπὴ: comparison with other parts of the sentence shows that the article, omitted in the mss., should come immediately before *πομπὴ*, not (where Reiske placed it) before *ἐπὶ Ληναίῳ*.

τοῖς ἐν ἄστει Διονυσίοις: the most important festival of Dionysos, held in Elaphebolion (approximately March). It is now generally agreed that it began on 10 Elaphebolion: the *προαγών*, held shortly before, was on 8 Elaphebolion (Ais. 3.67); 9 Elaphebolion was a day on which the Ekklesia often met, and on one occasion (270 B.C.) the calendar was stopped on that date for four more days, presumably to give extra time before the Dionysia began (*SEG* 14.65.4); but meetings on 10 Elaphebolion and subsequent days were rare. Cf. W. S. Ferguson *Hesp.* 17 (1948) 133–5 n. 46, W. B. Dinsmoor *Hesp.* 23 (1954) 307–9, Pickard-Cambridge *Festivals*^{2} 63–6, Mikalson *Calendar* 123–6. The festival continued until 15 Elaphebolion (**8** n. *τῶν Πανδίων*).

ἡ πομπὴ: a procession to the precinct of Dionysos (from what starting-point is not known), presumably on 10 Elaphebolion. Men and women, in the most splendid dress they could muster, walked or rode (cf. **180** *σκῦτος ἔχων*) escorting animals and carrying other sacrificial offerings for the god. Cf. Pickard-Cambridge *Festivals*^{2} 61–3.

οἱ παῖδες καὶ ὁ κῶμος: the 'fasti' of the city Dionysia (*IG* 2^2 2318, revised text in Pickard-Cambridge *Festivals*^{2} 104–6) record the victors of four contests every year in the order: *παίδων, ἀνδρῶν, κωμῳδῶν, τραγῳδῶν*. For the boys' and men's choruses the victor in each case is a tribe (not a poet), and it is clear that each of the ten tribes provided both a boys' chorus and a men's chorus, not just one or the other, for sometimes the same tribe won both those contests; cf. Pickard-Cambridge *Festivals*^{2} 75 n. 1. But in the law of Euegoros, where we should expect *οἱ ἄνδρες*, the text offers us *ὁ κῶμος*, 'the revel'. It would be imprudent to eject this as a scribal blunder, because the heading of the 'fasti' includes the word *κῶμοι*: but unfortunately that heading is so fragmentary that the meaning there is not clear. In our law, several possibilities must be considered:

1. The simplest is to take *ὁ κῶμος* as a name for the men's choruses.

Pind. *Ol.* 4.9 and *Pyth.* 5.22 show that a lyric chorus could be called *κῶμος*. Perhaps in early times this word was used for any kind of singing and dancing, and the contest of men's choruses, having been called *κῶμος* at a time when it was the only musical event in the festival, retained the name even after other varieties of chorus were added. Cf. P. Ghiron-Bistagne *Recherches sur les acteurs dans la Grèce antique* (Paris 1976) 225–7. But it is not easy to believe that this word could be used, in so formal a context as a law, to distinguish men's dithyrambic choruses from other choruses, particularly from *κωμῳδοί*.

2. Perhaps *καὶ οἱ ἄνδρες* has been lost from the text, an easy haplography. But then what was *ὁ κῶμος*? In this law everything else seems, as far as we can check, to be listed in chronological order; but no other event, either a 'revel' or anything else, is known to have occurred between the dithyrambic choruses and the plays.

3. Foucart *Rev. Phil.* 1 (1877) 175–9 proposes *οἱ παῖδες καὶ ὁ κῶμος* ⟨*καὶ ὁ ἀγὼν*⟩. Since the boys' and men's choruses at the Thargelia are both comprehended in *τῷ ἀγῶνι* in the next line of the law, one might expect the same word to be used for those choruses at the Dionysia. Then *οἱ παῖδες* must refer to another event: a hymn to Dionysos sung by a chorus of boys on the opening day of the festival. The only real evidence for that event is a decree of 185 B.C. honouring persons responsible for the performance of the festival, among them *τοὺς παῖδας τοὺς ἐλευθέρους καὶ τὸν διδάσκαλον αὐτῶν* (*IG* 2^2 896.59–63); this cannot refer to the contest of boys' choruses, for which there would have been ten *διδάσκαλοι*. Foucart's suggestion gives a plausible chronological sequence of events on the first day of the festival: *ἡ πομπὴ καὶ οἱ παῖδες καὶ ὁ κῶμος*—procession, religious ceremony, revelry. But two difficulties remain (besides the lack of early evidence for the boys' hymn): it is hard to take *οἱ παῖδες* as excluding the contest of boys' choruses, and it is hard to take *ὁ ἀγών* as excluding the contests of comedies and tragedies. (At the Thargelia there were no comedies or tragedies, so that *τῷ ἀγῶνι* covers the whole of the competitive section of that festival.)

Until further evidence is discovered, all these possibilities remain open.

οἱ κωμῳδοὶ καὶ οἱ τραγῳδοί: at the city Dionysia five comedians each presented one comedy, and three tragedians each presented three tragedies and one satyr-play. (The old view that the number of comedies was reduced to three for some years during the Peloponnesian War has been effectively challenged by W. Luppe *Philol.* 116 (1972) 53–75.) Since the law of Euegoros seems to observe chronological order, the phrase here is evidence that at this festival the comedies were performed before the tragedies. One passage has been thought to provide contrary evidence, Ar. *Birds* 786–9, where the chorus of birds points out an advantage of having wings: if a spectator in the theatre had wings, and then got hungry at the performances of tragedies, he could fly home for lunch and then fly back to 'us' again (*κᾆτ' ἂν ἐμπλησθεὶς ἐφ' ἡμᾶς αὖθις αὖ κατέπτατο*). But 'us' does not mean 'comedy' but 'the theatre'; *αὖθις αὖ* means that the hypothetical winged spectator returns to the same performance as he left, viz. the tragedies (which he left not because he disliked them, but because he was hungry). So we may accept that the comedies preceded the tragedies. Presumably the comedies were on 12 Elaphebolion and the tragedies on 13–15 Elaphebolion.

Θαργηλίων: the Thargelia were a festival of Apollo celebrated on 7 Thargelion (Plu. *Eth.* 717d). There were dithyrambic contests, but no plays; the ten tribes were arranged in pairs and provided five boys' choruses and five men's choruses. Ant. 6.11–13 gives an account of the preparations for the performance of a boys' chorus at the Thargelia; cf. Lys. 21.1, *AP* 56.3. On the festival in general see L. Deubner *Attische Feste* (1932) 179–98, H. W. Parke *Festivals of the Athenians* (1977) 146–9, E. Simon *Festivals of Attica* (1983) 76–9.

The older mss. here give *ὁ θαργηλιὼν*, the name of the month in the nominative; two later mss. give *θαργηλιῶνος*, the month in the genitive. Wolf deleted the article and changed the accent to give the name of the festival in the genitive. This seems the best emendation, even though it does produce a rather pointless change of construction from the temporal clause *ὅταν … ᾖ …* to the temporal dative *τῇ πομπῇ καὶ τῷ ἀγῶνι*: the nominative may have been introduced by someone who thought that the temporal clause ought to continue. Drerup *Jb. Cl. Ph.* Supp. 24 (1898) 302 proposes *καθὰ Θαργηλίων*: if this were right, it would imply that another law already existed about seizure of property at the Thargelia.

ἐνεχυράσαι: 'to distrain', to seize property in satisfaction of an unpaid debt. A creditor could do this in any of the following circumstances: (*a*) if the property had been awarded to him by the verdict of a jury or an arbitrator; (*b*) if the debtor himself, when the loan or lease was made, had agreed that certain property should pass to the creditor if the debt was not paid; (*c*) certain officials were entitled by law to seize property in satisfaction of debts owed to the state. Cf. Harrison *Law* 2.244–7.

ἕτερον, if right, means 'other than an *ἐνέχυρον*'; but Weil's proposal to emend *ἕτερον* to *ὁτιοῦν* (used in D.'s paraphrase in **11**) is attractive. In any case the phrase has a wider meaning than *ἐνεχυράσαι*, since it refers also to seizure of the actual money or object owed, not merely to other property seized instead.

μηδὲ τῶν ὑπερημέρων: strictly superfluous, since if payment was not overdue there would be no justification for a seizure anyway. The words are inserted in order to emphasize that this law forbids even acts which at other times would be justifiable.

ὑπόδικος ἔστω τῷ παθόντι: the victim may bring a private case (*δίκη*) claiming compensation or damages.

προβολαί: distinct from *ὑπόδικος τῷ παθόντι*: a *probole* was a public accusation which could be brought by any citizen, not merely by the victim.

ἔστωσαν, instead of *ὄντων*, does not prove the text spurious: **8** n. *παραδιδότωσαν*.

ἐν Διονύσου: **8** n. If the text is right, it implies that each of the four festivals with which this law is concerned was followed by a meeting of the Ekklesia in the precinct of Dionysos. But there is no other evidence for such a meeting after the Dionysia in Peiraieus, the Lenaia, and the Thargelia; so possibly the words *τῇ ἐν Διονύσου* are an incorrect gloss and ought to be deleted.

τῶν ἄλλων τῶν ἀδικούντων: the second article is redundant, and Spalding's conjecture *τῶν ἄλλως πως ἀδικούντων* may be correct.

γέγραπται can be taken as a reference to the law quoted in **8** and to similar laws about the Dionysia in Peiraieus, the Lenaia, and the Thargelia. It is not evidence that another law existed specifying more exactly the offenders at the Dionysia who were liable to *probole*.

11. ἐνθυμεῖσθε: better taken as imperative, as it must be in **197** and **209**.

τοὺς ὑπερημέρους εἰσπραττόντων: 'exacting ⟨payment from⟩ overdue debtors'. With this verb the accusative may be used both for the payment and for the payer (cf. the English verb 'charge'); alternatively the payer may be in the genitive with *παρά*, as in **155**.

ἢ βιαζομένων: a gratuitous addition by D. to suit his own case. The law of Euegoros refers only to seizure of property, not to other acts of violence.

ἐποιήσατε: in Athens every jury was considered equivalent to the whole citizen body. Therefore 'you', addressed to a jury, can mean all the citizens in the Ekklesia (as here) or it can mean another jury (as in **91**).

οὐ ... ὅπως: 'not only ... not' (LSJ *ὅπως* A.II.2), negativing *ᾤεσθε χρῆναι*. An interpolated *μή*, contrary to the idiom, is added in all the mss., though in S only as a later correction. As Σ (45 Dilts) points out, the passage is an *a fortiori* argument: if even claiming property is forbidden during a festival, all the more must assault and damage be.

ἄν: in an indefinite clause the *ἄν* which accompanies a subjunctive is sometimes irregularly retained with the optative when the clause is put into indirect speech, subordinate to a past verb. Cf. 30.6 *ἐπειδὰν* *τάχιστ' ἀνὴρ εἶναι δοκιμασθείην*, Isok. 17.15 *ἕως ἂν τἀληθῆ δόξειεν αὐτοῖς λέγειν*, Pl. *Phd.* 101d *ἕως ἂν τὰ ἀπ' ἐκείνης ὁρμηθέντα σκέψαιο*, S. *Tr.* 164, 687, X. *Hell.* 2.3.48, 2.4.18. Some editors delete *ἄν* in these passages, but that is hardly justifiable when the instances are so numerous.

λειτουργίαν: the earlier form *λητ-* and the later *λειτ-* were both in use in the time of D.; cf. Threatte *Grammar* 1.371. The mss. of D. regularly give the later form and there is no compelling reason to alter it; cf. N. Lewis *GRBS* 3 (1960) 180–1.

ἑλόντων ... ἑαλωκότων: in legal contexts *αἱρέω* and *ἁλίσκομαι* are used respectively of a successful prosecutor and an unsuccessful defendant. The verb *γίγνομαι* is used to refer to a transfer of property or power which is legally authorized but has not yet been put into effect; cf. W. E. Thompson *CQ* 33 (1983) 293–4.

12. εἰς τοσοῦτον ἀφῖχθε φιλανθρωπίας: 'you had reached such an extent of generosity'. This kind of phrase, including a partitive genitive of a noun denoting a type of behaviour, with a verb meaning 'advance' or 'arrive', is common in D.; there are further instances in **62**, **65**, **131**. Here the verb is best taken as pluperfect, referring to the condition of the Athenians at the time in the past when the law was made. If it is taken as perfect, emendation of *ἐπέσχετε* to *ἐπέχετε* (Dobree *Adversaria* 1.457) is necessary.

προσενέτεινε: *ἐντείνω* is used of inflicting blows; cf. LSJ *ἐντείνω* IV. *προσ-* means 'in addition' to the other maltreatment.

ἓν ... οὐδέν: 'not a single one'; cf. Hdt. 1.32.8 *ἀνθρώπου σῶμα ἓν οὐδὲν αὔταρκές ἐστι*, Th. 2.51.2 *ἕν τε οὐδὲ ἓν κατέστη ἴαμα*.

13–18. *Narrative of the events leading up to the Dionysia two years ago.*

The main purpose of this part of the oration is of course to catalogue Meidias' offensive acts, but at the same time D. gives prominence to his own public-spirited generosity as a volunteer chorus-producer. On the relevance of the contrast between the characters of D. and Meidias see p. 31.

13. ἐπειδὴ governs the whole of **13** (down to *ἔλαχον*); the main clause occupies **14**. D. wishes to recount the circumstances leading to his appointment as a chorus-producer (because they are to his credit), but primarily to direct attention to Meidias' attitude by making that the climax of the sentence.

The appointment of khoregoi was one of the first concerns of the arkhon after he entered office at midsummer (*AP* 56.3). For the city Dionysia each tribe had to nominate to the arkhon one khoregos for its men's chorus and one for its boys' chorus. Epigraphical evidence shows that each tribe provided both a men's and a boys' chorus, not merely one or the other (*IG* 2^2 2318.320–4, 3061; cf. Pickard-Cambridge *Festivals*[2] 75 n. 1). Therefore this passage and a similar phrase in Isai. 5.36, *τῇ μὲν φυλῇ εἰς Διονύσια χορηγήσας*, should not be taken as meaning that the tribe had to provide only one khoregos (as W. Wyse *The Speeches of Isaeus* 455 takes it). It is better to infer that being 'khoregos for the tribe' customarily meant being khoregos for the men, who were regarded as the tribe's main chorus. Probably Pandionis had succeeded in producing a nomination for its boys' chorus, which was a less expensive burden.

τῇ Πανδιονίδι φυλῇ: D. belonged to Paiania, an inland deme of the Pandionis tribe.

τρίτον ἔτος τουτί: 'two years ago', at the beginning of the year 349/8. For the form of words cf. 3.4 *τρίτον ἢ τέταρτον ἔτος τουτί* (of which A's reading here may be a reminiscence), 54.3, 56.4. The idiom may have originated as a parenthetic sentence, 'This is the third year', rather than as an accusative of time.

ἐπικληροῦν: the order of choice (not the individual names) of pipers was assigned to the khoregoi by lot. Cf. the last words of **13**, which show that *χορηγοῖς*, not *χοροῖς*, is the right reading here. Lot was used also for assignment of poets (Ant. 6.11).

τῶν ἐπιμελητῶν τῆς φυλῆς: each tribe appointed three *ἐπιμεληταί* (one from each trittys). Besides the duty of arranging for the nomination of khoregoi, they had charge of the tribe's funds, and arranged sacrifices and honorific inscriptions on behalf of the tribe. Cf. *IG* 2^2 1138 ff., and (for the number) *IG* 2^2 1151, 1152, 2818, *Hesperia* 5 (1936) 402 no. 10 lines 167–70, 32 (1963) 41 no. 42. These *ἐπιμεληταί* are to be distinguished from those elected to organize the procession for the Dionysia (**15**).

ἐθελοντής: D. by volunteering obviated the need for someone else to be compelled to serve.

κληρουμένων: genitive absolute, with *τῶν χορηγῶν* understood.

14. **ὡς οἷόν τε μάλιστ' ἀπεδέξασθε:** 'you welcomed as warmly as possible', an unusually strong sense for this verb (though 17.1 *ἀποδέχεσθαι σφόδρα* is similar); more often it is just 'not dispute'.

ὡς ἂν with participles: 'as you would if'.

παρ': 'throughout': cf. **1** *παρὰ πᾶσαν τὴν χορηγίαν*.

ἐπηρεάζων refers to deeds rather than speech; cf. **25**, where *ἐπήρεια* leads to *βλάβη*.

15. **ὅσα μὲν** ... is answered by *ἀλλ' ἃ* ... at the end of **15**; within this sentence *τῷ μὲν ἐπηρεαζομένῳ* ... is answered by *ὑμῖν δὲ* ... The pronoun *ὅσα* is an internal accusative with *ἠνώχλησεν*, while the clause τοὺς χορευτὰς ἀφεθῆναι is the object of ἐναντιούμενος. For the infinitive with that verb cf. Pl. *Ap.* 31d. The addition of μή (found by Lambinus in some ms.) is possible but not essential.

To enable them to rehearse, members of a chorus were exempt from military service until the festival was over; cf. 39.16 and MacDowell *Symposion 1982* (1985) 70–2. Presumably they (or the khoregos on their behalf) applied to the military officers of their tribe to be omitted from any call-up list which might be issued before the festival. What form Meidias' opposition took is not clear. Possibly there was a formal procedure, by which the military officers published a list of men who had applied for exemption and other citizens could object to the inclusion of particular names in it. Alternatively Meidias may just have had private conversations with the military officers, in which he tried to persuade them to refuse exemption to some of D.'s best singers.

προβαλλόμενος: 'being proposed'. The participle is better taken as passive, not middle, even though Meidias no doubt instigated someone to propose him. The middle voice is regularly used of proposing another person (And. 1.132, X. *An.* 6.1.25, Pl. *Laws* 755c). In our passage LSJ are wrong in saying (under προβάλλω B.I.4) that ἑαυτόν is the object of προβαλλόμενος: in fact ἑαυτόν is the object of χειροτονεῖν. There is no evidence that it was possible in Athens for a man formally to propose himself for election to office (in **13** D. volunteers to perform a liturgy, but that is a different matter), and D. 18.149 rather implies that it was not possible. There Aiskhines wants to become a pylagoros, because Philip (D. alleges) will pay him for it; hardly anyone is willing to support him, and yet he still does not propose himself, for the participle mentioning his proposal is indubitably passive: προβληθεὶς πυλάγορος οὗτος καὶ τριῶν ἢ τεττάρων χειροτονησάντων αὐτὸν ἀνερρήθη. So here we may take προβαλλόμενος as passive (and also προβέβληται in **200**). Meidias, however, was not elected; cf. **17** ἰδιώτης ὤν.

ἐπιμελητήν: in organizing the city Dionysia the arkhon was assisted by ten ἐπιμεληταί (not to be confused with the ἐπιμεληταί of the tribes, mentioned in **13**). They were elected by the Ekklesia (presumably affluent men were chosen) and had to pay the expenses of the procession out of their own pockets; but in later times they were selected by lot and the expenses were paid from public funds (*AP* 56.4). The change to appointment by lot was made not later than 328/7 (*IG* 2^2 354.15–16 οἱ λαχόντες ἐπιμεληταὶ τῆς εὐκοσμίας τῆς περὶ τὸ θέατρον), but in our passage χειροτονεῖν shows that in 349/8 the change had not yet been made. Thus Meidias' proposal was an offer to spend some of his own money. Although D. disparages it, it may have been no more selfish than D.'s own offer to be a khoregos.

ἐάσω: this passage is a favourite example of the rhetoricians' to illustrate the device of paralipsis; see the list of testimonia on p. 96.

παρίστη: *παρίστημι* is regularly used of putting either an idea or an emotion into a person's mind or heart; cf. **72**, **73**, 18.1, 19.333, 23.96, 23.103, etc.

ἃ ... ἀγανακτήσετε: for the accusative with *ἀγανακτῶ* cf. **123**, 8.55, 19.305, 24.88, 54.15. In some mss., because this construction was not understood, *ἃ* has been changed to *ἐφ' οἷς* or *ἀκούσαντες* has been added; the appearance of alternative remedies shows that *ἃ ... ἀγανακτήσετε* was the original reading. *-αιτε* for *-ετε* is a common corruption arising from late Greek pronunciation, and the insertion of *ἄν* is a subsequent attempt to correct the grammar; the future indicative must be right because D. actually is going to relate the facts.

16. ὑπερβολή: literally 'going beyond', a favourite word of D.'s in passages of moral condemnation. Sometimes it is not immediately clear what goes beyond what. A genitive may be objective; e.g. 19.66 *εἴ τις ἔστιν ὑπερβολὴ τούτου* means 'if anything goes beyond this'. More often a genitive is subjective, as here: 'the subsequent events go beyond'. Then the object has to be understood. Here it could be 'beyond those already mentioned'. But it is more probably 'beyond all reasonable limits', as in 27.38 *ταῦτ' οὐχ ὑπερβολὴ δεινῆς αἰσχροκερδίας;* 'are these acts not instances of profiteering going beyond everything?' For other instances of *ὑπερβολή* in this speech see **46**, **75**, **109**, **119**, **122**.

οὐδ' ἂν ἐπεχείρησα ἔγωγε: 'I should not even have tried', because there would have been no hope that anyone would believe so dreadful a thing. Note how D. makes the offences sound as bad as possible.

τότε ἐν τῷ δήμῳ: at the meeting of the Ekklesia immediately after the festival, when the evidence was fresh.

ἐσθῆτα: dress for the chorus to wear while performing, called *τῶν ἱματίων* in **25**.

ἱερὰν γὰρ ἔγωγε νομίζω ...: a tendentious definition. Σ (65 Dilts) points out that *χρησθῇ* may mean either 'worn' (at the festival) or 'worn out', but the former is probably what D. means.

τέως for *ἕως* is quite common in D.; cf. 2.21, 14.36, 19.326, 20.91, 23.108, 24.64, 24.80–1, 25.70, 29.43, 56.14, *Pr.* 21.4.

ἐποιησάμην: 'I got made' by the goldsmith.

ἐπεβούλευσεν, ὦ ἄνδρες Ἀθηναῖοι, διαφθεῖραι: a good example of the use of a vocative to create a pause between two words both of which the speaker wishes to emphasize.

μου may be construed with the accusatives *τὴν ἐσθῆτα* and *τοὺς στεφάνους*. It is not uncommon for this enclitic pronoun to follow the governing verb; cf. **17** *τὸν διδάσκαλον ... διέφθειρέν μου τοῦ χοροῦ*, **69**

τοῦ σώματος τὼ χεῖρε ... ἀποσχέσθαι μου, 6.6, 47.75, 53.14, 54.9. So the conjecture *μοι* (Iurinus) is unnecessary.

οὐ γὰρ ἐδυνήθη: the reason is not given. We are left to imagine Meidias trying single-handed to destroy a huge quantity of cloaks and crowns, and unable to get through them all before being interrupted by the goldsmith.

καὶ τοιοῦτον (S) is preferable to *καίτοι τοῦτό γε* (A) because (*a*) to reach a climax of indignation an adversative particle (*καίτοι*), though not logically impossible, is less appropriate than an accumulative particle (*καί*), and (*b*) it is a stronger statement to say that no one ever did anything of this kind (*τοιοῦτον*) than merely that no one ever did this particular thing (*τοῦτό γε*). Butcher adduces 9.16 and 20.117, but those passages are not helpful for solving the textual question, because in both of them the text is uncertain and the rhetorical arguments are different.

17. τὸν διδάσκαλον: the poet who composed the words and music as well as directing the chorus. D. means that Meidias bribed him to train the chorus badly.

μου: **16** n.

Τηλεφάνης: otherwise unknown. Presumably D. chose him as piper after the drawing of lots described in **13**.

ἀνδρῶν with *βέλτιστος*: 'an excellent fellow'; not with *ὁ αὐλητής*, 'the men's piper', which would need *τῶν* (cf. **18** *τῷ ἀγῶνι τῶν ἀνδρῶν*). Contrasted with this is *τὸν ἄνθρωπον* (the corrupted director), a good example of the tendency of *ἄνθρωπος* to be a more derogatory word than *ἀνήρ*. Cf. Dover *Morality* 282–4 on foreigners and slaves called *ἄνθρωποι*: here, however, the piper and the director are presumably both Athenian and the distinction between them is purely moral.

συγκροτεῖν: 'knock together' in the sense of getting men to work as a unit; in 2.17 well-trained soldiers are *συγκεκροτημένοι*. The use of an infinitive subordinated to another infinitive is a notable feature of D.'s style, especially with *οἴομαι δεῖν*: so also in **46**, **81**, **105**, **122**, **143**, **203**. It is studied by B. Gaya Nuño *Sobre un giro de la lengua de Demóstenes* (Madrid 1959).

οὐδ' ἂν ἠγωνισάμεθα, 'we should not have competed', means (as the rest of the sentence shows) that the chorus would have performed badly, not that it would not have appeared at all.

οὐδ' ἐνταῦθ' ἔστη ...: 'he did not even stop at that degree of *hybris*, but he had so much of it still left that ...'. After *ὥστε* there is much use of asyndeton, which helps to convey the impression that Meidias did one thing after another in rapid succession.

ἐστεφανωμένον adds nothing to the meaning of the sentence, but is inserted solely to stress the arkhon's dignity and thus make Meidias' offence sound worse. Each of the nine arkhons wore a wreath or crown when performing official duties; cf. **32**, **33**, Lys. 26.8, Ais. 1.19. (An exception is mentioned in *AP* 57.4: the basileus removed his crown when pronouncing the verdict in a trial held before the Areopagos or the ephetai.) For an arkhon to be deprived of his crown or to have it given back to him was synonymous with being removed from or restored to office; cf. 26.5, 58.27. No Attic author identifies the material of the arkhons' crowns. Several later texts say they were made of myrtle (schol. Ar. *Knights* 964, schol. Ais. 1.19, Pol. 8.86, Hesykh. μ 1918, Souda μ 1438), but these statements may all be derived from a misunderstanding of a statement by the historian Apollodoros (schol. Ar. *Frogs* 330 = *F. Gr. Hist.* 244 F140): he said that the thesmothetai were crowned with myrtle because 'the goddess' was connected with that plant, but this reference to Demeter must mean that they were crowned with myrtle at the Eleusinian festival, not all the time. In fact myrtle wreaths were donned by many ordinary people for various religious occasions (e.g. Ar. *Wasps* 861, *Thesm.* 37, 448); cf. P. G. Maxwell-Stuart *Wien. Stud.* 6 (1972) 145–61. The arkhons' crowns are likely to have been more distinctive and durable.

τοὺς χορηγούς: the chorus-producers of the other nine tribes. Probably D. did not really suffer any adverse consequence, or he would have said what it was.

ὀμνύουσι παρεστηκὼς τοῖς κριταῖς: on the appointment of judges at the Dionysia cf. Pickard-Cambridge *Festivals*[2] 95–7. They were selected by lot. Before each contest began, the arkhon drew one name from each of ten urns, each containing names of candidates previously selected from one tribe (cf. Isok. 17.33–4, Plu. *Kimon* 8.8). The ten names were called out (cf. **65** *καλουμένων*) and the ten men came forward from wherever they were sitting in the audience. Before taking the seats at the front reserved for them, they swore an oath to vote for the best performers. (Cf. Pl. *Laws* 659a. A comic chorus can say 'Don't break your oath, but vote for us!'; cf. Pherekrates 96, Ar. *Ekkl.* 1160.) But what was it that Meidias did? D. uses several phrases to describe the incident: Meidias stood beside the judges while they were taking the oath (**17**), he corrupted the judges in advance of the contest (**18**), he stood by while the judges were being called and dictated an oath to them when they swore (**65**). D. is hinting that Meidias whispered to the judges, trying to get them to swear not to award a prize to D.'s chorus. In hyp. ii 4 (text on p. 428) we read 'As Demosthenes says, while the judges were taking the oath to award the victory to the one that sang well, Meidias was nudging them and saying "Except Demosthenes!"' D. does not in fact say this explicitly, but the author of the hypothesis has interpreted the innuendo correctly. Really, though, D. cannot have known what Meidias whispered; if he had known, he would have made not an innuendo but a plain statement.

τὰ παρασκήνια φράττων, προσηλῶν: for the lack of a connective particle between the participles, **81** n. *τῇ δίκῃ, τοῖς νόμοις*: deletion of *φράττων* (Naber *Mn.* II 31 (1903) 412) is unnecessary. According to Harpokration (under *παρασκήνια*), Theophrastos in his *Laws* said that the *paraskenia* were the place assigned for preparations for the contest, and Didymos said that they were entrances on each side of the *orkhestra*. Greater weight must be given to Theophrastos, who was almost contemporary with D., but his explanation need not exclude the other. We may infer that at each side of the *skene* there was a structure, probably wooden, with a doorway; behind or within it choristers dressed and prepared for their performance, and then they passed through the doorway to enter the *orkhestra*. Archaeological evidence adds little, because most of the remains of the theatre belong to the later reconstruction attributed to Lykourgos. In this later theatre the *skene* had a section at each end projecting towards the audience, and modern scholars generally call those sections *paraskenia*, but there is no evidence that they were so named in ancient times, nor that the pre-Lykourgan *skene* had similar projections. However, a suggestion based on the scanty remains of the earlier theatre is offered by C. W. Dearden *The Stage of Aristophanes* (London 1976) 32, that the *paraskenia* were wooden corridors leading to the *skene* from the offstage area at each side, enabling an actor to reach the inside of the *skene* without being seen and then make his appearance from the *skene* door. This suggestion is based on two features of the archaeological evidence: a row of slots, which could have held upright wooden beams, which could have supported a second row of beams with panels between them; and the absence of any indication that there was a way of getting inside the *skene* from the stone hall which was directly behind it. Several objections may be raised: such corridors would have been unduly long and flimsy; since the surviving remains of the wall between the *skene* and the hall behind it do not rise above the ground level of the *orkhestra*, we do not know that there was no doorway through the wall above that level; and why should D., who was not presenting a play, have wanted his chorus to appear from inside the *skene*? Still, Dearden's hypothesis cannot be definitely disproved.

What was Meidias' purpose in blocking the *paraskenia*? Σ (76 Dilts) suggests two: to obstruct the entrance of D.'s chorus, who would then have to go round and appear through 'the entrance from outside'; and to drown the chorus's singing by the noise of hammering. The two explanations are mutually exclusive, since the one assumes that the nailing-up was done before the chorus's performance began, the other after; and scholars have generally preferred the first, which suits *φράττων* better. Cf. A. W. Pickard-Cambridge *The Theatre of Dionysus in Athens* (Oxford 1946) 24 n. 1. We probably have to assume that D.'s chorus was performing last, and the nailing-up was not done until the other nine choruses had already performed; otherwise they would have been impeded too. But whatever the details, it is unlikely that D. and his chorus were seriously disadvantaged, or he would not pass over the incident so quickly.

18. ἐν τῷ δήμῳ: in the Ekklesia, as recounted in **13–14**.

πρὸς with the dative regularly refers to the official presence of judges or magistrates; cf. 20.98, 22.28, 27.49, 39.22, 59.40.

ὑμεῖς ἐστέ μοι μάρτυρες: D. assumes that the jurors in the present trial all attended the Ekklesia and the theatre. For the rhetorical device of telling the jurors that they know the facts already, **1** n. *οὐδένα ... ἀγνοεῖν*. For the particular wording, calling the jurors witnesses, cf. And. 1.37 *καί μοι ὑμεῖς τούτων μάρτυρές ἐστε*, Lys. 10.1, 12.74, Gorg. *Pal.* 15.

καίτοι, usually adversative, is here positive, 'and indeed', adding a second premise to support the implied conclusion 'these statements are certainly true'; cf. Denniston *Particles* 563–4.

δικαιοτάτους: A adds *καὶ πιστοτάτους*. See pp. 50–1 on extra words in A's text.

προδιαφθείρας: 'having corrupted in advance'. This refers to the same incident as **17** *ὀμνύουσι παρεστηκὼς τοῖς κριταῖς*. This reading should be accepted, though it occurs only in later mss., where it may be a copying error for *προσ-*, which all the older mss. have. But *προσ-*, meaning 'in addition', would introduce a new allegation, which would need further explanation.

ὡσπερεὶ κεφάλαια: 'heads, as it were'. For the image cf. 27.7 *τὸ μὲν κεφάλαιον τῶν ἀδικημάτων*.

νενεανιευμένοις: sarcastic; Meidias was nearly fifty (**154**).

κρατούσῃ refers to actual superiority, *νικῆσαι* to formal recognition of it. D.'s chorus performed well, despite the nailing-up of the *paraskenia* and other interference, but was not awarded the victory because Meidias had corrupted the judges; cf. the end of **5**. D. has arranged his account of the festival so as to make this the climax, implying that he is unselfishly concerned more with the tribe's honour than with the assault on himself.

19–24. *Remarks about the scope and arrangement of the speech, and evidence of the incidents already narrated.*

This part of the speech has been at the centre of modern scholars' criticisms of the structure of the whole; see especially pp. 129–30 of

Goodwin's edition. The criticisms are misguided, because they demand a more strictly logical structure than is necessary in a speech to a popular audience. On the problem in general see p. 25; specific criticisms are mentioned at appropriate points in the notes below.

19. ἠσελγημένα: 1 n. *ἀσέλγειαν*.

διέξειμι: D. does not in fact go on to relate any further incidents concerning the Dionysia of 348; incidents related subsequently either are the same ones as have already been mentioned or do not concern that particular festival. But this does not prove that the speech is incomplete; it is just a rhetorical device to give the impression that Meidias' offences are very numerous.

ἔχω δὲ λέγειν καὶ ...: D. digresses for a few moments to indicate that an account of many other offences is yet to come. But he has not yet finished with the offences concerning the Dionysia; he will first call for witnesses' evidence of those, and go on to the other offences afterwards, as he explains in **21**. The arrangement seems to me clear enough, though Goodwin (pp. 129–30 of his edition) is muddled about it.

αὐτοῦ: deleted by Cobet *Misc. crit.* 504. But it is better to retain it, despite the fact that *τοῦ μιαροῦ τούτου* follows. *αὐτοῦ*, placed near the beginning of the list of offences, indicates their authorship; *τοῦ μιαροῦ τούτου*, at the climax, adds an abusive comment.

20. τῶν πεπονθότων οἱ μέν ... οἱ δ' ... εἰσὶ δ' οἳ ...: D. distinguishes three categories of Meidias' victims: some were too frightened to take any action at all; some prosecuted but lost their cases; some accepted compensation arranged privately.

τοὺς περὶ αὐτὸν ἑταίρους: explained more fully in **139**. There we find that Meidias' group of supporters includes men like Polyeuktos, Timokrates, and Euktemon who are paid to speak for him, and others (*μαρτύρων ἑταιρεία*) who nod in agreement when he speaks.

πλοῦτον: *καὶ ὕβριν* is added in A, followed by F and a corrector in P. But *ὕβρις* is not appropriate as a reason for not prosecuting Meidias for *ὕβρεις*: cf. p. 51.

ἠδυνήθησαν: exactly when the long augment *ἠ-* in the verbs *βούλομαι, δύναμαι*, and *μέλλω* came into use in Attic is uncertain. It seems not to occur in Attic inscriptions before 300 B.C.; cf. Meisterhans *Grammatik*[3] 169. Yet it is guaranteed by the metre in A. *Pr.* 206 *ἠδυνήθην*, Ar. *Frogs* 1038 *ἤμελλ'*, *Ekkl.* 597 *ἤμελλον*, Philippides 16 *ἠδύνω*, and so it is unnecessary to expel it from the text of D. Cf. A. Debrunner in *Festschrift für F. Zucker* (Berlin 1954) 85–110.

διελύσαντο: accepted money to refrain from prosecution; cf. **39**, **216**.

πάντας τοὺς ἄλλους: D. maintains that the assault on himself was an offence against the whole community; cf. **7** *καὶ εἰς τοὺς ἄλλους ἅπαντας*.

κληρονόμοι: 'heirs'. The sense of the metaphor is that Meidias has incurred many debts, which now fall due to be paid to the jury.

21. ὅσα ὑμεῖς: the offences concerning the Dionysia of 348 involved insolence to the whole community. A, followed by F and correctors in S and P, adds *ἠδίκησθε*, but the perfect does not fit in well after the aorist *ὑβρίσθην*. Cf. pp. 50–1 on extra words in A.

τὸν ἄλλον βίον: the rest of Meidias' life includes the rest of the offences mentioned in **19–20**. I see no justification for Goodwin's complaint that the relation of this passage to **19** is obscure.

πρώτην: since the goldsmith's testimony is not followed by others in **23**, this word has been taken as evidence that part of the speech is missing there. But it need not mean that; it may mean merely that this is the first testimony presented in the speech as a whole. Other witnesses are not needed at this point because, D. has claimed, the jurors saw for themselves what happened in the theatre (**18** *ὑμεῖς ἐστέ μοι μάρτυρες πάντες*). Cf. Erbse *Hermes* 84 (1956) 140–2.

22. By D.'s time it was the rule that a witness did not give his evidence orally at a trial; a written statement was made out beforehand, and at the trial the witness simply confirmed that the statement read out by the clerk of the court was correct. Thus, even if the trial of Meidias was never held, it is still possible that before it was abandoned the witnesses' statements had already been put in writing and could have been preserved along with the text of D.'s speech. But in fact the document in **22** does not look genuine, but appears to be an imaginative composition made in a later age; see the following notes.

Παμμένης: a name used in the Roman period, not known before the second century B.C.

ἔπερχος: a word found nowhere else and evidently corrupt. Since the sentence otherwise lacks a verb, the best way of emending may be to create a verb by adopting Meier's conjecture *ὑπάρχω*, 'I am' (LSJ *ὑπάρχω* B.I.5), or, I suggest, *ἐπέρχομαι*, 'I come forward to speak' (LSJ *ἐπέρχομαι* I.1.c). Alternatively we may adopt *ἔχω* from some later mss. and make the previous word into a description or identification of Pammenes. F in fact has *ἔπαρχος*, but there is no parallel for this word in the sense 'master craftsman'. (In Plu. *Cic.* 38.4, *τεκτόνων ἔπαρχος* translates the Latin *praefectus fabrum*, which is a different matter.) Another possibility is a demotic, such as *Ἐρχιεύς* (Buttmann, cf. Dobree *Adversaria* 1.457–8) or *Εἰτεαῖος* (A. Westermann *Untersuchungen über die in die attischen Redner eingelegten Urkunden* 72). Yet another, I suggest, is *ἔποικος*, meaning 'immigrant'; the author of this document may not have realized that the correct Attic term was *μέτοικος*.

καταγίγνομαι: a Hellenistic and later word, not Attic.

στέφανον ... ἱμάτιον: one crown and one cloak, to be worn by D. in the procession. But this is an error; D. himself says (**16**, **25**) that the goldsmith was making crowns and cloaks for the chorus. The forger of the document has read **16** carelessly and got the impression that only one crown and one cloak were concerned.

διάχρυσον: another non-Attic word.

23. **μὲν**: answered by *βούλομαι δὲ πρὸ τούτων εἰπεῖν* in **24**. To create expectation that tremendous crimes will shortly be revealed, D. keeps mentioning Meidias' offences against other people and keeps postponing the details. Goodwin fails to appreciate this rhetorical device and transfers the whole of **23** to a later part of the speech (between **127** and **128**), tearing *μέν* away from its *δέ*.

ἐν ἀρχῇ τοῦ λόγου must refer to the first sentence of the speech, but the words used here are really more like the latter part of **19**. D. will not have expected his listeners to remember the exact words of either passage.

ὕβρεις αὐτοῦ καὶ ἀτιμίας: *hybris* and dishonouring are related but not identical; see pp. 19–22. Cf. 18.205 *τὰς ὕβρεις καὶ τὰς ἀτιμίας, ἃς ἐν δουλευούσῃ τῇ πόλει φέρειν ἀνάγκη*. But the alternative reading *πονηρίας* (AF) could be right; cf. **19**.

24. **ἀκήκοα**: prosecuting speakers often claim to have heard what the defendants are going to say, e.g. 19.80, 19.202, 22.17, 23.110, 24.144, 38.19, 45.43, 54.38, 58.50, 59.119. Perhaps in some of these cases the speaker is really only guessing, but no doubt a good deal of genuine information did circulate in gossip about forthcoming trials. Cf. A. P. Dorjahn *TAPA* 66 (1935) 274–95.

κωλύσας: aorist for a future event which will be completed before that of the main verb. This accurate use of the aorist is preferable to A's reading *κωλύσων*.

οὗτος shows that *λόγος* here does not mean the speech as a whole but this part of it; cf. LSJ *λόγος* VI.3.d. In the next sentence *τούτῳ τῷ λόγῳ* has the same sense.

25–8. *The first of the objections that Meidias will make is that the wrong form of legal action is being used against him.*

On the legal procedure and the definition of the offence see pp. 13–23. Strictly D. is right to say that the function of a defendant is to answer the charge actually brought. Yet Meidias could reasonably retort that in that case D. ought not to have brought other offences into his speech.

25. πρῶτον μὲν: there is no responding δέ, and the second item is not reached until **29**; cf. Denniston *Particles* 382.

ἐπεπόνθειν: the normal 1st. sing. pluperfect ending in the fourth century.

δίκας ἰδίας: cases brought by the prosecutor for himself, in which the defendant, if he lost, would be required to pay financial or other compensation to the prosecutor. *δίκαι δημοσίαι* (which included the actions of *γραφή*, *φάσις*, *ἀπογραφή*, *ἀπαγωγή*, *ἔνδειξις*, *ἐφήγησις*, *δοκιμασία*, *εὔθυνα*, *εἰσαγγελία*, and *προβολή*) were those in which the prosecutor was acting on behalf of the whole community, and the penalty was payable to the state. In **25–8** a contrast is being drawn between private cases such as a *δίκη βλάβης*, which D. might have used against Meidias but did not, and *probole*, a public case, which he did use. But the contrast is confused by the mention of *hybris*. A prosecution for *hybris* was a *graphe*, a public case. Nevertheless it involved less publicity than a *probole*, because it was not heard in the Ekklesia; so here D. lumps *γραφὴ ὕβρεως* together with *δίκαι ἴδιαι*, with some legal inaccuracy, for the rhetorical purpose of setting against *probole* all the types of prosecution which he refrained from using. That does not mean that he is using the words *δίκαι ἴδιαι* in a special non-legal sense (as Goodwin maintains); there is no evidence that any such sense existed (for in 18.210, which Goodwin adduces, the reference is to ordinary private cases such as a *δίκη βλάβης*.) D. is using the words *δίκαι ἴδιαι* in the normal sense, and he knows perfectly well that a case for *hybris* is not really a *δίκη ἰδία*, as his words in **28** show (*καὶ δίκας ἰδίας ... καὶ γραφὴν ὕβρεως*).

προσῆκεν: it is normal to omit *ἄν* with this imperfect (as with *ἔδει* and *ἐχρῆν*) in a conditional sense; so also in **33**, **98**.

λαχεῖν: this verb, literally 'obtain', is regularly used of initiating a prosecution. Presumably the usage originated from the sense of obtaining a hearing before a magistrate; possibly the magistrate would use lots to allocate dates for trial to the various cases.

διαφθορᾶς ... ἐπηρείας ... ὧν: genitive for the cause of accusation.

βλάβης: sc. *δίκην*. A *δίκη βλάβης* was a case for damage to property, or for other acts causing financial loss; cf. MacDowell *Law* 149–53, H. Mummenthey *Zur Geschichte des Begriffs βλάβη im attischen Recht* (dissertation, Freiburg im Breisgau, 1971). On the problem of identifying 'the law of damage', **35** n.

οὐ μὰ Δί' οὐχὶ: an emphatic negative, used also in 19.285, 22.33.

δημοσίᾳ: 'publicly', a vague word which might be taken to mean either *δίκῃ δημοσίᾳ*, referring to any kind of public case, or *ἐν τῷ δήμῳ*, referring to the hearing of the *probole* in the Ekklesia.

κρίνειν: 'to put on trial', a term used for public cases rather than private ones; cf. LSJ *κρίνω* III.2.

τίμημα: assessment of a penalty by a jury. In a *probole* the prosecutor could propose any penalty, whereas in a *δίκη βλάβης* the compensation was always payment of the amount lost (single or double, as explained in **43**).

παθεῖν ἢ ἀποτεῖσαι: a common phrase in laws, e.g. **47**. It leaves open any kind of penalty, either non-monetary or monetary. Although mss. here and elsewhere usually give -*τι*-, inscriptions indicate that -*τει*- is correct in the future, aorist, and perfect of *τίνω* and its compounds, and in such names as *Τεισίας* (**62**). Cf. Meisterhans *Grammatik*³ 180–1.

26. προυβαλόμην ... προβαλέσθαι: Cobet *Misc. crit.* 505 prefers *προυβαλλόμην ... προβάλλεσθαί.* The present infinitive is given by S, the imperfect indicative only by later mss. Either aspect is possible, since the verb, like the noun *προβολή*, can refer either to the initiation of the case (past) or to all its proceedings (present); cf. p. 16. *βαλλ-* and *βαλ-* are often confused in mss.

ἐδικαζόμην: contrasted with *προυβαλόμην*, this verb implies a private case, just as *δίκη* does when contrasted with *γραφή*, *προβολή*, etc.

παρ' αὐτὰ τἀδικήματα: 'immediately after the offences', 'flagrante delicto' (LSJ *παρά* C.I.10), as in **215**, 18.13, 20.139, 37.2.

παρεσκευάζετο: A has *παρεσκεύαστο*. Since there is no other evidence to show whether the preparation of the dress was complete or still in progress, we cannot decide between the readings on grounds of sense. The words *ἐμοῦ συντελέσαντος αὐτὰ καὶ ἔχοντος παρ' ἐμαυτῷ ἕτοιμα* in **22** are not acceptable evidence because that document is spurious, and are unlikely to have influenced A, which does not contain it.

ἦν need not be emended to *ἦ*. *ἦν* as 1st. sing. is guaranteed by the metre in E. *Alk.* 655, *Ion* 280, Ar. *Wealth* 29, Men. *Sam.* 379, etc.; cf. E. Harrison *CR* 56 (1942) 6–9.

27. παρόντα ... ὄνθ': the prepositional prefix is often omitted in repetition of a compound verb, without change of meaning. On this phenomenon see C. Watkins *HSCP* 71 (1966) 115–19, R. Renehan *Greek Textual Criticism* (1969) 77–85 and *Studies in Greek Texts* (1976) 11–27 with Demosthenic instances on p. 113.

μὴ: the 'redundant' negative is added to the infinitive because *διακρουόμενον* implies a negative, like verbs meaning 'prevent', etc. It would be equally correct grammar to omit *μή* (cf. **128**), and A does so, but it is better to retain the less obvious but idiomatic usage given by S.

διακρουόμενον: literally 'knock away' and so metaphorically 'get rid of', 'evade', 'postpone'. Rare in other authors, it is a favourite expression of D.'s; cf. **128**, **186**, **201**, 19.33, 19.168, 19.258, 24.36, 34.13, 38.12, 39.37.

τούτοις: since we have *τὸ ... λέγειν* in the earlier part of the sentence, it is possible that another *τὸ* has been lost from the text before *τούτοις*, but it is not grammatically essential.

28. μὴ δὴ τοῦτο λέγειν αὐτὸν ἐᾶτε: similar wording is used in **40**, 19.82, 23.219, etc. These passages do not mean that a jury had authority to direct a speaker what to say and what to omit, but that the jurors might make such an uproar that it was hard for him to continue speaking on a particular point. He might nevertheless press on in defiance of them; cf. **40** *ἂν βιάζηται*. On jurors' *θόρυβος* generally see V. Bers in *Crux* 1–15.

ὁ νόμος: collective (as 'the law' may be in English), referring to numerous individual laws such as the one quoted in **47**.

ἐπὶ τῶν ἰδίων δικῶν: on the sense of the preposition, **2** n. *ἐπὶ τῶν ἄλλων*. Any payment of money or other penalty imposed in a private case went to the successful prosecutor, in a public case to the state.

ἐνέγκοι μοι: see app. crit. for variant readings. In D.'s time it appears that *οι* and *αι* were both current in the optative of this verb, and it is not possible to say confidently which he used. But *ἐνέγκοιμι* (or *-αιμι*) would require the emendation *διὰ τοῦτ'* earlier in the sentence.

29–35. *Meidias will beg not to be delivered over to Demosthenes. But the punishment of an offender is not in the hands of the prosecutor, but is a confirmation of the laws. Demosthenes was a chorus-producer at the time of the offence, and the laws give special protection to public officials, and at the time of the Dionysia.*

The main weakness in D.'s argument here is that the laws do not seem to have mentioned khoregoi specifically as being entitled to special protection.

29. τούτῳ πολεμῶ: Meidias will represent his quarrel with D. as a private feud, with which other people need not concern themselves. I find no reason to accept Humbert's comment that the phraseology attributed to Meidias here is colloquial.

φθέγξεται: this verb denotes sound rather than significance, and is normally contemptuous or sarcastic in D.; cf. **79**, **95**, **201**.

συνάγειν: *συλλέγειν* means the same and could be right (cf. **36**) but has less convincing support from the mss.

30. οὐδ' ἐγγύς: 'and not nearly', i.e. 'far from it'; so also in 18.12, 18.96, 37.38.

οὐδὲ γάρ: 'not even'. The argument runs: the jurors do not hand over an offender to the prosecutor for punishment, and do not even allow the prosecutor to decide what the punishment shall be (which would be a lesser concession). Of course jurors often do vote in favour of a prosecutor's proposal for conviction and punishment, and D. hopes that they will do so in the present case, but the proposal must be consistent with the law; the emphasis is on the indefinite *ὡς ἂν* ..., 'in any way in which a particular victim persuades you'.

ἐπ': 'with a view to' (not temporal, as Goodwin alleges).

ἀδικησομένοις: passive; so also in **220**. The -*θησ*- form is not used in Attic in the future of this verb.

οἱ νόμοι: personification of 'the laws' is common, the most elaborate instance being in Pl. *Kriton* 50–4.

31. ἀλλὰ μὴν ... γε marks progression to another point in the argument; cf. Denniston *Particles* 344.

32. τῶν θεσμοθετῶν τούτων: the demonstrative adjective is used because one thesmothetes is presiding at this trial; **3** n. *τις εἰσάγει*.

ἰδίαν qualifies only *δίκην κακηγορίας*, not *γραφὴν ὕβρεως*. For private cases, **25** n.; for prosecution for slander, **81** n.

ἐὰν δὲ θεσμοθέτην ...: the only precise evidence for this law. Arist. *Prob*. 952b 28–32 has a more general statement: *ἐὰν μέν τις ἄρχοντα κακῶς εἴπῃ, μεγάλα τὰ ἐπιτίμια, ἐὰν δέ τις ἰδιώτην, οὐδέν. καὶ καλῶς· οἴεται γὰρ τότε οὐ μόνον εἰς τὸν ἄρχοντα ἐξαμαρτάνειν τὸν κακηγοροῦντα, ἀλλὰ καὶ εἰς τὴν πόλιν ὑβρίζειν*. Aristotle may be assumed to have had Athens in mind, but his statement is not exactly correct about Athenian law, since one who slandered a private person might be prosecuted by *δίκη κακηγορίας*. D.'s statement can be accepted, up to a point: no doubt the law did specify disfranchisement as the penalty for anyone who attacked, physically or verbally, an official during performance of his duties; and no doubt it applied to all the nine arkhons. But did it extend beyond them? D. in **33** alleges that it applied to all to whom the state gave *τινὰ στεφανηφορίαν ἤ τινα τιμήν*. This must be too wide; see the notes there. If the law had clearly covered khoregoi, D. would have had it read out to the court. Probably only *ἄρχοντες* were mentioned in it.

ἄτιμος: 'disfranchised'. For a summary of what was meant by *ἀτιμία* see MacDowell *Law* 73–5; for a detailed discussion, Hansen *Apagoge* 54–98.

καθάπαξ: 'permanently'. Citizens disfranchised once for all are distinguished from state-debtors, whose disfranchisement ended when they paid their debts; cf. 25.30 *ἢ τοῖς ὀφείλουσι τῷ δημοσίῳ ἢ τοῖς καθάπαξ ἀτίμοις*, and Hansen *Apagoge* 68.

ἤδη: 'now', i.e. when he goes as far as that (LSJ *ἤδη* I.4); not to be deleted, as was proposed by Benseler *De hiatu* 109, who wished to remove the hiatus.

προσυβρίζει: *προσ-* is 'in addition' to the individual who holds the office. Aristotle makes the same point with *οὐ μόνον ... ἀλλὰ καὶ* ... in the passage quoted above.

ὑμέτερον κοινὸν ... τῆς πόλεως: both the crown and the title of an arkhon are the property of the state.

ὁ ... θεσμοθέτης: 'the word *θεσμοθέτης*'. One might expect *τό*: for the attraction to the gender of the word under consideration, cf. Lys. 10.16 *ἡ ποδοκάκκη αὕτη ἐστίν ... ὃ νῦν καλεῖται ἐν τῷ ξύλῳ δεδέσθαι*, Ais. 2.40 *ὁ κέρκωψ ... ἢ τὰ τοιαῦτα ῥήματα*.

33. πάλιν: often used by D. to mean 'also', introducing the next in a series of similar items; so in **44**, **64**, **141**, **171**.

τὸν ἄρχοντα: the chief arkhon, as in **13**.

ταὐτὸ τοῦτο: 'it is just the same' in parenthesis; *τὸν ἄρχοντα* is the object of *πατάξῃ*. So also in **39**.

ἐστεφανωμένον: while performing an official function; **17** n.

πατάξῃ τις ... εἴπῃ: S has *πατάξῃς ... εἴπῃς* here, and likewise *πατάξῃς* at 4.40 *κἂν ἑτέρωσε πατάξῃ τις*. 'You' would mean the same as 'one', but it seems less likely to be right, because it is awkward to have a 2nd. sing. expression among 2nd. pl. ones, and because with *ἄτιμος* and *ὑπόδικος* omission of *εἶ* would be less normal than omission of *ἐστίν*. Cf. Cobet *Misc. crit.* 505.

ἰδίᾳ ὑπόδικος: synonymous with *δίκην ἰδίαν φεύξεται* (**32**). This is not a normal sense of adverbial *ἰδίᾳ*, but it is clear here from the context.

ἄδειαν: in Athens *ἄδεια*, when given by the state to an individual, means immunity from prosecution for a specified act. This may be permission to perform an act which is otherwise illegal (e.g. 24.45–7, And. 1.77, *IG* I³ 52B (ML 58B) 16, 370 (ML 77) 15); or it may be pardon for an offence already committed, granted to the offender because, for example, he gives evidence about other offenders (e.g. And. 1.11–34, 2.23, Lys. 13.55, Th. 6.60.3). But D. cannot have meant to say that persons receiving immunity from prosecution for specified acts were all given special protection against assault and slander. Nor can we take *ἄδεια* here to mean some other kind of immunity, for the phrase is '*all* those to whom the state gives any immunity', and this cannot be interpreted so as to exclude the usual kind. A further reason for suspecting the text is that it offers *τινὰ* only twice with three accusative nouns. *ἄδειαν* should therefore be deleted; it is a gloss on the words that follow, written in a later period by someone, such as the author of Σ here (108 Dilts), using the word *ἄδεια* in the wider and non-Attic sense of 'privileged position'.

στεφανηφορίαν: vague, perhaps deliberately ambiguous. Different kinds of crown-wearing may be distinguished:

A. Some officials wore a crown as a badge of office. But the nine arkhons were perhaps the only such officials in Athens; that seems to be implied by a comment on a law in Ais. 1.19, "*μὴ ἐξέστω αὐτῷ τῶν ἐννέα ἀρχόντων γενέσθαι*", *ὅτι οἶμαι στεφανηφόρος ἡ ἀρχή*.

B. A crown was given to winners of athletic contests.

C. Occasionally a crown was given as an honour to someone who had performed distinguished services to Athens, as for example Ktesiphon proposed the conferment of a crown on D. in 336.

D. Anyone might choose to wear a crown on a religious or festive occasion.

D. has just explained the special legal protection given to category *A*. Similar protection may have been given to *C*, but obviously not to *D*. But was a khoregos in category *A*? Surely not, or D. would have had more to say about the significance of his crown. No doubt it was customary for a khoregos (and his chorus; cf. **16**) to wear a crown at a festival, but that will not legally have placed him in any category other than *D*. D.'s use of the word *στεφανηφορία* encourages his hearers to think that a khoregos enjoyed the same privileged status as an arkhon when really he did not.

τιμὴν: an even vaguer expression: who is to say what is an honour and what is not? There is in fact no further evidence to show what honours carried the privilege of special legal protection against assault and slander.

προσῆκεν: **25** n.

34. εἰ: introducing a fact as a ground of argument; cf. LSJ *εἰ* B. VI. The emendation *ἐπεὶ* (Cobet *Misc. crit.* 505) is unnecessary.

ἱερομηνίας οὔσης: in Athens *ἱερομηνία* does not mean 'sacred month' but refers to the particular days on which a festival is held. Cf. 24.29–32, where it refers to the one day of the Kronia (12 Hekatombaion).

καὶ τὸ: F, followed by correctors in S and P, has *καίτοι*, but even though *καίτοι* can be accumulative rather than adversative (cf. **18**), that is not satisfactory here, because it normally occurs at the start of a new sentence. Therefore we should read *καὶ τὸ* and understand with it *τὸν χορηγὸν ὑβρίζεσθαι*. This is possible only if *τοῦτο* likewise refers to the act, not merely to the title *χορηγός*. Therefore *ὄνομα*, inserted in A and other mss., is not correct. It must have been added by someone who took *ἐστὶ τῆς πόλεως* to be echoing the end of **32**.

χρὴ ...: D. forestalls the objection that the special law about offences at the festival is unnecessary and need not be enforced.

εὔορκα: 'in accordance with your oath'. The alternative *ἔνορκα* (A), 'in your oath', could be right: *τοὺς νόμους φυλάττειν καὶ χρῆσθαι* could be regarded as being in the jurors' oath, because the oath included the words ψηφιοῦμαι κατὰ τοὺς νόμους (24.149; cf. 18.121, 19.179, etc.).

35. ἣν ... ἣν ... ἣν: on the combination of asyndeton and repetition see Denniston *Style* 106–9.

βλάβης: as a legal offence, βλάβη always means damage to property, not injury to a person (which is αἴκεια or τραῦμα or ὕβρις), but it is used for various different kinds of damage: not only for physical damage to a piece of property, such as the damage which Meidias is alleged to have done to the cloaks and crowns at the goldsmith's (**25**), but also for causing financial loss or simply failing to repay a debt (e.g. 36.20, 37.22, 52.14). A private case (δίκη βλάβης) could be brought; the penalty is stated in **43** to have been payment of the amount of the damage if it was done unintentionally, of double that amount if it was done intentionally, and a phrase quoted from a law evidently refers to this distinction (23.50 ἄν τις καταβλάψῃ τινὰ ἑκὼν ἀδίκως). But otherwise we do not possess the text of the law or laws on this subject, and a difficulty arises from the apparent discrepancy between **35**, where D. speaks of 'the law of damage' in the singular, and **43**, where he speaks of 'all these laws about damage'. Was there just one law covering all kinds of damage, or several laws each dealing with a particular kind? This question has been considered by R. Maschke *Die Willenslehre im griechischen Recht* (1926) 114, H. J. Wolff *AJP* 64 (1943) 324, E. Ruschenbusch *ZSSR* 82 (1965) 306, H. Mummenthey *Zur Geschichte des Begriffs βλάβη im attischen Recht* (dissertation, Freiburg im Breisgau, 1971) 34–41. In my opinion **43** is decisive evidence that there were several laws. The problem then is to explain the singular νόμος in **35**. This sentence is a reference to particular statutes about βλάβη and αἴκεια and ὕβρις which existed earlier than the law about offences at the Dionysia. It is not correct that 'νόμος obviously is not meant in the technical sense of *statute*' (Wolff), nor that νόμος here means 'Inbegriff von Rechtssätzen' (Mummenthey). It is better to assume that, in addition to a number of laws about specific kinds of damage (e.g. damage caused to a neighbour's farm by an overflow of water, as in the case of Kallikles in D. 55, discussed by Wolff), there was also a general law about damage of unspecified kind (as Ruschenbusch maintains). To this law the phrase quoted in 23.50 may well belong. It will have been the original and basic law about βλάβη, meaning physical damage to property. Subsequent laws either added further kinds of act (such as causing financial loss) to those which were regarded as βλάβη, or specified the valuation of particular kinds of damage (e.g. 1000 dr. for any damage caused by an overflow of water; cf. 55.25). But D. here calls the general law about damage '*the* law', partly because it is the one which existed *πάλαι*, and partly because it is the one under which he could have prosecuted Meidias for damaging the cloaks and crowns.

πάλαι: the dates of the laws mentioned in this sentence are not known, but *πάλαι* need mean no more than that they existed in the fifth century.

αἰκείας: 'battery', hitting a person, for which a private case (*δίκη αἰκείας*) could be brought. D. 54 (*Against Konon*) and Isok. 20 (*Against Lokhites*) are speeches for the prosecution in such cases, and another case is described in 47.45–7. From the latter it appears that, if there was a fight, only the man who struck the first blow (*ὁπότερος ἦρξεν χειρῶν ἀδίκων*) was guilty of *αἴκεια*. It does not seem that the prosecutor had to prove that he had suffered injury, though of course if he had been injured that would help to prove that he had been hit. In the only known instances the penalty was payment of a sum of money, assessed by the jury, to the successful prosecutor (47.64, Isok. 20.16).

ὕβρεως: on this offence see pp. 18–22.

τοῦδε τοῦ νόμου ... ἱερὸν νόμον: the law quoted in **8**. D. calls it a sacred law simply because it was about the festival of Dionysos. He is not referring to some different law; if another law about the sanctity of the Dionysia had existed, he would undoubtedly have quoted it in full.

πᾶσι: D. will go on to relate other offences committed by Meidias, but 'all' is rhetorical exaggeration.

πότερα ...: the first part of the double question contains a deliberative subjunctive, the second part a potential optative, which requires the insertion of *ἄν* (as earlier editors saw, but G. H. Schaefer *Apparatus* 3.339 was the first to propose *κἄν*). The normal penalty for *βλάβη* (cf. **43**) or *αἴκεια* (cf. 47.64, Isok. 20.16) was the payment of a sum of money to the victim. The penalty for *hybris* could legally be anything to which the jury agreed (cf. **47**), and in **49** D. alleges that the death penalty was often imposed; thus it is not strictly logical to suggest that Meidias deserves a greater penalty than the one for *hybris*.

36–41. *Meidias will also adduce previous cases of assault in which the offenders have gone unpunished. But he should be punished as an example; besides, in the other cases there were extenuating circumstances, and the prosecutions were dropped for improper reasons.*

36. περιιόντ': the elided form with a single *ι* is given by both S and A here, and could be correct; cf. LSJ under *περίειμι* (*εἶμι ibo*).

συλλέγειν: cf. **23**. Apparently, while D. was cataloguing acts of *hybris* committed by Meidias, Meidias was cataloguing those committed by other people.

τούτους: object of *λέγειν*: 'that he was intending to mention these men ...'. Alternatively it might be the subject: 'that these men were going to give evidence ...'. But the subsequent course of the sentence (including *φασιν*, 'people say') does not suggest that the men were expected to speak in court themselves.

πρόεδρον: **9** n. Nothing is known of this incident except what D. tells us in **38–9**. The words *ποτέ φασιν* suggest that it had happened quite a long time ago.

Πολυζήλου: unknown, unless this is the Polyzelos who was arkhon in 367/6.

ἀφαιρούμενος: rescuing her from a man who was drunk (**38**). Nothing else is known of this incident either, but it was a more recent one (*ἔναγχος*). Σ (121 Dilts) says that the thesmothetai had the duty to patrol at night to prevent rape, but that is probably no more than a false inference from this passage.

ὡς ... ὀργιουμένους: after *ὡς* in the sense 'thinking that ...' the accusative absolute may be used instead of the genitive absolute, even when the participle is not impersonal. Cf. 54.32 *ὡς ὑμᾶς εὐθέως πιστεύσοντας*, 14.14, 18.276, Lys. 14.16, Ais. 3.142, Pl. *Rep*. 345e, 426c, etc. Meidias' argument would be one from precedent: others committing acts like his were not prosecuted or punished. In **38–9** D. argues that Meidias' case is different.

37. ὑπὲρ: usually no preposition is added to the genitive with *μέλει* or *φροντίζω*, but there are exceptions: cf. **39** *οὐχ ὑπὲρ ὑμῶν ... φροντίσας*, 1.2 *εἴπερ ὑπὲρ σωτηρίας αὐτῶν φροντίζετε*, 59.15 *μεμέληκεν αὐτῷ περὶ τούτων*. So it is better to retain *ὑπὲρ* here; SYP are more likely to have lost it by haplography after *εἴπερ* than AF to have inserted it wrongly.

τὸν ἀεὶ ληφθέντα: 'the man caught each time', a normal meaning of *ἀεί* with a participle; cf. **131**, **177**, **194**, **223**.

38. συγγνώμην: 'reason for being forgiven'; so also in **66**, **198**.

σκότους καὶ νυκτὸς: both genitives are temporal, but in effect the second explains the first, so that we could translate 'in the darkness of night'. Cf. **85** *ἑσπέρας οὔσης καὶ σκότους*.

ὀργῇ: for anger as an excuse for misconduct cf. **41**, where D. argues that only sudden loss of temper, not sustained anger, is an acceptable excuse.

φθάσας τὸν λογισμὸν ἁμαρτὼν ἔπαισεν: S and other mss. offer *φθάσαι τὸν λογισμὸν ἁμαρτὼν ἔφησεν*, and Wolf conjectures *φθασάσῃ τὸν λογισμὸν ἁμαρτεῖν ἔφησεν*. It is possible to defend *ἔφησεν* by regarding the whole passage as an account of the various defendants' statements, comparing *προφάσεις* in the previous sentence and *εἰπεῖν* in the following sentence. However, D. is really more concerned with facts than with speeches; and since the case of Polyzelos did not come to trial (**39**) and D. seems to have known of it only by hearsay (**36** *φασιν*) he may not have known what Polyzelos said about it.

ἐφ' ὕβρει: *ἐπί* with the dative may denote a purpose; so also in **72**, **92**.

ἐπὶ τούτου: **2** n. *ἐπὶ τῶν ἄλλων*.

39. ἐμοὶ καὶ τούτοις: dative of agent.

καθυφεὶς: *καθυφίημι* usually implies bribery or some other disreputable motive for abandoning a prosecution; cf. **151**.

ταὐτὸ τοῦτο: **33** n.

ἐρρῶσθαι πολλὰ τοῖς νόμοις εἰπὼν: 'saying many goodbyes to the laws', i.e. ignoring them. For the sarcastic metaphor cf. 19.248 *ἐρρῶσθαι πολλὰ φράσας τῷ σοφῷ Σοφοκλεῖ*, 5.22, 18.152. The meaning is not that he broke the laws, but merely that he took no action to enforce them.

40. εἰ μὲν ...: D. means this hypothesis to be rejected; he is maintaining that it is inappropriate to bring up other cases of assault. *ἐκείνων* means the proedros and the thesmothetes, who failed to prosecute when they ought to have.

πᾶν γὰρ τοὐναντίον: adverbial accusative. The rest of this sentence repeats the point made in **3**.

μὴ τοίνυν ἐᾶτε: **28** n.

δίκαιον refers not to the truth of the facts, but to the validity of the argument that Meidias ought to be let off because Polyzelos and the man who hit the thesmothetes were let off.

41. οὕτως: treating that argument as irrelevant.

ἔνεστ' implies futurity, 'there is no argument which he will be able to use'. It is therefore unnecessary to adopt *ἐνέσται* (FYP).

ἀνθρωπίνη: a 'human' excuse is one which justifies mercy (instead of strict justice) on the ground that the misconduct was due to a weakness to which human beings are prone, or to a misfortune which might befall anyone. Cf. 23.70 *οὐκ ἐπέθεντο τοῖς ἀτυχήμασιν, ἀλλ' ἀνθρωπίνως ἐπεκούφισαν ... τὰς συμφοράς*, 45.67 *ἡ τῆς χρείας ἀνάγκη φέρει τινὰ συγγνώμην παρὰ τοῖς ἀνθρωπίνως λογιζομένοις*, and Dover *Morality* 269–71.

μετρία: 'moderate', 'reasonable', showing that Meidias' conduct did not go beyond what was proper. *μετριότης* is the opposite of *ὕβρις*, and D. complains that Meidias is not *μέτριος* (**128**, **186**, **199**).

ὀργὴ: the excuse given for Polyzelos (**38**).

νὴ Δία: idiomatically used to mark a hypothetical reply; so also in **88**, **98**, **99**, **149**, **160**, **222**. The speaker's own retort is then often introduced by *ἀλλά*, but not by *ἀλλὰ μήν* (cf. Denniston *Particles*

10–11); therefore the reading ἀλλ' ἃ μὲν is preferable at the beginning of the next sentence.

τυχὸν: accusative absolute, 'if it chances', 'perhaps'; so also in 18.221, 25.88.

ἐξαχθῇ πρᾶξαι: 'is carried away ⟨by excitement⟩ to do'; a similar phrase in **74**.

ἔνι φῆσαι: 'it is possible for him to say'. Y's reading φήσαι (optative, without γ' ἔνι) would require the insertion of ἄν.

τοῦ μὴ μετ' ὀργῆς: sc. πράττεσθαι. The 'redundant' μή is analogous to the one found with verbs meaning 'prevent' or 'refrain', and we need not follow the corrector of O in deleting it. It is possible, of course, to act from anger sustained over a long period, but D. means that such anger is no excuse for an act which is otherwise wrong.

42–50. *The laws treat with special severity an offence committed intentionally or violently or with* hybris.

42. ὁπηνίκα, usually purely temporal, here has a causal implication. Cf. 18.14, where it is conditional.

δεῖ σκοπεῖν: this order is normal in D. (**154**, 18.62, 19.30, 20.25, etc.). σκοπεῖν δεῖ (SYP) is abnormal.

ὀμωμόκατε: on the jurors' oath, **4** n. ὀμωμοκὼς.

ἀξιοῦσι: the laws are the subject, personified as in **30**.

ἑκουσίως καὶ δι' ὕβριν: D. gets as close as he can to identifying *hybris* with committing any wrongful act intentionally. In fact the laws which he now adduces as parallels refer to intention or to violence, but not to *hybris*.

43. βλάβης: **35** n.

οὗτοι: the demonstrative adjective is used merely because the law of damage has already been mentioned. It does not imply that D. has copies of the laws in court.

ἵν' ἐκ τούτων ἄρξωμαι: 'to begin with these'. The phrase indicates to the hearer that there will be several examples, which could have been taken in a different order. Cf. 9.8 ἵν' ἐντεῦθεν ἄρξωμαι.

ἂν μὲν ἑκὼν βλάψῃ: sc. τις (which is in fact inserted in A). This is the only surviving statement of the distinction between the penalties for intentional and unintentional damage in Athenian law, but some words introducing this distinction in a law about damage are quoted in 23.50, ἄν τις καταβλάψῃ τινὰ ἑκὼν ἀδίκως. A fragment of a law about the Eleusinian Mysteries seems to refer to a similar rule: *IG* I[3] 6B.4–8 [τὰ] μὲν hακόσι[α h]απλεῖ, τὰ δὲ [hε]κόσια διπλ[εῖ].

τὸ βλάβος: 'the damage done', as distinct from βλάβη, which is the act of damaging, though this distinction is not always observed elsewhere.

ἐκτίνειν: 'pay compensation for', as in the law quoted in **8**. Since a prosecution for damage was a private case, the compensation was paid to the victim, not to the state.

πανταχοῦ: temporal, 'on every occasion'. So also in **153**, **161**.

βοηθείας: legal remedy, including compensation; **7** n. *βοηθῆσαι*. But there is a penological problem here which D. does not raise: it may be granted that the offender ought to pay more for intentional than for unintentional damage, but it does not follow that the victim ought to receive more in the one case than in the other.

οἱ φονικοὶ: sc. *νόμοι*. There were special courts and procedures for trying cases of homicide. A person accused of intentional killing of an Athenian was tried by the Areopagos. Other homicide cases were tried by the fifty-one *ἐφέται*: at the Palladion for unintentional homicide, or for the killing of a metic, alien, or slave, or for complicity in homicide (*βούλευσις φόνου*); at the Delphinion if the accused person maintained that the killing was lawful; at Phreatto for a second accusation of homicide against a person already exiled for a previous one. There were proceedings at the Prytaneion if death was alleged to have been caused by an unknown person, an animal, or an inanimate object. On all these see MacDowell *Athenian Homicide Law* (1963). The law or laws about homicide were attributed to Drakon and dated to the arkhonship of Aristaikhmos (probably 621/0). In 409/8 'Drakon's law about homicide' was newly inscribed on stone, and this inscription is partially preserved and can be supplemented from quotations in D. 23.22–82, 43.57. The inscription has been studied by R. S. Stroud, who published his new readings in *Drakon's Law on Homicide* (1968); his text is reproduced as *IG* 1^3 104 and ML 86. On the question whether it was really a seventh-century law remaining in force in the fifth and fourth centuries, see M. Gagarin *Drakon and Early Athenian Homicide Law* (1981); it is probable that some amendments had been added to Drakon's original law. D.'s use of the plural *οἱ φονικοὶ* does not imply that he is referring to other laws besides this inscription; Drakon's law could be called *νόμοι* because it contained a number of clauses (cf. 23.51).

ἐκ προνοίας: 'deliberately', 'on purpose'. The nature of premeditation in Athenian homicide law is discussed by W. T. Loomis *JHS* 92 (1972) 86–95; see also R. Maschke *Die Willenslehre im griechischen Recht* (1926) 53–63, MacDowell *Homicide* 58–60, Stroud *Drakon's Law* 40–1, E. Cantarella *Studi sull'omicidio in diritto greco e romano* (1976) 95–111, Gagarin *Drakon* 30–7, A. Biscardi *Diritto greco antico* (1982) 288–91. There are two problems. First, how is the expression *ἐκ προνοίας* related to the distinction between intentional and unintentional homicide (*φόνος ἑκούσιος* and *φόνος ἀκούσιος*, or *ἑκών* and *ἄκων*)? In English usage an act may be committed intentionally but without forethought, for example if a man suddenly loses his temper.

But in Athenian law this distinction is not made, and *ἐκ προνοίας* is simply a synonym of *ἑκούσιος*: indeed it is the normal term for intentional homicide, for which *ἑκούσιος* is more seldom used. In our present passage *ἐκ προνοίας* and *ἀκουσίως* are used as antonyms, and in other passages (23.45, 23.50, *AP* 57.3) it is clear that *ἐκ προνοίας* and *ἀκούσιος* (or *ἄκων*) are regarded as an exhaustive list; there is no such thing as homicide which is not *ἐκ προνοίας* and yet not *ἀκούσιος*. The fullest exposition is in Arist. *Ethika Megala* 1188b 29–38: taking the example of a woman who gave a man a drink which she thought was a philtre to make him love her, but which in fact killed him, Aristotle explains that such homicide is unintentional because it is *οὐκ ἐκ προνοίας*. Yet the woman did give the man the drink deliberately, and this brings us to the second problem: what kind of intention is meant by saying that a killing was done *ἐκ προνοίας* or *ἑκουσίως*? It is clear that, if the killer intends to kill, he kills *ἐκ προνοίας*, and if he intends no harm (like the woman in Aristotle's anecdote), he kills *οὐκ ἐκ προνοίας*, but what about the person who intends merely to harm without killing but does in fact kill? In 54.28 the speaker, Ariston, who accuses Konon of assaulting him, says that, if he had died from his injuries, Konon would have been tried by the Areopagos for homicide; that means that the charge would have been one of intentional homicide, even though Ariston does not allege that Konon meant to kill him. Likewise in Ant. 4 (the *Third Tetralogy*, an imaginary case) the accused man admittedly did not want to kill (4c.4) and yet has been prosecuted for intentional homicide (for the fourth speech of the tetralogy shows that he has gone into exile to avoid the death penalty). Therefore *ἐκ προνοίας*, in connection with homicide, means not necessarily 'intending to kill', but 'intending to harm'.

ἀποκτιννύντας: there appears to be no difference in meaning between *ἀποκτίννυμι* and the commoner *ἀποκτείνω*. But the form *-κτεινν-* (SA) should be rejected; cf. Herodian Gr. 2.539 (Lentz) *κτεινύω διὰ τῆς ει διφθόγγου καὶ δι' ἑνὸς ν κατὰ παράδοσιν· τὸ δὲ ἀποκτίννυται διὰ τοῦ ι καὶ διὰ δύο νν.*

θανάτῳ καὶ ἀειφυγίᾳ: death was the normal penalty for intentional homicide. But a defendant who expected to be found guilty could go into exile, either before the trial or after making the first of his two speeches in defence, and was then free to continue his life elsewhere (23.69, Ant. 5.13; cf. Ant. 2b.9, 4d.1). The present passage is the clearest surviving evidence that perpetual exile was not merely an unofficial evasion of the death penalty but was specified by law as an alternative to it. Cf. MacDowell *Homicide* 110–15, Gagarin *Drakon* 112–15.

δημεύσει τῶν ὑπαρχόντων: confiscation of property was imposed as an additional penalty, not as an alternative, to death or exile for intentional homicide (23.45, 24.7, Lys. 1.50). The property was sold by the poletai at a meeting of the Boule to anyone wishing to buy it (*AP* 47.2).

αἰδέσεως: respecting a person's rights, by not treating him as an outlaw, and thus 'forgiveness', 'pardon'. The noun is rare (*AP* 57.3 seems to be the only other instance), but the verb *αἰδέομαι* is used in this sense in the homicide law (43.57, restored in *IG* 1³ 104.13–16) and 23.72, 23.77, 37.59, 38.22. The penalty for unintentional homicide was exile, but it could be waived by the killed person's relatives or by members of his phratry. 'Pardon is to be granted, if there is a father or brother or sons, by all, or the one who opposes it shall prevail. And if these do not exist, pardon is to be granted by those as far as the degree of cousin's son and cousin, if all are willing to grant it; the one who opposes it shall prevail. And if there is not even one of these alive, and the killer did it unintentionally, and the Fifty-one, the Ephetai, decide that he did it unintentionally, then let ten members of the phratry admit him to the country, if they are willing. Let the Fifty-one choose these men according to their rank' (Stroud's translation of *IG* 1³ 104.13–19). **43** makes clear that in the fourth century *αἴδεσις* was possible only in a case of unintentional, not of intentional homicide, but it is disputed whether the inscription also means that. The condition that the homicide must have been unintentional seems to be emphasized by Drakon (if the wording is his) particularly in the provision for pardon granted by members of the phratry. One interpretation is that the victim's relatives had the power to pardon for intentional homicide in Drakon's time, but lost it by a later amendment of the law; so Gagarin *Drakon* 50–1, 137–40. The other interpretation is that pardon was already restricted to unintentional homicide, and Drakon was merely making clear that extending to the phratry the power to pardon did not authorize any breach of this restriction; so E. Heitsch *Aidesis im attischen Strafrecht* (Abhandlungen Mainz 1984.1). No other evidence resolves this problem, and it remains open.

φιλανθρωπίας πολλῆς: an overstatement. A person condemned for unintentional homicide, if the victim's relatives refused to pardon him, had to remain in exile, and thus was no better off than a person who had gone into exile for intentional homicide, except that his property was not confiscated.

44. τοῖς ἐκ προαιρέσεως ὑβρίσταις: D.'s own phrase for intentional offenders. The laws about damage and homicide did not, as far as is known, use the words *προαίρεσις* and *ὕβρις*.

ὀφλὼν δίκην: losing a private case, and therefore required to pay a sum of money to the prosecutor.

ἐξούλην: 'ejectment'. A person who won a case and was awarded a sum of money as damages, but then could not obtain the money (or property of equal value) from his opponent, could bring a case of ejectment (*δίκη ἐξούλης*) against him, as D. did against Meidias after the case of slander (**81**). If he won that case too, that entitled him to use force to seize the money due to him or property of equal value. The opponent, on losing the case of ejectment, had his name recorded in a public list (39.15), and was required to pay a fine to the public treasury also, as D. tells us here; the amount of the fine was the same as the amount owed to the prosecutor (P. Oxy. 221.14.10–15, Harp. *ἐξούλης*, Souda ε 1815; cf. Isai. 5.22–4). As with all debts to the state, he was disfranchised until he paid (And. 1.73). When D. says 'the law no longer made the ejectment private', he means only that a fine was payable to the state; the case was nevertheless a private one in the sense that it could be initiated only by the injured party. Details of this procedure (which was used also for other property claims, besides claims for damages awarded by a jury) have been much discussed; see Harrison *Law* 1.217–20, who gives references to earlier work.

προστιμᾶν ἐπέταξεν τῷ δημοσίῳ: 'ordered ⟨the jury⟩ to award ⟨a payment⟩ to the public treasury in addition'. D. provides no answer to this rhetorical question, but the answer implied is that, if a court has ruled that a man must pay something and he still does not pay it, he must be assumed to be withholding it deliberately, not from unawareness that he owes it. So this is another example of imposition of a severer penalty when an offence is intentional.

καὶ πάλιν ...: D. now moves to a different distinction: granted that an offence is intentional, there is a severer penalty if it is committed with violence.

τάλαντον ἓν ἢ δύο ἢ δέκα: these, especially the last, would be very large sums to lend or deposit.

ἀποστερήσῃ: the normal verb for omitting to repay money owed, with the intention of defrauding the creditor; so also in 34.27, 35.42, Ar. *Clouds* 1305, etc. Its sense is discussed at length by D. Cohen *Theft in Athenian Law* (Munich 1983) 10–33.

τι: omitted in most mss. by haplography before *τιμήματος*, but restored accidentally by dittography in Aa. The adjective *ἄξιον* seems to need the pronoun.

βίᾳ: this passage (with Harp. *βιαίων*, which is based on it) is our only evidence for the law that a person convicted of theft with violence had to pay not only compensation to the owner but also a fine of equal amount to the public treasury. Some other kinds of theft also led to severe penalties (cf. 24.113–14), but the reason for singling out theft with violence here is that violence is a feature of Meidias' conduct.

τῷ ἰδιώτῃ: the individual from whom the property was stolen. This makes an appropriate contrast with *τῷ δημοσίῳ*, and the emendation *τῷ ἑλόντι* (Cobet *Misc. crit.* 506), though it has some palaeographical plausibility, is unnecessary. (Harp. *βιαίων*, based on our passage, has *τῷ ἑλόντι*, but the reason for using it there is that *ὁ ἁλούς* precedes.)

45. ὁ νομοθέτης: the man who drafted the law, whoever he was. By and large, D. is right in saying that the reason why certain offences incur a penalty payable to the state and (in the next sentence) prosecution by any member of the public is that those offences are considered detrimental to the community in general. But theft with violence is not a clear-cut example, because for this offence only half of the payment went to the state, and (according to Harp. *βιαίων*, schol. Pl. *Rep.* 464e) the prosecution was by a *δίκη* and so not open to every member of the public.

τοὺς δὲ νόμους ἁπάντων: one of the functions of laws is to protect ordinary people against powerful individuals.

τὸν πεισθέντα: the man who agreed (cf. **44** *παρ' ἑκόντος*) to lend a sum of money; not, as Σ (136 Dilts) would have it, the thesmothetes and the proedros of **36–9**, since D. is here talking about theft and misappropriation of money and does not return to *hybris* until the next sentence.

βοηθείας covers both prosecution and exaction of penalties; only the latter (and only half of that) was public in a case of theft with violence, but both were public in a case of *hybris*, to which D. now turns.

αὐτῆς implies 'the particular offence with which we are concerned at present'.

τῷ βουλομένῳ: it was a normal feature of public cases that 'anyone who wishes' could prosecute; but D. omits to say that in some cases, including *hybris* (as we see from the text of the law in **47**), the right was restricted to Athenian citizens.

τὴν τιμωρίαν: the satisfaction of seeing his opponent punished.

τῶν τοιούτων: the genitive as regularly used for the offence with a verb of punishment.

ἐφ' ἑαυτῷ: *ἐπί* with a dative denoting a person is rare, and may here indicate either a cause, 'because of one's own injury', or a purpose, 'for one's own benefit'; cf. LSJ *ἐπί* B.III.1–2.

46. ὑπερβολῇ: **16** n. Here the point is that the legislator went beyond what was necessary or expected. For the use of *χράομαι*, cf. 18.212 *τοσαύτῃ γ' ὑπερβολῇ συκοφαντίας οὗτος κέχρηται ὥστε ...*

ὑπὲρ τούτου: 'on behalf of him'. The law permitted only Athenian citizens to prosecute for *hybris*. The slave could not himself prosecute, and Vince's translation 'he granted him the same right of bringing a public action' is wrong.

τὸ πρᾶγμα ὁποῖόν τι: compare the comment on the same law in Ais. 1.17: 'Perhaps someone may wonder, on first hearing it, why that phrase about slaves was added in the law of *hybris*. If you consider it, men of Athens, you will find that it is excellent. The legislator was not concerned on the slaves' behalf; it was because he wanted to accustom you to keep well clear of *hybris* against free persons that he added that *hybris* was not to be committed even against slaves. In a democracy he considered that a person committing *hybris* against anyone at all (ὅλως) was unacceptable (οὐκ ἐπιτήδειον) as a fellow-citizen.' Thus D. and Aiskhines concur in holding that the object of the law about *hybris* was not so much to protect the victims as to check and deter the offenders. Ath. 266f–267a tells us that the matter was mentioned also in lost speeches of Hypereides *Against Mantitheos* (fr. 145 Sauppe = 120 Kenyon, Jensen = 37.1 Burtt) and Lykourgos *Against Lykophron* (fr. 72 Sauppe = 74 Blass = X.12 Conomis).

ἐπέταξεν: this reading is supported by the use of the same verb in **44**, whereas ἐπέτρεψεν (SYP) is tautologous with ἐξεῖναι ('he did not permit it to be allowed').

οὐ γάρ ἐστιν, οὐκ ἔστιν: repetition for emphasis. The same phrase is used in 2.10, 4.46, 8.61, 19.296, 23.127; cf. Denniston *Style* 91.

οὐδὲν γὰρ οἷον ἀκούειν: introducing a quotation of *ipsissima verba*; cf. Ar. *Birds* 966 ἀλλ' οὐδὲν οἷόν ἐστ' ἀκοῦσαι τῶν ἐπῶν, Pl. *Gorg*. 447c οὐδὲν οἷον τὸ αὐτὸν ἐρωτᾶν.

47. Another document purporting to be the law of *hybris* is preserved in Ais. 1.16; it is completely different from the document which we have here. Which text, if either, is genuine? It does not seem to be open to us to accept both, because there was only one law, ὁ τῆς ὕβρεως νόμος (**35**, **46**, Ais. 1.15; in D. 54.24, although A has τοὺς νόμους τοὺς τῆς ὕβρεως καὶ τὸν περὶ τῶν λωποδυτῶν, other mss. do not have the second τοὺς and editors generally accept Dindorf's emendation τόν τε τῆς ὕβρεως). The same consideration enables us to reject the text preserved in Ais. 1.16; for, if there was only one law, it must have been a general one covering all kinds of *hybris*, whereas the text in Ais. 1.16 deals only with sexual violation of boys. That kind of *hybris* is relevant to the context in Aiskhines, and so it seems clear that a forger, unacquainted with the real law of *hybris*, has invented a text to fit that context. What he has overlooked is that some of the words with which Aiskhines introduces the law are actually quoted from it: διαρρήδην γέγραπται, ἐάν τις ὑβρίζῃ εἰς παῖδα ... ἢ ἄνδρα ἢ γυναῖκα, ἢ τῶν ἐλευθέρων τινὰ ἢ τῶν δούλων, ἢ ἐὰν παράνομόν τι ποιῇ εἰς τούτων τινά, γραφὰς ὕβρεως εἶναι πεποίηκεν καὶ τίμημα ἐπέθηκεν, ὅ τι χρὴ παθεῖν ἢ ἀποτεῖσαι. These words are plainly based on the text which we have in **47**, which must therefore be accepted as the genuine law.

Although one phrase appears to be spurious (see note on γραφὰς ἰδίας), nothing else in it conflicts with known facts about legal procedure or with the comments made by D. and Aiskhines on the law of *hybris*. Some objections raised by nineteenth-century scholars, especially Drerup *Jb. Cl. Ph.* Supp. 24 (1898) 297–300, are answered by Lipsius *Recht* 421–3.

The date and purpose of this law are uncertain and disputed. The main problem is its relationship to other laws about different types of personal violence, especially battery (αἴκεια) and rape. There are two approaches to this problem:

A. I have argued in *G&R* 23 (1976) 24–31 that the difference between *hybris* and the other offences lies in the offender's state of mind and intention. Exactly what state of mind is implied by *hybris* is itself disputed (see pp. 18–22 above); but the point here is that, in my view, a prosecutor for *hybris* would need, whereas a prosecutor for αἴκεια would not need, to provide evidence of the offender's state of mind and intention. This explains why the law of *hybris* existed concurrently with the other laws about personal violence, without being redundant.

B. E. Ruschenbusch *ZSSR* 82 (1965) 302–9 and M. Gagarin in *Arktouros* 229–36 regard *hybris* as merely a collective term for offences against the person. In their view, *hybris* is not a different offence from battery, rape, and so on; the purpose of the law of *hybris* is, rather, to make possible different (more severe) penalties than the fines specified in the other laws about those offences. Ruschenbusch sees this as a chronological development: the law of *hybris* was made in the time of Perikles and was intended to supersede the other laws. The objection to this is that the other laws in fact continued to exist in the fourth century. So Gagarin argues instead that the law of *hybris* was intended simply to provide an alternative procedure for prosecuting, by which a severer penalty could be proposed if the prosecutor was prepared to take the risk involved in using the *graphe* procedure (the risk of a fine of 1000 dr. if he failed to obtain one-fifth of the votes). But this view seems to me to take too little account of the evidence of the nature of *hybris* in D. and other authors, including Arist. *Rhet.* 1374a 13–15 (quoted on p. 20); I continue to think that *hybris* was a different offence from the others.

Discussion of the date of the law has turned partly on the interpretation of the phrase παράνομόν τι. Does it mean 'anything contrary to (written) law' or 'anything contrary to (conventional) rule'? In the fifth century νόμος was the usual word for a law, but in the sixth century a written law was usually called θεσμός and the sense of νόμος was wider. I have suggested (*G&R* 23 (1976) 25–6) that it would be tautologous and pointless to make a law forbidding what is contrary to law; thus παράνομόν τι should instead be given a fairly general sense, 'any improper behaviour', and the law was probably made in the sixth century, perhaps by Solon at the time when the *graphe* procedure was first introduced. Gagarin *Arktouros* 233–4 rejects this, and suggests the third quarter of the fifth century as the date of the law, because the procedure of γραφὴ παρανόμων came into existence in that period. But a γραφὴ παρανόμων concerned the proposal of a decree (ψήφισμα) inconsistent with a law, and that seems to me to have no relevance to the law of *hybris*; if the law of *hybris* belongs to an earlier period, παράνομος, like νόμος, will have had a somewhat different sense then. (On the meaning of παράνομος in the fifth century see Ostwald *Sovereignty* 111–29.) So I still prefer my earlier view, though the question remains open.

ἢ παῖδα ἢ γυναῖκα ἢ ἄνδρα: the order is evidently that of the degree of vulnerability to physical or sexual attack. Ais. 1.15 transposes γυναῖκα and ἄνδρα, but this and other minor variants may be due merely to his carelessness in quoting.

παράνομόν τι: a similar phrase occurs in the law about children and women whose adult male relatives have died: ἐὰν δέ τις ὑβρίζῃ ἢ ποιῇ τι παράνομον ... (43.75).

πρὸς τοὺς θεσμοθέτας: the thesmothetai were the magistrates in charge of *hybris* cases; cf. 37.33, 45.4, Isok. 20.2.

ὁ βουλόμενος Ἀθηναίων: in *graphai* and other public cases the right to prosecute for some offences was restricted to Athenian citizens (cf. 24.63, 59.16, Ais. 1.32) but for other offences was open to aliens also (cf. **175**, 24.105, 59.52, 59.66). It is not known on what principle this distinction was made, and discussion of this problem has not reached any satisfactory conclusion; cf. Lipsius *Recht* 243–4, Harrison *Law* 1.195 n. 1. That *hybris* belonged to the former category is shown by Ἀθηναίων here; cf. Isok. 20.2 περὶ δὲ τῆς ὕβρεως ... ἔξεστιν τῷ βουλομένῳ τῶν πολιτῶν ...

οἷς ἔξεστιν: 'those who are permitted'. The purpose of this phrase (used also in 24.63, 59.16, 59.52, Ais. 1.32) is to exclude those who have lost by disfranchisement the right to speak in court, like Straton (**95**). One form of partial disfranchisement was the loss specifically of the right to bring a certain kind of prosecution; cf. And. 1.76 ἑτέροις οὐκ ἦν γράψασθαι, τοῖς δὲ ἐνδεῖξαι.

ἡλιαίαν: this word has a smooth breathing in YP, a rough one in F, and none in S; but fifth-century inscriptions which show the aspirate in other words do not aspirate ἠλιαία (*IG* 1³ 40.75, 71.14, *ATL* D14.II.7). It seems to have meant originally 'assembly' (cf. Chantraine *Dictionnaire étymologique de la langue grecque* under ἁλής), but in classical Athens it was used only for a meeting of citizens to try a legal case, or for a building in which such meetings were held. It has generally been thought that, when Solon introduced trials by the people (*AP* 9.1), the Eliaia was simply the Ekklesia under another name. This view has been rejected by M. H. Hansen *Cl. et Med.* 33 (1981–2) 9–47, who on the basis of the plural *τὰ δικαστήρια* in Arist. *Pol.* 1274a 3 and *AP* 7.3 argues, in opposition to P. J. Rhodes *JHS* 99 (1979) 103–6, that the word from the start referred to more than one court; Hansen's view in turn is rejected by Ostwald *Sovereignty* 10–11 n. 29, while R. Sealey *The Athenian Republic* (1987) 60–70 adopts a compromise, arguing that the Eliaia was originally one court but not identical with the Ekklesia. However that may be, by the fourth century *ἡλιαία* was the normal word in laws for what were called in more popular usage *δικαστήρια* collectively, or for any one *δικαστήριον*, in which the jury consisted of citizens selected by lot. The same word was used to mean one particular building in Athens (e.g. 47.12). This was evidently the building in which the court of the thesmothetai sat; it was the largest court building, because some kinds of case at which the thesmothetai presided required very large juries, but its location remains uncertain. Cf. MacDowell *Law* 29–35 and Hansen's article cited above. In the present passage the words *τὴν ἡλιαίαν* provide no evidence for dating the law of *hybris*, because they could equally well mean either 'the (only) Eliaia' or 'the Eliaia (of the thesmothetai)'.

τριάκοντα ἡμερῶν: was it normal for a law to specify a time-limit within which the magistrate must bring an accused person to trial? It has been argued by M. H. Hansen in *Symposion 1979* (1981) 167–70 that a thirty-day limit was common, but I am not convinced by his argument. The other evidence is as follows.

1. There is an instance of a thirty-day limit in 24.63. But there it applies to accused persons held in custody, and its purpose is probably to restrict the period of imprisonment (like the 110–day rule in Scots law). This is not evidence for a time-limit in cases in which the accused person is not imprisoned.

2. In 42.13, where the speaker is maintaining that private agreements should be enforced, he says: *πολλάκις γὰρ ἔν τε τοῖς νόμοις γεγραμμένης τριακοστῆς ἡμέρας ἑτέραν ἡμῖν αὐτοῖς συγχωρήσαντες ἐθέμεθα, παρά τε ταῖς ἀρχαῖς ἁπάσαις καὶ δίκας καὶ κρίσεις ἀναβάλλονται τοῖς ἀντιδίκοις οἱ ἄρχοντες συγχωρησάντων ἐκείνων ἀλλήλοις*. The second half of this means that, after a magistrate has fixed a date for a trial, he sometimes postpones it if both disputants agree to the postponement; but it does not say that his discretion in fixing the original date is restricted by a time-limit. (Cf. **84**, where the arbitrator wishes to postpone the arbitration, but cannot because D. does not agree. There too there is no evidence that the date had to be within a time-limit.) The first half of the passage does mention thirty-day limits

prescribed by laws, but does not say what was thus limited. This is unlikely to be a reference to *δίκας καὶ κρίσεις*, because those are introduced in the second half of the passage as if the first half had not been concerned with them. I do not know what the application of these thirty-day limits was, but I suggest the possibility that they were time-limits for paying money owed; cf. **81** n. *ὑπερήμερον*.

3. In the second half of the fourth century there existed a category of 'monthly cases', *δίκαι ἔμμηνοι* (37.2, *AP* 52.2–3). It has been cogently argued by E. E. Cohen *Ancient Athenian Maritime Courts* (Princeton 1973) 23–59 that this term means that the cases were accepted by the magistrates every month. Hansen op. cit. prefers the older view that it means that the cases had to be tried within a month. But even if that were the correct interpretation of *δίκαι ἔμμηνοι* in the second half of the fourth century, it would not show that such a limitation was normal at the earlier period when the law of *hybris* was made.

4. Hansen also refers to *IG* 2^2 46.57 *τρι]άκοντα ἡμερ[*, but this is too fragmentary for any interpretation.

So, apart from cases in which accused persons were held in custody, it is not certain that there was any offence other than *hybris* for which the trial had to be held within thirty days of the initiation of the case. There may have been some reason for distinguishing *hybris* cases from others in this respect; possibly it was thought that a hybristic man might try to browbeat the thesmothetai into postponing his trial indefinitely. When Apollodoros prosecuted Phormion for *hybris*, the trial was delayed for a long time and in the end was not held at all (45.4), but that was probably due to legal objections.

ἀφ' ἧς: i.e. *ἀπὸ τῆς ἡμέρας ᾗ*.

ἂν ἡ γραφή: *ᾖ* may have been lost before *ἡ* (Markland's conjecture, reported by Taylor), but not necessarily, since it can easily be understood, as in Ant. 5.32, 6.8, Pl. *Rep.* 370e, 416d.

παραχρῆμα does not mean that the penalty was to be fixed without the usual speeches by the prosecutor and the defendant in favour of their respective proposals, but merely that those speeches were to follow immediately after the verdict was declared.

παθεῖν ἢ ἀποτεῖσαι: **25** n. Here *ὅτου* is constructed with *ἄξιος*, and then the infinitives are added to define the phrase more clearly.

γραφὰς ἰδίας: an unparalleled expression. It is also self-contradictory, since a *graphe* was one kind of *δίκη δημοσία*. Attempts have been made to save it by the hypothesis (which I formerly entertained myself; cf. MacDowell *Law* 129–30) that a *graphe* was called *ἰδία* if the prosecutor was himself the victim of the alleged offence. But there is no other evidence for that hypothesis, which merely creates a further problem in our passage: why should the penalty for failure to obtain one-fifth of the votes be specified for a prosecutor who claimed to have suffered from an act of *hybris* himself and not for a prosecutor who did not so claim? It is hard to believe that in the latter case the penalty was to be taken for granted, since this law is one which otherwise states explicitly all the normal features of the *graphe* procedure. A better solution is to excise *γραφὰς ἰδίας*. The words *κατὰ τὸν νόμον* imply that the prosecution procedure referred to here is simply what was laid down earlier in the law, where *γραφέσθω* was intransitive; so *γράφωνται* should likewise have no accusative. For criticism of earlier discussions of the phrase see Lipsius *Recht* 241–3 n. 12. How it got into the text is not clear, but it may not be accidental that the same word *ἰδίας* has intruded into the text of another document in **52**. I suggest the possibility that in the margin of these documents in an early copy of the oration someone wrote *ἰδίᾳ* in its late sense of 'separate', meaning that the documents were not part of the speech; cf. the use of this word in the margin of a verse text to indicate that a word is *extra metrum* (schol. Ar. *Clouds* 41). If this was once copied into the text, conversion to *γραφὰς ἰδίας* (and in **52** to *ἰδίας δεξιὰς*) would soon follow in an attempt to make sense.

χιλίας δραχμὰς: this rule, that a prosecutor who failed to proceed with his case or failed to obtain one-fifth of the jury's votes should be fined 1000 dr., was intended to discourage frivolous prosecution and sycophancy; cf. 53.1, 58.6, etc. It applied to most kinds of public case except *eisangelia* and *probole*. The rule that such a prosecutor was disfranchised (**103** n. *ἠτίμωκεν*) is not mentioned here; probably it was the subject of a separate law passed at a later date than the law of *hybris*.

ἐὰν δὲ ἀργυρίου τιμηθῇ ...: this sentence seems out of place. Its subject is the offender, not the prosecutor, but no word has been inserted to indicate the change of subject. It would be better placed directly after *παθεῖν ἢ ἀποτεῖσαι*. Either it has got out of order in the mss. or (perhaps more likely) it is a late addition to the law, whether made as an amendment at the time when the law was passed or at a later period. One might expect the passive *τιμηθῇ* to have the offence, rather than the offender, as its subject. But cf. 24.103 *ἄν τις ἁλῷ κλοπῆς καὶ μὴ τιμηθῇ θανάτου* ...: this shows that it is unnecessary to adopt Taylor's deletion of *τῆς ὕβρεως*.

δεδέσθω: imprisonment until the fine was paid was the rule also when a fine was imposed for failing to observe the restrictions of disfranchisement (24.105, *AP* 63.3) or in a mercantile case (33.1, 35.46–7, 56.4).

ἐὰν ἐλεύθερον ὑβρίσῃ means that, if the victim of *hybris* was a slave, imprisonment until the fine was paid was not automatic. But it could still be included in the proposal for the penalty in an individual case.

48. τί οὖν implies 'What need I say, then? The conclusion is obvious.' So also in **76**, **129**.

εἰς τοὺς βαρβάρους ἐνεγκὼν: carrying a copy of the text to read out to them. D. shows no interest in defining the countries from which slaves were imported, and we have little information about the sources of slaves at this period, but probably the main areas were Asia Minor and Thrace.

ὅτι here, as often, introduces direct speech.

49. εἰσὶν Ἕλληνές τινες: 'There do exist some Greeks ...' implies that other Greeks showed less consideration for slaves than the Athenians did. Similar points about the Athenians' lenient treatment of slaves are made in 9.3, X. *Ath.* 1.10.

ὑφ' ὑμῶν ... πρὸς ὑμᾶς: the alternative readings *ὑφ' ἡμῶν ... πρὸς ἡμᾶς* would mean that the imaginary speaker was one of the barbarians himself. That should be rejected because **50** *εἰ ... συνεῖεν οἱ βάρβαροι* implies that he was speaking in Greek rather than in their own language.

φύσει: D. does not mean that it is natural to hate foreigners, but that it is natural to hate people who have wronged oneself or one's ancestors. No doubt he is thinking of the Persian invasions of 490 and 480 as a principal cause of Athenian hostility to Asiatics. Cf. Dover *Morality* 281 on grievance and retaliation, rather than nature, as a cause of enmity between Greeks and barbarians.

ὅσων: the genitive goes with *τιμὴν* and the accusative is understood with *κτήσωνται*: 'not even those whose price they pay and ⟨whom⟩ they acquire as slaves'.

πολλοὺς: it is surprising if many free men were really put to death for treating slaves with *hybris*. No instance is known to us.

50. συνεῖεν: most barbarians did not know the Greek language.

δημοσίᾳ: 'by public decree', echoing the same word in **49** to emphasize the *quid pro quo*.

προξένους: official patrons of people from a foreign state (something like a modern consul). Normally a foreign state would appoint one Athenian as its *πρόξενος* in Athens; D.'s suggestion that such an appointment would be given to all Athenians is jocular.

παρὰ τοῖς βαρβάροις εὖ δόξαντ' ἂν ἔχειν: the argument is logically weak. The law's high reputation abroad is not a fact but merely D.'s hypothesis; and even if it were true, the usefulness of the law to barbarians would not prove it to be important to the Athenians.

51–5. *Meidias is guilty also of impiety, because Demosthenes was a chorus-producer at the time of the assault and as such was performing a religious function.*

On the definition of *asebeia*, and on the question whether Meidias' conduct could be regarded as impiety, see pp. 17–18.

51. ὕβριν ... ἀσέβειαν: *καταγιγνώσκω* takes the accusative of the decision, the genitive of the person or thing condemned. Meidias' behaviour is to be judged to be both *hybris* and impiety.

κἂν: *ἄν* goes grammatically with *ποιεῖν*, *καί* with *ἀσέβειαν*. This is not to be confused with Plato's use of *κἂν εἰ* as a fixed phrase for 'even if', which does not have another word intervening; in this sentence *ποιεῖν* requires *ἄν*.

κνισᾶν ἀγυιὰς: 'to make streets smell of sacrifice', religious terminology. The same phrase is used in Ar. *Knights* 1320, *Birds* 1233, Hes. fr. 224 Rzach = 325 Merkelbach and West, and in an oracle in D. 43.66, and cf. *κατ' ἀγυιὰς* and *κνισᾶν* in the oracles in **52**. Harp. *ἀγυιᾶς* quotes this passage and says that the reading should be not *ἀγυιὰς* but *ἀγυιᾶς*, accusative plural of *ἀγυιεύς*, referring to the pointed pillars set up in the streets in honour of Apollo (**52** n. *ἀγυιεῖ*). The scholiast on *Knights* 1320 adopts the same interpretation there: he explains *ἀγυιᾶς* as *τοὺς ἀγυιαίους θεούς*. Harpokration says that some people attributed the pillars to Dionysos, but that may be just a false inference from **51**, and elsewhere the pillars and the word *ἀγυιεύς* are connected with Apollo only. That does not suit the Aristophanes passages, where (certainly in *Birds* 1233, probably in *Knights* 1320) sacrifices to all gods are meant. So it seems better to reject Harpokration's opinion and to interpret the traditional phrase *κνισᾶν ἀγυιάς* as a general one, 'to fill the town with sacrifices', without specific reference to Apollo's pillars.

στεφανηφορεῖν: **33** n. *στεφανηφορίαν*. Of the categories distinguished there, *D* is meant here. Many people would wear crowns or wreaths of leaves for a festival.

52. Presumably the texts in **52–3** come from a collection, kept in Athens, of oracles received by the Athenian people. They are not all closely relevant to D.'s argument, but perhaps they are the most relevant that could be found. The fact that they are only marginally relevant helps to reassure us that the texts are genuine, since a forger inventing oracles for this speech would have composed texts which fitted the speech more exactly; but it is possible that whoever put the documents into *Meidias* after D.'s death has selected the wrong oracles from the collection, not the ones that D. actually intended to be read here. Their text is more corrupt than any other part of *Meidias*; perhaps the copies obtained from the collection of oracles already contained errors. In **51** D. mentions oracles from Delphi and from Dodona; those in **53** are headed 'oracles from Dodona', and so those in **52** may be assumed to be from Delphi. There is no evidence to show when they were originally delivered. The Ionic or Homeric dialect of the first oracle was traditional in verse and is not evidence that this oracle is older than the others. On the other hand, there is no reason to presume (as some scholars have done) that these oracles were recent at the time when D. was writing.

The first oracle is relevant to D.'s case insofar as it gives divine authority to a festival in honour of Dionysos; but it does not refer to activities in the theatre, but in the streets. It is not about the regular Dionysia, but instructs the Athenians to hold an extraordinary festival for Dionysos in thanksgiving for the harvest.

Ἐρεχθείδαισιν ... Πανδίονος: Erekhtheus and Pandion were legendary kings of Athens. For *Ἐρεχθεῖδαι* meaning 'Athenians' cf. Pind. *Isth.* 2.19, S. *Ai.* 202, E. *Med.* 824. Meier proposes the emendation *-ησιν*, perhaps rightly, though we cannot really tell how strictly the Delphic priests kept to traditional Ionic forms.

ἰθύνεθ': S has *ἰθύνετ'*, which may be correct. Fifth-century inscriptions which use the aspirate in other words omit it in *ἑορτή* (*IG* I³ 5.5, 386.157, *Hesperia* 1 (1932) 43–4 no. 1); cf. Threatte *Grammar* 1.500–1.

ὡραίων ... χάριν: if the text is right, this 'thanksgiving for ripe crops' to be held in 'broad-spaced streets' must consist of singing and dancing choruses in which all the people participate. Some editors, thinking this obscure, prefer to emend *χάριν* to *χορὸν*. Then *ὡραίων χορὸν* is 'a chorus of young men'. But that is inconsistent with *ἅμμιγα πάντας*, of which the point must be that everyone is to join in, and so further change is necessitated: *ὡραῖον Βρομίῳ χορὸν*, 'a seasonable chorus for Bromios' (Hemsterhuys in his edition of Lucian (1743) vol. 1 p. 54), or *ὡραίῳ Βρομίῳ χορὸν*, 'a chorus for youthful Bromios' (Meier).

ἅμμιγα: shortened form of *ἀνάμιγα*, as in A. *Th.* 239 (cod. M), S. *Tr.* 838. It is to be taken with *πάντας*, 'all indiscriminately', including slaves; cf. **53** *ἐλευθέρους καὶ δούλους*.

κάρη: Ionic accusative of *κάρα*, probably singular (each person crowns his own head).

The beginning of the second oracle indicates that it was given on the occasion of an epidemic. Most of it closely resembles a Delphic oracle given to the Athenians on an occasion when a 'sign' had appeared in the sky, the text of which is given in 43.66. It consists of routine advice for times of god-sent affliction. It does not pertain to the Dionysia, but it has a general relevance to D.'s case insofar as it shows that choruses have divine authority.

θύειν καὶ εὔχεσθαι: all the infinitives in this oracle are jussive.

καὶ Ἡρακλεῖ: *καί* is not used elsewhere within the lists of gods in this oracle and in the oracle in 43.66, and it seems best just to excise it. However, Σ **54** (164 Dilts) begins *ἐπειδὴ ἀνέγνω μαντείας, ἐν αἷς* *ἐγέγραπτο ὅτι ἔδει θύειν Ἡρακλεῖ ἀλεξικάκῳ καὶ κνισᾶν τὰς ἀγυιάς* ..., and on this evidence Radermacher (as reported by Sykutris) postulated a lacuna and conjectured *καὶ Ἡρακλεῖ ⟨ἀλεξικάκῳ καὶ⟩ Ἀπόλλωνι*. But it is possible that the scholiast's *ἀλεξικάκῳ* is merely his own addition; he is not quoting the oracle verbatim, and *ἀλεξίκακος* is a standard epithet of Herakles, used again in **Σ 66** (206 Dilts).

Ἀπόλλωνι: naturally given special prominence by his own oracle.

προστατηρίῳ: 'protector'. Hesykhios s.v. interprets it as 'standing before doors', but that is probably wrong, for these reasons: (*a*) it would make *Ἀπόλλων προστατήριος* virtually synonymous with *Ἀπόλλων ἀγυιεύς*, but the two functions seem to be distinguished in this oracle; (*b*) Artemis is called *προστατηρία* in A. *Th.* 449, but she did not stand before doors; (*c*) the oracle in 43.66, at the same point in the list of gods, has *Ἀπόλλωνι σωτῆρι*.

τύχας ἀγαθᾶς: genitive singular. This oracle contains several Doric forms, which are doubtless genuine features of the Delphic text.

ἀγυιεῖ: 'of streets'. Apollo, like Hermes and Hekate, had many shrines in the streets in front of houses. His image took the form of a pointed stone pillar (schol. Ar. *Wasps* 875 *κίονας εἰς ὀξὺ λήγοντας ὡς ὀβελίσκους*, schol. E. *Ph.* 631, Harp. *ἀγυιᾶς*). On an altar beside it incense was burnt or other offerings made; cf. S. fr. 341 Nauck = 370 Pearson and Radt, Helladios ap. Phot. *Bibl.* cod. 279 p. 535b 33–8. For further discussion of the pillar and altar see my note on Ar. *Wasps* 875.

καὶ κατ' ἀγυιάς: the oracle in 43.66 at the corresponding point has *καὶ τὰς ἀγυιὰς κνισῆν καί*. Perhaps our text should be so emended, especially since D. does use the phrase *κνισᾶν ἀγυιὰς* in his introductory remarks in **51**.

ἱστάμεν: Doric infinitive, used also in 43.66.

στεφανηφορεῖν, καὶ κατὰ τὰ πάτρια: editors adopt the Doric forms *στεφαναφορεῖν* and *καττὰ πάτρια*. But neither here nor in 43.66 do any mss. give those forms, and the two passages support each other against change. It is not a likely coincidence that the mss. should lose Doric forms by chance at exactly the same points in both texts. Instead we should conclude that the texts of Delphic oracles (at least those kept in Athens) did not necessarily exhibit the utmost rigour of the Doric dialect. *καί* is given here only by later mss., whereas all mss. give it before *μνασιδωρεῖν*: but the raising of hands is a gesture of thanksgiving which should be taken with that infinitive, not with *στεφανηφορεῖν*.

Ὀλυμπίοις: *καὶ Ὀλυμπίαις* is added in 43.66, and perhaps should be added here (so Blass). That this was a customary phrase is shown by the fun which Aristophanes makes of it: *Birds* 865–70 *εὔχεσθε* ... *ὄρνισιν Ὀλυμπίοις καὶ Ὀλυμπίῃσι πᾶσι καὶ πάσῃσιν*, *Thesm*, 331–4 εὔχεσθε τοῖς θεοῖσι τοῖς Ὀλυμπίοις καὶ ταῖς Ὀλυμπίαισι, καὶ τοῖς Πυθίοις καὶ ταῖσι Πυθίαισι, καὶ τοῖς Δηλίοις καὶ ταῖσι Δηλίαισι.

ἰδίας: on the intrusion of this word into the text of a document, **47** n. *γραφὰς ἰδίας*. It is pointless to say that the hands that the people are to raise are their own. Various emendations have been proposed: *ὁσίας* Wolf, *ἰθείας* Buttmann, *λιτὰς* Weil, *λαιὰς* Richards (*CR* 18 (1904) 13); Sauppe ingeniously conjectures *πάσαισι καὶ* (omitting *καὶ* before *κατὰ τὰ πάτρια*). But in 43.66 there is no word between *πάσαις* and *δεξιὰς* and it is better just to delete *ἰδίας* here.

ἀνίσχοντας: the older mss. have *-τες* here, but *-τας* in 43.66. The subject of the infinitives needs to be accusative.

53. The first oracle from Dodona was evidently delivered on an occasion when the Athenians had failed to send representatives to attend some festival at Dodona, and it tells them how to atone for the omission. It has nothing to do with the Dionysia or with choruses, and is really irrelevant to D.'s case, except that it reinforces the general point that proper observation of festivals is important.

ὁ τοῦ Διὸς σημαίνει: the masculine noun understood may be *χρησμός*. It is not *προφήτης*, because in classical times the prophecies at Dodona were delivered by women (Hdt. 2.55, Euripides ap. Page *Greek Literary Papyri* 1.112, Pl. *Phdr.* 244b, Ephoros (*F. Gr. Hist.* 70) F119, etc.); besides, a prophet does not himself signify but interprets the god's sign. H. W. Parke *The Oracles of Zeus* (1967) 84–6 suggests *κλῆρος*, meaning that the oracle was obtained by drawing lots; but the instructions here, especially the demand for a bronze table for a previous dedication, do not look like answers selected by lottery. For *σημαίνει*, used of a god at an oracle, cf. Herakleitos 93 *ὁ ἄναξ, οὗ τὸ μαντεῖόν ἐστι τὸ ἐν Δελφοῖς, οὔτε λέγει οὔτε κρύπτει ἀλλὰ σημαίνει.*

παρηνέγκατε: 'let pass', 'ignored'. This is an uncommon sense of *παραφέρω*, but it occurs also in Plu. *Arat.* 43.7 *παρήνεγκε τὸ ῥηθέν*. Spalding's conjecture *παρήκατε* (proposed on pp. 122–4 of his edition; cf. Cobet *Misc. crit.* 506) is therefore unnecessary. Weil prefers to keep *παρηνέγκατε* and translate 'transposed', meaning that the Athenians altered the date by a reform of their calendar (cf. Ar. *Clouds* 607–26); but that can hardly be right, since the Athenians would not need to be told to send envoys if they had already sent them, though on the wrong date.

κελεύει: as in the next oracle, the infinitives go with *σημαίνει*, and *κελεύει* should be dismissed as a gloss (though its omission in Vd is a lucky accident, afterwards undone by a 'correction'); cf. **9** n. *χρηματίζειν* and Cobet *Misc. crit.* 506.

ἐννέα: the point is probably that the number must be larger than usual. The normal number may have been three.

τούτους: accusative subject of the infinitive following *σημαίνει*. This construction (instead of the dative) is also found with other verbs in formal proclamations and instructions, e.g. Th. 4.97.4, X. *An.* 2.2.21, Ar. *Ekkl.* 684–6.

ναΐῳ: a conjecture by Buttmann (pp. 127–8 of his 1833 edition), but certainly right, because it is a regular title of Zeus at Dodona. There are numerous instances in the inscriptions from Dodona assembled by O. Hoffmann in *Sammlung der griechischen Dialekt-Inschriften* (ed. H. Collitz) 2 (1899) nos. 1557–98. The meaning was unknown even in antiquity, from which two different explanations have been preserved: (*a*) schol. *Iliad* 16.233 *ὁ δὲ Δωδωναῖος καὶ νάϊος· ὑδρηλὰ γὰρ τὰ ἐκεῖ χωρία*, implying derivation from *νάω* (cf. the Naiads) with the meaning 'of running water'; (*b*) *Lex. Rhet.* 283.22–5 *Πέριρος γὰρ ὁ Ἰκάστου παῖς, τοῦ Αἰόλου, ναυαγήσας διεσώθη ἐπὶ τῆς πρύμνης, καὶ ἱδρύσατο ἐν Δωδώνῃ Διὸς ναΐου ἱερόν*, implying derivation from *ναῦς*, with the meaning 'of the ship'. Cf. Parke *The Oracles of Zeus* 78 n. 36.

βοῒ σῦς: Radermacher's emendation of the mss.' nonsensical *βοιήσεις*: cf. *Gnomon* 16 (1940) 13. An alternative, less close to the mss., is *βοῒ δύο οἶς* (Dobree *Adversaria* 1.459).

Διώνῃ: the regular name of the consort of Zeus at Dodona; cf. 19.299 and the inscriptions in Hoffmann loc. cit.

καλλιερεῖν: 'to sacrifice well', avoiding any ill-omened acts or circumstances. The same verb is used in the oracle in 43.66. Sauppe's emendation is good. The mss.' phrase *καὶ ἄλλα ἱερεῖα* is too vague for this document, which gives precise instructions about numbers and types of victims; and the change in palaeographically easy, since *καλλ-* may have been mistaken for *κἄλλ'*. An alternative way of providing an infinitive is worth noting: Buttmann emends an earlier part of the sentence to *θεωροὺς ἕνεκα τούτου, τοὺς δὲ ἀπάγειν τῷ Διί*.

καθιστάναι: my conjecture for the mss.' *καί*. The accusative *τράπεζαν* needs a governing verb; it cannot be taken with *καλλιερεῖν*, which refers to sacrifices. Presumably the Athenians had previously dedicated a bronze statue or something similar, and they are now instructed to provide a table or plinth for it to stand on (cf. LSJ *τράπεζα* III.4). I postulate that in some copy of the text *καθιστάναι* was abbreviated or written small at the end of a line, and so was misread as *καί*.

The second oracle from Dodona authorizes a one-day festival of Dionysos, much like the first oracle in **52**. Perhaps (though this is pure speculation) that oracle from Delphi and this one from Dodona were both obtained on the same occasion, when the Athenians wanted advice about holding an extra festival.

δημοτελῆ ἱερὰ τελεῖν: Buttmann's emendations are generally accepted. Unknown to him, *δημοτελῆ* is actually in two of the later mss.

κρατῆρας: Humbert rightly emends the singular to the plural; cf. **52**. A single bowl of wine, for a public festival in which many perform (*χορούς*) and which all attend (*ἐλευθέρους καὶ δούλους*), is not credible.

κεράσαι: mixing wine and water in appropriate proportions.

ἐλευθέρους καὶ δούλους: cf. **52** *ἄμμιγα πάντας*.

Ἀπόλλωνι ἀποτροπαίῳ βοῦν θῦσαι: the mss. give these words in the middle of the preceding sentence, where they are clearly out of place. *θῦσαι* needs to be taken both with *Ἀπόλλωνι* ... and with *Διὶ* ...

54. καὶ ἀγαθαί: *καλαὶ* is inserted in A (followed by F and by correctors in Y and P). Cf. *Ep.* 1.16 *πολλὰς καὶ καλὰς κἀγαθὰς καὶ ἀληθεῖς ὑμῖν μαντείας ἀνῃρήκασιν*. But in both passages *καλαὶ* or *καλὰς* should probably be omitted, since *καλὸς κἀγαθός* is a social or moral expression normally applied only to persons. Cf. pp. 50–1 on extra words added in A.

ἐφ' ἑκάστης μαντείας: **2** n. *ἐπὶ τῶν ἄλλων*.

ἀεὶ: preserved only by A, but it looks idiomatic in this position and should probably be retained.

προσαναιροῦσιν: 'ordain in addition'. D. is trying to explain away the fact that none of the oracles which he quoted mentioned the particular festival with which he is concerned. His explanation is that the regular celebrations of the Dionysia, especially the choruses and crown-wearing, are implied by the general instructions to continue traditional observances.

55. καὶ οἱ χορηγοὶ: note how D. slips khoregoi into his argument here. They were not mentioned in the oracles quoted.

τὴν τῶν ἐπινικίων: 'for the day of his victory celebrations'. This was not an official event, but a winning khoregos would customarily hold a sacrifice and a party for his chorus after the festival; cf. Pl. *Symp.* 173a, Ar. fr. 433 Kock = 448 Kassel and Austin, D. 59.33 (the winner of a chariot race), Ar. *Akh.* 1155 (a khoregos who neglected to hold a celebration—but perhaps he was not victorious). At such a celebration it would be normal to wear a wreath or crown (**33** n. *στεφανηφορίαν*, category *D*); the reference here is not to a crown awarded as a prize, since the prize for a winning chorus was not a crown but a tripod. The present tense of *στεφανοῦται*, 'puts on a crown', has prompted some editors to emend *τὴν* to *τῇ*, but the change is unnecessary; *ἃς συνερχόμεθα* earlier in the sentence is an adequate parallel for the accusative.

ἐν τῷ τοῦ θεοῦ ἱερῷ: in the theatre of Dionysos; so also **74** *ἐν ἱερῷ*.

56–61. *Choristers are protected by law or by customary piety against interference while a festival is in progress; all the more should a man who assaults a chorus-producer be punished.*

The argument of this passage is essentially one from analogy, or *a fortiori*. D. is unable to refer to any law giving special protection to a khoregos. He therefore refers to a law giving special protection to members of a chorus, and infers that punishment should also be imposed on someone who has attacked a khoregos. But his argument is strictly invalid. The special law about choristers concerned challenges to their citizen status, and its purpose was evidently to minimize interruption and delay to the choral performances when an individual chorister was suspected of not being entitled to participate. There was no reason why this law should apply to khoregoi, not only because the identity of a khoregos was known and his status could be challenged long before the festival, but also because a khoregos did not himself perform and an attack on him would therefore not delay the performance. It is noticeable that D. does not have this law read out; perhaps the actual wording, if read out to the jury, would have made its irrelevance to his case obvious.

The details of the law are not easy to reconstruct. (The following account is abstracted from a slightly more detailed discussion in *Symposion 1982* (1985) 72–7.) It probably referred only to the city Dionysia, for schol. Ar. *Wealth* 953 tells us that it was at the Dionysia, not at the Lenaia, that choristers had to be Athenian citizens: *οὐκ ἐξῆν δὲ ξένον χορεύειν ἐν τῷ ἀστικῷ χορῷ . . .· ἐν δὲ τῷ Ληναίῳ ἐξῆν, ἐπεὶ καὶ μέτοικοι ἐχορήγουν*. Anyone who saw an alien about to perform in a chorus could object, and could 'lead him out' by summoning him to the arkhon; the wording which we find in **60** looks like legal terminology, *προσκαλέσασθαι πρὸς τὸν ἄρχοντα ἐξεῖναι, . . . εἰ ξένον τις ἐξαγαγεῖν ἠβούλετο*, and there is a vaguer reference to it in And. 4.20, *κελεύοντος δὲ τοῦ νόμου τῶν χορευτῶν ἐξάγειν ὃν ἄν τις βούληται ξένον ἀγωνιζόμενον*. But a rival khoregos might use this procedure unscrupulously to expel a good singer and thus give his own chorus a better chance of winning. So another clause of the law, or a separate law passed subsequently, added the penalties mentioned in **56**: a payment of 50 dr. for 'calling' a chorister, 1000 dr. for ordering him to sit down and take no part in the performance.

A difficulty is to distinguish the active *προσκαλέσαι* in **56–7** from the middle *προσκαλέσασθαι* in **60**. The middle is normal for a legal summons, and its meaning is clear: the accuser told a person to attend at the magistrate's office on a certain day on which the magistrate received charges and made arrangements for trials. But the arkhon surely did not receive charges on the days of the Dionysia; legal business was not done on festival days, and the arkhon would be busy with the festival arrangements. So *προσκαλέσασθαι πρὸς τὸν ἄρχοντα* (**60**) means that the accuser told the chorister to attend at the arkhon's office on a certain later date; but *ἐξαγαγεῖν* shows that, if such a summons was made, the accused chorister had to leave the chorus forthwith. The summons had the effect of an injunction not to perform, taking immediate effect, although the trial for the offence would not take place until later.

The active *προσκαλέσαι* is not used elsewhere for a legal summons (and the two instances of the simple *καλεῖν* in this sense in Ar. *Wasps* 483, 1418 are exceptional); and in **56** it is associated with another active verb, *σκοπεῖν*. These verbs evidently refer to activities carried out by the accuser himself, not by the arkhon. So the procedure appears to have been as follows. Anyone who saw in the orkhestra, or preparing to enter it, a chorister whom he suspected of being an alien, could accost him (*προσκαλέσαι*) and ask him questions (*σκοπεῖν*) such as 'Who is your father, and which is your deme?' If he thought the answers unsatisfactory, he could then summon the chorister to appear before the arkhon on the appropriate day (*προσκαλέσασθαι πρὸς τὸν ἄρχοντα*) and remove him from the chorus (*ἐξαγαγεῖν*), telling him to sit in the audience (*καθίζεσθαι*).

The financial deterrents mentioned in **56** fit this interpretation: removing a chorister stopped him performing at all, and 1000 dr. was payable for this serious interference; merely questioning him did not stop him performing, but still it incurred a small penalty of 50 dr. because it might unsettle him and so affect his performance to some extent. From the legal point of view, these payments are an interesting instance of an accuser's being required to pay money in advance of any trial. They are comparable in some respects to the fee or deposit (*πρυτανεῖα* or *παράστασις* or *παρακαταβολή*) payable in some other kinds of case. If the case went to trial and the prosecutor won, he would recover from the defendant the payment which he had made. The defendant would not be the chorister himself, but the khoregos who included an alien in his chorus; cf. Plu. *Phok.* 30.6 *νόμου γὰρ ὄντος Ἀθήνησι τότε μὴ χορεύειν ξένον ἢ χιλίας ἀποτίνειν τὸν χορηγόν*. Plutarch proceeds to relate an anecdote to illustrate the ostentatiousness of Demades: he recruited choristers who were all aliens, and brought the cash along to the theatre, 1000 dr. for each of his hundred choristers. The anecdote is dubious (could one khoregos present two choruses of fifty at the same festival? And even if he offered to pay, would the aliens have been allowed to perform?), but the statement about the law is acceptable: the convicted khoregos would have to pay 1000 dr. to the prosecutor, who would thus not be left out of pocket, while the state would retain the 1000 dr. paid by the prosecutor in advance. If the accuser did not proceed with a prosecution, or if he lost the case, he would not recover the payment which he had made; this would always be so if he only paid 50 dr. for questioning a chorister and did not remove him from the chorus.

For the procedure for removing a disfranchised citizen from a chorus, **59** n. *οὐδεὶς ἥψατο*.

56. οὐκ ἐδώκατε means that the Athenian people passed a law against it.

ἁπλῶς: 'just like that', without complication or restraint.

καλέσῃ: repetition of *προσκαλ-* without the prepositional prefix; **27** n. *παρόντα ... ὄνθ'*.

καθίζεσθαι: to sit in the audience, taking no part in the performance. *καθίζεσθαι* and *καθέζεσθαι* have the same meaning and are both well attested in Attic. In this speech the mss. give *-ιζ-* in **162**, *-εζ-* in **216** and **227**, and are divided in **119**. D. may have used both forms in different places. Σ **56** (168c, 169 Dilts) uses *-εζ-* in his comment, but that may be his own usage rather than D.'s.

ἀποτίνειν: the payment would have to be made at a later date, not on the spot, since a man would not normally carry 1000 dr. around with him.

λειτουργοῦντα τῷ θεῷ: 'serving the god', a sly use of vocabulary. Being a chorister was not a liturgy in the ordinary Athenian sense; the verb is therefore metaphorical. But being a khoregos *was* a liturgy; so the use of this verb encourages the hearer to think that the sentence is applicable also to D. as a khoregos. (In Hellenistic and Christian Greek this metaphor became the normal usage, so that now 'liturgy' regularly means service to God; but this was not normal in Attic.)

ἐξεπίτηδες: 'from an ulterior motive', to obtain some advantage for himself; so also in **187**, 31.13, Pl. *Gorg.* 461c, etc. A khoregos might think that his own chorus would have a better chance of winning if he could expel a good singer from a rival chorus, and might therefore (if there were no financial deterrent) accuse a chorister of being an alien without having any evidence that he was so.

57. εἶτα introduces the argument *a fortiori*: thrashing illegally is worse than accosting illegally. If the reading *αὐτὸν* (AF and a corrector in P) is correct, that adds another element to the *a fortiori* argument: interference with 'the chorus-producer himself' is worse than interference with a chorister. But possibly *αὐτὸν* should be omitted; cf. pp. 50–1 on extra words added in A.

δώσει: A and all other mss. except S insert *οὐ*. That could be correct; for *οὐ* reinforcing, not contradicting, a preceding *οὐδέ*, cf. 22.32, Ais. 3.78, and Denniston *Particles* 196–7. But the omission of *οὐ* here is confirmed by the testimonia (Minucianus and Maximus Planudes).

58. μηδὲν ἀχθεσθῆναί μοι: the usual phrase when a speaker is going to say something which he thinks may irritate his audience; cf.

10.54, 19.227, 23.144, Lys. 21.16, Lyk. *Leo.* 128, Pl. *Ap.* 31e. Here D. fears that some of the jury may think it objectionable that he should name two Athenians who have been disfranchised—even though they were guilty, and their convictions occurred many years ago. This is a notable instance of sensitivity (whether D.'s or the jury's) to other people's feelings.

ἐπὶ συμφοραῖς: a euphemism for legal condemnation; so also in the next few sentences *συμφορᾷ*, *ἀτυχίαν*, *ἠτυχηκώς*. Here *ἐπί* with the dative means 'involved in'; cf. 2.12 *ὄντων ἐπὶ τοῖς πράγμασιν*. Emendation to *ἐπὶ συμφορᾶς* (G. H. Schaefer *Apparatus* 3.364) or *ἐν συμφοραῖς* (Cobet *Misc. crit.* 506–7) is unnecessary.

ὀνομαστὶ: 'by name', instead of some less explicit allusion (cf. **206**, 24.132), whereas *ὀνόματι* means 'in name', as opposed to 'in fact' (cf. 40.1, 59.19). This justifies Reiske's emendation.

Σαννίων: nothing else is known of Sannion, except for an anecdote which was told by Demokhares (nephew of D.): on one occasion when Aiskhines was playing the part of Oinomaos in a tragedy by Iskhandros, he fell over, and Sannion the *χοροδιδάσκαλος* helped him to his feet (anon. *Life of Aiskhines* = *F. Gr. Hist.* 75 F6). From this, and also from *ἐμισθώσατο* (**59**), we may conclude that Sannion did not write tragedies himself, but trained choruses for plays written by others. He may have taken on the whole task of directing plays for authors unwilling or unable to do so (as Kallistratos did for Aristophanes), but more probably he just assisted as a *ὑποδιδάσκαλος* (for this term cf. Pl. *Ion* 536a, Phot. *ὑποδιδάσκαλος*) specializing in the choral parts of plays.

ἀστρατείας: although choristers were exempt from military service while their rehearsals were in progress (cf. **15**, 39.16), such exemption may not have been extended to a *ὑποδιδάσκαλος*: or Sannion may have been convicted of failing to perform it at some other time. The penalty for that offence was disfranchisement (*ἀτιμία*), as the present passage implies; cf. 24.105, 59.27, And. 1.74, Ais. 3.175–6. A disfranchised citizen was banned from all sacred precincts, including the theatre of Dionysos.

59. ἐμισθώσατο: 'hired'. Khoregoi drew lots for poets, but extra professional assistance was a matter for private arrangement.

φιλονικῶν: here and in **60** and **66** some or all mss. have *φιλονεικ-*, but *ι* and *ει* are often confused in mss. It is doubtful whether *φιλονεικ-* is ever a correct form (cf. LSJ under *φιλόνικος*); and 'eager for victory', not 'eager for contention', is the meaning required in all these three passages.

Θεοζοτίδης: a rare name. So we may identify him with Theozotides son of Nikostratos, named in *IG* xii.v 542.35 as a proxenos of

Karthaia in Keos, grandson of the Theozotides who took some part in public affairs in the late fifth and early fourth centuries; cf. Davies *Families* 222–3.

κωλύσειν: what was it that the rival khoregoi intended to prevent, but in the end allowed? Not the rehearsals, since the reference here is to the day of the performance. So these words are evidence that a *διδάσκαλος* or *ὑποδιδάσκαλος* might take part in the performance itself. Probably he was the coryphaeus.

οὐδεὶς ἥψατο: cf. **60** *ἐπιλαβόμενον τῇ χειρί*. These words show that to apprehend a man who had been disfranchised for failure to perform military service and was then seen in a sacred precinct, the correct legal procedure was *apagoge*: anyone who wished could arrest him and hand him over to the Eleven for custody and trial. The law is quoted in 24.105: *ἐὰν δέ τις ἀπαχθῇ, τῶν γονέων κακώσεως ἑαλωκὼς ἢ ἀστρατείας ἢ προειρημένον αὐτῷ τῶν νομίμων εἴργεσθαι, εἰσιὼν ὅποι μὴ χρή, δησάντων αὐτὸν οἱ ἕνδεκα καὶ εἰσαγόντων εἰς τὴν ἡλιαίαν, κατηγορείτω δὲ ὁ βουλόμενος οἷς ἔξεστιν*. Contrast **182**, where the procedure used against a man disfranchised as a debtor to the state and then seen in a jury is *endeixis*, denunciation to a magistrate. D.'s words in **59–60** (especially *δεῖν αὐτὸν ἐπιλαβόμενον τῇ χειρί*) seem not to allow for *endeixis* as a possibility in this case, but we do not know exactly where the line was drawn between disfranchised men liable to *apagoge* and those liable to *endeixis*. On that problem cf. Hansen *Apagoge* 94–6, but he does not discuss this passage.

τὸ συγκεχωρηκὸς: *τὸ* is absent in S and other mss., but A seems right to retain it, so that the participle is equivalent to an abstract noun, meaning that the avoidance of quarrels is an element of piety at a festival. Such participles are usually present, but cf. Th. 5.9.6 *ἐν τῷ ἀνειμένῳ αὐτῶν τῆς γνώμης*. Aristeides, perhaps under D.'s influence, uses *τὸ συγκεχωρηκός* in 1.171, 1.298, 1.771, 2.202 (Dindorf's pages).

τοσοῦτ' ἀπέχει τῶν χορηγῶν: the verb is impersonal: 'So far from the khoregoi, not even any of his personal enemies stops him.' Sannion's enemies were those who accused him of avoiding military service and got him disfranchised; they might have been expected to be even keener than the rival khoregoi to enforce the ban. With this reading, which is in S only, the grammar is somewhat compressed, but the point is clearly the right one. The alternative *τοσοῦτ' ἀπέχει τοῦ χορηγῶν τινος ἅψασθαι*, 'he is so far from touching any of the khoregoi', is an attempt to clarify the construction made by someone who missed the point about Sannion's position and thought that the point was merely that one ought not to hit a khoregos.

60. 'Αριστείδης, belonging to the Oineis tribe, could be the father of Eukles son of Aristeides of Akharnai, who was one of the dedicators of an inscription in Euboia (*IG* xii.ix 1242.18); and that may be the Eukles of Akharnai who was 59 years old in 325/4, when he served as a public arbitrator (*IG* 2² 1926.97). But Aristeides and Eukles are both common names, and so the identifications cannot be regarded as certain.

τοιοῦτον certainly means that Aristeides suffered the same penalty as Sannion, disfranchisement, and probably that he suffered it for the same offence, failure to perform military service.

ἡγεμὼν τῆς φυλῆς: in the context it is clear that this means 'leader of his tribe's chorus', i.e. the chief member of it, not the producer or director. *κορυφαῖος* cannot stand alongside *ἡγεμὼν* and is evidently a gloss which has got into the text. (This was first pointed out by Reiske under *κορυφαῖος* in the 'Index Graecitatis Demostheneae' of his edition.)

οἴχεται: metaphorical, as in 25.20.

εἶδεν: 'paid attention to'; cf. 45.64 *Φορμίωνα δὲ πάλιν ἑόρακεν καὶ τούτῳ γέγονεν οἰκεῖος*. The meaning is not that no khoregos realized that Aristeides was disfranchised, but that none took the advantage which his disfranchisement offered them.

ἐπιλαβόμενον τῇ χειρὶ: to seize hold of a chorister by force would obviously create a disturbance and antagonize the audience expecting to hear him. For the arrest of a disfranchised citizen, **59** n. *οὐδεὶς ἥψατο*. For the removal of an alien from a chorus, see p. 276. The treatment of a disfranchised Athenian appearing in a chorus was severer than that of an alien, because he had already been convicted by a court and was defying the court's sentence.

ἠβούλετο: for the augment, **20** n. *ἠδυνήθησαν*.

61. οὔκουν δεινόν ...: a characteristic idiom of Athenian rhetoric, going back at least to the fifth century (e.g. Ant. 1.12, E. *Andr.* 269, *Hek.* 592, Ar. *Knights* 875). What is presented as disgraceful is the inconsistency of two facts or possibilities, which are stated with *μέν* and *δέ* in either infinitival clauses, as here, or conditional clauses: 'is it not terrible that, whereas none of the khoregoi ..., Meidias ...?' The indignant tone makes it preferable to regard the sentence as a question introduced by *οὔκουν*, though the mss. give *οὐκοῦν* with no question mark; cf. **120** *πῶς οὐ δεινόν ...;* and Denniston *Particles* 432–3. The rhetorical effect is enhanced by much use of asyndeton within the sentence. For further comment cf. Pearson *Art* 107.

σχέτλιον: a synonym of *δεινόν*, and often paired with it, e.g. **104**, 19.146, 19.226, 20.156.

παρὰ τοῦτ': 'from this cause', by the removal of Sannion or Aristeides. The use of *παρά* to denote a sufficient cause ('this is enough to ...'; cf. LSJ *παρά* C.III.7) is found occasionally in earlier authors, e.g.

Ant. 3d.5, Th. 1.141.7, but becomes much commoner in the time of D.; cf. **96**, 4.11, 9.2, 18.232, 19.42, 19.263, etc. If *ταυτὶ* is included before this phrase (SYP), the meaning is 'would win those victories', but it is better omitted (AF); the deictic *-ί* is inappropriate for past events. Dobree *Adversaria* 1.460 suggests that it originated from *ταύτῃ*, a variant or gloss on *παρὰ τοῦτ'*.

ἀναλίσκοντας, ἀγωνιῶντας: it is not certain that this is an instance of two-limbed asyndeton (**81** n. *τῇ δίκῃ, τοῖς νόμοις*) since it is possible to take the second participle as dependent on the first, 'spending in keen rivalry'; but the assonance is rather in favour of taking them as an asyndetic pair.

προορᾶσθαι: the point of the middle voice is that the khoregoi foresaw that the audience's desires (to hear and see the performance) would have an adverse effect on themselves (if they hindered the performance).

ἰδιώτην ὄντα: not being a khoregos.

ἐπίτιμον: unlike Sannion and Aristeides, who were disfranchised.

62–9. *Meidias pursued a personal quarrel beyond proper limits.*

It is a familiar Greek notion that it is right to harm one's enemies; cf. Dover *Morality* 180–4. But here we have something different: D. is arguing that there are limits beyond which enmity should not be carried. He adduces Iphikrates and Khabrias as examples of proper restraint, but the principle emerges most clearly in **67**: it is reasonable and acceptable to *λυπεῖν* an enemy, but not to *ὑβρίζειν* him. Even though the distinction is not clearly defined, this is an advance on old-style Greek morality. It is because Meidias has overstepped this proper limit that he deserves to be punished, not by his enemy D. but by the jury representing Athens as a whole; his *hybris* has converted personal enmity into public offence.

62. ἐκεῖνον: 'the well known', like Latin *ille*; so also in **71**, 3.21, 18.219, 23.202. Iphikrates was the leading Athenian general in the first half of the fourth century.

Διοκλεῖ: Diokles son of Diokhares of Pithos was a rich man prominent in the first half of the fourth century; for detailed discussion see S. Lauffer *Historia* 6 (1957) 287–9, Davies *Families* 158–9. His deme Pithos seems to have been near Pallene, east of Athens (Ath. 234f), but he owned land at Nape in the silver-mine area (*Hesperia* 10 (1941) 16–17 no. 1 lines 48–9, 58) and his wealth probably came from mining. Besides the khoregia mentioned here, he performed two trierarchies (*IG* 2^2 1604.91, 1609.118), and his wife officiated at the Thesmophoria (Isai. 8.19). We have an epigram, composed possibly by himself, inscribed on the base of a dedication made perhaps from the first produce of a mine (*IG* 2² 4320). With the adoption of standard orthography and Lauffer's restorations it runs:

Διοχάρους Διοκλῆς με Πιθεὺς ἀνέθηκεν Ἀθηνᾷ
γαίας οἰκείας ἔρνος ἀπαρξάμενος.

Πιθεῖ, not *Πιτθ-*, is the spelling attested by inscriptions in the fourth century; cf. Threatte *Grammar* 1.545–6.

ἔτι πρὸς τούτῳ: emphasizing that the rivalry which occurred between Diokles and Teisias as khoregoi was additional to hostility already existing.

Τεισίαν: nothing else is known of this brother of Iphikrates, except the names of his two sons Timotheos (*Hesperia* 7 (1938) 92 no. 12) and Timarkhos (Ais. 1.157). On the spelling, **25** n. *παθεῖν ἢ ἀποτεῖσαι.*

πολλοὺς μὲν ἔχων φίλους: Polyainos relates an anecdote, in two slightly different versions (3.9.15, 3.9.29), to the effect that, when Iphikrates was prosecuted for treason after the battle of Embata was lost in 356, he was acquitted because the jurors were afraid of his armed supporters surrounding the court.

τιμῶν: when Iphikrates gave up military commands, he was awarded various honours, including the right to dine in the Prytaneion and the erection of a bronze statue of him on the Akropolis (23.130, Paus. 1.24.7). These were similar to the honours given to the tyrannicides Harmodios and Aristogeiton. The proposal to confer them was opposed by Harmodios, a descendant of the tyrannicide's family. The speech which Iphikrates made in reply was attributed to Lysias in later centuries, but Dion. Hal. *Lys.* 12 is probably right to reject that and attribute authorship of the speech to Iphikrates himself. Some of the surviving quotations from it fit D.'s comment that he was proud of himself, e.g. 'My actions are more akin to those of Harmodios and Aristogeiton than yours are' (Arist. *Rhet.* 1398a 20–2) and 'My pedigree begins with me, yours ends with you' (Plu. *Eth.* 187b).

63. **τὰς τῶν χρυσοχόων οἰκίας:** the plural is sarcastic. D. now lists the things which he accused Meidias of doing (**16–17**).

τῇ τῶν ἄλλων βουλήσει: the wishes of the audience; cf. **61** *τὰς ὑμετέρας βουλήσεις.*

στεφανούμενον refers not to the wearing of a crown during the festival by every khoregos, but to the winner's crown at his victory celebration; cf. **55**.

πολιτείᾳ: the whole organization of public life in Athens, including here the arrangements for festivals. The sentiment, that if you accept the advantages of living in a community you must also accept its adverse decisions, is the one developed in Pl. *Kriton* 50–4.

64. πάλιν: **33** n. For the use at the start of a sentence with no other connective word, cf. **197**.

Φιλόστρατον: Philostratos son of Dionysios of Kolonai was a well-to-do public figure. Wealth is indicated by the khoregia mentioned here, and he was also a member of a symmory (*IG* 2[2] 1622.773). He had no son, and left a substantial estate to his daughter's son Phainippos, whom he adopted (42.21, 42.27); in the 320s, after the death of Philostratos, Phainippos appeared rich enough to be the target of an *antidosis*, the subject of the speech *Against Phainippos* (D. 42). No particular political activity of Philostratos is known apart from the prosecution of Khabrias mentioned here; but he must have spoken often, since he is called 'Philostratos the orator' (42.21). He had been a friend of Lysias in his youth (59.22). Cf. Davies *Families* 552.

Κολωνῆθεν: there were two demes named Kolonai, one of the Leontis and one of the Antiokhis tribe, both perhaps in the vicinity of Pentelikon; cf. Lewis *BSA* 50 (1955) 12–17, W. E. Thompson *Hesperia* 39 (1970) 64–5. It is not known to which of them Philostratos belonged.

Χαβρίου: the distinguished general, whose achievements are summarized and praised by D. in 20.75–86.

τὴν περὶ Ὠρωποῦ κρίσιν: Oropos, on the borders of Attika and Boiotia, had been held by the Athenians for some years when in 366 it was taken by Oropian exiles with the help of Themison, tyrant of Eretria; the Athenians sent a force to recapture it, but the Thebans, invited as a neutral party to hold it temporarily, simply kept it (18.99, Ais. 3.85, X. *Hell.* 7.4.1, Diod. 15.76.1). In consequence Khabrias and the politician Kallistratos were prosecuted; the exact grounds of accusation are not known, but probably they were accused of *προδοσία* because they let the Thebans into Oropos. Besides Philostratos, Leodamas was one of the prosecutors (Arist. *Rhet.* 1364a 19–23; cf. D. 20.146–7 for his hostility to Khabrias). Kallistratos' speech in his own defence was much admired, and secured his acquittal; it is traditionally supposed to have implanted in the young D. the wish to become an orator (Plu. *Dem.* 5). Khabrias was supported by Lykoleon and Plato (Arist. *Rhet.* 1411b 6–7, Diog. Laert. 3.23–4); since his military career continued, it is clear that he too was acquitted.

θανάτου: death was a normal penalty for *προδοσία*, though it should probably be regarded as a penalty proposed by the prosecutors rather than fixed by law; cf. MacDowell *Law* 176–7.

νικῶντα: victory tripods of both Philostratos and his son-in-law Kallippos are mentioned in 42.22: *ἑκατέρου τρίπους ἀνάκειται, νικησάντων αὐτῶν Διονύσια χορηγούντων*. *IG* 2[2] 3035 may actually be a fragment of Philostratos' tripod; cf. the supplements proposed by

A. N. Oikonomides in *SEG* 19.194. The date of Philostratos' victory is not known, but it must have been before the death of Khabrias in 357, and probably not long before, since a khoregos for a boys' chorus had to be forty years old (*AP* 56.3, Ais. 1.11) and Philostratos is unlikely to have been born before about 400; cf. Lewis *BSA* 50 (1955) 24, Davies *Families* 552.

ὅποι μὴ προσῆκεν: such as goldsmiths' houses at night. The generalization causes the negative in the relative clause to be *μή*.

65. ἂν ἔχων εἰπεῖν: 'though I would be able to mention', if there were time, or if it were necessary. This is a shorter, more colloquial, substitute for a formal conditional sentence like S. *Phil.* 1047–8 *πόλλ' ἂν λέγειν ἔχοιμι πρὸς τὰ τοῦδ' ἔπη, εἴ μοι παρείκοι.* The phrase is common in D. both with and without *ἄν*. Blass (in his note on 18.100) formulated a rule that *ἄν* is inserted only when *παραλείπω* or a phrase of similar meaning precedes, not when it follows; but that is refuted by our passage and by 18.258 *πόλλ' ἂν ἔχων ἕτερ' εἰπεῖν περὶ αὐτῆς παραλείπω*, Isok. 12.262 *πόλλ' ἂν εἰπεῖν ἔχων ἔτι ..., ταῦτα μὲν ἐάσω*, cf. also D. 9.21 *πάνθ' ὅσα τοιαῦτ' ἂν ἔχοιμι διεξελθεῖν παραλείψω*. There are other instances with *ἄν* in **132** (probably; see note there), 3.27, 20.33; without *ἄν*, 10.75, 18.100, 18.138, 18.264; with *λέγειν* instead of *εἰπεῖν*, 18.50, 22.3, 22.46, 40.38.

ἔτι with *πολλούς*: 'many others'.

ἀκήκοα: 'I have heard of' (rather than 'I have heard'), as in 35.26 *οὐδὲ ἀκήκοα πώποτε πρᾶγμα μιαρώτερον.*

ἐκεῖνο foreshadows *οὐδένα ... παρεστηκότα ...*, instead of the commoner *τοῦτο* and *ὅτι ...*

ἐχθρῶν ἀλλήλοις: *γενομένων* is added by A and a corrector in P, but it is not essential; cf. pp. 50–1 on extra words in A.

καλουμένων ... ἐξορκοῦντα: for this allegation against Meidias, **17** n. *ὀμνύουσι παρεστηκὼς τοῖς κριταῖς*. The verb *καλουμένων* means that, when the lots were drawn to select the judges, the name of each man selected was called out and he came forward. The emendation *κληρουμένων*, proposed by Herwerden *Mn.* II 1 (1873) 308, is unnecessary.

ἐχθρὸν ἐξεταζόμενον: 'revealing himself as an enemy'. For the sense of *ἐξετάζομαι* (almost equivalent to *φαίνομαι ὤν*, but usually with a stronger sense of showing one's attitude deliberately: 'stand up and be counted') cf. 19.291 *συγκατηγόρει μετ' ἐκείνου σοῦ καὶ τῶν ἐχθρῶν τῶν σῶν εἷς ἐξητάζετο*, **127** (*συν-*), **161**, **190** (*συν-*), **202**; the last of these passages has *ἐξητάσθη*, so that probably all should be regarded as passive forms rather than middle.

66. φιλονικίᾳ rather than *-νεικ-*: **59** n. *φιλονικῶν*.

χορηγὸν ὄντα: subordinate to *ὑπαχθέντα* and so not linked to it by *καί*. Notice the contrast with *ἐκ προαιρέσεως*, 'on purpose'; the suggestion is that harm done by one competitor to another is done not for its own sake but for the sake of winning.

ἐλαύνοντά τινα: 'harassing someone'; the verb has the same sense in **111**, **131**, **135**, **173**. With this participle understand *ταῦτα πάντα καὶ τὰ τοιαῦτα ποιεῖν*, as subject of *ἐστιν*.

δύναμιν: A and F add *καὶ βίαν*, but power and violence are different, and if D. had wished to mention both he would probably have written *καὶ τὴν βίαν*. Cf. pp. 50–1 on extra words in A.

Ἡράκλεις: exclamatory vocative, used also in 9.31, 19.308, to express horror or indignation.

καὶ οὐχὶ ... οὐδὲ: a clear instance of the rule that 'and not' is *οὐδέ* when another negative precedes, *καὶ οὐ* when no negative precedes. *οὐχί* is more emphatic than *οὐ*: 'certainly not right'.

ὑμῖν: the people of Athens (not only the jury).

ὁ δεῖνα: 'some particular man', 'so-and-so'; so also in **141**. D. is generalizing from his own case, and *μοι* in effect means 'anyone'.

κἂν ἄμεινον ἀγωνίσωμαί τινος: the point is the same as in **5**, that D.'s chorus would have won if the judges had not been corrupted by Meidias.

διατελέσω: in earlier Attic authors the future of *τελῶ* and *καλῶ* is the same as the present, but by the time of D. the longer forms are coming into use, and there is no good reason to remove them by emendation. Cf. 8.14 *παρακαλέσειν*, 19.133 *ἐγκαλέσει* (in some mss. only), 23.123 *ἐγκαλέσουσιν*, Lyk. *Leo.* 17 *ἐπικαλέσεται*, 143 *ἐπικαλέσεται*, Ais. 1.174 *ἐκκαλέσεσθαι* (in some mss. only), Pl. *Rep.* 425e *διατελέσουσιν*. But the contracted future forms are found in **119** *καλῶ*, **167** *καλῶ*, 20.28 *συντελοῦσιν*, 24.159 *καλεῖ*, 55.17 *ἐγκαλεῖ*.

ἀλόγιστος: 'uncalculating', 'careless with money'; so also in 39.6.

ἄθλιός: 'foolish'; so also in **191**.

ἑκὼν added to *ἐθελήσειεν* emphasizes that the reference here is to voluntary expenditure, beyond the necessary minimum. A parsimonious khoregos may have been able to dress his chorus quite cheaply. But, for an effective show, the Athenians relied on the willingness of khoregoi to spend lavishly; this willingness will evaporate, says D., if khoregoi are not treated fairly.

67. πάντας: all khoregoi; not 'everyone' (Vince), because only the richest Athenians performed liturgies.

τῶν ἴσων καὶ τῶν δικαίων: neuter plural, serving as abstract nouns. There are similar instances of *ἴσα* in **96**, **112**, **188**, **211**.

αὐτῷ: 'himself', in apposition to *Μειδίᾳ*, contrasted with the preceding *ἐμὲ*. (Herwerden *Mn.* II 1 (1873) 308 misses the contrast, thinks *αὐτῷ* redundant, and emends the text.)

διᾶραι τὸ στόμα: perhaps a colloquial expression; cf. 19.112, 19.207 *οὐδὲ διῆρε τὸ στόμα.*

ἔχειν ἐμέ: accusative and infinitive added to *ἐξῆν* with dative: it was a possibility for Meidias that D. was not able even to open his mouth.

68. χορηγὸς ὑπέστην: 'undertook to be chorus-producer', as described in **13**. For the construction cf. Lys. 29.7 *ὑποστῆναι τριήραρχον.*

τῆς Ἐρεχθηίδος: sc. *χορηγόν*. Meidias' retort to this suggestion would no doubt have been that Erekhtheis, unlike Pandionis, had already selected a khoregos for its men's chorus.

69. ἐν ᾧ: 'in doing which'. This (instead of *ᾧ*) means that a compliment to the people, though not the main purpose of the act, would have been implied by Meidias' willingness to spend his money on a chorus for them.

οὐδ' ἐνεανιεύσατο: cf. **18** *τοῖς ἑαυτῷ νενεανιευμένοις*. Meidias imitated only the bad features of youth.

μανείς: nominative, agreeing with *ὅς*, even though it is governed by *νομίσαι*. Compare the usage with *λέγω* in the sense 'I mean' (**83** n. *ταύτης λέγω*); it is as if the word were a direct quotation from a person's speech or thought. This reading, though found only in S, should be preferred to *μανίᾳ* and *μανίαν*, which are easily explained as attempts to regularize the construction. Corruption probably occurred early in some copies of the oration: Dion. Hal. *Dem.* 9, quoting the passage as an example of D.'s 'unnatural' style because of its complexity of parenthesis, has *μανίαν ... φιλοτιμίαν*, and this reading is taken over from him by later rhetoricians.

The remark that it is madness to attempt the impossible is a commonplace which goes back to Gorg. *Pal.* 25 ... *μανίαν· μανία γάρ ἐστιν ἔργοις ἐπιχειρεῖν ἀδυνάτοις*. The quality of *φιλοτιμία*, by contrast, is a creditable one (**159** n.).

ὑπὲρ δύναμιν: D. means that he could not really afford the money which a chorus required.

τῶν ἱερῶν ἱματίων: cf. **16**.

τὼ χεῖρε, like *τελευτῶν*, is to be taken only with *τοῦ σώματος*, not with the earlier genitives; Meidias' interference with the chorus did not include laying hands on it.

μου: with the preceding genitive nouns. For its late position, **16** n.

70–6. *An offence such as Meidias' can have disastrous consequences if the victim retaliates.*

70. ἄλλως πως ἔχει τὴν ὀργὴν ... ἢ ὡς ...: cf. **127** for the construction (and similar instances with *τὴν γνώμην ἔχειν* in Th. 7.15.1, X. *An.* 1.3.6). 'Have your anger in any other way than as when ...' means 'be any less angry than you are when ...'.

οὐ: omitted in S, but a 'redundant' negative is common when *ἤ* follows a negative, e.g. 49.3 *οὐ περὶ πλείονος ἐποιήσατο ὁ πατὴρ περιουσίαν χρημάτων μᾶλλον ἢ οὐ Τιμοθέῳ ὑπηρετῆσαι*. Occasionally the preceding negative is merely implied, e.g. 49.47, where *τί ἂν ποτε βουλόμενος ...;* implies 'he had no reason to ...', and Th. 3.36.4, where *ὠμὸν τὸ βούλευμα* implies 'they ought not to ...'. In the present instance the negative clause follows, *οὐκ ὀρθῶς ἔχει*, but the words *εἰ ... τις ... ἄλλως* have already hinted at the negative idea 'you ought not to ...'.

εὐλάβειαν: D. claims credit for his discretion in not hitting back when Meidias assaulted him. Consequently no one was injured, but Meidias' offence should not be regarded as less serious on that account.

μερίδα: 'contribution' to the defendant's case; so also in **184**.

ἐπὶ τοῦ βοηθεῖν: in the circumstances of voting for him, by giving a verdict in his favour; **2** n. *ἐπὶ τῶν ἄλλων*, **7** n. *βοηθῆσαι*.

71. αἴρω: 'exaggerate'. For the metaphor cf. Ais. 2.10 *ἄνω τὸ πρᾶγμα ἐξάρας*. D. now proceeds to adduce two cases in which fatal consequences actually did ensue from retaliation, even though the insolence in those cases was less serious than Meidias' insolence to D.

ἴσασιν ἅπαντες: **1** n. *οὐδένα ... ἀγνοεῖν*. The details which D. now goes on to give, and especially *εὖ οἶδ' ὅτι γιγνώσκουσίν τινες ὑμῶν ὃν λέγω*, suggest that really the incident was not well known; it happened far from Athens on a private occasion many years ago. Euthynos and Sophilos are not otherwise known to us, and the description of the incident here is not very clear. Since both names are in the accusative, the sentence is formally ambiguous: which is the subject and which the object of *ἀμυνάμενον*? Σ (219 Dilts) says *ἀμφιβόλου δὲ ὄντος τοῦ ῥηθέντος εἴη ἂν ὁ μὲν Εὔθυνος ὁ λυπήσας, ὁ δὲ ἀποκτείνας ὁ Σώφιλος*, but that is the wrong answer, as Boeckh *Abh. Berlin 1818–19* 76–7 n. 1 shows, giving three reasons: (*a*) the last sentence of **71**, which must mean that Euaion killed Boiotos (cf. **73**), shows that the first name should be taken as the subject; (*b*) the point of the anecdote is that a young man, when stung by a supposed insult, felled a mighty pancratiast, whereas the reverse would have been less surprising; (*c*) *ἦν* shows that Sophilos is dead, whereas Euthynos seems to be mentioned as a man still living. He is identified as 'the well known man (**62** n. *ἐκεῖνον*) who was a wrestler', but the anecdote implies that he was young and obscure when he killed Sophilos; this can be explained by assuming that it was after that incident that he became well known, and that now he is old and has given up wrestling.

τὸν νεανίσκον: it is strange to identify Euthynos as '*the* young man', and I suggest that *τὸν* may be a corruption of *ὄντα*, 'when he was a young lad'. Weil conjectures *τὸν Νεμεόνικον*, but that weakens the anecdote by removing the contrast between a young inexperienced man and an older one.

καὶ Σώφιλον: Reiske and others delete *καί*, but it may be kept in the sense 'even': it is remarkable that a young man was able to kill so strong an opponent.

μέλας: 'dark' in complexion (but not a negro); cf. Pl. *Rep.* 474e.

τοῦτον: 'well, him', resuming *Σώφιλον* after the parenthesis.

οὕτως: 'merely'; cf. LSJ *οὕτως* IV. The point of mentioning the privacy of the occasion is that an insult is less wounding if few people witness it.

ὁ τύπτων: inserted by someone who thought it desirable to make the subject of *ᾤετο* clear, and intended it to mean the man who hit his opponent so hard as to kill him. But in fact it is confusing to insert the phrase, in that sense, before *ἀμυνάμενον* ... *ἀποκτεῖναι* has been reached, and it seems unlikely that D. wrote it; the subject of *ᾤετο* is clear without it to anyone following the train of thought, and so it is best to delete it as a gloss (Boeckh *Abh. Berlin 1818–19* 76–7 n. 1). An alternative possibility is *ὅτι τύπτων αὐτὸν ὑβρίζειν ᾤετο δεῖν* (G. H. Schaefer *Apparatus* 3.376); then the subject of this clause is Sophilos and the meaning is 'because he thought fit to insult him by hitting him' (for *ὑβρίζειν ᾤετο δεῖν* cf. **81**, **143**).

ἀμυνάμενον: the main verb of the clause governed by *ἴσασιν*.

ἴσασιν: repetition, instead of a particle, connects the sentences.

Εὐαίωνα: not otherwise known.

Λεωδάμαντος: a minor politician, active in the second quarter of the fourth century. Leodamas was a pupil of Isokrates (if we emend *Λεώδαμος* in Plu. *Eth.* 837d) and became an able speaker (20.146, Ais. 3.139). He is mentioned as an associate of Hegesandros (Ais. 1.69, 1.111), an ambassador to Thebes (Ais. 3.139), and an opponent of Khabrias (20.146–7); he prosecuted Khabrias and Kallistratos about the Oropos affair (**64** n.). In 355/4 he spoke in defence of the law of Leptines in opposition to D. (20.146–7). For his genealogy see Davies *Families* 523.

Βοιωτὸν: a rare name; so this may well be Boiotos son of Pamphilos, uncle of the man attacked in the speeches *πρὸς Βοιωτόν* (39.32, 40.23). He was still alive in 348 (to judge by the present tense *ἔστι* in 39.32); so, if the identification is correct, his death at the hands of Euaion was a recent event.

κοινῇ, if right, means simply that several men were sharing a meal. It cannot mean that the meal was 'a public banquet' (Vince), because **73** tells us that Boiotos was at dinner with only six or seven friends in a private house. Weil's conjecture *οἰκείων*, 'of friends', gives better sense and may be correct.

72. παρέστησε: **15** n. *παρίστη*.

ἡ ἀτιμία: for the connection between *hybris* and dishonour see pp. 18–22.

ὂν precedes *δεινόν* because it is emphatic: 'though it *is* terrible'. So also 9.55 *καὶ οὐχί πω τοῦτο δεινόν, καίπερ ὂν δεινόν*. The form of expression is a paradox or oxymoron, appearing self-contradictory: *δεινός* can mean either 'deserving indignation' or 'causing indignation', and the point here is that a blow (without *hybris*) deserves indignation but does not in practice cause it.

τῷ σχήματι ...: though presented as a generalization, this is in effect a description of the manner in which Meidias struck D. The reason why a blow is more offensive if struck with the fist and on the face is that such a blow is deliberately aimed; it is not accidental or hasty.

ἐπὶ κόρρης: in Homer *κόρση* is synonymous with *κρόταφος*, meaning the temple at the side of the forehead, e.g. *Iliad* 4.502. But in Attic *κόρρη* is synonymous with *γνάθος*, meaning the side of the face; cf. Hyp. fr. 97 Kenyon, Jensen = 36.1 Burtt. A blow *ἐπὶ κόρρης* can be either a slap with the flat of the hand (*ῥαπίζειν* in Hyp.; cf. Harp. *ἐπὶ κόρρης*) or a punch with the fist (cf. Theokr. 14.34 *πὺξ ἐπὶ κόρρας*). Plato regards the expression *ἐπὶ κόρρης τύπτειν* as 'rather coarse' (*ἀγροικότερον*, *Gorg.* 486c).

This passage is often quoted as an example of the effective use of anaphora and asyndeton. 'Longinus' *On the Sublime* 20 says that D. imitates the assailant, striking the jurors' minds with repeated blows: *οὐδὲν ἄλλο διὰ τούτων ὁ ῥήτωρ ἢ ὅπερ ὁ τύπτων ἐργάζεται· τὴν διάνοιαν τῶν δικαστῶν τῇ ἐπαλλήλῳ πλήττει φορᾷ.*

ἐπὶ τῆς ἀληθείας: the preposition has a temporal sense, and the noun refers to the action itself, as contrasted with later talk about it. Cf. 18.226 *ἐπὶ τῆς ἀληθείας, ἐγγὺς τῶν ἔργων*.

73. πρὸς Διὸς καὶ θεῶν: lengthier than the common *πρὸς θεῶν*, but it does not seem to be reserved, as one might expect, for particularly earnest requests. In this speech it recurs in **108**; there is another instance with *σκέψασθε δή* in 20.43.

παρ' ὑμῖν αὐτοῖς: 'in your minds'; so also in 22.22. Not 'among yourselves', meaning 'in discussion with one another', which would require *πρός* with an accusative.

ὀργὴν ... παραστῆναι: in effect the passive of *παρέστησε τὴν ὀργήν* (**72**).

ἐκείνῳ: not 'the well known', but 'the one I mentioned' (in **71**).

ὁ μέν: Euaion. The reasons why the assault on Euaion was less serious than the assault on D. are arranged in three groups: (*a*) the assailant was a friend, and was drunk; (*b*) the witnesses were few, were friends, and were likely to give Euaion their support; (*c*) the place was a private house, and Euaion went there voluntarily. All these are expressed by phrases attached to *ὁ μέν … ἐπλήγη*. The other side of the antithesis (*ἐγὼ δ' …*) is in corresponding groups: (*a*) D.'s assailant was an enemy, and was sober; (*b*) the witnesses were numerous, and included foreigners; (*c*) the place was a sacred precinct, and D. was required to go there by his duties as a chorus-producer.

γνωρίμου: 'an acquaintance', a friend less close than a *φίλος*. Cf. 18.284 *ξένος ἢ φίλος ἢ γνώριμος*.

ἐναντίον ἓξ ἢ ἑπτὰ ἀνθρώπων: dishonour is less if few people see it.

ἀνασχόμενον καὶ κατασχόνθ' ἑαυτὸν: conditional: 'if he acquiesced and restrained himself', as in fact he did not.

ἤμελλον: for the augment, **20** n. *ἠδυνήθησαν*.

εἰς οἰκίαν: contrasted with a sacred place (**74** *ἐν ἱερῷ*). A similar contrast is implied in And. 1.11 *τὰ μυστήρια ποιοῦντα ἐν οἰκίᾳ*.

μηδὲ, not *οὐδέ*, because it goes with the infinitive, not with *ἐξῆν*: 'it was possible for him not to go', not 'it was not permitted to him to go'. (Humbert's interpretation, 'une maison plus accueillante que respectable', appears to result from missing this grammatical point.) The contrast between Euaion and D. becomes strained here: it is not really fair to suggest that Euaion culpably exposed himself to assault by going out to dinner, for it was not to be expected that a friend would hit him there.

74. ἕωθεν: 'in the morning', a reinforcement of *νήφοντος*, because drinking parties were held in the evening.

ὕβρει καὶ οὐκ οἴνῳ: drinking wine is one of the possible causes of *hybris* (cf. p. 19); they are not mutually exclusive. But here D. means that Meidias' *hybris* was not caused by drinking on one occasion; it was a more lasting and more deliberate state of mind, and therefore more culpable. The same point is made about Ktesikles in the same words (**180**).

ξένων: foreigners often attended the Dionysia; cf. **217**, Ar. *Akh.* 502–6, Ais. 3.43.

ὑβριζόμην: Herwerden *Mn.* II 1 (1873) 308–9 suggests deleting this word as a gloss; it can be regarded as tautologous after *ὕβρει* and it is possible to understand *ἐπλήγην* instead. But it is better to retain the word; it is far enough away from *ὕβρει* for the tautology not to be noticeable, D. wants to emphasize that he was treated with *hybris*, and stylistically *ἐναντίον … ὑβριζόμην* responds to *ἐναντίον … ἐπλήγη* on the other side of the antithesis.

ἐμαυτὸν: where the subject of an infinitive is the same as the subject of the governing verb, it is usually omitted (and then any other word qualifying it, including adjectival *αὐτός*, goes into the nominative);

but if expressed for any reason (such as emphasis or contrast) it is accusative, not nominative as is sometimes thought, e.g. Pl. *Soph.* 234e *οἶμαι δὲ καὶ ἐμὲ τῶν ἔτι πόρρωθεν ἀφεστηκότων εἶναι.* Cf. Kühner-Gerth 2.30–2, and appendix vii in Goodwin's edition of *Meidias*. There is another probable instance at the end of **204**, where *σὲ* rather than *σὺ* is given by SFYP.

σωφρόνως, μᾶλλον δ' εὐτυχῶς: prudence would be due to D. himself, luck to the gods. By correcting himself D. gives an impression of being modest and pious, while actually suggesting to his hearers the thought that he himself does deserve credit for his restraint.

οὐδὲν, not *μηδὲν* (AF), because it negatives *ἐξαχθέντα* ('not induced to do anything'), not merely the infinitive.

τῷ δ' Εὐαίωνι ...: D. wishes to avoid seeming to criticize Euaion in case some friends of his are in the jury. Indeed the next sentence tells us that nearly half the jurors voted for Euaion's acquittal, even though he used no improper means of influencing them; so he had considerable sympathy and support.

75. οὔτε ...: the numerous negatives are emphatic, but also precisely organized. *οὔτε κλαύσαντα οὔτε δεηθέντα ... οὔτε φιλάνθρωπον ... ποιήσαντα* is a trio. Within the third member of the trio, *οὔτε μικρὸν οὔτε μέγα* is a pair. *οὐδ' ὁτιοῦν* is not connective, but a more emphatic alternative for *οὐδέν*, going with *φιλάνθρωπον*: 'doing the jurors no favour whatever, either great or small'. *φιλάνθρωπον* needs a pronoun; so the deletion of *οὐδ' ὁτιοῦν*, proposed by O. Jahn *Philol.* 26 (1867) 3, is wrong. Before *φιλάνθρωπον*, *οὐδὲ* would be as good as *οὔτε* (cf. **129**), but emendation, proposed by H. Richards *CR* 18 (1904) 13–14, is not necessary.

δεηθέντα τῶν δικαστῶν οὐδενός: of course Euaion, when speaking in his own defence, will have addressed a plea for acquittal to the jury as a whole; but, D. means, he did not canvass any juror individually before the trial (in the manner described comically in Ar. *Wasps* 552–8). By this period such canvassing had been made virtually impossible in ordinary courts, because the system of allotment made it impossible to know in advance which jurors would try which case. But Euaion will have been tried at the Delphinion by the ephetai, whose identity could be known. For the use of the word *δικασταί* to refer to the ephetai cf. Ant. 6.1. The Delphinion was the court for any homicide case in which the accused man admitted killing but said that he did it lawfully (23.74, *AP* 57.3), and there was a law exonerating a man who killed in self-defence (Ant. 4c.2); cf. MacDowell *Homicide* 70–81. Gagarin *GRBS* 19 (1978) 111–20 considers it more probable that, if the plea was simply self-defence, the case was tried by the Areopagos, but there is no explicit evidence for that view; I do not think it can stand against the evidence of Ant. 4c.2 unless one rejects Antiphon's *Tetralogies* altogether as evidence for Athenian law (as a few scholars do, but not Gagarin).

φιλάνθρωπον: probably a euphemism for payment of money, rather like 'consideration' in English; cf. 8.70, where *τοιαύτας ἄλλας φιλανθρωπίας* concludes a list of payments.

μὴ: one might expect *οὐχ*: cf. 25.43 *οὐδέν' εἶναι τοιοῦτον τίθεμαι*. But there are many exceptions to the rule that *οὐ* is the negative in indirect statements; cf. Goodwin *Moods and Tenses* §685.

76. τί οὖν: **48** n.

παράδειγμά γε πᾶσι: A substitutes *τοῦτον* for *γε πᾶσι*, perhaps in order to bring this passage into conformity with **227** *παράδειγμα ποιήσαντες τοῦτον τοῖς ἄλλοις*. In fact the end of a speech, as Σ (241a Dilts) points out, is a more normal place for a reference to setting an example. Here it marks the conclusion of D.'s argument about Meidias' offence at the Dionysia, before he turns to earlier incidents.

77–82. *The origin of the quarrel between Demosthenes and Meidias: the* antidosis *and the cases of slander and ejectment.*

On the course of the quarrel, see pp. 1–13. This part of the speech is what the rhetoricians call *παρέκβασις*, a digression which, as Σ (242a Dilts) says, supports the case without being strictly relevant to it; cf. pp. 30–1.

77. ποθεῖν ἀκοῦσαι: D. probably has no reason to think that the jurors desire anything of the sort. But he himself wishes to give an account of Meidias' past misconduct; and since this may be criticized for irrelevance, he plants in the jurors' minds the thought that they would like to hear it.

μεγάλου τινὸς: 'some serious punishment' to atone for harm done previously. For the sense of *προοφείλω* cf. Ant. 5.61 *προωφείλετο αὐτῷ κακόν*, E. *IT* 523, Ar. *Wasps* 3.

εἰπεῖν καὶ διηγήσασθαι, ἵν' εἰδῆθ' ὅτι ... φανήσεται: this sentence exhibits some redundancy of wording. Possibly D. is nervous about embarking on a passage which may appear irrelevant to the case.

τούτων has no clear antecedent, but evidently means the actions which gave rise to the hostility.

ὀφείλων δίκην: A has *τὴν μεγίστην ὀφείλων δοῦναι δίκην*, but the extra words are omitted not only by S but also by Σ (245 Dilts) in later mss. which have them in the text. Cf. pp. 50–1 on extra words in A.

βραχὺς: a similar apology is made in **160**, 3.23, 43.21, 48.4. A's word-order is preferable to S's, to avoid a sequence of short syllables.

78. ἡνίκα: no connective particle; the preceding sentences point forwards to the beginning of the narrative.

τὰς δίκας ἔλαχον τῶν πατρῴων: for the use of *λαγχάνω*, **25** n. The cases brought by D. against his guardians are the subject of the extant speeches orations 27–31 (*Against Aphobos* and *Against Onetor*). D. was seven and his sister five when their father, also named Demosthenes, died. On his deathbed the elder Demosthenes gave his wife with a dowry to be married to his nephew Aphobos; he gave his daughter with a dowry to be married to his nephew Demophon when she was old enough; and he appointed Aphobos, Demophon, and a friend named Therippides to be jointly guardians of the young D. and his property. The guardians took control of the property, but Aphobos did not marry the widow; and when D. came of age ten years later, they gave back to him much less than they had received. So D. prosecuted them. The date of the prosecutions was 364/3, and it seems to have been in the second half of that year, because 30.15 indicates that, when Aphobos was divorced from Onetor's sister in Posideon (approximately December), the trials had not yet taken place. The plural *τὰς δίκας* shows that there was a separate case for each defendant, but they would be tried on the same day or successive days.

μειρακύλλιον ὢν κομιδῇ: cf. **80** *νέος κομιδῇ*. He was about twenty, having been born in 384; on the problem of his date of birth, **154** n. *δύο καὶ τριάκοντα ἔτη*. He stresses his youth in order to make Meidias, an older man, seem more of a bully. *μειρακύλλιον* is a colloquial form; the other Attic instances are in comedy, e.g. Ar. *Frogs* 89, Men. *Dys.* 27.

οὐδ' ... οὐδὲ ... μηδὲ: in Attic prose *οὐδέ* (or *μηδέ*) is 'and not' when it follows another negative, 'not even' or 'not ... either' when it does not; cf. Denniston *Particles* 190. The sentence assumes a tone of scorn or sardonic humour: *οὐδ' εἰ γέγονεν εἰδὼς* is similar in effect to the English expression 'I didn't know him from Adam' (cf. 29.29 *τὸν Μιλύαν δ' οὐδ' ὅστις ἐστὶν οὐδεὶς ᾔδει*), and the parenthetic wish *ὡς μηδὲ νῦν ὤφελον* abruptly recalls the present situation. Cobet *Nov. lect.* 107 finds *οὐδὲ γιγνώσκων* a 'languidum additamentum' arising from a gloss on the preceding words. However, Dion. Hal. *Dem.* 58 quotes *τὸν Μειδίαν τοῦτον οὐκ εἰδὼς ὅστις ποτ' ἐστὶν οὐδὲ γιγνώσκων*: this is evidently an inexact quotation from memory, but it confirms the genuineness of *οὐδὲ γιγνώσκων*.

εἰς (not a dative) is normal to denote a point of time in the future. The reckoning of days is inclusive.

ἀδελφὸς: the article is required: *ὁ* is given by F, but the other mss. omit it. Probably D. would have spoken the article in crasis but omitted aspirates when writing; so if we put in breathings we should write *ἁ*-.

Thrasylokhos, the brother of Meidias, is first heard of in 367/6, when he renewed leases of two silver mines (*Hesperia* 10 (1941) 16 no. 1 lines 47–52; cf. M. Crosby *Hesperia* 19 (1950) 196–8). This implies that he had already held the leases for some time, and since this is earlier than any known activity of Meidias it seems likely that Thrasylokhos was the elder brother; cf. Schaefer *Demosthenes*[2] 2.89 n. 3, Davies *Families* 385. After his appointment as a trierarch in 364/3, mentioned here and in 28.17, he was a trierarch on at least two other occasions (50.52, *IG* 2^2 1629.753–4, 1631.121–2) besides carrying on other activities: he is recorded as the owner of a workshop in the mining area (*Hesperia* 26 (1957) 3 no. S2 line 12) and he is probably the Thrasylokhos mentioned in 50.13 as a mortgagee of property.

ἀντιδιδόντες τριηραρχίαν: 'because they were proposing to exchange ⟨property with regard to⟩ a trierarchy'. The verb *ἀντιδίδωμι* refers to a proposal or agreement to make an exchange, not necessarily to the carrying out of the exchange; cf. 28.17 (discussed below) and 42.1, where the perfect *τοὺς ἀντιδεδωκότας* refers to men who have agreed to an exchange of property but have not yet carried it out. The same is true of the noun *ἀντίδοσις*: cf. **156**, where *ἐξ ἀντιδόσεως* means 'as a result of a proposal for exchange' in a case in which no exchange of property took place, and 42.5, where *ἐποίουν οἱ στρατηγοὶ ... τὰς ἀντιδόσεις* means 'the generals arranged for the proposals for exchange to be made', not 'the generals arranged the carrying out of the exchanges'. In the phrase *ἀντιδιδόντες τριηραρχίαν* the form of expression, which must be idiomatic, is elliptical; 'property' is omitted, and instead the accusative is used to denote the liturgy with which the proposal is concerned. Likewise the noun *ἀντίδοσις* may have the liturgy added in the genitive; cf. X. *Oik.* 7.3 *ὅταν γέ με εἰς ἀντίδοσιν καλῶνται τριηραρχίας ἢ χορηγίας*, 'whenever they summon me to a proposal for exchange of ⟨property with regard to⟩ a trierarchy or a khoregia'.

The trierarchy was one of the most important types of liturgy. The duty of a trierarch was to arrange and pay for the maintenance of a ship in the Athenian navy and to command it if it was sent out to sea. However, he could pay someone else to command the ship at sea on his behalf, or even to perform the entire functions of the trierarch; cf. **80** *τὴν τριηραρχίαν ἦσαν μεμισθωκότες*. In the first half of the fourth century it was usual, though not invariable, for two men to be appointed jointly for each ship; cf. **154**, which shows that it was only a syntrierarchy (i.e. a half share of a trierarchy) which Thrasylokhos got D. to take over from him. In or before 357 the system was changed (**155** n.), but that change had not yet been made at the time of the incident recounted in **78–80**. Trierarchs were appointed by the strategoi, and it is likely that the appointments were normally made at midsummer, at the beginning of the official year; cf. Jordan *Navy*

65–7. But in the present instance it appears that Thrasylokhos had been appointed to a syntrierarchy in the winter; for that was when D.'s prosecution of his guardians took place (see note above on *τὰς δίκας* ...). We should conclude that because of some emergency additional trierarchs were appointed in the winter of 364/3. (D. in 4.36 comments on the Athenians' tendency to appoint trierarchs when they heard some bad news: *ἅμ' ἀκηκόαμέν τι καὶ τριηράρχους καθίσταμεν καὶ τούτοις ἀντιδόσεις ποιούμεθα.*) The emergency arose no doubt from the naval successes of Epameinondas in the eastern Aegean (Diod. 15.79.1).

The procedure of *antidosis* was the same for all liturgies. Liturgies were supposed to be performed by the richest men in Athens. A man appointed to a liturgy might seek to avoid it by claiming that there was some richer man, not exempt, who should be appointed in preference to him. The magistrate responsible for each liturgy fixed a date for proposals for exchange (*ἀντιδόσεις*), and anyone wishing to put forward a proposal had to summon his opponent (the man whom he wished to challenge) to appear before the magistrate on that day (42.5, X. *Oik.* 7.3). He then challenged his opponent to choose between two alternatives: either (if he admitted being richer) to take over the liturgy from the challenger, or (if he claimed to be poorer) to hand over the whole of his property in exchange for that of the challenger, who would then perform the liturgy himself. If the man challenged chose the second alternative, to exchange property, each had to produce within three days an inventory of his own possessions; and each could, if he wished, go with witnesses to the other's house or estate to observe what was there. If the man challenged failed to proceed with the exchange, the challenger could apply to the magistrate for a trial by jury (*διαδικασία*) to decide which of them was to perform the liturgy. Most of this information about the *antidosis* procedure comes from D. 42 (*Against Phainippos*), a speech composed for such a trial in the 320s. Cf. W. A. Goligher *Hermathena* 14 (1907) 481–515, Lipsius *Recht* 590–9, Harrison *Law* 2.236–8, MacDowell *Law* 161–4, V. Gabrielsen *Cl. et Med.* 38 (1987) 7–38.

The case in which D. was involved is also recounted, soon after the event, in the second oration *Against Aphobos* (28.17):

> ὡς γὰρ τὰς δίκας ταύτας ἔμελλον εἰσιέναι κατ' αὐτῶν, ἀντίδοσιν ἐπ' ἐμὲ παρεσκεύασαν, ἵν' εἰ μὲν ἀντιδοίην, μὴ ἐξείη μοι πρὸς αὐτοὺς ἀντιδικεῖν, ὡς καὶ τῶν δικῶν τούτων τοῦ ἀντιδόντος γιγνομένων, εἰ δὲ μηδὲν τούτων ποιοίην, ἵν' ἐκ βραχείας οὐσίας λειτουργῶν παντάπασιν ἀναιρεθείην. καὶ τοῦτ' αὐτοῖς ὑπηρέτησε Θρασύλοχος ὁ Ἀναγυράσιος· ᾧ τούτων οὐδὲν ἐνθυμηθεὶς ἀντέδωκα μέν, ἀπέκλεισα δ' ὡς διαδικασίας τευξόμενος· οὐ τυχὼν δὲ ταύτης, τῶν χρόνων ὑπογύων ὄντων, ἵνα μὴ στερηθῶ τῶν δικῶν, ἀπέτεισα τὴν λειτουργίαν ὑποθεὶς τὴν οἰκίαν καὶ τἀμαυτοῦ πάντα, βουλόμενος εἰς ὑμᾶς εἰσελθεῖν τὰς πρὸς τουτουσὶ δίκας.

Shortly before the trial of my cases against them [i.e. Aphobos and the other guardians] they got up an *antidosis* against me. The purpose was that, if I agreed to the exchange, I should be unable to litigate against them, because these cases too would become the property of the man who proposed to exchange with me; while if I did not do so [i.e. did not agree to the exchange], the purpose was that, by paying for a liturgy out of scanty resources, I should be utterly ruined. The man who obliged them in this way was Thrasylokhos of Anagyrous. Without thinking at all of these consequences I agreed to the exchange with him, but I excluded him [sc. from my property] because I wanted to obtain a *diadikasia*. But not obtaining one, since the dates [sc. of the trials of the guardians] were imminent, to avoid being deprived of my cases I paid the cost of the liturgy by mortgaging my house and everything I had, wanting to get my cases against these men tried in your court.

Although the two accounts (28.17 and **78–80**) give different details, there is no inconsistency between them and no reason to regard either as inaccurate. The ulterior purpose of the *antidosis* must have been as stated in 28.17, to free Aphobos and the other guardians from prosecution by D.; D.'s claims on his guardians counted as part of his estate, so that if he exchanged property with Thrasylokhos those claims would also pass to Thrasylokhos, who would then drop the prosecutions. Thus the guardians would keep D.'s estate, while D. would end up with only Thrasylokhos' property, which was presumably much smaller. Thrasylokhos would be rewarded by the guardians with some personal or financial or political favour; παρεσκευάσαν (28.17) implies improper inducement without making clear what form it took. Meidias' part in the scheme is not mentioned in 28.17, where it is irrelevant to the prosecution of Aphobos, but in the oration against Meidias himself he becomes the arch-plotter: **78** τὰ δ' ἔργα πάντ' ἦν καὶ τὰ πραττόμενα ὑπὸ τούτου. It may be true that he, as a friend of Aphobos and brother of Thrasylokhos, suggested the scheme and took the lead in carrying it out.

When D. was faced with the challenge, he must have had to weigh up conflicting considerations. If he could recover his whole estate from his guardians, he was rich, and would do better to accept the liturgy than forfeit his property; but his prosecutions of his guardians had not yet come to trial, and he could not be sure of winning them, or, even if he did win, of recovering all that was due to him. Without his estate he had little money to pay for a liturgy. He therefore determined to try to avoid both the exchange and the liturgy—for that is how we should interpret the most difficult sentence of 28.17, ἀντέδωκα μέν, ἀπέκλεισα δ' ὡς διαδικασίας τευξόμενος. The Phainippos case (D. 42) shows that, if a man agreed to an exchange of property but failed to carry it out, his opponent could take him to court for a *diadikasia*. In such a trial the jury decided which man was the richer and ordered him to perform the liturgy; cf. Isok. 15.5 ἔγνωσαν ἐμὴν εἶναι τὴν λειτουργίαν. That is what D. tried to contrive: he agreed to exchange property (ἀντέδωκα: see the first paragraph of this note for evidence that this verb means 'agreed to exchange' rather than 'exchanged'), but then refused to let Thrasylokhos enter his property (ἀπέκλεισα); he expected that Thrasylokhos would go to the strategoi and demand a *diadikasia*, and that the jury would decide that D., lacking his estate, was the poorer man and Thrasylokhos must perform the liturgy. This may indeed have been the normal way for a man to proceed if he was faced with an *antidosis* which he considered to be unreasonable; for D. expects a jury to understand his plan from that very brief sentence in 28.17, as if it needed no further explanation or apology.

Unfortunately, the plan did not work. Thrasylokhos and Meidias did not want an adjudication which might well have been in D.'s favour. What they wanted was that the exchange of property should go ahead. So they proceeded to D.'s house, and on finding it closed against them they did not go away to make application to the strategoi but simply broke down the doors. They assumed that they were entitled to enter because D. had formally agreed to exchange property with them (**79** ὡς αὑτῶν ἤδη γιγνομένας κατὰ τὴν ἀντίδοσιν); and D. appears to have conceded this, for although he later prosecuted Meidias for slander on this occasion (**81**) he does not seem to have prosecuted him for forcible entry or damage to his house. They also declared, assuming that they and not D. now had the right to claim D.'s estate from his guardians, that they would not go ahead with prosecution of the guardians (**79** τὰς δίκας ὡς αὑτῶν οὔσας ἀφίεσαν τοῖς ἐπιτρόποις). Possibly it was not until this stage was reached that D. realized that they were not concerned merely with the liturgy but were in collusion with the guardians; τούτων οὐδὲν ἐνθυμηθείς (28.17) means that he did not think of this at the start. From both his accounts of the next stage (28.17 and **80**) two things are clear. (*a*) His formal agreement to exchange property (28.17 ἀντέδωκα), even though he excluded Thrasylokhos from his house, had the effect of transferring to Thrasylokhos immediately all his legal rights to property, and this transfer could be invalidated only if either a *diadikasia* were held or D. performed the liturgy after all; this is shown by 28.17, 'not obtaining one [sc. a *diadikasia*], since the dates [sc. of the trials of the guardians] were imminent, to avoid being deprived of my cases I paid the cost of the liturgy'. (*b*) A *diadikasia* could not be demanded by a man who failed to carry out an exchange of property, but only by a man who claimed that his opponent had failed to carry it out; this is clear because, if D. had been entitled to insist on a *diadikasia*, he would obviously have done so. So, caught by this dilemma, D. resolved to undertake the liturgy. He mortgaged his house (28.17) to raise the necessary sum of 20 mnai (**80**); and once he had paid for the liturgy, all his legal rights in his own estate reverted to him and he was able to proceed with the prosecution of his guardians.

ἐκεῖνος: Thrasylokhos. D. expects the jurors to know who Meidias' brother is, but he adds the name later in the sentence in case there are any who do not.

79. τῶν οἰκημάτων: a number of rooms, and thus a fairly large house by the standards of the time. We may imagine an enclosed yard with several rooms opening off it; compare the Dema house described in *BSA* 57 (1962) 75–114.

ἔνδον: 'in the house' because she had not yet left it in marriage.

παιδὸς: adjectival with *κόρης*, 'a little girl'; cf. Ar. *Lys.* 595 *παῖδα κόρην γεγάμηκεν*. Actually she was only two years younger than D. (27.4) and thus about eighteen at the time of this incident.

οὐ γὰρ ἔγωγε ...: perhaps the language used by Meidias and Thrasylokhos really was improper, but D. makes the most of it by his paralipsis: he implies that he himself (*ἔγωγε* is emphatic) is more clean-mouthed than they, and he lets the jury assume the worst about them.

πάντας ἡμᾶς includes the slaves.

ῥητὰ καὶ ἄρρητα: so also 18.122 *βοᾷς ῥητὰ καὶ ἄρρητ' ὀνομάζων*, 22.61 *ῥητὰ καὶ ἄρρητα κακά*.

ἐξεῖπον: *ἐξ-* is made appropriate by what precedes; they spoke out what they ought to have kept in. Emendation is therefore not needed (*εἶπον* Wolf, *ἑξῆς εἶπον* Blass, *ἐπεῖπον* Weil).

ἀφίεσαν: a term for abandoning a prosecution or claim, used also in 36.25, 37.1, 59.30, etc.

80. ἔρημος ὢν: having no father or other elder to advise him; cf. **96** *ἐρημίαν*.

νέος κομιδῇ: **78** n. *μειρακύλλιον ὢν κομιδῇ*.

οὐχ ὅσα ...: D. naively assumed that, if he won the cases against his guardians, he would get back the whole amount of the estate which his father had left him. How much he did get back is uncertain, but Plu. *Dem.* 6.1 says that it was not even a small fraction; cf. Davies *Families* 133.

δίδωμι: D. sometimes uses the historic present for a single verb in a passage in which the other verbs are in past tenses. Cf. **83** *γίγνεται*.

τὴν τριηραρχίαν ἦσαν μεμισθωκότες: anyone who was appointed trierarch but did not want the trouble of maintaining and commanding the ship could, by private arrangement, pay someone else to perform all the duties on his behalf. An account of the duties may be read in the oration *On the trierarchic crown* (D. 51), which is an attack on some men who have let a contract for a trierarchy (51.7 *μεμισθώκασι τὴν λειτουργίαν*) made by another trierarch who has not done so. The price would be settled by mutual agreement, and so might vary from case to case; it would have to cover not only the expected outlay on repairs and equipment, but also some payment for the services of the contractor, who would have to go to sea in command of the ship. The sum of 20 mnai (one third of a talent) which Thrasylokhos and Meidias had paid, and for which D. now reimbursed them, was probably not excessive for a syntrierarchy; a few years later one talent was apparently the going rate for a full trierarchy (**155**). It is presumably on that evidence that Σ here (262 Dilts) assumes that Thrasylokhos had been appointed to one third of a trierarchy, not one half, but that is unlikely to be true; there is no other evidence for division of a trierarchy into three.

81. τῆς κακηγορίας: 'for the slander'. The article shows that the reference is to the incident already mentioned (**79**). *κακηγορία*, as a legal offence, was not just any use of bad language; the law listed particular expressions which were forbidden. Our information about it comes mainly from the oration *Against Theomnestos* (Lys. 10), which is a speech for the prosecution in such a case. It shows that the prohibited expressions included *ἀνδροφόνος*, *πατραλοίας*, *μητραλοίας*, and *τὴν ἀσπίδα ἀποβεβληκέναι*. (Lys. 10.6–9). Disparaging any citizen's work in the Agora was also forbidden (D. 57.30). Such allegations were actionable only if false; showing that a statement was true was sufficient defence (23.50, Lys. 10.30). The case was a private *dike*, and the penalty on conviction was a payment of 500 dr. to the prosecutor (Lys. 10.12, Isok. 20.3). D. claimed 1000 dr. from Meidias (**88**, **90**); possibly the amount of the fixed penalty had been increased since the time of Lys. 10, but more likely D. claimed double compensation because he alleged that Meidias had made two slanderous statements or had slandered two persons, his mother and his sister. Cf. Lipsius *Recht* 646–51, MacDowell *Law* 126–9.

ἐρήμην: 'deserted', meaning 'I won the case by default'. *ἔρημος* is the legal term meaning that a trial or arbitration is formally decided against one party because he fails to attend the hearing. Elsewhere in this speech we find the form *ἔρημον* as feminine (**85**, **87**, **92**, in all of which *δίκην* or *δίαιταν* is understood), but *ἐρήμην* is used in other speeches (e.g. 33.20, 39.18, 55.6) and so need not be emended here. How Meidias lost this case is described in detail in **83–101**, from which we see that he did not admit that his conviction was valid.

ἀπήντα: the usual verb for attending a hearing, used also in **84**, **90**, **92**.

λαβὼν ... καὶ ἔχων: 'getting and having him'. There is little difference in meaning between the two participles; strictly the first refers to the moment at which the payment became overdue and the second to the ensuing period in which it remained unpaid, but the distinction hardly seems worth making here. So the conjecture ἐξὸν proposed by Herwerden *Mn.* II 1 (1873) 310 may possibly be correct, giving the sense 'though I could have done so, I never touched any of his property'.

ὑπερήμερον: the legal term for an overdue debtor; so also in **10**, **11**, **89**. It is not known what length of time was allowed to a person who was condemned in a private case to make a payment to his opponent. Possibly it was thirty days; cf. 42.13, with my comment on that passage in **47** n. *τριάκοντα ἡμερῶν*. In 47.49–51 we read of an instance in which the time-limit was extended by mutual agreement. Once the time-limit had expired, the creditor could seize from his opponent either the money owed or items of property of equal value; cf. the law of Euegoros (**10**) which prohibits such seizures during certain festivals. D. now implies that he deserves approval for not resorting to self-help in this way. Instead he used a further legal procedure, *δίκη ἐξούλης*. For that procedure, **44** n. *ἐξούλην*.

πάλιν with *λαχών*, not with *ἐξούλης*: 'bringing another case against him, ⟨this time⟩ for ejectment'.

οὐδέπω καὶ τήμερον: the same phrase occurs in **91**, **157**, 30.33, 37.46, 45.76. Since the occasion of the alleged slander in 364/3 about seventeen years had passed. D. cannot really have been pressing the case for all that time. But there must have been no time-limit for claiming money awarded by a court, so that he can speak of the matter as not yet closed. Meidias' devices for causing delay doubtless included *ὑπωμοσίαι* and *παραγραφαί* (**84** n.).

τῇ δίκῃ, τοῖς νόμοις: asyndeton for a series of only two items is common when a word is repeated, e.g. **72** *ταῦτα κινεῖ, ταῦτ' ἐξίστησιν*. Without such repetition it is less common, but is sometimes used for two items which are considered to go together naturally as a pair. Other instances in D. are **17** *φράττων, προσηλῶν*, **91** *ἄνω κάτω* (where A inserts *καὶ*), **135** *ἀπειλεῖς πᾶσιν, ἐλαύνεις πάντας* (helped by the repetition of 'all', though in a different case and not as the first word of the phrase), **177** *τοὺς νόμους, τὸν ὅρκον*, 3.31 *χρήματα, συμμάχους* (where A inserts *καὶ*), 18.67 *τὴν χεῖρα, τὸ σκέλος*, 18.94 *δόξαν, εὔνοιαν* (where A adds a third item, *τιμὴν*), 18.100 *ναυμαχίας, ἐξόδους*, 18.234 *ὁπλίτην δ', ἱππέα* (where A inserts *ἢ*), 19.190 *σπονδῶν, ἱερῶν*, 19.220 *Εὔβοιαν, Ὠρωπόν*, 19.299 *ἔξωθεν ..., ἔνδοθεν ...*, 47.30 *τῷ καιρῷ, τῇ χρείᾳ*. **61** *ἀναλίσκοντας, ἀγωνιῶντας* is a doubtful instance; see note there. Denniston *Style* 105 understates the frequency of this phenomenon.

εἰς τοὺς φυλέτας: a renewed reference to Meidias' interference with D.'s chorus at the Dionysia; cf. the beginning of **19**.

82. ὡς οὖν ταῦτ' ἀληθῆ λέγω: a standard phrase for introducing witnesses; cf. **93**, **107**, **119**, **121**, **167**, **174**. Grammatically it may be regarded as *oratio obliqua* subordinate to *μάρτυρας*: 'witnesses that what I say is true'. Σ (271 Dilts) expands the statement thus: *ὅτι ὕβρισε κἀμὲ καὶ τὴν μητέρα καὶ τὴν ἀδελφήν, ὅτι θύρας κατέσχισαν, ὅτι τὰς εἴκοσι μνᾶς ἔλαβον αὐτοὶ μηδὲν ἀναλώσαντες*. This guess at the content of the testimony may be wrong, but what it does prove, as Christ *Abh. Bay. Akad.* 16.3 (1882) 197 points out, is that the scholiast did not have in his text of D. the document which now appears in our mss.

Like the other witnesses' statements found in this oration, this document should be rejected as spurious. *οἴδαμεν* (for *ἴσμεν*) and *κρίσιν λελογχότα* (for *δίκην εἰληχότα*) are non-Attic forms, and an Athenian would probably have said *τῇ δίκῃ* where the text offers *τῇ κρίσει*. So the document should be regarded as a composition of a later period; the names given to the witnesses are fictitious, and *ἔτη ὀκτώ* is not to be used as evidence for the date of D.'s *δίκη ἐξούλης*.

Θορίκιος: the mss. give *θεωρίσκος*, but a Hellenistic or Roman writer is not likely to have thought that that was an Athenian demotic; the error should be attributed to a later copyist.

83–101. *The arbitration in the case of slander and Meidias' attack on the arbitrator.*

83. ὑπερηφανίαν: a term of moral condemnation, but (like *ἀσέλγεια* and unlike *ὕβρις*) not a legal term. The earliest instance of the adjective *ὑπερήφανος* refers to those sons of Heaven and Earth who had fifty heads and a hundred arms (Hes. *Th.* 149), which suggests the possibility that its original meaning was 'monstrous' in a physical sense. Not inconsistent with that would be its later use as a colloquial term of praise, something like English 'terrific' (in comedy, Alexis 261.6, Philippides 27; probably we should interpret it in this way in Pl. *Phd.* 96a, *Symp.* 217e, *Gorg.* 511d). But in the great majority of instances both the noun and the adjective express strong disapproval of a man's behaviour. *ὑπερηφανία* is often associated with *ὕβρις*, as in **83** and **195** (cf. *Iliad* 11.694–5, where the form is *ὑπερηφανέοντες*, and also Solon fr. 4.34–7 West (ap. D. 19.255), Pind. *Pyth.* 2.28, D. 24.121, Pl. *Symp.* 219c, Arist. *Rhet.* 1390b 33), or with *ἀσέλγεια*, as in **137** (cf. Isok. 7.53; also the spurious letter of Philip, D. 12.15); and it is not easily distinguished from them. Like them it is commonly caused by good fortune (Isok. 12.32, 12.196, Arist. *Rhet.* 1391a 33, Men. fr. 252 Kock = 218 Koerte) and accompanies wealth and luxurious living (**96**, **195**, Pl. *Laws* 691a, Arist. *Rhet.* 1390b 33), such as having too grand a house (13.30, Pl. *Kritias* 112c) and excessive indulgence in food (Diphilos 32.20 Kock = 31.20 Kassel and Austin) or sex (Ais. 1.70). It may include boasting of one's own ancestry (Pl. *Tht.* 175b) and over-confidence (18.252), such as the Persians displayed when they invaded Greece (Pl. *Menex.* 240d, Isok. 4.89). But perhaps the core of it in fourth-century usage is the scorn of ordinary people by one who considers himself superior, who is *μισόδημος* and *μισάνθρωπος* (Isok. 15.131). Thus D. contrasts *ὑπερηφανία* with *τὸ τῶν πολλῶν εἷς εἶναι* (**96**); it is the attitude of one to whom everyone else is *καθάρματα καὶ πτωχοὶ καὶ οὐδ' ἄνθρωποι* (**198–9**). If we can distinguish it from *ὕβρις* and *ἀσέλγεια*, the difference may be that *ὑπερηφανία* does not in itself include action; it is 'arrogance' rather than 'aggressiveness' or 'bullying'. At the end of the classical period Theophrastos makes it the subject of one of his character sketches (*Char.* 24); for him, it seems to be no more than giving one's own personal convenience priority over other people's. But in D. the sense is clearly stronger than that.

ταύτης λέγω ...: a parenthesis inserted to make clear that, of the two cases mentioned in **81**, D. here means the case of slander, not the case of ejectment. One might expect *ταύτην*, but instead this word retains the case which *τῆς δίκης* has in the main clause; cf. 8.24 *τούτων τῶν τὴν Ἀσίαν οἰκούντων λέγω*, 19.152 *λέγω δὲ Φωκέων καὶ Πυλῶν*, 57.24 *λέγω φράτερσι, συγγενέσι, δημόταις, γεννήταις*.

γίγνεται: in this narrative D. several times switches to the historic present for one or two verbs and then back again to a past tense.

διαιτητὴς: 'arbitrator'. A prosecution for slander, like many kinds of private case, was submitted to the four judges of the defendant's tribe (*φυλή*); these were the officials sometimes called collectively 'the Forty'. Unless the case was a minor one involving no more than 10 dr., the tribe judges had to refer it to a public arbitrator. All Athenian citizens served as arbitrators in their sixtieth year (the forty-second year after registration in a deme), and cases were assigned to them by lot. After hearing the arguments and evidence, the arbitrator gave a decision, which was final if the prosecutor and the defendant both accepted it; but if either appealed against it, the case went for trial by jury. The system, which was evidently intended to save the time and expense of a trial by jury in cases in which a settlement could be reached without it, is described in *AP* 53; D.'s narrative here is the fullest surviving account of an actual case. For modern discussion see T. D. Goodell *AJP* 12 (1891) 319–26, Lipsius *Recht* 226–33, Bonner and Smith *Justice* 1.346–53, 2.97–116, Harrell *Arbitration*, Harrison *Law* 2.66–8, MacDowell *Law* 206–11, Rhodes *Comm. on AP* 587–96, Ruschenbusch *Symposion 1982* (1985) 31–40.

Στράτων Φαληρεύς: Straton is not otherwise known. Phaleron was a deme of the Aiantis tribe. Meidias belonged to Erekhtheis, and one might have expected arbitrators to hear cases for members of their own tribes. But evidently that was not how the system worked; after all, in any particular year the tribes would not necessarily each produce the same number of either arbitrators or defendants. Cf. 47.12 *ἡ μὲν γὰρ δίαιτα ἐν τῇ ἡλιαίᾳ ἦν, οἱ γὰρ τὴν Οἰνηίδα καὶ τὴν Ἐρεχθηίδα διαιτῶντες ἐνταῦθα κάθηνται*: that passage seems to imply that each arbitrator was allocated to one tribe for the whole year (so Lipsius *Recht* 227, Rhodes *Comm. on AP* 594). If that is right, it appears that Straton was one of the group of arbitrators picked by lot at the beginning of the year to arbitrate cases brought against members of Erekhtheis, and also that the place where the arbitration proceedings were held for this case was the Eliaia. (On the Eliaia, **47** n. *ἡλιαίαν*.)

πένης does not mean that Straton was destitute, but that he, unlike Meidias, had to work for a living. For the meaning of *πένης* (contrasted with *πτωχός*, which does mean 'destitute') cf. Ar. *Wealth* 552–4:

> *πτωχοῦ μὲν γὰρ βίος, ὃν σὺ λέγεις, ζῆν ἐστιν μηδὲν ἔχοντα·*
> *τοῦ δὲ πένητος ζῆν φειδόμενον καὶ τοῖς ἔργοις προσέχοντα,*
> *περιγίγνεσθαι δ' αὐτῷ μηδέν, μὴ μέντοι μηδ' ἐπιλείπειν.*

Cf. M. M. Markle in *Crux* 267–71. However, Markle is wrong when he says (*Crux* 287–8 n. 40) that Straton had become poor through the persecution of Meidias. **83** means that he was already poor when he was appointed arbitrator. Note the implication of *μέν* that a man who is poor normally deserves less respect than one who is not; so also in **95**. Although exceptions may occur, the normal expectation is that poverty causes a person to behave worse; cf. E. *El.* 375–6 *ἀλλ' ἔχει νόσον πενία, διδάσκει δ' ἄνδρα τῇ χρείᾳ κακόν*, and Dover *Morality* 109–10.

ἀπράγμων: not taking an active part in public affairs. In forensic speeches this normally means 'reluctant to litigate'; cf. **141**, where *ἀπραγμοσύνη* is a reason for failure to prosecute. It is generally a compliment, linked with such favourable terms as *μέτριος*: cf. 42.12 *καὶ μέτριον καὶ ἀπράγμονος εἶναι πολίτου μὴ εὐθὺς ἐπὶ κεφαλὴν εἰς τὸ δικαστήριον βαδίζειν*, 36.53, 40.32, 47.82, 54.24. For the antonym *φιλοπραγμοσύνη* cf. **137**. These terms have been much discussed; see especially V. Ehrenberg *JHS* 67 (1947) 46–67, Gomme *HCT* 2.121–2, Dover *Morality* 187–90. But **83** is a not quite typical instance: the structure of the sentence with *μέν* and *δέ* shows that *ἀπράγμων*, like *πένης*, is here a term of disparagement. Ignorance of legal matters is a weakness in a man required to serve as an arbitrator. Another passage in which *ἀπραγμοσύνη* is regarded as a weakness, though a venial one, is 58.24, where it is pointed out that ignorance of the law may lead one to transgress it unintentionally.

ἄλλως δ' οὐ πονηρός: 'but in other ways not bad', i.e. having no other weaknesses. Goodwin takes *ἄλλως* as 'moreover', 'besides'; but that is probably wrong, because it would require us to take *πένης* as a favourable term.

ὅπερ ...: in saying that Straton's virtue has destroyed him, D. means no more than that his verdict in D.'s favour caused Meidias to accuse him, as recounted in **84–7**.

τὸν ταλαίπωρον: 'the wretched man', expressing D.'s sympathy for Straton; likewise **104** *τῷ ἀθλίῳ καὶ ταλαιπώρῳ*. In these phrases the definite article does not perform its normal identifying function but is just exclamatory in effect.

84. οὗτος refers back, and so makes a connective particle unnecessary.

ἐπειδή ποθ': 'when eventually', hinting at a considerable lapse of time. Cf. 24.13 *ἐπειδή ποτ' ἐπαύσανθ' οὗτοι βοῶντες.*

ἡ κυρία: 'the appointed day', the day fixed by the arbitrator for giving his verdict. It seems that an arbitrator would normally hold a preliminary meeting with the disputants (cf. 54.29 *ἐν τῇ πρώτῃ συνόδῳ πρὸς τῷ διαιτητῇ*), and then fix a day for the formal arbitration, on which the disputants presented their arguments and evidence and the arbitrator pronounced his verdict. A disputant who could not attend on that day could apply for a postponement; otherwise, as we see from the rest of this sentence, the arbitrator could postpone it only with the agreement of both disputants.

διεξεληλύθει: 'had run their course', 'had been completed'. For this sense of the verb cf. Pl. *Laws* 805b *τὸν μὲν λόγον ἐᾶσαι διεξελθεῖν, εὖ διελθόντος δέ* ... The use of the pluperfect shows that the reference is not to proceedings on the appointed day (*ἡ κυρία*) but to proceedings which had been completed before that day arrived. (Thus Harrell *Arbitration* 30 is wrong in saying that here 'we read of *παραγραφή* and *ὑπωμοσία* before an arbitrator, brought apparently on the very day of award'.) D. means that Meidias expected the verdict to go against him, and therefore had tried every possible method of putting it off.

τὰκ τῶν νόμων: 'the procedures laid down in the laws'. Cf. 24.28 *τὸν ἐκ τῶν νόμων χρόνον.*

ὑπωμοσίαι: 'oaths on his behalf'. If a litigant was unable to attend a trial or arbitration, he could apply for a postponement, sending a relative or friend to swear an oath that he was ill or absent from Attika. Such an oath is called *hypomosia*, in which *ὑπ-* is best interpreted as meaning 'by proxy' or 'deputizing'. The middle verb *ὑπόμνυμαι* is used both with the oath-taker as subject, meaning 'swear an oath on another's behalf', e.g. 48.25 *ὑπωμοσάμεθα ἡμεῖς τουτονὶ Ὀλυμπιόδωρον δημοσίᾳ ἀπεῖναι στρατευόμενον*, and with the litigant as subject, meaning 'have an oath sworn on one's behalf', e.g. 47.45 *ὁ* *μὲν Θεόφημος παρεγράφετο καὶ ὑπώμνυτο*. If the opponent in the case wished to oppose postponement, he could take a counter-oath denying that the litigant was ill or absent from Attika; thus 48.25 (just quoted) goes on *ὑπομοθέντος δὲ τούτου ἀνθυπωμόσαντο οἱ ἀντίδικοι*. For other instances of *hypomosiai* see 39.37, 47.39, 58.43, Hyp. *Eux*. 7, fr. 202 Kenyon, Jensen = D.7 Burtt.

παραγραφαί: the usual meaning of *paragraphe* in Athenian law is 'counter-prosecution', a procedure by which a defendant tried to bar an action against him by prosecuting the prosecutor for proceeding in a way contrary to law; D. 32–8 are orations composed for such cases. But here it is generally thought that we have something quite distinct, a procedure for applying for postponement of an arbitration, the *paragraphe* being simply the written statement of the plea of illness or absence abroad which was the subject of a *hypomosia*. This, the view of Schoemann, is most fully expounded by Bonner and Smith *Justice* 2.91–6; it is adopted also by Lipsius *Recht* 836 n. 22, Harrell *Arbitration* 30, H. J. Wolff *Die attische Paragraphe* (1966) 8 n. 5, MacDowell *Law* 208. But it is rejected by G. M. Calhoun *CP* 14 (1919) 20–8, who holds that *paragraphe* here has its usual sense of counter-prosecution; and Ruschenbusch *Symposion 1982* (1985) 31–4 follows Calhoun, insofar as he accepts that a *paragraphe* in the sense of counter-prosecution could be brought before an arbitrator, though he does not discuss the present passage. The other evidence is as follows:

A. In D. 47 the speaker (a trierarch) relates that he and Theophemos came to blows and each prosecuted the other for battery (*αἴκεια*). Both cases went for arbitration. In the case in which the speaker prosecuted Theophemos, *ἐπειδὴ ἡ ἀπόφασις ἦν τῆς δίκης, ὁ μὲν Θεόφημος παρεγράφετο καὶ ὑπώμνυτο* (47.45); but in the case in which Theophemos prosecuted the speaker, *οὐ παραγραφομένου ἐμοῦ οὐδ' ὑπομνυμένου* the case went on to a jury (47.39; cf. 47.45 *ἐγὼ δὲ πιστεύων ἐμαυτῷ μηδὲν ἀδικεῖν εἰσῄειν εἰς ὑμᾶς*). It is clear that, according to the speaker, Theophemos used delaying tactics to which the speaker himself scorned to resort. But *ἐπειδὴ ἡ ἀπόφασις ἦν* is a difficult phrase: *ἀπόφασις* means the arbitrator's declaration of his verdict (cf. 33.21, and *ἀποφαίνειν* in **85**, etc.), but what time is 'when the declaration of the verdict *was*'? Strictly it ought to mean after the verdict had been declared, but the phrase seems hardly tolerable in any sense, and may need some emendation such as ⟨*ἐγγὺς*⟩ *ἦν*. If it means that Theophemos began his delaying tactics before the arbitrator gave his verdict (as is generally assumed), we can accept that those tactics were the same as those used by Meidias in **84**, but we are no nearer knowing what they were.

B. In *Lex. Cant*. under *μὴ οὖσα δίκη* we find a report of what Demetrios of Phaleron said on the matter (*F. Gr. Hist*. 228 F13):

Δημήτριος ὁ Φαληρεὺς ... λέγει ... ἐνίους δὲ ἀσθενὲς τὸ δίκαιον ἔχοντας καὶ δεδοικότας τὴν καταδίαιταν χρόνους ἐμβάλλειν καὶ σκήψεις οἵας δοκεῖν εἶναι εὐλόγους, καὶ τὸ μὲν πρῶτον παραγράφεσθαι, εἶτα ὑπόμνυσθαι νόσον ἢ ἀποδημίαν, καὶ τελευτῶντας ἐπὶ τὴν κυρίαν τῆς διαίτης ἡμέραν οὐκ ἀπαντῶντας [so Dobree: *ἀπάντων τὰς* cod.] *ὅπῃ ἂν δύνωνται ἀντιλαγχάνειν τὴν μὴ οὖσαν τῷ ἑλόντι*. This clearly indicates three separate delaying tactics: (*i*) *παραγράφεσθαι*, whatever that was, (*ii*) *ὑπόμνυσθαι*, (*iii*) non-attendance on the day of the verdict. The use of *τὸ μὲν πρῶτον ... εἶτα* is not consistent with the view that *paragraphe* and *hypomosia* were inseparable parts of a single application for postponement.

C. Polydeukes 8.60: *ὁπόταν τις παρὰ διαιτηταῖς παραγραψάμενος καὶ* [so ABL: *ἢ* FS] *ὑπομοσάμενος νόσον ἢ ἀποδημίαν εἰς τὴν κυρίαν μὴ ἀπαντήσας ἐρήμην ὄφλῃ, ἐξῆν ἐντὸς δέκα ἡμερῶν τὴν μὴ οὖσαν ἀντιλαχεῖν*. The disagreement of mss. over the crucial conjunction makes it impossible to be sure whether the author regarded *paragraphe* and *hypomosia* as alternative procedures or two parts of a single procedure. But anyway he is confused. He thinks that the man who had an arbitration decided against him in his absence (*ἐρήμην ὄφλῃ*) and wished to apply to have it set aside on the ground that his absence had been unavoidable (*τὴν μὴ οὖσαν ἀντιλαχεῖν*: for this procedure see **86**) could do so only if he had previously presented a *paragraphe* and/or a *hypomosia*; but that cannot be right, because attendance might be prevented by a sudden cause when no application for postponement had been made, and in fact the oath required to support such an application was not a previous *hypomosia* but came after the action of *ἀντιλαχεῖν* (**86**). It seems that Polydeukes must have misunderstood his source of information (which may indeed have been none other than the passage of Demetrios reported in *Lex. Cant.*), and no reliance should be placed on his account.

D. Σ here (281b Dilts): *πολλάκις γὰρ οἱ ἀντίδικοι {οὐ} παρεγράφοντο "οὐ πρὸς διαιτητήν με κρίνεσθαι δεῖ, ἀλλὰ πρὸς ἄρχοντα ἢ θεσμοθέτην, οὐ σὲ δεῖ κατηγορεῖν, ἀλλ' ἕτερον, οὐ νῦν, ἀλλ' αὖθις" καὶ τὰ τοιαῦτα*. This is virtually the same as Calhoun's interpretation, taking *paragraphe* as an objection to the prosecutor's procedure. However, it may be based merely on a reading of the *paragraphe* orations (D. 32–8); it is not certain that the scholiast had other evidence about Athenian law not available to us.

I conclude that *A* and *C* get us nowhere, while *B* and (less reliably) *D* are rather against Schoemann's view. Also against it perhaps is *πάντα* in **84**; although D. may well be using rhetorical exaggeration in this passage, *πάντα* does seem unsuitable if the words *ὑπωμοσίαι καὶ παραγραφαί* refer to only a single kind of procedure. On the other hand Bonner and Smith object that the plural *παραγραφαί* is fatal to

Calhoun's view, because a defendant objecting to the legality of a prosecution brought only one *paragraphe*, however many objections it included. However, I agree with Calhoun's suggestion (rejected by Bonner and Smith) that D. may here be using a scornful rhetorical plural; and another possible explanation is that before the arbitrator, where proceedings would be less formal than before a jury, Meidias did raise different objections on different days. Although certainty is impossible, I am now inclined to think (departing from my earlier view) that we should regard *paragraphe* as a separate delaying tactic from *hypomosia*, and that there is no strong reason why it should not have its usual sense of 'objection to the legality of the prosecution' or 'counter-prosecution', as Calhoun maintains. If that is right, Meidias submitted to the arbitrator one or more written objections to the legal procedure which D. was using against him, the arbitrator was persuaded by D. to reject them, and Meidias did not insist on appealing to a jury about them.

ἐπισχεῖν ἐδεῖτό μου: the subject is *ὁ Στράτων*. Formally the arbitrator himself must have been the person to authorize an adjournment, but the wording here implies that this could not be done without the prosecutor's agreement. *ἐπισχεῖν* evidently means a longer delay than adjournment to the following day, which Straton suggested next as a lesser concession. But it does not mean abandoning the case altogether (Vince's translation 'abandon' is wrong); cf. **12** *τὸ λαμβάνειν δίκην ἐπέσχετε ταύτας τὰς ἡμέρας*. For another reference to the possibility of postponement of the verdict in an arbitration, cf. 47.14 *ἀναβαλέσθαι κελεύων τὴν δίαιταν εἰς τὴν ὑστέραν σύνοδον*.

τῆς δ' ὥρας ... ὀψέ: 'late in the day'. Cf. X. *Hell.* 2.1.23 *τῆς ἡμέρας ὀψὲ ἦν*, Makhon 441 (Gow) *ὀψὲ τῆς ὥρας*. It may have been standard practice for an arbitrator to wait until sunset for an absent disputant to appear (so Harrell *Arbitration* 33); cf. 49.19, where a witness fails to appear and the arbitrator gives his verdict *ἑσπέρας ἤδη οὔσης*.

κατεδιῄτησεν: *καταδιαιτάω* is the verb for 'convict in an arbitration', *ἀποδιαιτάω* (**85**, etc.) for 'acquit in an arbitration'. In past tenses of these verbs the mss. vary between the double and the single augment, but the double is commoner and was probably D.'s normal usage. The verdict had to be against Meidias because he failed to attend. An arbitrator, besides pronouncing his verdict orally, put it in writing (*AP* 53.2).

85. ἑσπέρας οὔσης καὶ σκότους: a picturesque detail; cf. p. 33.

τῶν ἀρχόντων: 'the ⟨relevant⟩ magistrates', the tribe judges. For *οἱ ἄρχοντες* referring to the tribe judges cf. 45.87. It is not known where their office was, nor even whether all the Forty used a single office or the four judges for each tribe had a separate office. The context shows that, when an arbitrator gave a verdict, he had to deliver it to the tribe judges immediately. The litigants did not have to accompany him (for D. goes on to say that he got information about these events from a bystander).

τὴν ἔρημον δεδωκότα: 'having delivered ⟨to the tribe judges⟩ the ⟨verdict pronounced against Meidias⟩ in absence'; **81** n. *ἐρήμην*.

οἷός τε ἦν: 'he was able to', meaning 'he had the impudence to'. *τε* was originally omitted in S, and is therefore removed from the text by Blass; but *τὸ μὲν οὖν πρῶτον οἷος ἦν* ..., 'at first he was the sort of man to ...', is not suitable, because it implies that Meidias' character was later reformed.

πείθειν: 'to try to persuade', 'to urge', the common inceptive sense of the present tense.

ἀποφαίνειν: 'to declare'. This is the correct verb for an arbitrator declaring his verdict, and is used in **96**; cf. 33.19–20, 33.33, 54.27. So the alternative reading *ἀποφέρειν* (SFYP) should be rejected.

μεταγράφειν: to change the written record of Straton's verdict.

αὐτοῖς includes Straton as well as the magistrates, as **96** shows.

86. οὐδετέρους: neither Straton nor the tribe judges.

τί ποιεῖ;: a striking use of a rhetorical question to enliven a narrative.

καὶ: emphasizing, not connective: 'just look at ...!'

ἀντιλαχὼν: when an arbitrator gave judgement against a disputant in his absence (*ἔρημος*), the disputant could within ten days apply to the tribe judges to have the judgement set aside (*ἀντιλαγχάνειν*). He had to take an oath that his absence had been unavoidable, and the case was called 'non-existent' (*μὴ οὖσα*) and went afresh to arbitration (whether to the same arbitrator is not known). Cf. **90** *τὴν μὴ οὖσαν ἀντιλαχεῖν*, 39.38, Lys. 32.2, and the passages from *Lex. Cant.* and Pol. 8.60 already quoted (in note on **84** *παραγραφαί*). Polydeukes proceeds: *ἐξῆν ἐντὸς δέκα ἡμερῶν τὴν μὴ οὖσαν ἀντιλαχεῖν, καὶ ἡ ἐρήμη ἐλύετο, ὡς ἐξ ἀρχῆς ἐλθεῖν ἐπὶ διαιτητήν. εἰ δὲ μὴ ἕλοι* [so L: *ἕλη* ABFS] *τὴν μὴ οὖσαν, ὀμόσας μὴ ἑκὼν ἐκλιπεῖν τὴν δίαιταν, κύρια τὰ διαιτηθέντα ἐγίνετο· ὅθεν ἐγγυητὰς καθίστασαν τοῦ ἐκτίσματος.* The latter sentence is diffucult. *ἕλοι* should mean 'obtain a conviction' and is not applicable to a defendant. Perhaps we should read *τελοίη*: 'If he did not carry through the "non-existent" case by taking the oath that ...'; that would then be a description of what in fact happened in the case of Meidias. Meidias applied to have the arbitrator's judgement set aside, but then failed to attend at the proper time to take the oath that his absence from the arbitration had been unavoidable (*οὐκ ὤμοσεν*). So he was reported as not having taken the oath (*ἀνώμοτος ἀπηνέχθη*), and the arbitrator's judgement against him automatically became final and binding (*κυρίαν*). We may infer that *τὴν μὴ οὖσαν ἀντιλαγχάνειν* was the only kind of appeal permitted against an arbitrator's judgement given against a disputant in his absence (*ἔρημος*); appeal to a jury was not allowed. The last five words of Pol. 8.60 mention the appointment of guarantors (*ἐγγυηταί*) but fail to make clear at what stage that was done; D. says nothing to suggest that Meidias did it.

φυλάξας: 'awaiting'. D. implies that Meidias could have accused Straton earlier, but deliberately refrained from doing so in order to catch him unawares and accuse him on a day when he was absent.

τὴν τελευταίαν ἡμέραν τῶν διαιτητῶν: the last day on which the arbitrators met, not necessarily the last day of the year. The words *τὴν τοῦ Θαργηλιῶνος ἢ τοῦ Σκιροφοριῶνος γιγνομένην* are clearly a scholiast's alternative suggestions about when the last day may have been; they should not be attributed to D., who must have known when the last day was. (Cf. Lipsius *Recht* 232 n. 46. Blass prefers to delete only *Θαργηλιῶνος ἢ τοῦ* and *γιγνομένην*, but that is unlikely to be right, because if D. had written *τὴν τοῦ Σκιροφοριῶνος* no gloss would have been wanted. Spalding retains the whole phrase and interprets it as a reference to the *ἕνη καὶ νέα*, a day belonging to both the old and the new month; but that is wrong, both because it would require the conjunction to be *καί* instead of *ἢ* and because the *ἕνη καὶ νέα* in fact was regarded as belonging to the old month only, e.g. *IG* 2^2 916.10.) Presumably the main business at most of the arbitrators' meetings was to assign cases to them for arbitration, but it would be no use assigning a case to an arbitrator so near the end of the year that he had no hope of completing it before he ceased to hold the office. Hearing an accusation against an arbitrator would be an additional item, not arising at every meeting. D.'s comments make sense if we assume that there was a meeting every month, but the meeting in the twelfth month (Skirophorion) was badly attended because new cases were not assigned to arbitrators in that month and consequently it was usually just formal. Σ (297 Dilts) considers that the arbitrators had to undergo *euthynai*, but that should be dismissed as a wrong guess; if it were right, surely all the arbitrators would have attended. (The scholiast is clearly ignorant, for he thinks that Skirophorion is the eleventh month of the year.)

The procedure for accusing an arbitrator is recounted briefly in *AP* 53.6: *ἔστιν δὲ καὶ εἰσαγγέλλειν εἰς τοὺς διαιτητάς, ἐάν τις ἀδικηθῇ ὑπὸ τοῦ διαιτητοῦ, κἄν τινος καταγνῶσιν, ἀτιμοῦσθαι κελεύουσιν οἱ νόμοι· ἔφεσις δ' ἔστι καὶ τούτοις.* This kind of *eisangelia* should be distinguished from two other kinds: *eisangelia* for treason and other serious offences, which was initiated by a denunciation to the Ekklesia or the Boule, and *eisangelia* for maltreatment of an orphan or heiress. It is clear from both **86–7** and *AP* 53.6 that the accusation was received and the verdict given by the whole body of arbitrators for the year. (Emendation of *διαιτητάς* to *δικαστάς* in the text of *AP*, which Harrison *Law* 2.68 is inclined to accept, is shown by **86** to be wrong; cf. Rhodes *Comm. on AP* 595.) Both texts agree also that the penalty for an arbitrator found guilty of misconduct of an arbitration was fixed at disfranchisement, to which **87** adds expulsion from the list of arbitrators. Only *AP* adds that appeal to a jury (*ἔφεσις*) was permitted, but cf. note on **91** *ἐχαρίσασθε*.

D. nowhere says what was the accusation which Meidias made against Straton. Most probably he alleged that Straton had given judgement in the arbitration without informing him of the date on which he intended to do so.

87. τὸν πρυτανεύοντα: the chairman of the arbitrators, probably selected from the arbitrators by lot.

δοῦναι τὴν ψῆφον: 'to put the matter to the vote', a regular expression for the action of a chairman, often found in inscribed decrees, e.g. *IG* 2^2 109b.16, 222.24. Contrast the different sense in **188**.

παρὰ πάντας τοὺς νόμους: 'all the laws' means only 'all the laws about the accusation of arbitrators'; but in what way did Meidias infringe those laws? D. must mean that it was illegal to try someone who was not present and had not been summoned to attend. But the law about arbitrators probably said nothing about a summons because it assumed that all arbitrators would be present anyway at the arbitrators' meetings; so Meidias may have claimed that it was Straton's own fault if he was absent, even though in practice absence from the arbitrators' last meeting was common. It is also possible that the usual procedure was for the accusation to be received at one meeting and the verdict given at the next meeting, but in the present case Meidias would have argued that the verdict must be given at this meeting because it was the last before the arbitrators demitted office.

κλητῆρα οὐδ' ὁντινοῦν ἐπιγραψάμενος: 'writing on the charge-sheet the name of not a single summons-witness'. In an ordinary case the prosecutor had to give the names of two men who, if the defendant failed to attend, could testify that the summons had been delivered to him (e.g. 53.14). But if no summons was required for accusing an arbitrator (see previous note), there would naturally be no summons-witnesses.

κατηγορῶν ἔρημον: 'making an accusation for which the defendant was not present'; **81** n. *ἐρήμην*. Cf. Pl. *Ap.* 18c *ἐρήμην κατηγοροῦντες ἀπολογουμένου οὐδενός*.

οὐδενὸς παρόντος: an exaggerated way of saying that the attendance was small.

ἐκβάλλει καὶ ἀτιμοῖ: 'expelled and disfranchised', a vivid way of saying that Meidias caused the arbitrators to vote for expulsion and disfranchisement. For this form of expression, attributing the penalty to the accuser, cf. 37.49 *ἀτιμῶσαι ζητεῖς*, And. 1.58 *ἔδησεν αὐτούς*, Lys. 10.22, 30.11, Isai. 5.19. Expulsion from the list of arbitrators would not have any practical effect when it was already the end of the year (though it may have been considered shaming), but disfranchisement was more serious, as D. goes on to emphasize (**95**).

εἷς Ἀθηναίων: 'an Athenian citizen'. Throughout this passage D. stresses that it was an Athenian whom Meidias bullied, not merely an alien (which would have been thought less serious); cf. **88** *ἀνδρὸς πολίτου*, **90** *Ἀθηναίων ἕνα*, **95** *πολίτης ὤν*.

καθάπαξ: **32** n. This does not contradict the statement of *AP* 53.6 that a convicted arbitrator could appeal; it states the penalty imposed when the appeal was rejected or not made.

καὶ οὔτε λαχεῖν ...: not a statement of Straton's disabilities when disfranchised, but a sweeping and sarcastic conclusion about how dangerous it is for anyone to do anything which gives offence to Meidias; not 'it is unsafe for him ...' (Vince), but 'it is dangerous to prosecute Meidias when he wrongs you (as D. did), or be appointed arbitrator for him (as Straton was), or even walk along the same street!'

88. ἐπεβούλευσε, instead of just *ἐβούλετο*, implies malicious scheming. S has *ἐπεβούλευσαι*, but the augment and accent show that this is just a slip for *ἐπεβούλευσε*, and someone has written *ε* above the *αι*. The alternative reading in the margin of S takes it as *ἐπιβουλεῦσαι* and adds *ὥστε* to make sense of it; but that would require emendation of *παθὼν* to *ἔπαθε*.

ὑπερφυές: always an unfavourable term in D., linked with *δεινόν* again in 19.225.

μηδέν: sc. *δεινὸν καὶ ὑπερφυές*: not 'nothing at all', because he did suffer condemnation to pay 1000 dr., as D. immediately points out.

θεάσασθε: possibly D. wrote *θεάσασθαι* (Spalding), continuing the construction after *δεῖ*: Byzantine pronunciation made corruption of *αι* to *ε* easy. But probably he has now dismissed *δεῖ* and is using the more forceful imperative.

τὴν ἀσέλγειαν καὶ ...: cf. the first sentence of the speech.

τῶν ἐντυγχανόντων: cf. the end of **87**: it is dangerous even to walk along the same street as Meidias.

νὴ Δί': **41** n. Here D. is using irony in suggesting that someone might say that the penalty imposed on Meidias was a heavy one. Each 'change of speaker' in the imaginary debate (the figure of speech called hypophora) is marked by *ἀλλά*: cf. **98** and Denniston

Particles 10–11. Thus *φαίη τις ἄν* (**89**) is not strictly necessary, and G. H. Schaefer *Apparatus* 3.389 suggests that it is a marginal explanation wrongly inserted into the text.

χιλίων ... δραχμῶν: for the amount of the penalty in the case of slander, **81** n. *τῆς κακηγορίας*.

89. δάκνει: 'bites', meaning 'annoys' (LSJ *δάκνω* III); cf. **148**, if the conjecture *δακνόμενος* is correct there.

ὑπερημέρῳ: **81** n. The hypothetical interlocutor means that Meidias, because he was treated wrongfully by Straton (in that Straton did not notify him of the date of the arbitration), not only lost the case but did not even know that he had lost it and therefore did not make the required payment, thus becoming an overdue debtor liable to seizure of property or to a *δίκη ἐξούλης*. (The meaning is not, as most editors say, that Meidias knew he had lost the case but subsequently, because of his indignation at being treated unjustly, forgot about the date when the payment was due; for D.'s next sentence, saying that Meidias knew about the verdict on the day when it was given, would not be an answer to that.)

ἀλλ' αὐθημερὸν ...: D. replies again to the hypothetical objector, referring to the incident recounted in **85**. If Meidias had really not known the date of the arbitration, he would not have turned up in the evening; and since he did turn up and knew what verdict had been given, he had no excuse for letting the payment get overdue.

τὸν ἄνθρωπον: Straton.

ἐκτέτεικεν: for the spelling *-τει-*, **25** n. *παθεῖν ἢ ἀποτεῖσαι*.

ἀλλὰ μήπω τοῦτο: sc. *λέγωμεν*. Cf. similar phrases in 18.99, 19.200, 57.45. D. will return in **91** to Meidias' failure to pay, after further comment in **90** on the disfranchisement of Straton.

90. τὴν μὴ οὖσαν ἀντιλαχεῖν: **86** n. *ἀντιλαχὼν*.

δήπου, 'surely', perhaps betrays doubt. A man applying for an arbitration to be reopened, after a verdict had been given against him in his absence, had to take an oath that he had been unable to attend; cf. **86** *οὐκ ὤμοσεν*, Pol. 8.60 *ὀμόσας μὴ ἑκὼν ἐκλιπεῖν τὴν δίαιταν*. We have no evidence that it was an acceptable alternative for him to swear that, though he could have attended, he was not informed of the date. If the law did not in fact specify that as an alternative, it might have been disputed whether *τὴν μὴ οὖσαν ἀντιλαχεῖν* was permissible in Meidias' case.

πρὸς ἐμὲ τὸ πρᾶγμα καταστήσασθαι: 'to raise his action against me'. The translation 'settle it with me' (LSJ *καθίστημι* A.II.2.b) is not correct; *τὴν μὴ οὖσαν ἀντιλαχεῖν* is not a method of ending litigation but of reopening it.

ἀτίμητον: 'unassessed', meaning that the penalty was fixed by law and did not have to be assessed by the arbitrator or the jury. Thus Meidias knew that, if the case were reopened, there was no risk of his having to pay more than 10 mnai (1000 dr.). For the amount of the penalty for slander, **81** n. *τῆς κακηγορίας*.

μήτε λόγου ... τυχεῖν: 'not to obtain speech' means that Straton had no opportunity to speak in his own defence, because he was tried in his absence, as recounted in **86–7**. For the sense of *λόγου τυχεῖν*, 'to get a chance to speak', cf. 18.13 *οὐ γὰρ ἀφαιρεῖσθαι δεῖ τὸ προσελθεῖν τῷ δήμῳ καὶ λόγου τυχεῖν*. Here *μήτε ἐλέου* is written in the margin of S, while in **209**, where the other mss. have *ἐλέου*, S has *λόγου*, but there is no particular reason why D. should have used the same word in both places.

ἐπιεικείας: 'fairness', implying reasonably generous treatment, not insisting on the letter of the law; cf. Arist. *EN* 1137b 11–12 *τὸ ἐπιεικὲς δίκαιον μέν ἐστιν, οὐ τὸ κατὰ νόμον δέ*. D. here seems to come near to conceding that Meidias' proceedings against Straton were strictly correct according to law, though he denies it in **87** and **91**.

καὶ τοῖς ὄντως ἀδικοῦσιν: guilty men receive a fair trial and are allowed to speak in their own defence. But *συγγνώμης* is now forgotten or ignored, for it is obviously not true that all offenders are pardoned.

91. ἠτίμωσεν: **87** n. *ἐκβάλλει καὶ ἀτιμοῖ*.

ἠβουλήθη: for the augment, **20** n. *ἠδυνήθησαν*.

ἐχαρίσασθε: every Athenian jury is regarded as equivalent to the whole people of Athens. Therefore a speaker addressing one jury can refer to another jury's verdict as a verdict which 'you' gave, even though there may really have been not a single juror in common between the two. But a meeting of arbitrators is not regarded as equivalent. If it had been, appeal from it would not have been possible; but *AP* 53.6 says that those condemned at an arbitrators' meeting can appeal. By saying that 'you' did this favour to Meidias, D. can only mean that Straton appealed to a jury against the arbitrators' verdict and the jury confirmed his disfranchisement. This fact, which D. has almost but not quite completely concealed, makes his whole complaint that Meidias treated Straton unfairly a good deal less convincing. It is not satisfactory to avoid this conclusion by emending the text to *τοῦτ' ἐχαρίσαθ' αὑτῷ*, 'he did himself this favour' (Dobree *Adversaria* 1.461; cf. Σ (313 Dilts) *φαίνεται γὰρ αὐτὸς τὸ πᾶν συσκευάσας*); for even if we thus removed any reference to condemnation by a jury, we should still have to admit that Straton could have appealed to a jury if he had wished, and if he did not do so it was probably because he thought a jury would decide against him.

γνώμην ... ἐνέπλησεν: cf. Th. 7.68.1 *ἀποπλῆσαι τῆς γνώμης τὸ θυμούμενον*.

αὑτοῦ: with *γνώμην*. For the placing of the genitive pronoun after the verb, cf. **16** *μου*.

ἐκεῖν': 'the thing I mentioned earlier'. This is explained by *τὴν καταδίκην ἐκτέτεικε*, which is in apposition to the preceding phrase and therefore has no connective particle.

τὴν καταδίκην: the payment which Meidias had been condemned by Straton to make. For this sense cf. 47.51 *ἐκέλευον ἐπὶ τὴν τράπεζαν ἀκολουθοῦντα κομίζεσθαι τὴν καταδίκην*.

οὐδὲ χαλκοῦν: 'not a penny'. There is a similar rhetorical use of *χαλκοῦς* in 42.22. Meidias claimed that he owed nothing, on the ground that the conviction of Straton for misconducting the arbitration invalidated the verdict in the arbitration; cf. the first sentence of **92**.

δίκην ἐξούλης: the case mentioned in **81**.

παρ-: 'amiss', 'miserably'. Editors who translate 'as an accessory' or 'on a side issue' are making too much of the prefix, which is simply equivalent to the English 'mis-'. Cf. Ar. *Wasps* 1228 *παραπολεῖ βοώμενος*, 'you'll be shouted to death!', where 'as an accessory' does not fit the context. (This is different from the cook's comment on his art in Dionysios com. 2.35, where the preceding *ἀπολέσῃς* makes clear that *παρ-* means 'as well'.)

ἄνω κάτω: 'up and down' in confusion. For the asyndeton, **81** n. *τῇ δίκῃ, τοῖς νόμοις*.

92. γνῶσιν: 'verdict', the normal sense of this noun in D. Two verdicts were given: Straton's verdict against Meidias (*ἣν δ'* ...) and the arbitrators' (and jury's) verdict against Straton (*τὴν μὲν* ...). Meidias chooses to regard the verdict against himself as invalid and the verdict against Straton as valid. He evidently assumes that the later verdict overrides the earlier one, because it shows that the earlier one was given illegally. D. ignores that argument, implies that Meidias' view has no grounds other than his own advantage, and argues that the earlier verdict is the one more soundly based in law, because it was preceded by a summons and the later verdict was not.

ἀπρόσκλητον: **87** n. *παρὰ πάντας τοὺς νόμους*.

αὐτὸς implies that Meidias has decided this himself, without having legal authority for it.

προσκληθείς, εἰδώς, οὐκ ἀπαντῶν: Meidias did not attend the arbitration when Straton gave his verdict (**84**), but in what sense had he received a summons (*προσκληθείς*) to attend it? For the sake of the contrast with the condemnation of Straton by the other arbitrators, which was *ἀπρόσκλητον*, D. blurs two distinct stages of the slander case: (*a*) D.'s summons to Meidias to appear before the tribe judges when he initiated proceedings, which is correctly called *προσκαλεῖσθαι* and undoubtedly did take place; (*b*) Straton's notification to Meidias of the date on which the verdict would be given. It was this latter notification which, as it seems, Meidias complained had not been given, with the result that he was absent; but it was not, as far as is known, called *προσκαλεῖσθαι*. Thus the impression given by the juxtaposition of *προσκληθείς* and *οὐκ ἀπαντῶν*, that Meidias ignored a formal summons, is spurious.

ἄκυρον ποιεῖ: Meidias does not pay the 1000 dr., so that the verdict has had no effect.

παρὰ τῶν ἔρημον καταδιαιτησάντων αὐτοῦ: a rhetorical generalization; Straton is the only such person. For *ἔρημον*, **81** n. *ἐρήμην*.

λαμβάνειν ... λαβεῖν: disfranchisement continues for life, but death is a single event.

νόμων καὶ δικῶν καὶ πάντων στέρησις: paraphrase of *ἀτιμία* (not a separate penalty). A disfranchised man was deprived of 'laws and cases' in the sense that he could not speak in a court and thus could not prosecute anyone who wronged him. He also lost various other rights: to enter temples or the Agora, to hold public office, and so on. But *πάντων* is an exaggeration: disfranchisement did not include loss of personal property.

93. ἀλλὰ μὴν: cf. **107**, **119**, **167**, and Denniston *Particles* 346: 'Demosthenes and Isaeus (not, I think, any of the other orators, who prefer *καί*, *οὖν*, etc. in such cases) often use *ἀλλὰ μήν* to mark the transition from a statement to the calling of evidence in support of it'.

The document is another forgery, compiled in a later age by someone who has read **81–92** and added two fictional names.

οἴδαμεν: a non-Attic form for *ἴσμεν*, used also in the spurious documents in **82**, **121**.

τοῦ κακηγορίου: the mss. have *κατ-*. The author of the document must have written *κακ-*, but *τοῦ κακηγορίου* is a non-Attic form for *τῆς κακηγορίας*.

ἑλομένους: a factual error. Cases were distributed to public arbitrators by lot; only private arbitrators were chosen by the disputants. This confusion recurs in the spurious law in **94**; probably both documents were composed by the same person.

ἡ κυρία τοῦ νόμου: for 'the day appointed by law' an Attic writer would probably have written *ἡ κυρία ἐκ τοῦ νόμου* (conjectured here by Wolf and by Dobree *Adversaria* 1.461) or *ἡ κυρία ἡ ἐκ τοῦ νόμου*.

γενομένης δὲ ἐρήμου κατὰ Μειδίου: another odd phrase. An Attic writer would probably have written *καταδεδιῃτημένης ...*

ἄρχοντας: the author of the document has taken this out of **85**, and does not really know what officials are meant. A genuine witnesses' statement would have been more specific, e.g. τῇ Ἐρεχθηΐδι δικάζοντας.

καταβραβευθέντα: another non-Attic expression; cf. Σ (319 Dilts). The right word would be either κατηγορηθέντα or εἰσαγγελθέντα (cf. *AP* 53.6).

94. The law instituting public arbitration and laying down its procedures, mentioned also in Lys. fr. 16 Thalheim, was made in 400/399; cf. MacDowell *RIDA* 18 (1971) 267–73. But why does D. call for it to be read out here? In **83–92** he clearly assumes that his hearers are already familiar with the normal procedure of arbitration. There are two parts of the narrative which might be illuminated by knowledge of the exact terms of the law: did Straton follow the correct procedure for notifying Meidias of the date of the arbitration and for giving a verdict against him when he failed to attend, and did Meidias follow the correct procedure for accusing Straton of misconducting the arbitration? But D. makes no attempt to use the text of the law to answer those questions. Indeed he makes no use of the law at all: the reading is neither preceded nor followed by any analysis or singling out of points favourable to his argument, such as we have in **9**, **11**, **45–6**. Without such commentary by D. the reading of the law appears pointless. Perhaps he intended to add comments but never did; cf. p. 26.

Furthermore, the document which the mss. (SFYP) preserve is plainly about private arbitration (by which disputants make their own arrangements to refer a question to an arbitrator of their own choice), not public arbitration; thus it is quite irrelevant to D.'s speech and cannot be the text which he wanted read. Some scholars hold that it is nevertheless a genuine Athenian law, and that someone (not D., but an editor who had access to a collection of Athenian laws) has by mistake put into the text the law about private arbitration instead of the law about public arbitration; so Drerup *Jb. Cl. Ph.* Supp. 24 (1898) 304, Lipsius *Recht* 222 n. 6, A. Steinwenter *Die Streitbeendigung durch Urteil, Schiedsspruch und Vergleich nach griechischem Rechte* (1925) 60–1. But that view is rightly rejected by K. Latte *Gnomon* 2 (1926) 211, L. Gernet *REG* 52 (1939) 391 n. 3 = *Droit et société dans la Grèce ancienne* (1955) 104 n. 7. (Harrison *Law* 2.65 n. 1 tries unsuccessfully to have it both ways.) Besides the unconvincing features mentioned in the following notes, the document omits or glosses over several matters which must have been specified in the real law about private arbitration. (*a*) The law must have permitted appointment of more than one man as arbitrator; cf. 27.1, 33.14, Isai. 5.31. (*b*) The law must have required the disputants to make a formal agreement (*συνθῆκαι*) about who the arbitrator was to be and what question he was to decide; cf. 33.14–19, 34.18. (*c*) The law must have required the arbitrator to take an oath before giving his verdict; cf. 52.30, Isai. 5.32. (*d*) The law must have specified what was to happen if, when the verdict was given, one of the disputants failed to abide by it. So the document before us should be rejected as spurious. Σ (320 Dilts) makes suggestions about what the law of arbitration will have contained, showing that the scholiast did not have any text of this law in his copy of the oration.

συμβολαίων ἰδίων: it is not clear what the writer meant by this phrase. 'Private contracts' is too narrow to be an accurate definition of the scope of private arbitration, which could be used for disputes not involving contracts. But if it just means 'private relations' the phrase is too vague to serve any purpose in a law.

διαιτητὴν ἑλέσθαι: the repetition of these words appears so pointless that it is probably right to delete them on their second occurrence as a copying error, rather than to attribute the redundancy to the author of the document.

δ' ἕλωνται: Reiske's necessary correction of *βούλωνται*.

κατὰ κοινόν: in fact private arbitrators were not always chosen jointly. Sometimes each disputant chose an equal number; cf. 33.14, Isai. 5.31.

μενέτωσαν: the use of the imperative ending *-τωσαν* is not in itself proof of the document's inauthenticity (**8** n. *παραδιδότωσαν*). But, for the sense of abiding by a decision, normal Attic usage is the compound *ἐμμένειν* rather than *μένειν ἐν* (e.g. 27.1, 33.15, Ar. *Wasps* 524, Men. *Epit.* 237).

μεταφερέτωσαν: *με* is written above the line in O, and this gives a more suitable form for the sense 'transfer' than the older mss.' *καταφερέτωσαν*. But neither seems likely in an Athenian law, which would more probably have said something like *μὴ ἐξεῖναι δίκην λαχεῖν περὶ τῶν αὐτῶν* ...

τὰ κριθέντα: another odd expression. One would rather expect simply *τὴν δίαιταν κυρίαν εἶναι* (cf. And. 1.87).

95. ἑστάναι: to stand in the court without speaking. This is one of two passages (the other is 59.26–7) which show clearly that a disfranchised citizen was not permitted to testify in a lawcourt. The rule is curious to modern eyes; the Athenians evidently considered it more important to degrade a disfranchised man than to obtain all relevant evidence at a trial. Yet Straton could, it appears, enter the court as long as he did not speak. Some ancient critics apparently could not believe this, for Σ (323 Dilts) says *ὠβέλισται δὲ καὶ ταῦτα· οὐδὲ ἐπιβῆναι γὰρ τοῖς τοιούτοις ἐφεῖται*, 'This is obelized too; for such men are not permitted even to set foot ⟨in a court⟩'. But the following lines, which presuppose Straton's silent appearance, are so Demosthenic in quality that the passage must be genuine, and we must accept that Straton could appear in court without speaking. How is this to be reconciled with the law excluding disfranchised citizens from the Agora (Ais. 3.176, etc.)? Hansen *Apagoge* 62 takes our passage as evidence that the exclusion of *ἄτιμοι* from the Agora was not always strictly enforced; if that is right, *δήπουθεν*, 'I suppose', could be a sign that D. is not sure that Straton will be allowed to infringe the exclusion rule on this occasion. But it does not seem likely that one exclusion rule (from the Agora) would have been relaxed when the other (from testifying) was not. So I prefer to take the passage as evidence that the lawcourts were not technically part of the Agora. I take *δήπουθεν* not as indicating any genuine doubt, but as sarcastic, similar to **87** *τὴν αὐτὴν ὁδὸν βαδίζειν*: D. pretends to think that Meidias is so hostile to Straton that possibly he will not even allow him to stand in silence.

οὗτος: D. must be supposed to have paused while Straton was stepping forward. Now he is standing before the court.

πένης: **83** n. Here *ἴσως* reinforces the concessive sense of *μέν*: 'perhaps' implies 'I admit', not doubt about the fact.

πολίτης: a citizen who was disfranchised remained nevertheless a *πολίτης*, not a *ξένος*.

ἐστρατευμένος ... εἰργασμένος: chiasmus, making a rhetorical pair out of Straton's life: he did good and did not do harm. An unkind critic might point out that the only good done by Straton which D. has managed to discover is routine military service, which was compulsory.

ἐν τῇ ἡλικίᾳ: *οἱ ἐν ἡλικίᾳ* is the usual phrase for 'men of military age'; cf. 1.28, 4.7, 10.40, 13.4, etc. But here the meaning is not 'campaigns of military age' but 'campaigns during *his* military age', for which the article is needed and I have therefore restored it; cf. 24.126 *τὰ ἐπιτηδεύματα τὰ ἐν τῇ ἡλικίᾳ*.

τῶν ἄλλων: not really *all* others; **92** n. *πάντων στέρησις*.

φθέγξασθαι: 'utter a sound'; **29** n. *φθέγξεται*. This word and *ὀδύρασθαι* are used for rhetorical effect, to develop the pathos of Straton's situation.

δίκαια: Straton, if permitted to speak, would not really be likely to say that he had been treated justly. But again D. is aiming at a rhetorical effect: Straton is not allowed even to say what is right.

96. πλούτου: there is no evidence that money really had anything to do with Straton's condemnation. But wealth is traditionally linked with *hybris*, and D. wishes to exploit the notion of the rich man oppressing the poor man.

ὑπερηφανίας: **83** n.

παρὰ means that being poor, friendless, and one of the many was sufficient cause for Straton to be treated insolently by Meidias; **61** n. *παρὰ τοῦτ'*.

ἐρημίαν: lack of friends to support him; cf. **80** *ἔρημος*.

τὰς πεντήκοντα δραχμὰς: the bribe offered by Meidias for reversing the verdict in the arbitration (**85**).

ἀπέφηνεν: **85** n. *ἀποφαίνειν*.

97. τὸν οὕτως ὠμόν: 'a man as cruel as this'. The whole of **97** is expressed in a generalizing manner; note the negatives *μήτε ... μηδενὸς* which give *ποιούμενον* the generic sense 'a man who does not ...'. D. implies that the verdict about Meidias will be taken as a precedent for other offenders; this leads up to the last words of **97**, *οὐ παράδειγμα ποιήσετε;* Note also the unobtrusive skill with which D. in this sentence passes from Straton's case to the present one (at *ὑβρίζοντα λαβόντες*), giving the impression that everything recounted in **83–96** supports a verdict against Meidias in this trial.

ποιούμενον: the verb *καταψηφίζομαι* normally takes a genitive; the mss. give it an accusative in 24.65, but *τοῦτον* there is generally emended. Perhaps *ποιουμένου* (Buttmann) should be adopted here; but editors have objected to the sound *-ου οὐ*, and it may be better to place a comma after *ποιούμενον* and understand *λαβόντες* from the previous sentence. Many other emendations have been proposed, including transfer of *ἀφήσετε* to follow *ποιούμενον* (Cobet *Misc. crit.* 509), deletion of *ἀφήσετε* (Butcher), deletion of the whole passage *καὶ μήτε ... ποιούμενον* (G. H. Schaefer *Apparatus* 3.395), change of *καὶ* to *καὶ ταῦτα* (Gebauer *De hypotacticis* 34) or to *κἂν ἴδητε* (H. Demoulin *Rev. Instr. Publ. Belg.* 42 (1899) 382–3), or the insertion of another verb, such as *ποιουμένου ⟨προνοήσεσθε⟩* (Weil) or *καὶ ⟨ἐλεήσετε⟩* (Sykutris).

For the series of questions beginning with *οὐ*, cf. 4.44 *οὐκ ἐμβησόμεθα; οὐκ ἔξιμεν αὐτοὶ ...; οὐκ ἐπὶ τὴν ἐκείνου πλευσόμεθα;*

98. τί φήσετε; implies 'What reason will you give for acquitting him?'

νὴ Δί': **41** n. The replies which D. here attributes to an imaginary interlocutor are obviously bad reasons for acquitting Meidias; thus he implies that it is impossible to think of any good ones.

ἀλλὰ μισεῖν ὀφείλετ' ...: cf. 24.170 *ἀλλὰ μισεῖν ὀφείλετε τοὺς τοιούτους, ὦ ἄνδρες Ἀθηναῖοι, μᾶλλον ἢ σῴζειν*, 22.64 *ἀλλὰ μισεῖν δικαιότερον διὰ ταῦτά σ' ὀφείλουσιν ἢ σῴζειν*. The use of *ὀφείλω* with an infinitive, meaning 'ought', is less rare in Attic prose than might be supposed from a reading of LSJ *ὀφείλω* II.1; in D. there are further instances in **166**, 2.8, 23.97, 23.183, *Pr.* 45.2.

ἀλλ' ὅτι πλούσιός ἐστιν: the second suggestion by the hypothetical interlocutor. For the use of hypophora cf. **88–9**, **148–9**, and Denniston *Particles* 10–11. For wealth as a cause of *hybris* cf. p. 19.

ὄν: for the participle in *oratio obliqua* with *εὑρίσκω*, cf. 19.332, 20.44.

ἀφελεῖν τὴν ἀφορμήν: to confiscate his property.

προσῆκε: for the omission of *ἄν*, **25** n.; it is unnecessary to adopt *προσήκει* (A) as most editors do.

99. The third reason for acquitting Meidias which D. attributes to the hypothetical interlocutor (*νὴ Δία*) is pity for his children, who will suffer undeservedly if Meidias is condemned. Some 75 years earlier the practice of bringing weeping children into court to persuade the jury to acquit their father was satirized by Aristophanes in *Wasps* 976–84, where the dog Labes is on trial and the appearance of his weeping puppies makes the juror Philokleon burst into tears; cf. also *Wasps* 568–74, And. 1.148, Pl. *Ap.* 34c, Lys. 20.34–5. Yet **99**, **186**, and **188** show that it still persisted in D.'s time, presumably because defendants found that it was effective; and there is an even later instance in Hyp. *Phil.* 9.

κλαιήσει: this form is given by all the mss. and there seems to be no need to emend it to either *κλαήσει* or *κλαύσεται*. There is no relevant epigraphical evidence; *κλαιήσει* is in the papyrus of Hyp. *Dem.* 40 (first century A.D.). Of course the weeping ought to be done by the children; D. is sarcastic in suggesting that Meidias will weep himself.

φέρειν: 'bear up against', 'endure'. This might seem to imply that pity is not required if an unjust punishment is endurable, and *ἀφαιρεῖν* (a variant reading in S) is probably a conjecture intended to avoid that implication. But emendation is unnecessary; D. is not trying to state the minimum requirements which justify pity, but the cases in which pity is most obviously appropriate.

τοῦδε: Straton, who is still standing before the court. The use of pronouns in this passage may be confusing to a reader, but in delivery of the speech D.'s gestures would have made the references clear. In my translation I have substituted names for some of them in order to clarify the sense.

οὐδ' ἐπικουρίαν ἐνοῦσαν ὁρᾷ: 'they see also that no remedy is possible'. A man disfranchised because he failed to pay a fine could regain his rights (*ἐπίτιμος γενέσθαι*) by paying it, but Straton's disfranchisement was permanent (cf. **87** *καθάπαξ ἄτιμος*).

ἁπλῶς οὕτως: 'just like that', equivalent to *καθάπαξ* in **87**. For the addition of *οὕτως* to *ἁπλῶς* cf. Pl. *Prot.* 351c, *Gorg.* 468c.

ῥύμῃ: 'force', 'impetus', a notable metaphor.

100. ταῖς ἐσχάταις συμφοραῖς: for *συμφορά* used of a legal penalty cf. **6**, **58**, **96**, **99**. Whereas in English 'the extreme penalty' means death, ἔσχατος can refer to other penalties; another passage where it refers to disfranchisement is 59.53 ταῖς ἐσχάταις ζημίαις.

τούτῳ δ': 'well, with *that* man ...'. For the duplication of δέ with a demonstrative summing up the preceding clause, cf. 8.3 ὅσα δ' ἐχθρὸς ὑπάρχων τῇ πόλει ... πειρᾶται προλαβεῖν ..., περὶ τούτων δ' οἴομαι ..., and other instances cited by Denniston *Particles* 183–5.

οὐδὲ συνοργισθήσεσθε: although the negative in a conditional clause is normally μή, οὐ is sometimes used where it is felt to apply to an individual word rather than to a clause as a whole. When that word is the verb, the effect is commonly to denote a definite negative act (οὔ φημι 'I deny', οὐκ ἐῶ 'I forbid', οὐκ ἐθέλω 'I refuse', etc.) rather than a mere absence of action. So here D. means that, if the jurors vote for the acquittal of Meidias, that will be a clear refusal of sympathy to Straton. οὐδέ, 'not even', implies that it would have been better still if Straton could somehow be relieved of his disfranchisement.

101. Substantially the same comparison of life to an ἔρανος is made both in **101** and in **184–5**. Surely D. did not mean to include both passages in his speech; and since **184–5** is expressed rather better (cf. notes below on ἀξιοῦν and οἷον ἐγώ τις οὑτοσί), probably that is the revised version and D. intended to omit **101**. On the relevance of these passages to the problem of the composition of the oration, see pp. 26–7. Some phrases from **184** are added to **101** either in the text or in the margin of some mss. (see app. crit.), but do not appear in the text as originally written in S and A; this should be taken to mean that those phrases were not in the text of **101** in the archetype, but were added later by people who thought that the two passages ought to be alike.

The word ἔρανος developed and changed its meaning over the centuries; for a collection and discussion of the evidence see J. Vondeling *Eranos* (Groningen 1961, in Dutch with a summary in English). In the earliest period an ἔρανος is a friendly dinner arranged on a reciprocal basis: either the friends take turns at providing a dinner (ἀμοιβαῖα δεῖπνα in Pi. *Ol.* 1.39) or one man provides the dinner and each guest brings him a present as payment (as, according to Pherekydes, *F. Gr. Hist.* 3 F11, Polydektes gave a feast and each guest brought him a horse, except Perseus who brought the Gorgon's head instead). Such dinners continued to be held in the fifth and fourth centuries (Ar. fr. 408, Arist. *EN* 1123a 22), but in D.'s time ἔρανος usually means a friendly loan raised by contributions. A man in need of money for some purpose would collect contributions from a number of friends, and repay them by instalments over a period. The commonest purpose of such a loan may have been to provide capital for a business, such as the perfume-shop in Hyp. *Ath.* (where ἔρανοι are mentioned in 7–11). We also hear of ἔρανοι to ransom a person from slavery, as in the cases of Neaira (59.31 ἔρανον εἰς τὴν ἐλευθερίαν συλλέγουσα) and Nikostratos (53.8 ἀπεκρινάμην αὐτῷ ὅτι ... χιλίας τε δραχμὰς ἔρανον αὐτῷ εἰς τὰ λύτρα εἰσοίσοιμι). Aristonikos collected contributions to enable him to end the disfranchisement which he incurred by being in debt to the state (18.312 τὸ συνειλεγμένον εἰς τὴν ἐπιτιμίαν); this too was doubtless an ἔρανος.

D. uses in **101** and **184–5** several words which are regularly found in connection with ἔρανοι. συλλέγειν is the word for collecting contributions, used in Ant. 2b.9 ἔρανον παρὰ τῶν φίλων συλλέξας, D. 18.312, 53.11–12, 59.31, Thphr. *Char.* 22.9. πληρωτής is the word for a contributor, used in Hyp. *Ath.* 7 τοὺς χρήστας καὶ τοὺς πληρωτὰς τῶν ἐράνων, ibid. 9, D. 25.21. The verb for contributing to an ἔρανος is usually εἰσφέρειν, used in 53.7–8, 61.54, Ar. *Lys.* 651, Isok. 11.1, Pl. *Symp.* 177c, *Laws* 927c, Philemon 213.14, Thphr. *Char.* 15.7, 17.9, etc. Sometimes the simple verb φέρειν and the noun φορά refer to the contribution by a lender, as in D. 10.40, 25.58, Thphr. *Char.* 15.7; sometimes to the borrower's repayment by instalments, as in D. 25.21, Hyp. *Ath.* 11, Lys. fr. 1.4 Thalheim = xxxviii 4 Gernet and Bizos.

Since the point which D. wishes to make is 'What you do to other people, they should do to you in return', one might expect that the use he would make of the terminology of an ἔρανος would be to say 'Whatever you lend to other people, they should repay to you'. But in fact he does not say that; instead he says 'Whatever you lend to other people, they should make a similar loan to you on another occasion' (ταὐτὰ εἰσφέρειν, ἐάν που καιρὸς ἢ χρεία παραστῇ). This is easier to understand when one realizes that an ἔρανος was a loan made without charging interest, and it was considered generous to make this kind of loan instead of demanding interest in the usual way. The point is clear in Thphr. *Char.* 17.9, where the grumbler (μεμψίμοιρος) ought to be gratified at receiving an ἔρανος but is not: καὶ ἐράνου εἰσενεχθέντος παρὰ τῶν φίλων καὶ φήσαντός τινος "ἱλαρὸς ἴσθι", "καὶ πῶς;" εἰπεῖν, "ὅτι δεῖ τἀργύριον ἀποδοῦναι ἑκάστῳ, καὶ χωρὶς τούτων χάριν ὀφείλειν ὡς εὐεργετημένον;", 'And when a loan has been made to him by his friends and someone says "Enjoy it!", "How can I?" he says, "I've got to repay the money to each of them, and be grateful to them besides, as if they'd done me a good turn!"'

Perhaps the earliest passage in which ἔρανος is used metaphorically of doing someone a service is a remark attributed to Thales, οἵους ἂν ἐράνους ἐνέγκῃς τοῖς γονεῦσι, τοιούτους αὐτὸς ἐν τῷ γήρᾳ παρὰ τῶν τέκνων προσδέχου (Diels–Kranz *Die Fragmente der Vorsokratiker*[6] 1.64 lines 5–6); however, it is doubtful whether ἔρανος had yet acquired the sense of a financial loan in the time of Thales. Instances of the metaphor in the fifth and fourth centuries include E. *Supp.* 363, Ar. *Lys.* 651–5, Th. 2.43.1, X. *Kyr.* 7.1.12, Isok. 10.20, 11.1, 14.57, Dikaiogenes 4, Alexis 280, Pl. *Symp.* 177c, *Laws* 927c, Lyk. *Leo.* 143, D. 10.40, 25.21–2, 61.54, *Ep.* 5.5, Arist. *Pol.* 1332b 38–41. It is used ironically of doing harm in D. 59.8, where Theomnestos and Apollodoros have set about getting their own back on Stephanos: *τούτῳ δὲ δικαίως τὸν αὐτὸν ἔρανον ἐνεχειρήσαμεν ἀποδοῦναι.* Our passage (**101** and **184–5**) is unique in using the metaphor neutrally, to refer to good and bad deeds alike.

ἀξιοῦν, 'think it right', is not particularly appropriate, because some of the conduct concerned, especially the bad conduct, would be unpremeditated; the point is rather that everyone does in fact contribute by his behaviour to other people's lives, whether he intends to do so or not. The word is absent from **184**; that probably means that D. wrote it in the first draft of the passage and then, realizing its unsuitability, omitted it in the second.

αὑτοῖς: contributing to an *ἔρανος* is regarded as being for one's own benefit, because it entitles one to receive contributions if one needs them later; it is a kind of insurance. The dative has the same sense in Pl. *Laws* 927c *ἔρανον εἰσφέροντα ἑαυτῷ τε καὶ τοῖς αὑτοῦ.*

παρὰ: 'throughout'; cf. **1** *παρὰ πᾶσαν τὴν χορηγίαν.*

οἷον ἐγώ τις οὑτοσὶ: the other three words serve to show that *ἐγώ* is a hypothetical instance: 'For example, here is someone, say myself, ...'. In **185** D., realizing that it is hardly tactful to take himself as the paragon of good conduct, substitutes *τις ἡμῶν* in his second draft.

ἅπασι προσήκει: no connective particle. D. often uses asyndeton in a pair of sentences of which the first states some circumstances (actual or hypothetical) and the second the consequence (actual or to be expected); the first is equivalent to 'if ...' or 'when ...', but instead of using a subordinating conjunction D. makes the construction paratactic. The second is sometimes a rhetorical question. In such a pair either the first sentence or the second or both may lack a connective particle. Here *οἷον* serves to introduce *ἐγώ* ..., but *ἅπασι* ... has asyndeton. In the immediately following pair of sentences there is asyndeton both at *ἕτερος* ... and at *τούτῳ* ... Other instances in this speech include **179** *οὐδ' οὕτω* ...· *ἐπιβολὴν* ..., **187** *ἐμοὶ* ...· *διὰ τοῦτ'* ...;, **220** *μισεῖ* ...· *ἆρ' οὖν* ...; Cf. Denniston *Style* 118–19.

ταὐτὰ εἰσφέρειν: 'to lend him the same amounts'. Not 'to pay him back' (King and Goodwin), because in the vocabulary of *ἔρανοι* the verb *εἰσφέρειν* is used of contributing to a loan, not of repaying it; see p. 323.

που: *του* (S) could be correct, but D. does not usually attach an objective genitive to *καιρός*.

βίαιος: sc. *ἐστίν*. A adds *ὠμός*: cf. pp. 50–1 on extra words in A.

οὔθ' ὅλως ἄνθρωπον: Meidias considers men like Straton to be not human beings but 'rubbish' (*καθάρματα*); cf. **185**, **198**.

σεαυτῷ: see note on *αὑτοῖς* above.

συλλέξασθαι: usually active in connection with *ἔρανοι*, but there is another instance of the middle in Lys. fr. 1.4 Thalheim = xxxviii 4 Gernet and Bizos.

102–22. *The attempts of Meidias to get Demosthenes prosecuted for desertion and for homicide.*

D. now jumps to a much later date. The incidents described in this part of the oration occurred after the occasion when Meidias hit him in the theatre. They were probably in 348/7; see p. 9 on their place in the sequence of events.

102. δεινότερα: in fact it is not true that more serious accusations will be made later in the speech, but the remark is rhetorically effective at this point and will not be checked or remembered later by the jurors. The whole sentence serves, as Σ says (351 Dilts), to magnify both what has preceded and what will follow.

τιμᾶν αὐτῷ τῶν ἐσχάτων: 'to make an assessment for him at the utmost amount', i.e. 'to condemn him to the extreme penalties'. *τῶν ἐσχάτων* does not have a specific meaning (**100** n.), but could refer here to death and confiscation of property. The genitive of value or price is normal with *τιμάω* in the sense 'assess'; so also in **151**, **152**, **182**.

τῶν μετὰ ταῦτα: 'the rest of my speech' (not 'the later events').

πεποίηκε: Meidias 'has made' the accusations in the sense that he has committed acts which cause accusations to be made.

103. λιποταξίου: the procedure for prosecution (which was *graphe*) and the penalty (which was disfranchisement) were the same for desertion (*λιποτάξιον*) as for failure to attend when called up for military service (*ἀστρατεία*); cf. 15.32, And. 1.74, Ais. 3.175–6. For the ground of Euktemon's prosecution of D. see p. 9.

τὸν κονιορτὸν Εὐκτήμονα: 'dusty Euktemon', evidently the man's regular nickname; cf. **139**. Compare *κυρηβίων*, 'bran-man', a nickname of Epikrates (19.287, Ath. 242d). We have no evidence to show how he acquired the nickname, and it is probably a mistake to try (as several editors do) to explain it by reference to the events mentioned in **103**. Σ (355 Dilts) suggests that it means 'easily persuaded', because dust is easily blown about by the wind; but that is just a guess based on the preceding words *λίαν εὐχερῆ*.

Euktemon was a common name, and it is a problem whether this holder of it can be identified with others mentioned elsewhere. But since his deme was Lousia (mentioned later in **103**), it is reasonable to accept that *Χαρίας Εὐκτη*[, who was the representative of Lousia in the Boule of 303/2 (*IG* 2^2 1746.33 = *The Athenian Agora* 15 no. 62.192), was his son. If so, then he himself was surely identical with *Εὐκτήμων Χαρίου*, who was an Athenian *ναοποιός* at Delphi in the years 346–37 (*Fouilles de Delphes* 3(5) 91.23, cf. 19.74, 19.96, 48.1.13; the identification is made by J. Sundwall in *Klio* 5 (1905) 131–2). Euktemon son of Aision, honourably mentioned in **165**, was clearly a different man.

ἐάσω: a notable instance of paralipsis: in saying that he will say nothing about the incident, D. not only tells his listeners all that he wishes them to know about it but also escapes the obligation to provide evidence for his allegation.

ἀνεκρίνετο: the active of this verb is used of the magistrate who conducts the inquiry (e.g. 48.31), the middle of the prosecutor who gets the magistrate to conduct it (e.g. 53.17). When a prosecution was initiated, the magistrate (in most kinds of case) held an inquiry (*ἀνάκρισις*) before fixing a date for the trial; he would put questions to the prosecutor and the defendant to elucidate the charge and ensure that the correct legal procedure was followed. If the prosecutor failed to attend the inquiry, the prosecution lapsed; that is evidently what happened in Euktemon's prosecution of D. Cf. Harrison *Law* 2.94–105, MacDowell *Law* 240–2.

συκοφάντης need not be given any very specific interpretation. It is simply an abusive word for a prosecutor, implying that he had no good reason for prosecuting.

ἐκκέοιτο πρὸ τῶν ἐπωνύμων: a prosecutor's charge was written by the clerk on a wax-covered tablet and displayed in public; cf. 58.10 *ἐξέκειτο πολὺν χρόνον ἡ φάσις*, Ar. *Clouds* 770–2, *Wasps* 349, 848, Isok. 15.237. Our passage alone shows that the place where the charges were displayed was in the Agora, in front of the statues of the heroes after whom the ten tribes were named; for that monument see Thompson and Wycherley *Agora* 38–41, U. Kron *Die zehn attischen Phylenheroen* (Berlin 1976) 228–32.

Λουσιεὺς: Lousia was a city deme of the Oineis tribe.

κἂν προσγράψασθαι ...: a sarcastic joke, suggesting that Meidias takes pride in acts which normal people think disgraceful. The allegation is D.'s invention, not an attested fact. The middle voice is used to mean that Meidias would have got the magistrate's clerk to make an addition to the official notice; the conjecture *προσγράψαι* (Cobet *Misc. crit.* 509) is unnecessary.

γέγραπται: Jahn *Philol.* 26 (1867) 3 and Herwerden *Exerc. crit.* 169, *Mn.* II 3 (1875) 136 propose to delete this word, leaving *ἐγράψατο* to be understood. But its tense defends it: a glossator would simply have repeated ἐγράψατο, but D. has moved into the perfect because he is thinking of Meidias' (alleged) complacency at the result of his action ('Look what I've achieved!').

ἐφ' ᾗ: sc. γραφῇ.

ἠτίμωκεν αὑτὸν οὐκ ἐπεξελθών means that Euktemon's failure to proceed to the trial of his charge against D. led automatically to his disfranchisement. This passage is one of the principal pieces of evidence for the problem, not yet satisfactorily solved, of the extent of disfranchisement as a penalty for frivolous prosecution. It is known that a prosecutor in a *graphe* and some other kinds of public case, if he failed to proceed with the case or failed to obtain one-fifth of the jury's votes, was fined 1000 dr. (cf. **47**). Was disfranchisement (mentioned in **103**, And. 1.33, Hyp. *Eux.* 34) an alternative or an additional penalty? A fragment of Thphr. *Laws* makes clear that it was additional: Ἀθήνησιν οὖν ἐν τοῖς δημοσίοις ἀγῶσιν, ἐὰν μὴ μεταλάβῃ τις τὸ πέμπτον μέρος, χιλίας ἀποτίνει, καὶ ἔτι πρόσεστί τις ἀτιμία, ὥστε μὴ ἐξεῖναι μήτε γράψασθαι παρανόμων μήτε φαίνειν μήτε ἐφηγεῖσθαι· ἐὰν δέ τις γραψάμενος μὴ ἐπεξέλθῃ, ὁμοίως (text from the combined evidence of Σ 22.3 (13b Dilts) and *Lex. Cant.* under πρόστιμον: cf. also Harp. under ἐάν τις, Pol. 8.53). Furthermore Theophrastos seems to mean that this disfranchisement was not total, but partial: the frivolous prosecutor did not lose all his citizen-rights, but merely the right to bring similar prosecutions in future. That is confirmed by D. 26.9 ὅταν τις ἐπεξιὼν μὴ μεταλάβῃ τὸ πέμπτον μέρος τῶν ψήφων, ἐφ' οἷς οἱ νόμοι κελεύουσι τὸ λοιπὸν μὴ γράφεσθαι μηδ' ἀπάγειν μηδ' ἐφηγεῖσθαι, and by And. 1.76 which, without mentioning their offence, says that there were men whose partial disfranchisement took the form of a prohibition to γράψασθαι or to ἐνδεῖξαι, and by D. 57.8, where Euboulides got less than one-fifth of the votes as prosecutor in a *graphe* and was subsequently a member of the Boule, and by our present passage, where Euktemon is now disfranchised (**103** ἠτίμωκεν is perfect, implying that the disfranchisement is still valid) and yet is apparently able to give evidence as a witness in court (**139**); cf. Hansen *Apagoge* 63–5. We should therefore reject the suggestion of U. E. Paoli *Studi di diritto attico* (Florence 1930) 322–3, followed by Harrison *Law* 2.83, that the only kind of disfranchisement suffered by frivolous prosecutors was that which applied to all citizens condemned to pay fines, viz. total disfranchisement which ended as soon as the fine was paid.

Yet further problems remain. Some passages mention the fine of 1000 dr. without giving any hint that the frivolous prosecutor was subject to partial disfranchisement in addition (**47**, 22.21, 22.26–7, 23.80, 24.3, And. 4.18, Pl. *Ap.* 36a–b), and Androtion, who got less than one-fifth of the votes as prosecutor in a *graphe* for impiety (24.7), was subsequently one of the prosecutors in a *γραφὴ παρανόμων* (24.14 *γράφονται τὸ ψήφισμα*). So it seems that the prohibition to prosecute in a *graphe* in future was not invariably imposed on a prosecutor in a *graphe* who failed to proceed or failed to get one-fifth of the votes. Perhaps it was imposed only if the offence was committed three times, as Genethlios said, according to Σ 22.3 (13a Dilts). Genethlios, who lived in the third century A.D., may have been only guessing; but the guess is plausible, since disfranchisement was likewise imposed only after the third offence on defendants convicted of giving false evidence or proposing illegal decrees (And. 1.74, Hyp. *Phil.* 11–12, Antiphanes 196.14). A different problem is raised by And. 1.33, where Kephisios, if he fails to get one-fifth of the votes, will suffer a form of disfranchisement which will prohibit him from entering the temple of Demeter and Kore. We can only conclude that the law about disfranchisement of frivolous prosecutors had complexities which cannot be fully elucidated on our present evidence.

104. Now D. refers to the attempt to incriminate him in connection with the killing of Nikodemos by Aristarkhos. But his account of the murder is incomplete, because he expects the jurors to remember it (it occurred only a year ago, and must have caused a sensation at the time) and also because he wants to avoid creating an association between the murder and himself in their minds. More details are given, from a standpoint hostile to D., by Aiskhines in his speech *Against Timarkhos*, where he alleges that D. habitually preyed on rich young men. 'After spending his inherited property, Demosthenes went around the city hunting rich young orphans whose fathers had died and whose mothers were managing their property. I will pass over many instances and mention only one of his victims. He spotted a rich estate not being well managed: a woman had charge of it, a conceited and silly person, and a half-crazy young orphan was handling the property, Aristarkhos son of Moskhos. Demosthenes pretended to be his lover, and invited the lad to enter into this affectionate relationship with him. Then he filled him with false hopes that before long he would become a leading orator (he provided a list of names [sc. of other men taught oratory by him]), and became his instigator and instructor in activities of such a kind that, in consequence, Aristarkhos is an exile from his native land, Demosthenes, who was entrusted with the money for his upkeep in exile, has robbed him of 3 talents, and Nikodemos of Aphidna has been violently killed by Aristarkhos; the poor man had both eyes gouged out and his tongue cut off—the tongue with which he spoke openly because he relied on the laws and you Athenians to protect him. And then, although you put Sokrates the sophist to death because it was found that he had trained Kritias, one of the Thirty who subverted the democracy, is Demosthenes going to get comrades acquitted, when he exacts revenge like that from private individuals and democrats for speaking freely?' (Ais. 1.170–3). He recurs to the subject more briefly in his *Embassy* speech: 'You were prosecuted for desertion and got off because you bribed the prosecutor, Nikodemos of Aphidna, whom you later joined Aristarkhos in killing' (Ais. 2.148); 'You entered the prosperous house of Aristarkhos son of Moskhos, and destroyed it; you were entrusted with 3 talents by Aristarkhos when he was exiled, and robbed him of the money for his upkeep in exile, though you were not ashamed of the reputation, which you claimed, of being an admirer of his youthful beauty—it wasn't true, for genuine love has no room for villainy' (Ais. 2.166). (The last passage especially is notable for the implication that love and dishonesty are incompatible; cf. K. J. Dover *Greek Homosexuality* (1978) 46–7.)

Little is known of Nikodemos. Ais. 2.148 says that he prosecuted D. for desertion. It seems unlikely that this was a separate prosecution from Euktemon's (**103**). Perhaps Nikodemos supported Euktemon, or perhaps he merely talked of prosecuting and did not proceed, though we cannot know whether it is true, as Aiskhines says, that D. paid him to drop the prosecution. We should accept that it was Euktemon, not Nikodemos, who formally initiated the prosecution for desertion, because the charge quoted in **103** gives his name.

As for Aristarkhos, we should think of him as a young adult, not a boy, at the time of these events; for a minor in Athens had to have a male relative as legal guardian, whereas Ais. 1.171 says that no one but Aristarkhos and his mother controlled the property. We may reconstruct the events as follows. D. befriended the young man; whether attracted by his beauty or his money, we cannot say on the basis of Aiskhines' hostile evidence. No doubt he promised to teach him oratory and help him to become a politician. Then Aristarkhos killed Nikodemos. Why? The killing cannot have been accidental, if Aiskhines' description of the multilation of the body has any truth in it. Did he really commit murder out of love and loyalty to D., because Nikodemos accused D. of desertion? He may have been mentally disturbed (Ais. 1.171 ἡμιμανής), but that would not make him any the less guilty of murder in Athenian eyes. The penalty for intentional homicide could be either death or exile, if the accused left Athens before the end of the trial (**43** n. θανάτῳ καὶ ἀειφυγίᾳ); and it was the duty of the relatives of the killed man to prosecute.

This was the point at which Meidias intervened. He too was friendly with Aristarkhos, and indeed Aristarkhos had previously interceded with D. to urge him to drop his dispute with Meidias (**117**). Now Meidias approached the relatives of Nikodemos and tried to get them to prosecute D. instead of Aristarkhos for homicide. What reason or evidence against D. he produced is unknown (because D. naturally avoids mentioning that), but anyway his attempt failed. Probably it was clear that Aristarkhos was the murderer. So Meidias then distanced himself from Aristarkhos by a remarkable denunciation. The death of Nikodemos was raised for discussion in the Boule. Why and by whom we are not told; homicide was not normally a matter for the Boule, but on this occasion someone must have declared that the killing of Nikodemos was a matter of public importance. Meidias stepped forward and demanded the execution of Aristarkhos (**116**). But the Boule apparently took no action, and on the next day Meidias was again chatting with Aristarkhos and denying that he had ever accused him (**119**). Unless D. is simply lying here, Meidias' behaviour does seem to have been even more capricious than usual.

Aristarkhos went into exile, either before or during the trial at which he was pronounced guilty of homicide. We cannot check the allegation that D. defrauded him of 3 tal. at this point (Ais. 1.172, 2.166). D. himself was never prosecuted for the killing, though the gossip that he was to blame for it, because he somehow advised or instigated Aristarkhos, was repeated by his enemies for the rest of his life (cf. Dein. 1.30, 1.47, besides Aiskhines).

One more problem remains: was Euboulos involved in these events (as Longo *Eterie* 96–7 maintains)? The only evidence that he was is in three scholia, presumably by the same scholiast. At **102** and **104** he calls Nikodemos a supporter and friend of Euboulos: *ἦν γὰρ οὗτος τῶν περὶ τὸν Εὔβουλον ἐσπουδακότων* (352 Dilts), *Εὐβούλου τοῦ πολιτευομένου ὢν εἰς τὰ μάλιστα φίλος* (364 Dilts). At **205** he associates Euboulos with the homicide charge: *τὴν φήμην λέγει τὴν περὶ τοῦ φόνου· Εὔβουλος γὰρ ἦν ὁ κατηγορῶν τοῦ Ἀριστάρχου καὶ συναιτιώμενος τὸν Δημοσθένην* (687 Dilts). If it is true that Euboulos was the man who prosecuted Aristarkhos for homicide, he must have been a relative of Nikodemos. He would then be included in **104** *τοῖς τοῦ τετελευτηκότος οἰκείοις*, and an Athenian jury would have known that, I find it hard to believe that D. would have said in this manner that Meidias tried to bribe Euboulos. Besides, it is not clear where the scholiast could have found this information; elsewhere the scholia on D. do not usually seem to draw on historical sources other than the orations which we still have, but no surviving oration links Euboulos with Nikodemos. I am therefore inclined to think that the scholiast has postulated this link on his own conjecture, in an attempt to explain the events as part of a rivalry between D. and Euboulos. I doubt that Euboulos really had anything to do with the matter.

δεινόν ... καὶ σχέτλιον ...: **61** n. This passage has much pleonasm for rhetorical effect; **1** n. *ὕβριν*.

κοινὸν: affecting the community in general, not just D. personally.

ἔμοιγ' goes syntactically with *δοκεῖ*, but is placed between *κοινὸν* and *ἀσέβημα* to space out those two words and so enable them both to be emphasized.

ἀσέβημα, οὐκ ἀδίκημα: the rhyme is reminiscent of Gorgias, who is fond of nouns in -*ημα*, e.g. *Helen* 19 *εἰ δ' ἐστὶν ἀνθρώπινον νόσημα καὶ ψυχῆς ἀγνόημα, οὐχ ὡς ἁμάρτημα μεμπτέον ἀλλ' ὡς ἀτύχημα νομιστέον*. Cf. **130** *κακουργήματα ... ἀσεβήματα*.

Homicide was regarded by the Greeks primarily as an offence against the victim and his family, but also as an offence against the gods; thus killing is sometimes called *ἀσέβεια* (e.g. Ant. 4c.6). This usage is sometimes extended in two ways:

A. A person who associates with a killer in a friendly manner, especially by going into the same house, may be regarded as *ἀσεβής*. This applies particularly to relatives of the killed person, since it is their duty to bring the killer to justice; thus a man was actually prosecuted for *ἀσέβεια* because he went into the same house with his nephew, who was alleged to be a parricide (22.2). In **120** D. says it was *ἀσεβές* for Meidias to go into the same house with Aristarkhos after denouncing him for homicide, though in this case Meidias was not a relative of Nikodemos and so did not have a duty to prosecute.

B. A person who accuses an innocent man of homicide may be regarded as *ἀσεβής*. Here the logical basis may be that such an accusation causes a killing, if the innocent man is executed. Cf. Ant. 4b.7 *καθαρῷ μέν μοι τῆς αἰτίας ὄντι φόνον ἐπικαλοῦντες, ἀποστεροῦντες δέ με τοῦ βίου ὃν ὁ θεὸς παρέδωκέ μοι, περὶ τὸν θεὸν ἀσεβοῦσιν*. But in **104** D. blurs this logic: when Meidias has accused him of killing Nikodemos, he calls that accusation *ἀσέβημα* even though he would not have been executed but would have gone into exile as Aristarkhos did (**105** *ἐξόριστον* and *ἐφ' αἵματι φεύγειν*, **115** *ἐκβαλεῖν ἐκ τῆς πατρίδος*).

τούτῳ: dative of agent, as often with a perfect passive.

περιιὼν: **36** n. *περιιόντ'*. Cf. Phryn. com. 3.4 *κατὰ τὴν ἀγορὰν περιιόντες*.

τοῖς τοῦ τετελευτηκότος οἰκείοις: the duty of prosecution for homicide fell upon the deceased's relatives; cf. MacDowell *Homicide* 8–32.

105. πρὸς οὓς: best taken as equivalent to *τούτους πρὸς οὓς* (rather than *πρὸς τούτους οἷς*), because *πρός* is often used with *λέγω*, rarely with *αἰσχύνομαι*. 'He was not abashed even at the very men whom he was addressing', the relatives of the murdered man; much less was he afraid of anyone else's opinion.

ὅρον: 'limit': Meidias would stop at nothing until he had destroyed D.

ἀνῃρῆσθαι: this verb has already been used earlier in the sentence (ἀνελεῖν), and Naber *Mn.* II 31 (1903) 417 suggests that here D. wrote ἀνηρπάσθαι (cf. **120**). But the repetition is unobtrusive and emendation is hardly justified.

παρεθῆναι: 'be let off' without penalty. For this sense of *παρίημι* cf. 23.85 *τούτους μὲν ἀθῴους παρῆκε*.

ἡλωκέναι: there is no need to emend to *ἑαλωκέναι*, because *ἡλ-* is supported by the metre in two passages of fourth-century comedy, Antiphanes 204.7 and Xenarkhos 7.17.

ἐφ' αἵματι φεύγειν: 'be exiled for homicide'; cf. ML 43.2 *φεύγεν τὴν ἐπ' αἵμ[ατι φυγήν]*. The alternative translation 'be prosecuted for homicide' is less likely to be right here, because the rhetorical crescendo requires something more, not less, than the preceding *ἡλωκέναι*.

προσηλῶσθαι: nailing to a board (*σανίς* or *τύμπανον*) was a method of executing criminals. There is a comic example in Ar. *Thesm.* 930–46 (with a reference to a nail in 1003). In real life, Menestratos was executed in this way when condemned as a killer (Lys. 13.56), and other instances (not in Athens) are mentioned in Hdt. 9.120.4, Arist. *Rhet.* 1385a 10–11, Plu. *Per.* 28.2. Cf. A. D. Keramopoullos *Ὁ ἀποτυμπανισμός* (Athens 1923), Bonner and Smith *Justice* 2.279–82, MacDowell *Homicide* 111–13.

106. αὐτόχειρά μου: 'my destroyer'. *αὐτόχειρ* may be used of the perpetrator of any act (e.g. **60**), but it is most often used of a killer (e.g. **116**) and in that sense can take a genitive of the person killed; cf. 20.158 *αὐτόχειρ' ἄλλον ἄλλου γίγνεσθαι*, Ant. 5.47, Isok. 4.111, Men. *Sam.* 561. But the word does not necessarily refer to killing, and D. here is using it in an ambivalent manner for rhetorical effect. He never actually alleges that Meidias' activities could really have led to his own death (notice **105** *μόνον οὐ προσηλῶσθαι*, and in **106** the list *τὰ λοιπὰ πάντα ... τὰς ἐλπίδας* does not include 'life'), but by using the word *αὐτόχειρ* he makes it sound as if Meidias were a murderer.

τοῖς Διονυσίοις: temporal (defining *τότε*) but not precise, since the alleged damage to the cloaks and crowns occurred not at the festival but before it.

τὴν πόλιν, τὸ γένος: 'attacking my city and my family' means trying to deprive D. of them. He would be unable to live in Athens if he were pronounced guilty of homicide.

τὴν ἐπιτιμίαν: D. would be disfranchised if he were found guilty of desertion.

ἕν: 'just one', viz. the attempt to get D. convicted of homicide.

μηδὲ ταφῆναι ... οἴκοι: this seems to be the only evidence that a person exiled for homicide could not be brought back to Athens for burial when he died. But there is evidence that this prohibition applied to other serious offenders; cf. X. *Hell.* 1.7.22, Lyk. *Leo.* 113, Plu. *Eth.* 834b. *μηδὲ* is used rather than *οὐδὲ* (A) because the negative goes with the infinitive only, not with *προσυπῆρχεν*.

διὰ τί ...: the argument is 'What justification has Meidias for persecuting me like this, when the proceedings which I was taking against him were perfectly reasonable?'

βοηθεῖν αὑτῷ: **7** n. *βοηθῆσαι*.

προσκυνεῖν: Greeks regarded with scorn the oriental practice of obeisance to an autocratic ruler. All three instances of this verb in D. (the others are 19.314, 25.37) are contemptuous.

107. προσεξείργασται: *προσ-* means 'in addition to' the attack on D. at the Dionysia; so also in **109**.

The document here, like the other statements attributed to witnesses in this oration, is spurious; see the following notes on *κέρματα* and *γραφὴν*.

'Αφιδναῖος: the forger of the document is knowledgeable enough to have discovered (probably from Ais. 1.172, 2.148) not only the name of Nikodemos but also his deme, and so has given the same demotic to one of the relatives whom he has invented. That the other relative is given a different demotic is not absurd; he may be supposed to be, say, a maternal uncle of Nikodemos.

τὸν 'Αρίσταρχον: a dative of the person prosecuted is usual with *ἐπεξέρχομαι*. But there are instances of the accusative (Ant. 1.11, Lys. 31.18); so this is not in itself proof of the document's spuriousness.

κέρματα: 'cash', 'small change'. In Attic this word is virtually confined to comedy, and is a colloquial word for small coins. It is not suitable in a formal document as a general word for money; this is a blunder by the forger of the document.

γραφὴν: another blunder. A prosecution for homicide was a *δίκη φόνου*. Whether such a thing as a *γραφὴ φόνου* was ever possible has been disputed; cf. most recently Hansen *Apagoge* 108–12, Gagarin *GRBS* 20 (1979) 322–3. But even if it was possible, it was a procedure for use by accusers other than the deceased's relatives and is not appropriate here.

παραγράψασθαι: *παρα-* here implies substitution of one name for another, as in 39.31 *πατρὸς ἄλλου σεαυτὸν παραγράφειν*.

108. ἐν ὅσῳ: 'while', implying that it takes the clerk a little time to find the document. The law is eventually read at **113**. The various documents which an orator would wish to have read out during his speech must have been collected and given to the clerk in advance, but perhaps they were not arranged in the sequence in which the orator would call for them. Nevertheless, since usually a speech contains no material to cover the gap while the clerk is searching, a passage as long as **108–12** cannot really have been needed for that purpose here; *ἐν ὅσῳ … λαμβάνει* is just a rhetorical pretext for slipping in some further matters at this point. Cf. 18.219–21, also apparently spoken while the clerk is preparing to read a document.

δεηθεὶς ὑμῶν: a standard form of words in sentences saying 'I shall tell you …, after first making a request of you …'; cf. 4.13, 10.54, *pr.* 21.4, 50.1.

πρὸς Διὸς καὶ θεῶν: **73** n.

ἐφ' οἷς: equivalent to *ἐπὶ τούτοις ἅ*. For *ἐπί* and dative with *χαλεπῶς φέρω*, cf. 23.170 *τῶν δὲ Θρᾳκῶν ἁπάντων χαλεπῶς ἐνεγκόντων ἐπὶ τούτοις*, 45.82, 54.15.

τὴν λειτουργίαν: D.'s service as a khoregos.

τούτοις τοῖς μετὰ ταῦτα: 'at these subsequent events', described in **102–7**. For the dative (without *ἐπί*) with *χαλεπῶς φέρω*, cf. *pr.* 41.1, X. *Hell.* 5.1.29, *An.* 1.3.3.

109. ὑπερβολήν: **16** n. Here 'What going-beyond of shamelessness is there?' means 'How can shamelessness go beyond this?'

ποιήσας … ἀδίκως τινά refers to the incidents before and at the Dionysia.

ἀναλαμβάνειν: 'atone for', as in S. *Ph.* 1248–9 *τὴν ἁμαρτίαν αἰσχρὰν ἁμαρτὼν ἀναλαβεῖν πειράσομαι.*

ἐκβαλών τινα alludes to the attempt to get D. exiled for homicide. The attempt did not succeed, but that does not mean that we must adopt *ἐκβάλλων* (S); for this clause (with the future verb *εὐδαιμονιεῖ*) is one expressing purpose, not fact.

τῆς περιουσίας: 'his affluence', which enables him to defeat his enemy (by bribing people to prosecute him). For the sense of *περιουσία*, cf. **159**. For the genitive of cause with *εὐδαιμονίζω*, cf. 19.67.

110. οὐδὲν ἐμοὶ προσήκουσαν: for the sense of *οὐ προσήκων*, meaning that the accusation ought to have been made against someone else, cf. **111** *οὐδὲν ἐμοὶ προσῆκεν*, Ant. 5.2 *τῆς αἰτίας τῆς οὐ προσηκούσης*.

τὸ πρᾶγμα: 'the outcome', the conviction of Aristarkhos for the murder.

ἐγράψατο: actually Euktemon, not Meidias, was the prosecutor, according to **103**; so this verb should be interpreted as 'he got a prosecution brought'. For the double accusative cf. 18.251, 58.36, 59.1.

τρεῖς αὐτὸς τάξεις λελοιπώς: the three occasions when Meidias was guilty of desertion, according to D., were when he donated a trireme to avoid service with the cavalry (**162**), when he avoided service on his trireme by staying in Athens for the Dionysia (**163**), and when he eventually joined the ship to avoid another call-up of cavalry (**164**). That these acts did not legally constitute desertion is indicated by the fact that Meidias was not prosecuted for it, though D. says that he ought to be (**166**).

τῶν ἐν Εὐβοίᾳ πραγμάτων: the genitive goes with *αἴτιός εἰμι*. For the expedition to Euboia in 348 see pp. 5–7; for Meidias' speech about it see pp. 9–10.

μικροῦ: 'almost' (LSJ *μικρός* III.2); so also in **216**.

παρῆλθέ με: 'eluded me', i.e. 'I forgot'; cf. 19.234 *μικροῦ γ', ἃ μάλιστά μ' ἔδει πρὸς ὑμᾶς εἰπεῖν, παρῆλθεν*. Note the rhetorical hypocrisy involved in putting such a remark into the text of a speech written out before delivery; but this kind of hypocrisy is histrionic rather than immoral.

ξένος καὶ φίλος: after the failure of the expedition to Euboia Ploutarkhos was no doubt execrated in Athens; so, to disparage Meidias, D. emphasizes and perhaps overstates the connection between them.

κατεσκεύαζεν: the verb means 'fabricate', implying that the allegation was untrue, and the imperfect is conative, 'he tried to make out that ...'; similarly in **115**, **134**.

πρὸ τοῦ governs everything to the end of the sentence; *γενέσθαι φανερὸν* is equivalent to *φαίνεσθαι* and is followed by a participle.

111. βουλεύειν: 'to be a member of the Boule' for 347/6. On Meidias' accusation of D. at the *dokimasia* see p. 10.

ὑπέρδεινον: 'very alarming'. The same compound is used in 20.47 in the slightly different sense 'very shocking' (omitted in LSJ).

περιέστη: 'was converted'. Cf. 37.10 *τὸ πρᾶγμά μοι περιεστηκὸς εἰς ἄτοπον*.

δοῦναι πραγμάτων: *δίκην* is easily understood, and is probably an intrusive gloss in those mss. in which it appears in the text.

τῶν ἐρημοτάτων: '⟨one⟩ of the most friendless'. Cf. **80** *τότε παντάπασιν ἔρημος ὢν καὶ νέος κομιδῇ* (many years earlier).

112. εἰ γὰρ ... δεῖ: D. makes a show of hesitancy about introducing what is actually one of the main themes of the speech.

τῶν νόμων: sharing in the laws means having legal protection; the rest of **112** explains how legal protection against the rich is inadequate. Cf. **205** *μηδὲ τῆς κοινῆς τῶν νόμων ἐπικουρίας ἀξιῶν ἐμοὶ μετεῖναι*, 23.86 *ὥσπερ γὰρ τῆς ἄλλης πολιτείας ἴσον μέτεστιν ἑκάστῳ, οὕτως ᾤετο δεῖν καὶ τῶν νόμων ἴσον μετέχειν πάντας*. The alternative reading *τῶν ὁμοίων* (AF) is not preferable; although *ἴσος* and *ὅμοιος* are often linked, they are synonyms, so that here, if *ὁμοίων* were right, we should expect *τῶν ἴσων καὶ ὁμοίων*: cf. Th. 5.79.1 *ἐπὶ τοῖς ἴσοις καὶ ὁμοίοις*, 1.27.1, 1.145, 4.105.2, X. *Hell.* 7.1.45.

λοιποῖς: *πολλοῖς* (Π11AF) may be right, but is more likely than *λοιποῖς* to be an intrusive gloss; cf. *τῶν δ' ἄλλων ἡμῶν* later in the sentence.

ἡμῖν: Taylor's emendation should be accepted; the partitive genitive *ἡμῶν* would imply that the rich were also included among 'us', which would be contrary to D.'s aim of distinguishing Meidias and the rich from the jury and himself. Cf. **138** *τοῖς ἄλλοις ἡμῖν*, 18.324 *ἡμῖν δὲ τοῖς λοιποῖς*.

οὐ μέτεστιν, οὔ: repetition for emphasis, similar to **46** *οὐ γὰρ ἔστιν, οὐκ ἔστιν*. For the repetition of *οὔ* alone, cf. 19.97, 19.186, 19.255.

χρόνοι: the date for a trial was fixed by the arkhon or other magistrate responsible for the type of case concerned. There were some cases for which the trial had to be held within thirty days (**47** n. *τριάκοντα ἡμερῶν*). In other cases there was no time-limit, as far as we know (though presumably anyone who considered that a magistrate had delayed a case unreasonably could make an accusation against him at his *euthyna*). Here D. means that a rich man could persuade a magistrate, either by bribery or by more indirect favours, to delay the trial of a case in which he was the defendant.

ἕωλα ... ψυχρὰ ... πρόσφατος: a notable metaphor from food, perhaps fish. Compare the slightly different metaphor used in 25.61 for a fight between two prisoners: *νεαλὴς δὲ καὶ πρόσφατος ὢν ἐκεῖνος περιῆν αὐτοῦ τεταριχευμένου καὶ πολὺν χρόνον ἐμπεπτωκότος*, 'the other man, who was newly caught and fresh, got the better of him, who was smoked and had been in jail a long time'. As in that passage, so also here the metaphor is clearer when conveyed by two adjectives rather than one; so the deletion of *καὶ ψυχρὰ*, proposed by Jahn *Philol.* 26 (1867) 3, is not desirable.

ἄν τι συμβῇ: a euphemism for being prosecuted.

μάρτυρες: in **139** D. names some of the men (*μαρτύρων ἑταιρεία*) who, he alleges, were paid to testify in support of Meidias.

συνήγοροι: supporting speakers in court (not to be confused with other kinds of *συνήγορος*: cf. my note on Ar. *Wasps* 482). Such men were normally supposed to be relatives or friends of the defendant, supporting him because of their personal connection. To accept payment for being a *συνήγορος* was an offence, subject to prosecution by *graphe* (46.26); but D. here implies that it happened nevertheless.

οὐδὲ τἀληθῆ μαρτυρεῖν ἐθέλοντας: a witness who was reluctant to testify could be compelled to do so (58.7). Two procedures are known: a man who failed to attend as a witness when summoned could be prosecuted for *λιπομαρτύριον* (49.19); one who did attend but refused to testify was subject to a procedure called *κλητεύειν*, leading to a fine of 1000 dr. (59.28, Ais. 1.46, Lyk. *Leo.* 20); cf. Harrison *Law* 2.138–43. Here *ὁρᾶτ'* may indicate that D. expected (when he was writing the speech) to have to resort to one of these procedures at the trial.

113. θρηνῶν: a mainly poetic verb, not used elsewhere by D.; it is used here to give the effect of a wry joke. *ταῦτα* is its object (LSJ is wrong to classify this passage under *ἀπεῖπον* IV.1; it ought to be under IV.3.f), and the whole sentence means that D. is cutting his words short, refraining from speaking at length about corruption in the lawcourts: 'one would wear oneself out, bewailing these matters'.

ὥσπερ ἠρξάμην: 'as I began' seems illogical. The point is probably not that D. himself read out the first words of the law (as Weil suggests), but that the reading by the clerk is part of D.'s presentation of the law to the jury.

The document preserved here may be accepted as genuine, but it must be distinguished from other laws specifying other penalties for particular sorts of bribery: one about magistrates found guilty at their *euthyna* (*AP* 54.2, Dein. 1.60, Hyp. *Dem.* 24), one about speakers in the Ekklesia (Dein. 2.17), one about bribery of jurors, members of the Boule, or *συνήγοροι* (D. 46.26), and one about bribery to secure acquittal on a charge of simulation of citizenship (*δωροξενία*, *AP* 59.3). The law in **113** is more general than the others, specifying *ἀτιμία* as a penalty for any Athenian who either accepted or offered bribes in any circumstances to the detriment of the people or of an individual citizen. (*ἀτιμία* as a penalty for bribery is mentioned also in And. 1.74, Ais. 3.232.) I have discussed these laws in *RIDA* 30 (1983) 57–78, where I argue that the law in **113** is an old one, perhaps of the sixth century, because its last words *καὶ τὰ ἐκείνου* have no meaning unless *ἄτιμος* is given its old sense of outlawry, rather than its later sense of disfranchisement. (A similar phrase, *ἄτιμον εἶναι καὶ παῖδας {ἀτίμους} καὶ τὰ ἐκείνου*, occurs in the homicide law attributed to Drakon, quoted in 23.62.) Outlawry was a severe penalty: the offender was deprived of legal protection, so that there was no legal redress if he and his family were killed or injured or if his property was stolen or damaged. In the course of the fifth century *ἄτιμος* came to be interpreted as meaning merely 'disfranchised', so that the words *καὶ τὰ ἐκείνου* were no longer significant, and the penalty for bribery was less severe than when the law was made. The other laws were therefore added at various times in order to impose further penalties for the worst kinds of bribery. Here D. does not refer to the other laws, because the old general law is the only one applicable to Meidias' alleged attempt to bribe the relatives of Nikodemos to prosecute D. for homicide. For discussion of the prevalence of bribery in this period see F. D. Harvey in *Crux* 76–117.

᾿Αθηναίων: nothing is said here about metics or other aliens offering bribes. Perhaps they were dealt with in another law.

λαμβάνῃ: one would expect an object. Perhaps *τι* should be restored in the text.

ἐπαγγελλόμενος: promising to pay after the performance of the favour requested, whereas *διδῷ* refers to payment in advance.

τρόπῳ ἢ μηχανῇ ᾑτινιοῦν: similar phrases occur in other legal documents: 24.150 *οὔτε τέχνῃ οὔτε μηχανῇ οὐδεμιᾷ*, 59.16 *τέχνῃ ἢ μηχανῇ ᾑτινιοῦν*, Th. 5.18.4 *μήτε τέχνῃ μήτε μηχανῇ μηδεμιᾷ*, *IG* 1³ 40.22–3 *οὔτε τέχνει οὔτε μεχανêι οὐδεμιâι*. It is possible that *τρόπῳ* should be emended to *τέχνῃ*.

It is noticeable that the reading of the law is not followed by any comment or explanation of how it applies to Meidias; cf. p. 26.

114. ἀσεβὴς: because he accused an innocent man of homicide (**104** n. *ἀσέβημα*). Not because he allowed a man he believed guilty of homicide to perform sacred rituals, for the point is that he did not believe D. guilty; cf. **115** *εἴ γ' εἶχε ..., ταῦτ' ἂν εἴασεν;*

ἂν ὑποστὰς: 'who would consent' if occasion arose.

ἀλλ' οὐδ' ὁτιοῦν: 'not a bit', an idiom found also in 25.5. Cf. Denniston *Particles* 23–4.

εἰσιτητήρια: the mss. give *εἰσιτήρια* here and in 19.190, but the longer form is the one attested by Athenian inscriptions, *IG* 2² 17.22–3, 689.20, 975.4, 1011.34, 1315.7. For the accusative with *ἱεροποιῆσαι*, cf. *IG* 2² 1749.80. The inaugural rites were held at the beginning of the new year (midsummer 347) when D. became a member of the Boule (cf. **111**). Presumably the Boule itself elected one or more of its members to carry out the sacrifice; cf. Rhodes *Boule* 130.

115. ἀρχιθεωροῦντα: the function of an *ἀρχιθέωρος* was to lead the Athenian representatives at one of the international festivals (Olympic, Pythian, Nemean, or Isthmian games) and pay their expenses; this, like being a khoregos, was a liturgy. D. evidently performed this liturgy at the Nemean games in 347. There is some doubt about the spelling: Attic inscriptions of the fourth and third centuries offer two instances of *ἀρχεθέωρος* (*IG* 2² 1534.170, 1635.34), two of *ἀρκεθ-* (*IG* 2² 365a.10, 365b.7), and one of]*εθ-* (*Hesperia* 37 (1968) 375 no. 51 line 26). But that is perhaps not enough to prove that D. did not write *ἀρχιθ-*, which is epigraphically attested elsewhere.

ταῖς Σεμναῖς θεαῖς: a name used at Athens for the Erinyes or Eumenides, the avengers of homicide. Their shrine was on the Areopagos, near the entrance to the Akropolis (Th. 1.126.11, Paus. 1.28.6).

ἱεροποιὸν: hieropoioi were officials (not priests) appointed to perform a religious ceremony. Some were appointed by lot (e.g. *IG* 1³ 82.17–23, *AP* 54.6–7), but here *αἱρεθέντα ἐξ Ἀθηναίων ἁπάντων* means that those for the *Σεμναὶ θεαί* were elected by the Ekklesia. Their number was not fixed (cf. Phot. *ἱεροποιοί*): on this occasion there were three, on another ten (Dein. fr. 8.2 Conomis = 4 Burtt).

τρίτον αὐτὸν, 'with two others', does not mean that D. was superior to his two colleagues (cf. K. J. Dover *JHS* 80 (1960) 61–77), but *καταρξάμενον τῶν ἱερῶν* may imply that they gave him precedence.

ταῦτ' ἂν εἴασεν;: pollution (*μίασμα*) might be spread by a man guilty of homicide if he associated with other people, and especially if he participated in religious rites. Anyone who perceived a religious obstacle would be expected to object, and the proceedings would then be stopped. (Cf. the comic instance in Ar. *Akh.* 169–73.) So the fact that Meidias did not object to D.'s participation in public rites shows that he did not really believe him guilty of homicide.

116. οὐδὲ καθ' ἕν: 'in not a single respect'. The same phrase is used in Th. 2.87.7.

δι' ἐμὲ: having failed to get D. personally prosecuted for homicide, Meidias as a second-best tried to urge on the prosecution of Aristarkhos, so that D. as Aristarkhos' friend would be blackened by association. The sense of *δι' ἐμέ*, 'for the sake of attacking me', is the same as in 18.13 *Κτησιφῶντα μὲν δύναται διώκειν δι' ἐμέ*. The accusative is preserved by a scholiast on Hermogenes; it is written above the line, probably as an emendation, in two of the later mss. of *Meidias* (BT).

ἐσυκοφάντει: 103 n. *συκοφάντης*. Here the word is even vaguer: Meidias was not prosecuting (the prosecutors were the relatives of Nikodemos) but merely making allegations, as explained in the next sentence.

τὰ μὲν ἄλλα σιωπῶ: another paralipsis, to give an impression that Meidias made other attacks on Aristarkhos besides those mentioned.

περὶ τούτων καθημένης: the Boule was already discussing the death of Nikodemos before Meidias spoke. It could discuss any matter of public concern, but it did not have authority to decide homicide cases (which were tried by the Areopagos and other special homicide courts), and D. fails to explain here who had raised the subject of Nikodemos in the Boule and why.

παρελθὼν: there is no other evidence that Meidias was a member of the Boule this year. More likely he was not, but was present as a member of the public listening to the debate, and requested permission to address the Boule on the ground that he could contribute to the subject under discussion (like Andokides in Lys. 6.33).

οὐκ ἀποκτενεῖτε;: D. makes Meidias sound absurdly precipitate (execution first, arrest afterwards). In fact an alleged killer could not legally be arrested at his own house, but only if he entered sacred and public places; nor could he be put to death without trial (23.80). If Meidias really said what D. attributes to him here, he was speaking wildly, not proposing a proper legal action. Cf. MacDowell *Homicide* 130–40, E. M. Carawan *GRBS* 25 (1984) 111–21.

117. αὕτη: when an epithet follows the article, a demonstrative may stand between that and its noun. Cf. **119** *τοὺς ἐν τῇ βουλῇ τούτους λόγους*.

κεφαλή: 'person', used in this sense by D. only in abusive phrases. Cf. 18.153 *ἡ μιαρὰ κεφαλὴ … αὕτη* and the vocative in **135**, **195**, 19.313. Masculine participles follow, according to sense rather than grammar.

ἐξεληλυθὼς … παρ' Ἀριστάρχου: 'though he had come out of Aristarkhos' house', i.e. had visited Aristarkhos. Anyone who really believed a man guilty of homicide would not enter his house for fear of pollution. The same argument, to prove an accusation of homicide insincere, is used in Ant. 6.39 *συνῆσάν μοι καὶ διελέγοντο ἐν τοῖς ἱεροῖς, ἐν τῇ ἀγορᾷ, ἐν τῇ ἐμῇ οἰκίᾳ, ἐν τῇ σφετέρᾳ αὐτῶν καὶ ἑτέρωθι πανταχοῦ.* Since the dangerous thing was to enter the house, not to leave it, it is at first sight surprising that D. does not say *εἰσεληλυθὼς … παρ' Ἀρίσταρχον*: the reason is probably that he wants to convey a picture of Meidias going almost straight from the house to the Boule.

τὰ πρὸ τούτου: 'during the period before this'; cf. *τὰ πρὸ τούτων* in 18.188, 18.222. Reiske's conjecture *πρὸ τοῦ* is not needed.

ὅτ' ηὐτύχει: before Aristarkhos was accused of killing Nikodemos, which was his 'misfortune' (cf. **104**, **122**). For the occasion when friends of Meidias tried to persuade D. to drop the prosecution see **151**. For the augment *ηὐ-* (A), rather than *εὐ-* (SFYP), see Threatte *Grammar* 1.384–5.

διαλλαγῶν: 'reconciliation'. S has *ἀπαλλαγῶν*, 'withdrawal'; cf. **151** *ἀπαλλαγῆναι*. But *ἀπ-* does not fit well with *πρὸς τοῦτον*. (Pl. *Laws* 915c *ἐὰν μὴ … ἀπαλλάττωνται πρὸς ἀλλήλους τῶν ἐγκλημάτων* is different, because there *ἀπ-* is to be taken with the genitive.) Cf. **119** *πρὸς ἐμὲ … τὰς διαλύσεις*, 38.6 *πρὸς ὃν αὐτοῖς ἐγένονθ' αἱ διαλλαγαί* (where S has *δι-* and other mss. have *ἀπ-*).

ἐφ' οἷς ἀπόλωλεν: 'for which he has been ruined'. Not 'died'; Aristarkhos was not put to death, but went into exile (Ais. 1.172).

οὐδ' οὕτως: 'not even in those circumstances' ought Meidias to have attacked Aristarkhos.

118. τῆς λοιπῆς φιλίας: 'friendship in future'. Perhaps, as suggested by Dobree *Adversaria* 1.462 and Cobet *Misc. crit.* 510, D. wrote *τοῦ λοιποῦ* (cf. 4.15 *οὐκέτι τοῦ λοιποῦ*, 19.137, 19.149, 23.188, *pr.* 1.2, 21.4) and copyists have mistakenly assimilated it to the gender of *φιλίας*.

τούτῳ γε: sarcastic; what would be offensive in anyone else may be forgiven in Meidias.

ὁμωρόφιος: one should avoid sharing a roof with a killer. The same word is used in similar contexts in 18.287, Ant. 5.11. The passage is quoted by H. Stephanus in his *Thesaurus* under ἅλς with different readings: 'in Or. C. Mid. ita loquitur, *Εἰ δὲ ἁλῶν μὲν κοινωνήσας καὶ ὁμωρόφιος γενόμενος ὡς οὐδὲν εἰργασμένῳ φανήσεται·* ita enim legitur h. l. in optimis et vetustissimis codd.' No ms. that I have collated has *κοινωνήσας*, but A after correction has *αλων* (without breathing or accent) and *γενόμενος*. Presumably Stephanus saw a ms. derived from A in which *κοινωνήσας* had been added by conjecture to make sense of *αλων*. Sharing of salt would be a symbol of friendship, but in fact *λαλῶν* is more likely to be the correct reading here, because in **119** D. says *διείλεκτο ἐκείνῳ* and makes no mention of a meal.

δεκάκις ... ἀπολωλέναι: for the rhetorical statement that the accused deserves death several times over, cf. **201**, 19.110, 19.131, 24.207.

119. τῇ προτεραίᾳ ...: 'on what was, when he said that, the previous day'.

οὐκ ἔχον ἐστὶν ὑπερβολὴν: 'is not outdone', 'sets a record'; **16** n. *ὑπερβολή*. The words *ἔχον ἐστὶν* are not to be taken together as a two-word verb (like English 'is walking'). The phrase is *ἔχον ὑπερβολήν*, meaning 'second', 'inferior'; this phrase is emphasized by having an unemphatic word, *ἐστὶν*, placed after its first word. Cf. 2.26 *οὔτ' εὔλογον οὔτ' ἔχον ἐστὶ φύσιν τοῦτό γε*, 'that is neither reasonable nor natural'; 25.54 *δεινῶν γὰρ ὄντων, οὐ μὲν οὖν ἐχόντων ὑπερβολήν*, 'for although they are extraordinary, indeed unsurpassed'.

ἀκαθαρσίας: 'impurity', involvement in the pollution of homicide.

ἐφεξῆς οὑτωσὶ καθιζόμενος: 'sitting down close to him, like this'. Elsewhere in D. *ἐφεξῆς* is always temporal; but here the deictic *οὑτωσὶ*, which is surely to be accompanied by a gesture (perhaps pointing at the jury, as Σ (420 Dilts) suggests), makes the local sense certain. (LSJ is therefore wrong to classify this passage under *ἐφεξῆς* II.3.) *καθεζόμενος* (AF) may be correct; **56** n. *καθίζεσθαι*.

τούτους: **117** n. *αὕτη*.

κατ' ἐξωλείας: the regular phrase for an oath invoking destruction on oneself if one is not telling the truth; cf. 23.67 *διομεῖται κατ' ἐξωλείας αὑτοῦ καὶ γένους καὶ οἰκίας*, 57.22.

περὶ αὐτοῦ: it is not necessary to adopt *κατ' αὐτοῦ* (A). Cf. 24.127 (126 in some editions) *περὶ μὲν τοῦ πατρὸς αὐτοῦ οὐδὲν ἂν φλαῦρον εἴποιμι*, 40.48.

φλαῦρον (A), rather than *φαῦλον* (S), is normal in the sense 'disparaging'; cf. **208**, 20.13, 20.102, 22.12, 24.127.

καὶ τούτων sums up everything from the beginning of **119**. *καὶ* ('also') is omitted in S, but should probably be retained here, as in **121**.

τοὺς παρόντας: with past reference, 'those who were present at the time'.

καλῶ does not mean that D. will himself call the witnesses, but that he will get the clerk to do so; cf. **167**, and **10** *ἀναγνῶναι*. This sentence has prompted the insertion of *ΜΑΡΤΥΡΙΑ* or *ΜΑΡΤΥΡΕΣ* here in most mss., but in fact D. does not request the clerk to call the witnesses until **121**. Probably the word has been inserted by copyists, though Sykutris suggests that D. wrote it himself before deciding to add **120**.

120. πῶς οὐ δεινόν: **61** n. *οὔκουν δεινόν*. Here the actions of Meidias which D. regards as disgraceful because of their inconsistency are (*a*) calling a man (Aristarkhos) a murderer, (*b*) treating him as innocent of murder. Both are stated twice; in the first statement (*λέγειν ... ἀπομνύναι*) the contrast is marked not by the usual *μέν* and *δέ* but by *καὶ πάλιν*, 'and then again'.

ἀσεβές: **104** n. *ἀσέβημα*.

ὡς φονεύς: sc. *ἐστί τις*.

κἂν μὲν ἀφῶ ...: here D. leaves Aristarkhos and reverts to the attempts to prosecute himself, developing the point already made in **105**. He is alleging that Meidias' motive for getting him prosecuted for desertion or homicide was not that he thought him guilty of either offence, but to deter or prevent him from going ahead with his prosecution of Meidias. (The speaker of Ant. 6 alleges that the prosecution of himself for homicide has a similar motive.) D. expresses the motive in an effectively sarcastic manner, marked by *ὡς ἔοικ'*, which in effect means 'according to Meidias'.

προδῶ: 'betray', 'fail to preserve'. Here, as in **3** and **40**, D. regards the vote against Meidias, when the *probole* was considered by the Ekklesia, as an expression of the Athenians' wish that he should prosecute Meidias, so that it would be disloyal of him not to do so.

ἀνηρπάσθαι: like *ἀπολωλέναι*, 'be ruined', 'be destroyed' in a general sense; so also in **124**, **125** (*προ-*), 9.47, 10.18, 59.8. (LSJ *ἀναρπάζω* II.1 is wrong to say 'to be carried off to prison'. Imprisonment was not the penalty for either desertion or homicide.)

αὖ τοὐναντίον: both here and at 22.5 the mss. are divided between this reading and *αὐτὸ τοὐναντίον*. Corruption of either reading to the other, by haplography or dittography of *το*, would be very easy, and could have been produced independently by more than one copyist. But *αὖ* is supported by **37** *ἐμοὶ δ' αὖ τοὐναντίον*, **129** *πάλιν τοὐναντίον*.

ἄν: Blass's addition seems necessary to give the sense 'I think that, if I had let him go, I should have been a deserter from the post of justice'. Goodwin defends the omission of *ἄν* by saying that 'it would have been said' is implied. But that can hardly be right; this sentence does not state what Meidias or anyone else would have said, but D.'s own opinion (*ἐγὼ ... οἶμαι*). Humbert's translation 'si je renonce à le poursuivre, je crois que j'abandonne mon poste' is forbidden by the tenses of the verbs; it would require *ἂν ... ἀφῶ*, as in the previous sentence.

τάξιν: fairly often used by D. as a metaphor for duty (cf. Ronnet *Style* 161–2), but particularly apt here because the word has just been used in the literal, military sense.

φόνου δ' ἂν ...: 'and that it would have been reasonable for me to accuse myself of homicide'. D. means that life without justice is virtually not life at all, as the next clause explains. The notion of prosecuting oneself for killing oneself seems a rather contrived one; D. has introduced it to complete the rhetorical contrast with Meidias' attempts to get D. prosecuted for desertion and homicide.

ἦν ... βιωτὸν: *ἄν* is normally omitted with *ἦν* and an adjective expressing obligation or propriety, as with *ἔδει, ἐχρῆν*, and *προσῆκεν* (**25** n.). Cf. 24.141 *ἡγούμενος ἀβίωτον αὐτῷ εἶναι τὸν βίον τοῦτο παθόντι*, 'thinking that his life would not be worth living if he suffered that'.

121. Although this document is less obviously spurious than some of the other witness-statements, it does have one or two suspicious features (see the following notes), and should be assumed to have been invented by the same forger as the others on the basis of the information which D. gives in the oration.

Λυσίμαχος Ἀλωπεκῆθεν: the names invented for the witnesses are ordinary Athenian ones, but by making Lysimakhos belong to the deme Alopeke the forger has, whether by accident or by intention, produced the name of several real men; for the family of the famous Aristeides 'the just', which belonged to Alopeke, used this name for some of its members, including the Lysimakhos who appears in Pl. *Lakhes*.

καιροὺς: the general sense 'period' (instead of 'critical times') is not normal in Attic. The earliest instances seem to be *AP* 23.2, 26.1, 41.1.

εἰσαγγελία: perhaps a mistake made by the composer of the document about Athenian legal procedure. In the fourth century the use of *eisangelia* seems to have been restricted to certain offences, not including homicide (cf. MacDowell *Law* 184–6); however, not everyone accepts that it was so restricted (cf. Rhodes *JHS* 99 (1979) 107–8). But possibly the word is not intended technically here; the meaning may be merely 'information was given'.

Νικόδημον: the composer of the document has discovered elsewhere the name of the murdered man; cf. **107**.

οἴδαμεν: a non-Attic form, used also in the spurious documents in **82**, **93**.

μηδένα: *μή* in an indirect statement governed by *λέγω* is unusual in Attic, but not unknown; cf. Pl. *Rep.* 346e, X. *Symp.* 4.5.

φαῦλον: 119 n. *φλαῦρον*. Here it is impossible to say whether *φαῦλον* is the usage of the composer of the document or an error by a copyist.

ἀξιοῦντα ... ὅπως: in Attic *ἀξιῶ* normally takes an infinitive rather than *ὅπως* or *ἵνα*, but it takes *ἵνα* in another spurious document in 18.155.

122. ὑπερβολή: 16 n.

ἀτυχοῦντα: being accused of murder, the converse of **117** *ηὐτύχει*.

ἐῶ γὰρ εἰ φίλον: 'for I do not raise the question whether ⟨he accused⟩ a friend'. Yet another instance of paralipsis; by this parenthesis D. suggests that Meidias, besides everything else, infringed the obligations of friendship to Aristarkhos.

χρήματ' ἀνήλισκεν: Meidias' alleged attempt to bribe the relatives of Nikodemos to accuse D. of homicide along with Aristarkhos (**104**).

123–7. *Meidias' offences against Demosthenes concern everyone.*

This passage is a comment on the significance of the whole account of the quarrel between Meidias and D. (**13–122**). In sum, there are two kinds of argument to justify treating Meidias' conduct in this quarrel as offensive to the community, not just to D. personally: all ordinary citizens will suffer if rich men are allowed to offend with impunity, and at the Dionysia D. was performing a public religious function.

123. ἔθος, 'habit', and the generalizing plural *τοῖς ὑπὲρ αὑτῶν ἐπεξιοῦσι* imply that Meidias has treated many people in this way, but D. has not in fact given any instances except himself.

τὸ ... περιιστάναι κακά, 'to set trouble around', is in apposition to *τὸ τοιοῦτον ἔθος καὶ τὸ κατασκεύασμα* and explains that phrase. The whole of this is the object of the infinitives which follow; *ἀγανακτεῖν* does not usually govern an accusative (except neuter pronouns), but the construction is not difficult here because *βαρέως φέρειν* does so.

δικαίως: compare the placing of *ἀδίκως* at the end of **122**. Both adverbs are placed last in their phrases for emphasis.

οὐκ negatives the combination *ἐμοὶ μὲν ..., ὑμῖν δὲ ...*, and then the positive point is made by *ἀλλὰ πᾶσιν ...*: 'indignation should be shown, *not* by me only and not by you, far from it, but by everyone'.

τοῦ ... ῥᾳδίως κακῶς παθεῖν: 'easy maltreatment'. *κακῶς παθεῖν* is a standard phrase which here is itself modified by another adverb. This genitive is constructed with *ἐγγύτατα*, whereas *ὑμῶν* is a partitive genitive with *οἱ πενέστατοι καὶ ἀσθενέστατοι*.

ποιήσαντας: 'after doing it', meaning *ὑβρίσαντας* (LSJ *ποιέω* B.I.4). The *τοῦ* before *ποιήσαντας* governs everything down to *μισθώσασθαι*.

πράγματα refers primarily to prosecutions (LSJ *πρᾶγμα* III.4), such as Meidias' attempts to get D. prosecuted by Euktemon and by the relatives of Nikodemos.

εἰσιν ἐγγυτάτω: the repetition, with the pointless variation from *-τατα* to *-τάτω*, is dull and otiose, and it seems right to delete the phrase as an explanatory gloss.

124. οὐδὲ τὸν ἐξείργοντα ...: a notable example of D.'s ability to express a complex point effectively by the use of infinitives and participles. *τὸ δίκην ... λαμβάνειν παρ' αὐτοῦ* is the object of *ἐξείργοντα*, while *τὸν ἐξείργοντα ... ποιεῖν* is the accusative and infinitive forming the object of *νομίζειν*.

δέει καὶ φόβῳ: merely pleonasm for emphasis; no distinction between the words is intended. So also 23.103 *φόβον καὶ δέος*.

ἰσηγορίας here refers to the right to make an accusation by legal procedure.

μετουσίας: the plural of this noun is unparalleled (but the regular use of the plural of *συνουσία* is comparable): 'occasions when we participate in ...'

ἴσως: 'perhaps', not implying that there is any doubt about the fact, but conceding a point to an imaginary objector.

διεωσάμην: D. induced Euktemon and the relatives of Nikodemos not to prosecute him.

καὶ ἄλλος τις ἄν: sc. *διώσαιτο*: 'and so might someone else', implying that there would not be many such persons.

ἀνήρπασμαι: **120** n. *ἀνηρπάσθαι*.

δημοσίᾳ: 'by public decision', viz. the verdict in the present trial.

φοβερὸν: a thing which people are frightened to do, because of the likely legal consequences. For the sense cf. 20.158 *ὁ Δράκων φοβερὸν κατασκευάζων ... τό τιν' αὐτόχειρ' ἄλλον ἄλλου γίγνεσθαι*. The reading *φανερὸν* (YP) is doubtless due to someone who did not see the point of *φοβερὸν*.

ἀποχρῆσθαι: 'misuse', apparently the earliest instance of this compound in this sense, for which *καταχρῆσθαι* is more usual.

125. δόντα λόγον: 'after rendering an account', a common metaphor for standing trial, and thus synonymous with *ὑποσχόντα κρίσιν*. This sentence says what anyone in Meidias' position ought to do: he ought to answer the charges against himself first (aorist participles) and only then (*τότ'*) bring charges against his opponents. (Goodwin wrongly says that this is the answer to *τί ποιήσετε;* (**124**). That is a rhetorical question and receives no answer.)

καὶ τότ' ... ὁρᾷ τις: deleted by Sykutris as a commentator's paraphrase of words in the previous clause. *τότ'* cannot be taken with *οὐ* *προαναρπάζειν*, because *προ-* is contrasted with the earlier *τότ'*: 'then, not before'.

ἀσχάλλειν: a mainly poetic verb, used here for sarcastic effect.

126. πάντ' ἐπιβουλευόμενος τρόπον: 'being plotted against in every way'. It may be right to insert *ὡς* or *ὅσα* before these words, as Herwerden *Mn.* II 1 (1873) 310–11 conjectured. Otherwise *ὅσα* must be understood from the beginning of the sentence, as an internal accusative with *ἐκπέφευγα*, 'how many escapes I have had'.

παραλείπω: paralipsis again. It is unlikely that D. really had further complaints about Meidias as serious as those already mentioned.

ἐφ' ὅτῳ: throughout this sentence *ἐπί* is 'in relation to', 'as far as ... is concerned'.

ἡ φυλή: the Pandionis tribe, whose khoregos D. was.

τὸ δέκατον μέρος ὑμῶν: 'one-tenth of the Athenian people'; cf. Ais 3.4 *ἡ προεδρεύουσα φυλή, τὸ δέκατον μέρος τῆς πόλεως*. The article is omitted by S and some editors, but it is the regular idiom to include it; cf. 14.23 *τὸ τρίτον μέρος*, 18.266 *τὸ πέμπτον μέρος*, 20.115, 22.3, 22.61, etc.

ὁτιδήποτ' ἐστίν: 'whatever it is'. The point is not that the details of divine law are unimportant, but that human beings cannot fully comprehend them. Cf. A. *Ag.* 160 *Ζεὺς ὅστις ποτ' ἐστίν* ...

συνηδίκηται: with this otiose repetition, rightly deleted, compare *εἴσιν ἐγγυτάτω* at the end of **123**.

127. οὐχ ὡς ... τὴν ὀργὴν ἔχειν: 70 n. *ἄλλως πως ἔχει τὴν ὀργὴν ... ἢ ὡς* ...

ἡμῶν: the variant readings of the mss. are best explained by the hypothesis that D. here wrote *ἡμῶν* meaning 'me'. For a list of instances of *ἡμεῖς* used in the sense of the singular in D. and other orators, see Wyse's note on Isaios 7.37.

ὄντος μόνον: most editors adopt A's order *μόνον ὄντος*. But the order given by most mss. is preferable: the unemphatic *ὄντος* is placed so as to space out the emphatic *ὑπὲρ ἡμῶν* and *μόνον*. Cf. **128** *εἰς ἐμὲ ἀσελγὴς μόνον*.

συνεξεταζομένους: 65 n. *ἐχθρὸν ἐξεταζόμενον*.

συνηγόρους: 112 n.

δοκιμαστὰς: 'approvers'. D. means that Meidias is one of a number of men who are all liable to behave as he does.

128–31. *Meidias' offences against other people.*

D. speaks as if he were turning to an entirely fresh topic. In fact he has already described Meidias' alleged ill treatment of Straton and of Aristarkhos, because those incidents arose out of the quarrel between Meidias and himself. Presumably he is now going on to incidents which did not involve himself at all; but we do not know what most of them were, because the list of them read out at the end of **130** has not been preserved, and the fuller account which D. in **130** offers to give was evidently not written out but would have been extemporized if necessary when the speech was delivered. The only one of these incidents on which he expatiates in the written text is Meidias' speech criticizing the cavalry (**132–5**). That speech was insulting and annoying, but not illegal; and the same may well have been true of all the incidents referred to in this passage.

128. ἀσελγὴς μόνον: so both S and A, whereas other mss. reverse the order. Cf. **127** *ὑπὲρ ἡμῶν ὄντος μόνον*.

129. τί οὖν;: **48** n. The point of the question that follows lies in *δεινότερα*, not in *ἀγανακτεῖς*: 'You are indignant, but have you suffered worse treatment than all the others?'

τὸ παρ' ἀμφοτέρων ἡμῶν ὕδωρ: the time which D. and Meidias have at their disposal. Speeches in lawcourts were timed by a water-clock (*κλεψύδρα*); for *ὕδωρ* used to mean the amount of time available cf. 19.57 *ἀναστὰς ἐν τῷ ἐμῷ ὕδατι εἰπάτω*, 45.47 *οὐ γὰρ ἱκανόν μοι τὸ ὕδωρ ἐστίν*, 18.139, 19.213, 27.12, 29.9, 59.20, etc. For a public case, such as this one, the total amount of water was 11 amphoreis, of which one-third (44 khoes) was for the prosecutor, one-third for the defendant, and one-third for the speeches about the assessment of the penalty if the defendant was found guilty (Ais. 2.126, 3.197). A water-clock inscribed with the name of the Antiokhis tribe, found in the Agora excavations, holds 2 khoes and takes 6 minutes to run out (cf. S. Young *Hesperia* 8 (1939) 274–84, Thompson and Wycherley *Agora* 55); thus each of the main speeches in a public case was allowed 132 minutes, if the lawcourt water-clocks had plug-holes of the same size as that one. Cf. Rhodes *Comm. on AP* 719–28.

D. is alleging that other misdeeds of Meidias, not yet mentioned, are so numerous that to recount them would use up not only the rest of his time but all the time allotted for Meidias' speech as well. The suggestion is fanciful, and that may be why D. adds *πᾶν ... προστεθέν* to make the meaning clear; it is not necessary to delete *τό τ' ἐμὸν ... προστεθέν*, as proposed by Jahn *Philol.* 26 (1867) 3–4 and Herwerden *Mn.* II 1 (1873) 311.

οὐκ duplicates the preceding *οὐδ'* and does not cancel it; cf. Denniston *Particles* 196–7.

130. ὑπομνήματα: 'memoranda', a summary list of Meidias' offences, after which D. will give a full account (*λέξω*) of individual offences as required.

βουλομένοις ὑμῖν ᾖ: this idiom (LSJ *βούλομαι* II.4) is common in

D.; cf. 10.46, 16.3, 18.11, 23.195, 24.19, etc. For the offer of a choice to the audience, cf. 23.18 *δίκαιον δ' ἐστὶν ἴσως ἔμ' ... αἵρεσιν ὑμῖν δοῦναι τοῖς ἀκουσομένοις, τί πρῶτον ἢ τί δεύτερον ἢ τί τελευταῖον βουλομένοις ἀκούειν ὑμῖν ἐστιν*. The jurors are expected to shout out their preference, and the passage implies that D. is ready to depart from his prepared speech and extemporize.

τόπος: 'field', sphere of activity. For this sense (not distinguished by LSJ) cf. Ais. 3.216, where *τόπος* refers to Aiskhines' various kinds of activity, including politics, making public speeches, time spent away from public life, and time spent in gymnasiums. Goodwin's translation 'no spot on earth' is less satisfactory; D.'s point is not that Meidias has travelled extensively, but that he has committed different kinds of offence. Vince translates 'not a single passage', but that sense of *τόπος* does not seem to be paralleled in Attic; besides, there is not much point in saying that there is no passage in a list of offences that does not contain many offences.

The list is presumably read out by D. himself, not by the clerk (though *ἀναγνώσομαι* is ambiguous; **10** n. *ἀναγνῶναι*), and thus is really part of the speech. Yet it has not been preserved; perhaps D. wrote it out as a separate document for taking into court.

131. τὸν ἀεὶ προστυχόντ' αὐτῷ: 'a person who came into contact with him on any occasion'. For the use of *ἀεί* with a participle, cf. **37**, **177**, **194**, **223**.

ἡγεῖθ': SYP have *ἡγεῖ θ'* as two words. This may be derived from an earlier ms. in which *ἡγεῖσθ'* was written in error and the *σ* was then erased. But *ᾤετ'* later in the sentence makes it certain that *ἡγεῖθ'* is right here.

ὡς ἐμοὶ δοκεῖ emphasizes the sarcasm of this sentence.

λαμπρὸν: 'distinguished', 'striking' (not 'brilliant', since intellectual cleverness is not involved). This adjective and its noun *λαμπρότης* occur several times in this part of the speech, often with a sarcastic tone (cf. **153**, **158–9**, **174**). We may infer that Meidias had used the word of himself, and D. is now flinging it back at him.

ἄξιον θαύματος: 'worthy of admiration'; for this phrase cf. Pl. *Symp.* 221c, E. *Hipp.* 906, Hdt. 1.185.3, 3.113.1, 4.199.1, 7.135.1. *θαύματος* is an emendation made by Herwerden *Mn.* II 1 (1873) 311 and R. K. Boekmeijer *Adnotationes criticae in oratores Atticos* (Groningen 1895) 18. The mss. have *ἄξιον θανάτου*, which would mean that Meidias wanted to do something which would justify his execution; but that does not fit the sequence of thought in this ironic sentence, which requires after *λαμπρὸν* and *νεανικὸν* another expression which is normally used for praise. Corruption to *θανάτου* was probably assisted by the phrase *θανάτου ... ἄξια* in **130**. Other conjectures: *ἄξιον {θανάτου}* G. H.

Schaefer *Apparatus* 3.421, *ἄξιον αὐτοῦ* Buttmann, *ἄξιον ἑαυτοῦ* Weil, *ἄξιον ἐπαίνου* Naber *Mn.* II 31 (1903) 418–19.

ἔθνος: 'class' of Athenian society; not 'race', because later in the sentence *ὑμῶν* (addressed to a jury) shows that only Athenian citizens are in question. D. is already thinking of the *ἱππεῖς*, to be mentioned in the next sentence.

ἀβίωτον ...: Vince quotes here the description of Hotspur in *1 Henry IV* 2.4: 'he that kills me some six or seven dozen of Scots at a breakfast, washes his hands, and says to his wife "Fie upon this quiet life! I want work."'

132–5. *Meidias' attack on his fellow-cavalrymen.*

Meidias made this speech shortly after his return to Athens in the late winter or early spring of 348; for its place in the chronological sequence of events see p. 6. It is mentioned again in **197**.

132. ἂν εἰπεῖν ἔχων: **65** n. Here the mss. are divided between the inclusion and the omission of *ἄν*, but Π2 (second century A.D.) seems to have included it: though only]*νειπε̣*[is preserved, the length of line indicates that *ἔχων* must have followed, not (as in AF) preceded.

Ἄργουραν: we are told by Harpokration s.v. that this was a town in the territory of Khalkis. No other ancient author gives any information about its location, but it is the subject of an excellent article by D. Knoepfler in *BCH* 105 (1981) 289–329. From D.'s account it is clear that it was a port reasonably accessible from Attika, therefore on the south-west (not the north-east) coast of the country of Khalkis. The purpose of the expedition was to support the authority of Ploutarkhos in Eretria; Phokion and the infantry went to Tamynai, in the north-east part of the country of Eretria, where no doubt the opponents of Ploutarkhos were based, and the function of the cavalry is likely to have been to guard the approaches to the city of Eretria; Argoura should therefore be in the area of the Lelantine plain, near the frontier of Khalkis and Eretria. These considerations point to the harbour and site now known as Lefkandi. Knoepfler finds confirmation of this identification in the name itself, which he derives from *ἀργ-* meaning 'white'; for Lefkandi is *Λευκάντι*, so named after its white cliff. His argument from etymology is weaker than his argument from Athenian strategy, but together they are convincing. C. Bérard in a subsequent article, *Mus. Helv.* 42 (1985) 268–75, does little more than refine the details of Knoepfler's topography, arguing for a landing-point about one mile further east than Knoepfler's.

παρ' ὑμῖν: in the Ekklesia; cf. **197** *εἰς τὴν ἐκκλησίαν*. The phrase just means 'in the presence of the Athenians' and does not imply that all the jurors in this trial necessarily attended that meeting of the Ekklesia; on the contrary, *δήπου* implies that some may know about it only by hearsay.

ἐκ Χαλκίδος: from the country of Khalkis, in which Argoura was situated. The phrase does not necessarily mean that Meidias had gone to the city of Khalkis.

ὄνειδος: predicative. Since Meidias was the Athenian *πρόξενος* of Ploutarkhos (**200**), he was probably annoyed that the cavalry left Euboia without taking action to support Ploutarkhos; but D.'s words here and in **197** imply that his criticism was directed not against Phokion's strategy but against the misconduct of the cavalrymen themselves.

ἐλοιδορήθη: active in meaning; so also in 9.54, 54.5. D. never uses the aorist active or middle form of this verb.

Κρατίνῳ: from the context it seems likely that Kratinos was the hipparch in command of the cavalry on the expedition to Euboia. (Σ (466 Dilts) states that Kratinos was a hipparch, but this statement is probably a deduction from the text of D., not based on independent evidence. The note is anyway confused, since Kratinos is also called *στρατηγὸς τοῦ πεζοῦ*.) He is not otherwise known to us, but D. seems to assume that he is a well known figure; so he may tentatively be identified with a Kratinos who proposed several decrees at this period (*IG* 2^2 109.b8, 134.6, 172.3–4).

τῷ νῦν ... μέλλοντι βοηθεῖν αὐτῷ: 'the man who is now intending to support him', as a supporting speaker at this trial; **7** n. *βοηθῆσαι*. The words *τῷ νῦν* are written above the line in S and are omitted in all other mss. They may have been copied from another ms. now lost. Without them, we must read *αὑτῷ* (S) and translate 'when he was about to defend himself' against Meidias' criticism in the Ekklesia. But surely D. was present in the Ekklesia, and would not have used the words *ὡς ἐγὼ πυνθάνομαι*, 'as I am told', to refer to that occasion. So it is preferable to retain *τῷ νῦν* and take *ὡς ἐγὼ πυνθάνομαι* as a reference to information about the defendant's plans (as in 19.182, 40.45).

ταῦτα: Meidias' speech in the Ekklesia, just described. So *πράττειν* has an imperfect sense, 'that he was doing this'; Vince is wrong to translate 'what degree of wickedness and recklessness may we expect from him now?'

133. ὄνειδος: a 'boomerang' device of rhetoric: D. turns against Meidias a word which Meidias himself had used.

τὴν σκευὴν ἔχοντες: Athenian soldiers provided their own arms and equipment. No doubt it sometimes happened that some individuals reported for duty inadequately equipped, but D.'s point here is a different one: it is not that the soldiers brought what was needed (which would be ἣν ἔδει), but that they brought only what was suitable (ἣν προσῆκε) without unnecessary luxuries.

τῶν ἐξιόντων, sc. εἶναι: 'to be part of the expeditionary force'.

ἐκληροῦ: the cavalry was organized in ten troops, one for each tribe, each commanded by an elected phylarch (*AP* 61.5). From this passage it appears that, when a certain number out of the ten troops was to be sent on an expedition, lots were drawn to decide which troops should go; cf. Lys. 16.16 for a similar procedure for the tribal contingents of hoplites. Each phylarch must have drawn the lot on behalf of his troop; so the singular ἐκληροῦ is best interpreted as meaning that Meidias was the phylarch of his tribe in 349/8. If so, he was not a hipparch that year. There were only two hipparchs (4.26, *AP* 61.4); Meidias had held that office in a previous year (**171–4**), and when D. sarcastically calls him ὁ ἵππαρχος Μειδίας (**166**) that must refer merely to the fact that he boasted of having been a hipparch in the past. This interpretation is confirmed by the reference to his συνάρχοντες in the plural (**197**); a phylarch had nine colleagues, but a hipparch had only one. (The suggestion that Meidias was a phylarch is made by Knoepfler *BCH* 105 (1981) 291 n. 5, but he misses the significance of ἐκληροῦ.) D.'s assertion that Meidias hoped that his troop would not be selected is doubtless his own presumption, since Meidias will not have said so aloud; it may be true nevertheless.

ἀστράβης: a kind of chair mounted on the back of a mule or donkey. Σ (469a Dilts) gives the following definition: εἶδος καθέδρας, παρὰ τὸ μὴ στροβεῖσθαι μηδὲ στρέφεσθαι. ἔστι δὲ ἐπὶ πλεῖστον εἰς ὕψος ἀνῆκον, ὥστε τῶν καθεζομένων ἀνέχειν τὰ νῶτα. χρῶνται δὲ αὐτῷ μάλιστα αἱ γυναῖκες. We may think of it as halfway between a saddle and a howdah. (LSJ translate 'an easy padded saddle', but there is no evidence for padding.) The word is used not only for the chair, but also for the animal equipped with it; cf. Helladios, reported in Phot. *Bibl.* 533a: ὅτι τὴν ἡμίονον, ἐφ' ἧς ὀχούμεθά, φησιν, ἀζύγου οὔσης, Ἀθηναῖοι καλοῦσιν ἀστράβην· ἔνθεν λέγεται καὶ ὁ ἀστραβηλάτης. λέγεται μέντοι ἀστράβη καὶ τὸ κατασκεύασμα, ὡς Δημοσθένης ἐν τῷ κατὰ Μειδίου λόγῳ· "ἐπ' ἀστράβης ὀχούμενος ἀργυρᾶς". (The definition given by Herodian Gr. 1.308.7 (Lentz) is εἶδος ἁμάξης. This is puzzling, and may be a mistake. The other evidence, especially Helladios, implies that the rider is on the mule, not on a wheeled cart drawn by the mule.) In all other passages where the rider is identified it is either a woman (Makhon 389 Gow, Alkiphron 2.3 (=4.18.17 Schepers, Benner and Fobes); cf. A. *Supp.* 285) or an invalid (Lys. 24.11, Luc. *Lex.* 2). Thus D. is accusing Meidias of effeminate self-indulgence. Cf. Erbse *Ausgewählte Schriften* 423–6.

ἀργυρᾶς: a chair of solid silver would be uncomfortable to sit on and heavy for a mule to carry. We should therefore envisage a chair made of a lighter material, perhaps wickerwork, and take *ἀργυρᾶς* as meaning merely 'with silver ornamentation', as in Hdt. 9.82.2 *κλίνας τε χρυσέας καὶ ἀργυρέας*.

The alternative reading *ἐξ Ἀργούρας τῆς Εὐβοίας* was already current in ancient times (see p. 50), but it should be rejected, because D. is describing Meidias' journey from Athens to Argoura, not his departure from Argoura. Besides, D. would not need to tell his audience that Argoura was in Euboia; he has already mentioned Argoura in **132** without considering that explanation necessary. Nor should we accept *ἀργυρᾶς τῆς ἐξ Εὐβοίας*, 'silver, the one from Euboia'; that would require the unlikely hypothesis that Meidias had a well-known silver riding-chair which he had imported from Euboia on some previous occasion (taking silver to Athens was like taking coals to Newcastle) and now took back to Euboia, and even for that we should expect the wording to be *ἐπὶ τῆς ἀργυρᾶς ἀστράβης τῆς ἐξ Εὐβοίας* (cf. **158** *ἐπὶ τοῦ λευκοῦ ζεύγους τοῦ ἐκ Σικυῶνος*). There is no evidence that Helladios and Menander, who read *ἀργυρᾶς*, had *τῆς ἐξ Εὐβοίας* in their text. Corruption of *ἀργυρᾶς* to *Ἀργούρας* doubtless occurred very early in the history of the text (it was an easy error after *Ἄργουραν* in **132**), and the reference to Euboia is obviously a gloss added to *Ἀργούρας*, but how did Euboia get into mss. which had *ἀργυρᾶς*? Perhaps by contamination (i.e. the gloss was copied out of a ms. which had *Ἀργούρας* into one which had *ἀργυρᾶς*). Alternatively *ἀργυρᾶς*, after being lost from the text, may have got back into it by a second error which reversed the first; cf. **164**, where *Ἀργούρας* is right but the scribe of S at first wrote *αργυρας* by mistake. Cf. H. Diels and W. Schubart *Didymos: Kommentar zu Demosthenes* (Berlin 1904) pp. xlvii–xlviii, Knoepfler *BCH* 105 (1981) 328 n. 164.

χλανίδας: woollen cloaks of superior quality, and therefore a mark of affluence and luxury; cf. 36.45 *σὺ μὲν χλανίδα φορεῖς ... καὶ τρεῖς παῖδας ἀκολούθους περιάγει ..., αὐτὸς δ' ἐκεῖνος πολλῶν ἐνδεὴς ἐστιν*, Ar. *Wasps* 677, Men. *Dys.* 257, Poseidippos com. 31, etc.

κυμβία: a kind of small cup, long and narrow, according to the note of Didymos on our passage, quoted by Harpokration s.v.: *εἶδός τι ἐκπώματος τὸ κυμβίον. φησὶ δὲ Δίδυμος ἐπίμηκες αὐτὸ εἶναι καὶ στενὸν καὶ τῷ σχήματι παρόμοιον τῷ πλοίῳ ὃ καλεῖται κυμβίον*. According to Dorotheos, quoted by Ath. 481d, they had no base or handles: *γένος ποτηρίων βαθέων τὰ κυμβία καὶ ὀρθῶν, πυθμένα μὴ ἐχόντων μηδὲ ὦτα*. They were used at ordinary meals by the well-to-do, but were valuable enough to be worth stealing (47.58). We hear of *κυμβία* ornamented with gold (Alexis 95), and they were perhaps always made of metal rather than pottery. They are discussed at length in Ath. 481d–482e.

κάδους: large jars of a kind normally used for wine. We should assume that Meidias' *κάδοι* (like those in Ar. *Akh.* 549, *Peace* 1202) were full, not just empty jars.

πεντηκοστολόγοι: in Athens there was a two-per-cent tax on imports and exports (34.7, 35.29–30, 59.27). D.'s meaning is probably that, when Meidias embarked at Peiraieus, the tax collectors thought that his gear was merchandise being exported without payment of the tax, and so tried to confiscate it. But the imperfect verb implies that the confiscation was not carried out; D. is relating a joke rather than sober fact, and the next sentence makes clear that he does not vouch for its truth. Knoepfler *BCH* 105 (1981) 328–9, following Weil and others, suggests that the tax collectors were rather those of Khalkis, taxing imports at Argoura; that is possible, and it would fit well with the reference in the next sentence to the point of disembarkation, but there is no other evidence that Khalkis had a two-per-cent tax and officials called *πεντηκοστολόγοι*.

εἰς ταὐτὸν: 'to the same place'. The hoplites may be assumed to have disembarked at Porthmos, the port nearest to Tamynai; so Knoepfler *BCH* 105 (1981) 291.

134. εἰ introduces an admitted fact, as in 43.60 *ἀλλ' εἰ Θεόπομπος τετελεύτηκεν ὁ τουτουὶ πατήρ, οἱ νόμοι οὐ τετελευτήκασιν*. An alternative reading *εἷς* is written in the margin of S, making a paratactic contrast with *πάντας*: 'a single individual mocked you: you attacked them all'. But the emphasis on 'one' seems inappropriate when *Ἀρχετίων ἤ τις ἄλλος* implies that there may have been several mockers.

Ἀρχετίων: unknown. (Σ (474 Dilts) merely offers an explanation based on what D. says here.) Presumably he was a member of the cavalry and the mockery occurred at Argoura; cf. *οἱ λοιποὶ τῶν στρατιωτῶν* below.

πάντας ἤλαυνες: Meidias' speech in the Ekklesia criticizing the cavalry (**132**). *κατηγόρεις* in the next sentence refers to the same occasion; the ensuing optative shows that it is correctly accented as imperfect, not present.

κακῶς ἤκουες: 'you were called bad names' (LSJ *ἀκούω* III.1). The phrase is treated as a unit, to which another adverb, *δικαίως*, is applied.

τουτουσὶ: the present jury.

εἰ δὲ μὴ ...: a striking 'heads I win, tails you lose' argument: if the tales about Meidias' self-indulgence were invented, it was his own fault for being the sort of man to whom such tales were thought appropriate.

κατεσκεύαζον: **110** n. *κατεσκεύαζεν*.

ἐπέχαιρον: this compound is used especially of rejoicing that someone has got the penalty which he deserves; cf. 9.61 *τὸν δ'* (sc.

Euphraios, who was imprisoned) *ἐπιτήδειον ταῦτα παθεῖν ἔφη καὶ ἐπέχαιρεν.*

ἐκ τῶν ἄλλων ὧν ἔζης: 'from the rest of your life'; cf. **196** *ἐξ ὧν ζῇς*, 18.130 *ἀπ' αὐτῶν δ' ὧν αὐτὸς βεβίωκεν*, 18.198 *ἐξ ὧν ζῇς*. In those passages *ὧν* stands for *τούτων ἅ*, but here *ὧν* is merely attracted from the accusative by the genitive antecedent. A similar sense is expressed by the passive verb with a nominative pronoun in **151**, 22.77 *οἷα σοὶ βεβίωται*, 22.78 *οἷα τούτῳ βεβίωται.*

135. ἀπειλεῖς πᾶσιν, ἐλαύνεις πάντας: the reference is still to Meidias' speech in the Ekklesia, but the use of the present tense is perhaps not merely 'historic present' referring to the one past incident, but implies that Meidias keeps on behaving in this way. For the asyndeton, **81** n. *τῇ δίκῃ, τοῖς νόμοις.*

μή with future indicative: the relative clause includes an idea of purpose or intention.

τὸ δὴ σχετλιώτατον ... refers to the statement which follows, rather like a descriptive heading. Cf. 31.14 *ἔπειτα τὸ δεινότατον* ..., Ais. 3.161 *καὶ τὸ πάντων δεινότατον* ..., 3.232 *καὶ τὸ πάντων ἀτοπώτατον* ...

κεφαλή: **117** n.

κατηγόρεις: as in **134**, this is better accented as imperfect than as present. It does not mean that Meidias initiated a formal prosecution of every cavalryman who went to Euboia, but only that he made sweeping criticisms of the whole force.

ἔφριξεν: a mainly poetic verb, but used by D. also in 18.323, 51.9.

136–42. *Meidias' aggressiveness and supporters deter his victims from prosecuting him or testifying against him.*

D. is trying to convince the jury that the present trial is the time to make a stand against bullying, and the opportunity to punish Meidias for his misconduct in general, not merely for his treatment of D. This theme becomes increasingly prominent towards the end of the oration.

136. κατηγορεῖται (SA) is preferable to *κατηγοροῦνται* (FYP), because the subject of this verb in the passive is normally the charge, not the accused person (though the latter is possible, e.g. And. 1.7).

καταψευδομαρτυροῦμαι: unlike *κατηγορέω* this verb in the passive always has the person as subject; cf. 33.37, 45.1, Isai. 5.9, Pl. *Gorg.* 472a. *κατα-* just means 'against'; there seems to be no reason for LSJ's translation 'to be borne down by false evidence'.

137. πάντας ... εἰδέναι: **1** n. *οὐδένα ... ἀγνοεῖν.*

τὸν τρόπον ... καὶ τὴν ἀσέλγειαν καὶ τὴν ὑπερηφανίαν: not really three separate items, since the other two nouns define the first: 'his bullying and arrogant character'. On hendiadys in Greek, including this and other examples from D., see D. Sansone *Glotta* 62 (1984) 16–25.

ὧν: equivalent to *τούτων ἅ*. 'Some are surprised at what they have not heard' means that they are surprised at the omission. Thus *πάλαι* means 'during a considerable part of my speech'. An alternative interpretation is 'some have long been viewing with amazement what they know of themselves, without needing that I should tell them' (King); but it is less appropriate to say that people are amazed at Meidias' behaviour, when D. has just said that it is familiar to everyone.

οὐδὲ ... μαρτυρεῖν ἐθέλοντας: **112** n.

φιλοπραγμοσύνην: readiness to attack others, whether by force (cf. 1.14, 4.42) or by legal proceedings (cf. 39.1). This is a pejorative word; **83** n. *ἀπράγμων*.

ὀρρωδοῦντας: Taylor's conjecture should be accepted. Corruption to *ὁρῶντας* was easy, because that verb is common and has been used earlier in this sentence, but D. is less likely in fact to have used the same verb twice in one sentence without rhetorical point. The variant *δεδιότας* or *δεδοικότας* is no doubt a gloss which has got into the text; but it is a gloss on *ὀρρωδοῦντας*, not on *ὁρῶντας*.

ἀφορμήν: wealth; cf. the end of **98**.

κατάπτυστον: the first prose instance of this adjective. Earlier it occurs in tragedy, and a Middle Comedy instance (Anaxilas 22.6) is paratragic. So D.'s use of it should be regarded as slightly old-fashioned and formal, not colloquial. He uses it again in **167**, **171**, 18.33, 18.43, 18.196, 19.15. It is applied to D. himself by Dein. 1.15, perhaps to mock his use of the word.

138. ἐπ' ἐξουσίας καὶ πλούτου: for this sense of *ἐπί*, referring to the circumstances which make an action possible, cf. 9.61 *ἐπ' ἐξουσίας ὁπόσης ἐβούλοντ' ἔπραττον*, and **2** n. *ἐπὶ τῶν ἄλλων*.

τεῖχός ἐστι: 'is a wall' of defence against prosecutions, a striking metaphor not found elsewhere. The military metaphor is continued by *ἐξ ἐπιδρομῆς*, *προβέβληται*, *μισθοφόροι*.

ἂν: inserted because *ἐξ ἐπιδρομῆς* is equivalent to a conditional clause, 'if he were attacked'.

αὑτὸν: 'oneself'. Not 'him'; this sentence is a generalization. Then *ἐπεὶ ... οὗτος ...* gives Meidias as an example: *περιαιρεθεὶς ... δώσει* is concessive in effect, and *νῦν δ' ...* gives the main point, the protection which Meidias' money buys.

εἰ δ' ἄρα: 'or if he does, after all'. For the sense of *εἰ ἄρα* see Denniston *Particles* 37–8. Omission of the verb, to be understood from the previous verb, is common with *εἰ δὲ μή*, rarer when *εἰ* or *ἐάν* is not accompanied by *μή*, but cf. **222**, E. *Hipp*. 508, Pl. *Euthd*. 285c.

ἐλάττονος ἄξιος ... τοῦ μικροτάτου: 'worth less than the smallest', i.e. 'negligible'.

παρ' ὑμῖν: 'in your judgement'; the following words show that D. is thinking of the reception which will be given to Meidias by future juries.

139. τούτου προβέβληται: 'is set in front of him' to protect him. D. evidently expects one or more of these men to speak in support of Meidias.

Πολύευκτος: the son of Timokrates. Nothing earlier is known of him, but in this same year, 347/6, he was the proposer of an addition to a decree proposed by Androtion to honour the rulers of Bosporos (*IG* 2^2 212 (Tod 167) 65–6). He is mentioned again in D. 42.11 (*Phainippos*) as a friend of Phainippos.

Τιμοκράτης: a politician active in the 360s and 350s. He is known to us mainly from D. 24, a speech for the prosecution in a *γραφὴ παρανόμων*, in which he is attacked as an associate of Androtion and proposer of a law intended to protect Androtion. There are also inscriptions recording his victory in an Olympic chariot race (*IG* 2^2 3127) and his service as a trierarch (*IG* 2^2 1609.83, 1609.89–90). Cf. Davies *Families* 513–14.

Εὐκτήμων ὁ κονιορτός: **103** n.

μισθοφόροι continues the military metaphor, but also implies that Meidias pays these men to support him at trials.

πρὸς ἔτι ἕτεροι τούτοις: with these words D. moves on to men who do not make speeches but merely give silent support. All the older mss. give *προσέτι* as one word, but that is an adverb; the dative *τούτοις* demands the preposition *πρός*. For the intervention of *ἔτι* between *πρός* and the dative cf. *πρὸς ἔτι τούτοισι* in Hdt. 1.64.2, 3.65.7, 9.111.2. In S -*έτι* was at first omitted, though subsequently added by the original scribe, and in F *τούτοις* precedes *ἕτεροι*: Blass combines these variants and prints *πρὸς τούτοις ἕτεροι*. Herwerden *Mn.* II 3 (1875) 136 simply deletes *τούτοις*.

συνεστῶσα: 'banded together'; cf. LSJ *συνίστημι* B.III.1. The phrase suggests a conspiracy.

ἑταιρεία (S) or *ἑταιρία* (other mss.)? The two forms are constantly confused in mss., and there is no Attic passage in which either is guaranteed by metre. But note the Ionic form *ἑταιρηίην* in Hdt. 5.71.1, and the inscription of the law of Gortyn has *ἑταιρείαι* at 10.38; in both those passages the meaning is 'group of comrades' (rather than the abstract 'comradeship'), which is also the meaning here. So *ἑταιρεία* may be preferable here, but *ἑταιρία* remains possible.

The word is used especially, though not exclusively, of a group having a political as well as a social purpose. For discussion of such groups see G. M. Calhoun *Athenian Clubs in Politics and Litigation* (Austin 1913), F. Sartori *Le eterie nella vita politica ateniese del VI e V secolo a.C.* (Rome 1957), C. P. Longo *"Eterie" e gruppi politici nell'Atene del IV sec. a.C.* (Florence 1971). But in the present case 'club' is not an appropriate translation; Meidias' supporters here are parasites, hoping to get money or other favours from him in return for their support. On the question whether their association with him had a political function see p. 12.

ἐπινευόντων: 'nodding assent'. They do not testify formally as witnesses, but when Meidias and other rich men make speeches (perhaps in the Ekklesia as well as in the courts) they nod their heads to give the impression that they can confirm what is said; they are *κόλακες*.

οὐδὲν ὠφελεῖσθαι: 'they do not make anything'. D. skilfully kills two birds with one stone: Meidias' supporters are corrupt and hope to be rewarded by him, and Meidias is too mean to pay them.

φθείρεσθαι πρὸς τοὺς πλουσίους: 'go to the rich to be corrupted'; cf. Plu. *Phok.* 21.3 *Ἅρπαλος μετὰ χρημάτων πολλῶν ... τῇ Ἀττικῇ προσέβαλε, καὶ τῶν εἰωθότων ἀπὸ τοῦ βήματος χρηματίζεσθαι δρόμος ἦν καὶ ἅμιλλα φθειρομένων πρὸς αὐτόν*, *Eum.* 14.1, *Ant.* 24.1.

140. **καθ' ἑαυτὸν ὅπως δύναται**: without supporters and without wealth.

συλλέγεσθε, imperative: 'unite!' by all voting against Meidias.

141. **τάχα**: 'perhaps' (not 'soon'), introducing an attempt to forestall an argument which Meidias may use. Cf. 19.134 *τάχα τοίνυν ἴσως καὶ τοιοῦτος ἥξει τις λόγος παρὰ τούτων*, and similar phraseology in **191**, 20.18, 38.25, 45.83.

δή: sarcastic or scornful.

τὰ καὶ τά: 'such-and-such'; cf. LSJ *ὁ* A.VII.2.

μέν: there is no *δέ*: the contrast is introduced by *μέντοι* in **142**.

βοηθεῖν αὐτῷ: **7** n. *βοηθῆσαι*.

ἀπραγμοσύνη: **83** n. *ἀπράγμων*.

ἀπορία: 'lack of means'. Although prosecuting in Athens did not involve heavy expenses, it required time for collecting evidence, composing a speech, and attending preliminary and final hearings, and a poor man might be unable to afford to take so much time away from work.

142. **ἀλλ' ὡς οὐ ...**: cf. **28** *ἀλλ' ὡς οὐ πεποίηκεν ἃ κατηγόρηκα ...* Here *ἅ* is attracted into *ὧν* by the proximity of its antecedent *τούτων*. The alternative reading *ἐφ' οἷς* gives an abnormal construction with *κατηγόρηκα*.

ἀπολωλέναι πολὺ μᾶλλον: sc. *προσήκειν ἡγοῦμαι*. The reading *ἐστι δίκαιος* is a gloss added to provide a construction for the infinitive.

τηλικοῦτος: 'so powerful'; so also in **207**. Legally Meidias would be correct in arguing that any offences which he had committed against others should not affect the verdict in the case brought by D., but D.'s reply is that it is justifiable to take the other offences into account because Meidias by overbearing power has prevented separate cases being brought for them.

πᾶσιν: dative of agent with *τιμωρητέος*.

143–50. *Contrast between Meidias and Alkibiades. Even Alkibiades was punished for his insolence, though he had far greater merits and better excuses than Meidias.*

For a comparison with the treatment of Alkibiades in other fourth-century orations, especially Lys. 14 and Isok. 16, see M. Nouhaud *L'Utilisation de l'histoire par les orateurs attiques* (Paris 1982) 292–7. D.'s account contains a number of mistakes (see the following notes), but his purpose is not to compose a biography of Alikibiades but only to mention points which enable him to disparage Meidias by contrast.

143. λέγεται: D. speaks as if he knew about Athenian history only from hearsay, not from written accounts. That does not prove that he had not read Thucydides; but he is anxious to avoid seeming too learned for his audience. Cf. 3.21 *ἀκούω … τὸν Ἀριστείδην ἐκεῖνον, τὸν Νικίαν, τὸν ὁμώνυμον ἐμαυτῷ, τὸν Περικλέα*. But whereas those four statesmen are mentioned with the article, as if their names at least were well known, Alkibiades is here introduced without the article and with an indication of when he lived, as if the jury might never have heard of him.

κατὰ: 'during the time of'; cf. LSJ *κατά* B.VII.1. So also in **146**.

εὐδαιμονίαν: used also in 3.26 to refer to the time of the Athenian empire in the fifth century.

ποίων τινῶν: the combination *ποῖός τις* is common, whereas the alternative reading *ποίων τιμῶν* would require the following words to be 'from the people'.

τοῖς πολλοῖς ὑμῖν: D. contrasts ordinary citizens with prominent individuals like Meidias or Alkibiades (not with other cities: King wrongly comments 'Other people might do it, but not Athenians').

144. πρὸς πατρὸς μὲν Ἀλκμεωνιδῶν … πρὸς δὲ μητρὸς Ἱππονίκου: actually it was the mother of Alkibiades, not his father, who was an Alkmeonid (Lys. 14.39, Plu. *Alk.* 1.1); and it was his wife, not himself, whose father was Hipponikos (Isok. 16.31, Plu. *Alk.* 8.3). Thus D. wrongly says about the famous Alkibiades what was actually true of his son, who was also named Alkibiades; he has carelessly misapplied to the former something which he has heard or read about the latter.

ὑπὸ τῶν τυράννων ... ἐκπεσεῖν: according to Hdt. 1.64.3 the Alkmeonids went into exile when Peisistratos gained power permanently as tyrant about 546. However, Kleisthenes, who became the leading member of the family, is now known to have been arkhon in Athens in 525/4 (ML 6c.3); so either the family did not leave Athens until after that date or it had more than one period of exile before returning with the Spartans to overthrow Hippias, who had succeeded his father as tyrant, in 510.

ὑπὲρ τοῦ δήμου στασιάζοντας: the reforms of Kleisthenes in 507 were the foundation of Athenian democracy, but there is no evidence that the Alkmeonids had a democratic policy before the expulsion of Hippias. D. is probably guilty of anachronism here.

δανεισαμένους χρήματ' ἐκ Δελφῶν: the financial arrangements between the Alkmeonids and Delphi are connected with the rebuilding of the temple, which had been destroyed by fire some years earlier; but the subject is obscure and much disputed. The other evidence is Hdt. 2.180, 5.62–3, Pind. *Pyth.* 7.10–12 with schol. (quoting Philokhoros, *F. Gr. Hist.* 328 F115), Isok. 15.232, *AP* 19.4 (followed by schol. Ar. *Lys.* 1153), and Σ here (497–8 Dilts). For discussion see Jacoby's commentary on *F. Gr. Hist.* 328 F115, W. G. Forrest *GRBS* 10 (1969) 277–86, K. H. Kinzl *Hermes* 102 (1974) 179–90, Rhodes *Comm. on AP* 236–7. There are essentially two versions of the story. According to Herodotos, the Alkmeonids, after entering into a contract to rebuild (or to complete the rebuilding of) the temple, used their own money to make the building more splendid than the contract required (and also, it was alleged, to bribe the Pythia); in gratitude for this, the Delphians supported the efforts to expel the tyrants from Athens. According to the other version, given or implied by most of the other sources, the Alkmeonids used for the expedition against the tyrants some or all of the money which they were paid in advance for rebuilding the temple; they repaid it by completing the rebuilding after the tyrants had been expelled. Apart from D. (and Σ 498 Dilts, which is merely based on D.; Kinzl is not justified in attributing independent authority to it) the only source to mention a loan is Isok. 15.232, who says: *Κλεισθένης ἐκπεσὼν ἐκ τῆς πόλεως ὑπὸ τῶν τυράννων, λόγῳ πείσας τοὺς Ἀμφικτύονας δανεῖσαι τῶν τοῦ θεοῦ χρημάτων αὐτῷ, τόν τε δῆμον κατήγαγεν* ... This can be interpreted as a variety of the second version, rather than as an entirely different version of the story: the Amphiktyons agreed to allow the Alkmeonids to use for an expedition against the tyrants money given them for use in rebuilding the temple, on condition that they repaid it by completing the building afterwards. D. may have taken this point directly from Isok. 15 (*Antidosis*), written only a few years before, rather than from any independent source; Forrest's assertion that both used the lost *Atthis* of Kleidemos is mere speculation. But the exact truth about the Alkmeonids' financial dealings at Delphi cannot now be known.

Ἱππονίκου: Hipponikos belonged to the aristocratic clan of Kerykes, and was one of several successive members of his family to hold the positions of δᾳδοῦχος (torchbearer at the Eleusinian Mysteries) and πρόξενος for Sparta at Athens. Otherwise his only known public services are that he was secretary of the Boule in 445/4 or 444/3 (*IG* I³ 455.5) and a general in 426/5 (Th. 3.91.4; cf. D. M. Lewis *JHS* 81 (1961) 119–21); the statement that he died as a general at Delion in 424 (And. 4.13) probably results from confusion with Hippokrates, but is not certainly wrong (cf. A. R. Burn *CQ* 4 (1954) 139). His father Kallias was more distinguished: he led the Athenian embassy which made the Peace of Kallias with Persia in 449 (D. 19.273 calls him *Καλλίαν τὸν Ἱππονίκου ταύτην τὴν ὑπὸ πάντων θρυλουμένην εἰρήνην πρεσβεύσαντα*), and he was a member of the embassy which made the Thirty Years' Peace with Sparta in 446 (Diod. 12.7). The son of Hipponikos, also named Kallias, is known mainly as a man of wealth and culture, and is a prominent character in Plato's *Protagoras*, Xenophon's *Symposium*, and Andokides' *On the Mysteries*; but he held public offices too, as a general in 391/0 (X. *Hell.* 4.5.13) and as an ambassador to Sparta on three occasions (X. *Hell.* 6.3.4). On the whole family, especially the rise and fall of its wealth, see Davies *Families* 254–70.

οἷς has as antecedent the whole phrase *Ἱππονίκου καὶ ταύτης τῆς οἰκίας* (though even *ταύτης τῆς οἰκίας* alone, because it refers to a number of men, might be followed by a masculine plural pronoun). All mss. except S have *ἧς*, but we find the dative *ᾧ* in the similar phrase in **143** and *αὐτῷ* in **145**, and after *οἰκίας* corruption is more likely to have occurred to *ἧς* than from it.

145. θέμενος τὰ ὅπλα: 'bearing arms'; cf. LSJ *τίθημι* A.II.10.b.

δὶς μὲν ἐν Σάμῳ, τρίτον δ' ἐν αὐτῇ τῇ πόλει: campaigns in which Alkibiades is known to have taken part personally are Poteidaia in 432, Delion in 424, the Sicilian expedition in 415, and the naval operations from 411 to 407, including the battle of Kyzikos; but there was no battle at Samos or at Athens in which he can have taken part. D. appears to have read some account of his activities in 411–07 which said that he was elected general twice at Samos and once at Athens (viz. the elections mentioned in Th. 8.82.1 and X. *Hell.* 1.4.12, together with another at Samos not attested in any extant text) and misunderstood it as meaning that those were the places where he fought in battles.

Ὀλυμπίασιν ἀγῶνες: at the Olympic games in 416 Alkibiades provided seven entries for the chariot race, three of which came first, second, and fourth (Th. 6.16.2; cf. Dover's note there, *HCT* 4.246–7). The plural ἀγῶνες should imply that he competed at Olympia on at least one other occasion; but D. may be using it rhetorically without knowing of more than one race, or he may be confusing races elsewhere with those at Olympia. Alkibiades won victories also at the Pythian and Nemean games, and he had two pictures painted to celebrate them (Plu. *Alk.* 16.7, Paus. 1.22.7, Ath. 534d).

καὶ στέφανοι: omitted in S (though added in the margin by the second hand) and A. Hesitantly I retain it on rhetorical grounds: without it καὶ νῖκαι seems too short a phrase to make an effective end to the clause.

στρατηγὸς ἄριστος: this accords with the usual view of Alkibiades' military talent, of which the earliest extant expression is Th. 6.15.4 κράτιστα διαθέντι τὰ τοῦ πολέμου. In modern times this assessment has been disputed; for some recent views see H. D. Westlake *Individuals in Thucydides* (Cambridge 1968) chapter 12, E. F. Bloedow *Alcibiades Reexamined* (*Historia* Einzelschriften 21, 1973), Dover *HCT* 5.423–7.

λέγειν ... δεινότατος: it is difficult to find evidence to confirm this very favourable estimate of Alkibiades' oratory. No samples of it are preserved, except perhaps the one word ἀποβήσεται in Ar. fr. 198. Thucydides' version of his speech in favour of the Sicilian expedition (Th. 6.16–18) is hardly admissible as evidence of Alkibiades' style, though of course it is an undeniable fact that he succeeded in convincing his hearers on this and other occasions (e.g. Th. 8.86.5); cf. Dover and Andrewes in *HCT* 4.246, 5.287. Otherwise we have only Plutarch's comment, of which the most important part is some information drawn from Theophrastos: 'That he was an able speaker is attested both by the comic poets and by the ablest of the orators, who says in *Against Meidias* that Alkibiades, besides his other qualities, was also very clever at speaking. But if we trust Theophrastos, a man as learned and as fond of listening as any of the philosophers, Alkibiades was the most competent of all at thinking up the necessary ideas, but, while searching not only for what ought to be said but also for the words in which it ought to be expressed, not having a ready supply of them, he often stumbled, and fell silent in the middle of his speech, and paused because words excaped him, pulling himself up and pondering' (Plu. *Alk.* 10.4). If Theophrastos had reliable evidence for his description, it is likely that D.'s praise is overstated.

146. κατ' ἐκεῖνον: 'in his time'; cf. 18.95 τῶν καθ' ὑμᾶς πεπραγμένων. This sense of κατά is not common, and FYP add τὸν χρόνον to produce an easier phrase.

οὐδενὸς: genitive of price; cf. **160** οὐδενὸς πράγματος οὐδ' ἔργου. Herwerden *Exerc. crit.* 169 conjectures οὐδενὸς ⟨ἕνεκα⟩.

φυγάδα: Alkibiades was condemned to death in his absence in 415 after the mutilation of the Hermai and the profanation of the Mysteries (Th. 6.61.7); and after being pardoned and elected general, he was deposed from the generalship in 406 after the battle of Notion and stayed away from Athens for the rest of his life (X. *Hell.* 1.5.16–17). On neither occasion was exile (or disfranchisement; SFP add *ἄτιμον καὶ* above the line or in the margin) the legal penalty, though it was the practical effect.

Δεκέλειαν: the Spartans fortified Dekeleia as their base in Attika in 413. D. omits to say that it was Alkibiades, in exile, who advised them to do so (Th. 6.91.6), but he implies that Alkibiades, if retained as general, could have averted this and the other disasters.

τὰς ναῦς ἁλῶναι: at the battle of Aigospotamoi in 405.

συγχωρῆσαι: deleted by Cobet *Misc. crit.* 511. It could be an erroneous repetition after *συνεχώρησαν ὑβρίζειν* earlier in **146**; cf. **126** *συνηδίκηται*.

147. τοσοῦτον ... ἡλίκον: so also in 25.40, but the meaning seems not to differ from *τοσοῦτον ... ὅσον*.

Ταυρέαν ἐπάταξεν: a fuller account of the incident is given by And. 4.20–1. Alkibiades and Taureas were rival khoregoi at a contest of choruses of boys; Taureas tried to remove from Alkibiades' chorus a boy who was allegedly an alien (in accordance with the law mentioned in **56–61**; see note there), but Alkibiades struck Taureas and pushed him away, though the whole audience was looking on; the judges nevertheless awarded the first prize to Alkibiades' chorus. Though the authenticity of And. 4 is suspect, this anecdote corresponds perfectly with **147** and can be accepted. Plu. *Alk.* 16.5 adds no further information and is probably derived from D. This Taureas may well be the Taureas who was a cousin of Leogoras the father of Andokides (And. 1.47), though 'certainly to be identified' (Davies *Families* 29) is perhaps too strong a statement. He may also be the Taureas who owned a palaestra (Pl. *Kharm.* 153a).

ἔστω implies a concession to an opponent's argument, as in **118**. The whole of **147** is a passage of hypophora, with opposed points and answers; hence the lack of connective particles.

χορηγῶν γε χορηγοῦντα: D. implies that rivalry in a competition is some excuse for misconduct.

τόνδε τὸν νόμον: the law quoted in **8**. On its date, **8** n.

εἶρξεν Ἀγάθαρχον: the painter Agatharkhos, son of Eudemos, came from Samos (Harp. *Ἀγάθαρχος*) to Athens by the time of Perikles, and was one of the artists who worked on the buildings on the Akropolis; he prided himself on working fast, whereas Zeuxis prided himself on working slowly (Plu. *Per.* 13.3). Vitruvius 7 *praef.* 11 says that he was the first person to make scenery, for a tragedy by Aiskhylos; but that seems to conflict with the statement that Sophokles introduced scene-painting (Arist. *Poet.* 1449a 18–19), and from the evidence of extant plays it appears unlikely that there was in fact any scene-painting in the fifth century; cf. A. L. Brown *Proc. Camb. Phil. Soc.* 210 (1984) 1–17. The occasion when Alkibiades imprisoned him was notorious, but the three accounts differ in detail. Plu. *Alk.* 16.5 says simply that Alkibiades shut him up and, after he had painted the house (sc. with pictures on the interior walls), let him go with a present. And. 4.17 gives a fuller version: Alkibiades invited him to his house and then compelled him to paint it; though he tried to excuse himself and pleaded that he was already engaged in other contracts, Alkibiades threatened to tie him up unless he got to work painting at once; it was only in the fourth month that he eluded the guards and escaped, and even then Alkibiades had the impudence to go to him and complain that he had left the work unfinished. Σ (506 Dilts) is quite different: Agatharkhos was caught in intercourse with the concubine of Alkibiades, who thereupon imprisoned him. One might dismiss this as the scholiast's invention, if it were not that D. himself seems to imply it by saying that what Alkibiades did was justifiable. In Athenian law any man who caught another man in sexual intercourse with his wife, mother, sister, daughter, or concubine kept with a view to free children could subject the offender to any treatment he wished, including immediate death or imprisonment until he paid compensation or provided sureties for payment (Lys. 1.25, D. 23.53, 59.65–6). This appears to have been the only offence for which a free man could legally be imprisoned in a private house (as distinct from the public prison), and it must therefore be what D. insinuates by the words *τι πλημμελοῦντα*. We cannot be sure which of the three accounts is true, but it is possible to accept the greater part of all of them by supposing that Alkibiades caught Agatharkhos with his concubine and refused to let him go until he painted the house.

ὥς φασιν: at this point Π9 has *ως φα*[...........]*τεροι οπε̣*[. The space would apparently accommodate about twelve letters. Possibly D. wrote *ὥς φασιν ὑπὲρ αὐτοῦ ἕτεροι· ὅπερ* and scribes erred by haplography, jumping from *ὑπὲρ* to *ὅπερ*.

ὅπερ: the antecedent is the whole of *εἶρξεν ... πλημμελοῦντα*.

τοὺς Ἑρμᾶς περιέκοπτεν: two religious scandals occurred in Athens in 415, the mutilation of the Hermai and the profanation of the Mysteries. From our contemporary informants, And. 1 and Th. 6, it is clear that Alkibiades was associated only with the Mysteries, not with the Hermai; cf. MacDowell *Andokides: On the Mysteries* (Oxford 1962) 192–3, Dover *HCT* 4.280–1. But by the time of D., it appears, this distinction was no longer made: the words *τοὺς Ἑρμᾶς περι-*

ἔκοπτεν are a concession made in his argument, and it does not occur to him that the statement may be untrue.

ἱερά: all mss. have *ἱερὰν ἐσθῆτα* except S. Σ (508 Dilts) attributes *ἱερά* to 'the ancient text' (cf. p. 50) and points out that it covers more than just dress. In fact it would hardly be convincing for D. to claim that clothes were more important than the stone images of a god; his argument requires a contrast between *ἀφανίζειν* and *κόπτειν* only, avoiding any differentiation between the objects of the two verbs. We should therefore regard *ἱερὰν ἐσθῆτα* and *τοὺς Ἑρμᾶς* as glosses; *περικόπτειν* (all mss. except S) is probably part of the gloss, being the standard verb for the mutilation of the Hermai (Th. 6.27.1, And. 1.34, etc.). The fact that the scholiast on Hermogenes quotes the passage with *ἱερὰν ἐσθῆτα* and *περικόπτειν τοὺς Ἑρμᾶς* confirms that the alternative readings were current in antiquity.

κόπτειν: the prepositional prefix is often omitted in repetition of a compound verb; **27** n. *παρόντα ... ὅνθ'*.

οὐκοῦν: used to move from the major to the minor premiss of the argument; cf. Denniston *Particles* 439. The obvious conclusion 'Therefore Meidias is more guilty than Alkibiades' is left unstated.

ἐξελήλεγκται τοῦτο ποιῶν: D. did not even allege, let alone prove, that Meidias destroyed totally the clothes and crowns prepared for the chorus; cf. **16** *οὐ μέντοι πᾶσάν γε*. To defend his consistency we must stress the present aspect of *ποιῶν*: Meidias began the process of destruction, and intended to complete it, though he did not. The more one thinks about the argument that this was a worse offence than the mutilation of the Hermai, the less convincing it becomes; but D. does not give the jury much time to think about it.

148. ἐποίησεν δακνόμενος: all mss. have *ἐνδεικνύμενος*, giving the inept sense 'displaying these things to whom'. The question at issue here is what reason or excuse Meidias had for his conduct that was comparable to the excuses which Alkibiades had. In S a later hand has added in the margin *ἐπήρθη ταῦτα ποιεῖν*, and Σ (510 Dilts) offers the following expansion (taking examples from **147**): *ὑπὸ τίνος ἠδικημένος; ποίου χορηγοῦ παροξύνοντος; ποίου ζωγράφου κεκινηκότος εἰς ἀγανάκτησιν;* Clearly the author of this note had in the text not a word meaning 'showing' but a word meaning 'provoked'. I believe that the solution is to read *δακνόμενος*, taking *τίσι* as neuter, 'annoyed by what'. For the metaphorical use of this verb cf. **89**; it is uncommon enough to stimulate a scholiast to gloss it. I suggest that the error originated when an early scribe omitted *-ποιησε-* by haplography, jumping from *ε* to *ε*.

καλὸν ... θεμιτὸν ... ὅσιον: these terms are not easy to define precisely. The structure of the sentence, with the first *μηδὲ* meaning 'not even', shows that *καλόν* is stronger than the others; if a person does what is *καλόν*, he deserves praise, whereas if he does what is *θεμιτόν* or *ὅσιον* he merely avoids blame. *ὅσιον* is always a religious word, meaning that a thing is not forbidden by the gods. *θεμιτόν* refers to what is morally right; this may or may not have a religious basis, according as the speaker believes or does not believe that moral rules are made by the gods. For the use of these words in a prosecutor's assertion that the defendant deserves no mercy, cf. 25.81 ... *ἔλεον, συγγνώμην, φιλανθρωπίαν· ἀλλὰ τούτων γ' οὔθ' ὅσιον οὔτε θέμις τῷ μιαρῷ τούτῳ μεταδοῦναι.*

μηδαμόθεν: a scornful reference to the disreputable parentage of Meidias (**149**) by contrast with the aristocratic ancestry of Alkibiades (**144**).

μή τί γε: sc. *λέγε*, 'do not mention at all', and thus an idiomatic phrase for 'much less'. It is a particular favourite with D. who uses it also in 2.23, 8.27, 19.137, 22.45, 22.53, 24.165, 54.17.

κοινῇ ... ἰδίᾳ: 'concerning the community ... concerning individuals'. Not 'in public speeches ... in private conversation' (Goodwin and Vince); the whole sentence is about public speeches.

149. γένους ἕνεκα: at this point Σ (517a Dilts) says: *τοῦτο συνῆπται τοῖς παρασήμοις, πρὶν διορθωθῆναι τὸν λόγον· τὰ δὲ μέσα ἐνετέθη.* The phrase *τοῖς παρασήμοις* refers to the obelized passage **143–7** (cf. pp. 47–8), so that the scholiast means that **148** was previously omitted, until the speech was 'corrected'. But that does not mean that **148** is spurious (*γένους ἕνεκα* must be part of the series of suggested answers introduced by the question *τίνος γὰρ ἕνεκα;*), but rather that a copyist omitted **148** in error and a reviser correctly restored it.

ἀπορρήτους, literally 'not to be spoken', can mean either 'secret' or 'obscene'. Here D. exploits the two senses: really the birth of Meidias (if the story about it is true) was merely secret, but the use of this adjective and the allusion to tragedy may make some of the jurors think of incestuous births like those of the children of Oedipus. Cf. And. 1.129, referring to the matrimonial affairs of his opponent Kallias: *τίς ἂν εἴη οὗτος; Οἰδίπους, ἢ Αἴγισθος; ἢ τί χρὴ αὐτὸν ὀνομάσαι;*

There is no other evidence for the story that Meidias was a supposititious child, and it is unlikely to be true. No attempt was made to challenge his legitimacy or citizenship in the courts, as far as we know; and his parents will not have been desperate for a son, since they already had another son, Thrasylokhos. It was common for an Attic orator to attack his opponent's origin; the reason why the attack on Meidias' origin takes the form of alleging that he was a supposititious child may be, as Σ (517b Dilts) suggests, that his family was an important one, so that his birth could be disparaged only by saying that he did not belong to it.

τὰς τούτου γονάς: the use of the plural of this noun for a single birth occurs in tragedy (E. *Hipp.* 1082, *Ion* 328). D. may be imitating tragic usage here, as a joke. The repetition of the article suggests 'affected hesitation' (Goodwin), either before mentioning what is unspeakable or before launching into a poetic phrase.

δύο τὰ ἐναντιώτατα: S and A both omit the article, but the normal idiom is to include it in this phrase; cf. **196**, Isai. 1.22, Gorg. *Pal.* 25, Th. 3.42.1. This sentence inflates the audience's curiosity, which is then punctured by the joke at the end of **149**.

βελτίω: having a father of higher social status. D. implies that the father of Meidias was a slave of foreign extraction; cf. **150** *βάρβαρον*.

ταύτης τῆς τιμῆς: 'at that price', the price she paid for Meidias. *τῆς ἴσης τιμῆς* (AF) and *τῆς αὐτῆς τιμῆς* (Dobree *Adversaria* 1.464) are two attempts to clarify the phrase, but neither is necessary.

150. **καὶ γάρ τοι**: 'consequently', an oratorical expression particularly common in D.; cf. Denniston *Particles* 113–14.

τῶν οὐ προσηκόντων ἀγαθῶν: property inherited from his supposed father, whose son he really is not. So also 9.31 *τὰ μὴ προσήκοντα*, 39.34 *οὐ προσήκουσιν*.

πατρίδος τετυχηκὼς: Athenian parentage entitled Meidias to Athenian citizenship.

δοκεῖ softens *ἁπασῶν*: D. does not claim sure knowledge of every city. But it may well be true that Athens had a more comprehensive code of laws than any other city at this time.

ἕλκει καὶ βιάζεται: the object is not Meidias, as most editors assume (that would imply 'He can't help it, poor chap', which is far from D.'s meaning), but innocent citizens like D. The sense of *ἕλκειν* is to catch hold of someone manually, to stop him going away or make him go in a different direction; cf. **216**, **221**, 37.42, 54.20.

ὅπερ ἐστίν: 'which they are'; the antecedent of *ὅπερ* is *ἀλλοτρίοις*, and the subject of *ἐστίν* is *τὰ παρόντα*. The insinuation, that the speaker's opponent's misbehaviour shows that he finds his circumstances alien, is used also in the first oration *Against Boiotos*: *δόξεις εἰς ἀλλότρι' ἐμπεσὼν ὡς οὐ προσήκουσιν οὕτω χρῆσθαι* (39.34).

151–70. *The services which Meidias claims to have done to Athens, compared with those done by Demosthenes.*

151. **ἃ** (A) is preferable to *ὧν* (placed after *ἀναιδεῖ* in S) because attraction of the relative pronoun out of the nominative case is not common (though cf. 57.46 *ἐξ ὧν ἄρτι μεμαρτύρηται μεμαθήκατε*) and is awkward when *ὧν* is separated from the preceding genitives by other words. If ὧν is read, it should probably be placed immediately after *ὄντων*.

ἔνιοί μοι προσιόντες: not all at once, but several individuals on different occasions. That is shown by the optative in *ἐπειδή με μὴ πείθοιεν*, 'each time that they failed to persuade me'.

καθυφεῖναι: **39** n.

οὐκ ἐτόλμων λέγειν: D. cunningly conveys the impression (without giving any evidence) that Meidias' own supporters agree that he is guilty.

ἥλωκεν ἤδη ...: imagining the situation if the jury votes against Meidias. (Not a reference to the earlier vote in the Ekklesia, which did not lead to a penalty.) This kind of hypothesis is usually introduced by *καὶ δή* (for another instance without *καὶ δή* cf. E. *Andr.* 334), and is often, as here, followed in asyndeton by a question asking 'What will happen next?' (cf. 39.8, E. *Med.* 386, Gorg. *Pal.* 7). For the form of *ἥλωκεν*, **105** n.

τίνος: genitive of price; **102** n. *τιμᾶν αὐτῷ τῶν ἐσχάτων*.

τριηραρχίας ἐρεῖ καὶ λειτουργίας: Meidias will argue that it will be to the Athenians' advantage to let him off, because he will use his money to perform services for them. The trierarchy was itself a type of liturgy (**78** n.).

152. ἀγεννὲς: in D. this word has lost its earlier sense of 'low-born' and is a fairly vague word for moral inadequacy, usually referring as here to cowardice. Cf. 11.2 (about military resistance to Philip) *χρὴ μήτ' ὀρρωδεῖν ὑμᾶς τὴν ἐκείνου δύναμιν μήτ' ἀγεννῶς ἀντιταχθῆναι πρὸς αὐτόν*.

μάλιστα μὲν ..., εἰ δὲ μή: 'best of all ..., but failing that'. So also in 14.32, 18.103, 19.101, 20.25. The posing of alternatives here implies that D. has not yet made formally his proposal for a penalty, and thus that the proposed penalty was not stated in the charge (contrast Ar. *Wasps* 897); cf. Harrison *Law* 2.63–4.

153. ἡμεῖς οἱ λειτουργοῦντες does not mean that there was a permanent official list of men who performed liturgies. It just means that Meidias and his associates were among the richest men in Athens, so that they were often called on to perform liturgies, and they wanted to be rewarded by favourable treatment in other ways.

Ruschenbusch *ZPE* 31 (1978) 275–84, 59 (1985) 237–40, 69 (1987) 75–81 argues that there was a list of 300 men, each possessing property worth 4 tal. 1000 dr. or more, who were liable to liturgies. I argue in *CR* 36 (1986) 438–49 that this view should be rejected, and that there was no official list or limit; it was only in practice that all liturgies were performed by the same few hundred men, because the richest available were always selected. The main reasons against

Ruschenbusch's view are: (*a*) a law said that no one was exempt from the trierarchy except the nine arkhons (20.27–8); (*b*) 300 men would not be enough to perform all the liturgies required, at least in some years, while still allowing the statutory interval of one year between liturgies (20.8); (*c*) the *antidosis* procedure (**78** n.) implies that a man appointed to a liturgy could challenge anyone else to take it on whom he thought to be richer, not merely those on a restricted list.

οἱ προεισφέροντες: the tax called *εἰσφορά* was a levy on property paid by Athenian citizens and metics. It was not annual; it was imposed only when the Ekklesia so decided, usually to raise money for a specific purpose such as a naval expedition. The arrangements for collecting it in the time of D. have been discussed by modern scholars for more than a century, but there is still disagreement about the details. For a good account of different theories see R. Thomsen *Eisphora* (Copenhagen 1964), but Thomsen's own theory is not acceptable either; cf. G. E. M. de Ste. Croix *CR* 16 (1966) 90–3, P. A. Brunt *JHS* 86 (1966) 245–7. For more recent discussion see Ruschenbusch *ZPE* 31 (1978) 275–84, 59 (1985) 237–40, 69 (1987) 75–81, Rhodes *AJAH* 7 (1982) 1–19, P. Brun *Eisphora, syntaxis, stratiotika* (Paris 1983), MacDowell *CQ* 36 (1986) 438–49.

It is agreed that a new system began in 378/7, the arkhonship of Nausinikos (22.44, cf. Polyb. 2.62.6–7). Most scholars accept that a main point of the new system was the grouping of the taxpayers into symmories, because Philokhoros says that the Athenians were divided *κατὰ συμμορίας* in that year (*F. Gr. Hist.* 328 F41; cf. J. J. Keaney *Historia* 17 (1968) 508–9). The number and size of the symmories at that time is not known; but I accept (though Rhodes rejects) the view of Ruschenbusch that after 357 the same symmories were used for *eisphora* as for naval contributions (**155** n.), and thus that there were 1200 payers of *eisphora*, arranged in 20 symmories after 357 and in 100 symmories after 354. Individual taxpayers were allocated to symmories by the strategoi (39.8), who tried to combine rich men with less rich ones, so that the total wealth of the members was approximately the same in each symmory; the payments made by each individual were supposed to be in proportion to the amount of his property (**157**, 27.7–9).

The purpose of *προεισφορά* was to secure the prompt payment of *eisphora*. It was a liturgy: as a service to the state a rich man advanced from his own resources the total amount of *eisphora* due from a number of people, and subsequently collected the sums due from the individuals. The collection might give him some trouble, and if he failed to collect all the sums he was left out of pocket himself. In 42.25 'the three hundred' (*οἱ τριακόσιοι*) seems to be an alternative name for *οἱ προεισφέροντες*, and Isai. 6.60 mentions some men making payments of *eisphora* 'among the three hundred', implying that the 300 were a special category within the class of payers of *eisphora*: thus it appears that *proeisphora* was paid by 300 men. The Isaios passage is the earliest reference to them; it was written in 364 but refers to payments made over a period of some years before. Yet it seems that *proeisphora* cannot date from the reorganization of *eisphora* in 378, because Androtion was appointed to collect arrears of *eisphora* *ἀπὸ Ναυσινίκου* (22.44), a task which would have been the concern only of *οἱ προεισφέροντες* when once they had come into existence. So probably *proeisphora* was introduced in the late 370s.

Another obscure question is how the payers of *proeisphora* were appointed. At the time of an emergency in 362 the Athenians voted that there should be a collection of *eisphora* and that the members of the Boule, on behalf of the members of the demes, should nominate those who were to pay *proeisphora* (50.8). This is most easily interpreted as meaning that usually the payers of *proeisphora* were nominated by the demes, but on this occasion the members of the Boule belonging to each deme were to act on behalf of their fellow-demesmen in order to save time. It shows that at this date there was no standing list of payers of *proeisphora*, and they were appointed afresh each time there was to be a collection of *eisphora*. But by 340, when on the proposal of D. the whole of the liability for the trierarchy and naval maintenance costs was transferred to the 300 (Ais. 3.222, Dein. 1.42, Hyp. fr. 134 Kenyon, Jensen = 43.1 Burtt), the 300 were the 'leaders, seconds, and thirds of the symmories' (18.103; **157** n. *ἡγεμὼν συμμορίας*). This indicates that before 340 the old system of appointment of payers of *proeisphora* by demes had been abandoned; instead the three richest members of each of the 100 symmories paid *proeisphora* for their symmory. But no evidence shows whether this change had already been made when *Meidias* was written.

ἡμεῖς οἱ πλούσιοί ἐσμεν: some editors emend this clause (Blass deletes *ἐσμεν*, Weil changes *οἱ* to *οἳ*) to show that it gives not a third kind of public service but an explanation of the other two. But no change is necessary, because Meidias (according to D.) did regard his wealth as a matter for boasting or pride; cf. **198** *πλουτεῖ μόνος*.

154. ἔτη περὶ πεντήκοντα ἴσως ἢ μικρὸν ἐλάττω: the date of Meidias' birth is not known from other sources, and it is in the interest of D.'s argument here to make him seem as old as possible. Since the date of this oration is 347/6, we should infer that Meidias was born later than 397/6, perhaps as much as five years later. (Davies *Families* 386 wrongly says 'about 400'.)

ἐλάττω: *ἔλαττον* is adopted by several editors; but *ἐλάττω* (neuter plural) is acceptable with *ἔτη* and there is no clear authority for ἔλαττον, for the accent on S's reading ἐλάττον suggests that it is a corruption of ἐλάττω, and the reading of Π7 is uncertain.

δύο καὶ τριάκοντα ἔτη: this text conflicts with other evidence about D.'s date of birth. His *dokimasia* was in the summer of 366 (30.15: immediately after an event which occurred in Skirophorion of the year when Polyzelos was arkhon). That puts his birth in either 385/4 or 384/3. Scholars dispute whether a man's *dokimasia* occurred in his eighteenth year (R. Sealey *CR* 7 (1957) 195–7) or after his eighteenth birthday was past (Rhodes *Comm. on AP* 497–8), but probably it could be either; the Athenians did not reckon young men's ages by birth certificates but, at least in some cases, by their physical development (cf. Ar. *Wasps* 578 with my note there), and it is likely that a man would pass through his *dokimasia* as soon as he looked adult, regardless of whether the precise date of his eighteenth birthday had been reached. (For discussion of the date of a man's *dokimasia* see M. Golden *Phoenix* 33 (1979) 25–38.) Dion. Hal. *Amm.* 1.4 has a different chronology, by which D. was born in 381/0 and entered his seventeenth year in 364/3, but that is clearly incompatible with what D. himself says (30.15) and must be wrong; since Dionysios believed (wrongly) that *Meidias* was written in 349/8 (see p. 10 n. 4), a possible explanation is that he calculated D.'s date of birth from the figure δύο καὶ τριάκοντα ἔτη in **154**. The writer of the *Lives of the Ten Orators* places the birth of D. in 385/4 (Plu. *Eth.* 845d: the year of Dexitheos). There is no strong reason to reject that date; I do not find plausible the suggestion that it is based not on evidence but merely on a calculation from the date of D.'s *dokimasia* (Davies *Families* 126, following Hoeck), since the writer appears rather to be taking 385/4 as a known date *from* which D.'s age may be calculated. The other evidence is in his orations about his guardians, where we read that he was seven when his father died (27.4) and was under the charge of his guardians for ten years (27.6, etc.); the period of guardianship began before Aphobos went as a trierarch to Kerkyra (27.14), which must have been in 375 (X. *Hell.* 5.4.63–6). This evidence is most easily interpreted as meaning that D. was born in 384/3 and passed through his *dokimasia* before his eighteenth birthday; yet it can be stretched to fit in with the dating of his birth in 385/4, if we suppose that the guardianship lasted ten years and a few months, and that he was nearly eight when his father died (but calls it 'seven' to increase the effect of pathos). So I am inclined to accept that his birth was late in 385/4, while admitting the possibility that it could have been early in 384/3. But in either case he cannot have been aged thirty-two in 347/6. This figure cannot be reconciled with the other evidence and must be corrupt. (The attempt by Sealey *REG* 68 (1955) 100 to save it, by postulating that this part of the speech was written in 352/1, is unsuccessful; cf. Davies *Families* 386.) It is most likely that he was thirty-seven and that he wrote the figure in the form *ΔΔΔΠΙΙ*, which became corrupted to *ΔΔΔΙΙ*. If it is true that this corrupt figure was the cause of Dionysios' mistake about the date of D.'s birth, the corruption must have occurred early. Plutarch too, writing a century later, had it in his copy of the oration (Plu. *Dem.* 12.3).

ἐτριηράρχουν: for D.'s first trierarchy see **78–80** and notes there. When a man passed his *dokimasia*, he had to spend the next two years serving as an ephebe. Thus, after his *dokimasia* in 366, the first year in which D. was eligible for a trierarchy was 364/3, and that is the sense of *εὐθὺς ἐκ παίδων ἐξελθών*.

σύνδυο ἦμεν οἱ τριήραρχοι: during the Peloponnesian War, when the navy was often in action, service as a trierarch for a whole year came to seem an excessive burden, and from then on it became usual for two men to be appointed, each paying half of the ship's expenses and commanding it for six months. But this was not invariable; there were still some men who served as sole trierarchs, presumably because they volunteered to do so. Cf. Jordan *Navy* 70–3.

οἴκων: 'properties'; *ἦν* is understood. For the sense of *οἶκος* cf. MacDowell in *CQ* 39 (1989) 10–11. But *ἐκ τῶν ἰδίων* could stand without it here (cf. **11**) and a verb may seem preferable. A has *ἐδαπανῶμεν*, which may be a gloss; it is not the best verb to govern *τἀναλώματα*. Modern conjectures: *ἦκεν* (Dobree *Adversaria* 1.464), *διῳκοῦμεν* (G. H. Schaefer *Apparatus* 3.444), *ἦν* (Sykutris).

τὰς ναῦς ἐπληρούμεθ' αὐτοί: a trierarch had to recruit a crew for his ship and pay it. Exceptionally, in the emergency in 362, citizens were listed by demes for compulsory service in the navy; but even then Apollodoros says that few of the conscripts turned up at his ship, and so he hired sailors at his own expense after all (50.6–7).

155. τηνικαῦτα: 'only at the time ...'. The new system of paying for the maintenance of ships by symmories began in 357 or not much earlier (see next note). An inscription records that Meidias had been the trierarch of the *Olympias* (*IG* 2² 1612.291). Since the same inscription records a syntrierarchy of Demokhares and Theophemos (*IG* 2² 1612.313–14) and they are known to have shared a trierarchy in a year earlier than 357/6 (47.22, with the date in 47.44), it has been thought that Meidias too must have been a trierarch in an earlier year, and thus that D. is not telling the truth here. But the argument is not valid, because Demokhares and Theophemos may have shared a trierarchy more than once. (Friends may often have served together in this capacity. The same inscription tells us of Antidoros and Aristolokhos, who shared trierarchies on five different ships, *IG* 2² 1612.352–9.) A lie by D. about the period when Meidias was first a trierarch would very easily have been refuted by Meidias; so it is more likely that D. is telling the truth here.

διακοσίους καὶ χιλίους πεποιήκατε συντελεῖς: the reform of Athenian naval finance can to some extent be reconstructed from allusions in contemporary orations and from the fragments of the inscribed naval records of the period, but despite much modern discussion considerable disagreement persists. The foundations of modern study of the subject were laid by A. Boeckh *Urkunden über das Seewesen des attischen Staates* (1840), and a good summary with bibliography for the next century is given by H. Strasburger in *RE* 7.A.1 (1939) 106–16. More recently cf. M. Amit *Athens and the Sea* (1965) 103–15, Jordan *Navy* 61–93, Wankel *Kranz* 554–72, Ruschenbusch *ZPE* 31 (1978) 275–84, 59 (1985) 237–49, 69 (1987) 75–81, Rhodes *AJAH* 7 (1982) 1–19, Cawkwell *CQ* 34 (1984) 334–45, MacDowell *CQ* 36 (1986) 438–49. Here I concentrate on the points which arise in **155**.

It is agreed that the reform was brought about by a law proposed by Periandros, which concerned the organization of symmories (47.21). This law was in force in 357/6 (47.44); so it was probably passed in 358/7, or possibly a year or two earlier. As our present passage makes clear, the main point of the reform was to make 1200 persons, grouped in symmories, contribute to the cost of maintaining ships, which had previously been borne entirely by one or (more usually in the fourth century) two trierarchs for each ship.

The question arises whether that means that the 1200 served as trierarchs. Many modern writers assume that it does, taking *τριήραρχος* to be (at least in the fourth century) a term of merely financial significance; but in my view a distinction should be drawn between a trierarch and a contributor (*συντελής*). A contributor just paid money, but a trierarch took charge of a ship, as the second half of the word itself suggests; he organized the equipping and manning of it, and he commanded it when it went to sea. Admittedly the distinction tended to become blurred when a trierarch chose to employ a deputy to perform his functions (**80** n. *τὴν τριηραρχίαν ἦσαν μεμισθωκότες*), for then he too did nothing but pay money. There are some passages in the orators in which the distinction is not clearly drawn (e.g. Hyp. fr. 134 Kenyon, Jensen = 43.1 Burtt), and D. complains in 18.104 about rich men who called themselves not trierarchs but contributors (yet this does imply that D. thinks it wrong to confuse the words). Nevertheless a distinction must have existed; for the list of 1200 contributors included rich heiresses and orphan boys, estates belonging to cleruchs and corporations, and men physically disabled, who could not perform the duties of a trierarch (14.16). Another passage which points to this distinction is one in which the speaker says that he was the supervisor of his symmory and a trierarch, implying that other contributors to his symmory were not trierarchs (47.21–4); and the later part of our present sentence in **155** shows that trierarchy was a liturgy, giving exemption from other liturgies, whereas being a contributor was not. We should therefore conclude that the 1200 were not trierarchs. Trierarchs presumably continued after 357 to be appointed by the same procedure as before (by the strategoi, who simply selected the richest men up to the number required that year, exempting those who had served in the preceding year); the change was that they no longer had to pay the whole cost themselves, but received financial help from the contributors.

εἰσπραττόμενοι τάλαντον ταλάντου μισθοῦσι: the contributors allocated to a particular ship for a particular year had to pay for the maintenance of the ship. The cost varied according to the amount of repairs which the ship happened to need, and D. gives the figure of one talent merely as a typical example. The trierarch ought, of course, to collect from the contributors only the amount which the maintenance actually cost. It was then his duty to organize the maintenance and to command the ship; if he chose to employ a deputy to perform those functions (**80** n. *τὴν τριηραρχίαν ἦσαν μεμισθωκότες*), it was for him to pay the deputy out of his own pocket. D.'s allegation is that Meidias (and other trierarchs) exacts from his contributors a total of one talent, saying that this is the amount required for the ship's maintenance, although in fact one talent is enough to cover not only the maintenance but the employment of a deputy trierarch as well; he lets a contract at this figure (*ταλάντου* is genitive of price), by which the contractor undertakes to carry out the maintenance of the ship and perform all the functions of the trierarch too, so that Meidias does nothing and pays nothing.

οὗτοι: a vague rhetorical plural, meaning Meidias and men like him.

πληρώματα ἡ πόλις παρέχει means that, departing from the previous practice (**154**), the state now recruits and pays crews; cf. Isok. 8.48 *τοὺς δὲ πολίτας ἐλαύνειν ἀναγκάζομεν*. However, there continued in fact to be some occasions when the state provided only 'empty ships' (3.5, 4.43), for which the trierarchs and contributors must have had to supply crews.

σκεύη δίδωσιν: sails, rudders, anchors, ropes, etc. But here, even more than in the previous phrase, D. seems to be overstating the difference between the periods before and after 357 (cf. Cawkwell *CQ* 34 (1984) 342–3). Even before then a trierarch could usually draw such gear from the dockyard; it was exceptional, and a matter for boasting, if he provided gear for his ship at his own expense (47.23, 50.7, 50.34, 51.5). No evidence shows what change had really been made; possibly more or better gear was now provided.

τῶν ἄλλων λειτουργιῶν ἀτελεῖς: 'exempt from the other liturgies'. For the genitive cf. **166** *ἀτελῆ στρατείας*. The accusative *ἀτελεῖς* shows that *αὐτούς* is to be understood; F's reading *ἀτελέσι*, agreeing with *ἐνίοις*, produces a sequence of six short syllables, which D. probably wished to avoid.

It was the law that no one could be required to perform two liturgies in the same year or consecutive years (20.8, 50.9, *AP* 56.3; Isai. 7.38 indicates that at an earlier period a two-year interval between liturgies could be claimed). Thus the richest men, who were the first choice for appointment as trierarchs except when they were exempt, were never required to perform festival liturgies (20.19), though they could volunteer to do so. D.'s complaint is that Meidias (and some other men), performing a trierarchy at no cost to himself by the contrivance described earlier in the sentence, can then claim exemption from other liturgies for that year and the next on the ground that he is a trierarch.

156. τραγῳδοῖς: the occasion when Meidias was khoregos for a set of tragedies is not mentioned elsewhere. *ποθ'* here and *τότε* a few lines later suggest that it was some years ago. Presumably it was at the city Dionysia; if it had been at a lesser festival, D. would have said so in order to depreciate it.

αὐληταῖς ἀνδράσιν: the occasion when D. was khoregos for a men's dithyrambic chorus was of course at the Dionysia in 348, when Meidias hit him. Such a chorus could be called *κύκλιος χορός* or *ἀνδρικὸς χορός* or simply *ἄνδρες* (Lys. 21.1–2 contains instances of all three expressions), but *αὐληταῖς* is an inept gloss added by someone who thought that the simple *ἀνδράσιν* required explanation. The chorus did not in fact consist of pipers, but of dancers accompanied by a single piper (cf. **13**). There is a similar error in hyp. i (see p. 426), which suggests the possibility that Libanios was responsible for the gloss here.

πλέον ἐστὶν πολλῷ: it would naturally cost more to dress a dithyrambic chorus of fifty than a tragic chorus of fifteen. Even if a khoregos for tragedies had also to pay something for the actors' dresses and for stage properties, that would be outweighed by the custom of dressing dithyrambic choruses very magnificently; D. ordered gold crowns for his chorus (**16**). But a tragic khoregos could spend a large amount if he wanted to be lavish. The speaker of Lys. 21.1–2 claims to have spent 3000 dr. on a tragic chorus at the Dionysia, 2000 dr. on a men's chorus at the Thargelia, and 5000 dr. (including the cost of a victory tripod) on a men's chorus at the Dionysia the next year.

ἐθελοντὴς νῦν: cf. **13**. *νῦν* is 'on this recent occasion', two years ago.

ἐξ ἀντιδόσεως: induced to perform the liturgy by a challenge, made by the man originally appointed to perform it, that he should either perform it or exchange property (**78** n. *ἀντιδιδόντες τριηραρχίαν*). D. suggests that Meidias deserves no credit for a liturgy which he performed only because he was forced to do it.

εἱστίακα τὴν φυλήν: a *ἑστιάτωρ* was a man who provided a meal for the members of his tribe at a festival. This was a liturgy, for which the members of each tribe nominated someone in the same way as they nominated khoregoi (20.21, 39.7). Σ (Patm.) 20.21 says that it was performed at the Dionysia and the Panathenaia.

Παναθηναίοις κεχορήγηκα: at the Panathenaia there were dithyrambic choruses and choruses of dancers (*πυρριχισταί*); D.'s words could refer to either. The speaker of Lys. 21.1–4 claims that he spent 800 dr. as khoregos for dancers at the great Panathenaia, 300 dr. for a dithyrambic chorus at the lesser Panathenaia, and 700 dr. for dancers at the lesser Panathenaia. We are not told why choruses at the Panathenaia cost less than at the Dionysia and the Thargelia; most likely they were smaller.

οὐδέτερα: plural, where the singular might be expected, referring to single acts; so also 42.2 *οὐδέτερα τούτων ἐποίησεν*.

157. ἡγεμὼν συμμορίας: for the symmories or groups of payers of *eisphora*, **153** n. *οἱ προεισφέροντες*. The ten years for which D. was the leader of his symmory were 376–66, while he was a minor and his guardians had charge of his property; cf. 28.4 *Ἄφοβος μετὰ τῶν συνεπιτρόπων ... ἡγεμόνα με τῆς συμμορίας καταστήσας*. At that time symmories were used only for *eisphora*; they did not begin to be used for naval contributions until 357 or not long before. The leader of a symmory was the richest member of it (Harp. *ἡγεμὼν συμμορίας*) and therefore paid the highest amount. At a later period the three richest members of each symmory became liable to pay *proeisphora* (**153** n.), but D. did not have to do that: the orations against Aphobos, which include figures for the amounts received and expended by the guardians from D.'s estate, mention payments of *eisphora* (27.37) but say nothing of payments of *proeisphora* or receipts of *eisphora* from members of the symmory. So when D. boasts here that he was leader of a symmory, his point is simply that he paid a large amount of *eisphora*. When he came of age, his guardians claimed that they had paid 18 mnai of *eisphora* from his estate (27.37). For the rate at which they paid cf. 27.7 *εἰς γὰρ τὴν συμμορίαν ὑπὲρ ἐμοῦ συνετάξαντο κατὰ τὰς πέντε καὶ εἴκοσι μνᾶς πεντακοσίας δραχμὰς εἰσφέρειν, ὅσονπερ Τιμόθεος ὁ Κόνωνος καὶ οἱ τὰ μέγιστα κεκτημένοι τιμήματ' εἰσέφερον*. This sentence is best interpreted as meaning that D.'s guardians agreed on his behalf to pay 500 dr. = 5 mnai of every 25 mnai, i.e. one-fifth of the total amount of *eisphora* levied on the symmory, which was as much as Timotheos and other very rich citizens paid. Perhaps it was normal, either by law or by custom, for the leader to pay one-fifth of a symmory's *eisphora*, without calculating precisely the ratio between his property and the other members'.

ἴσον: not referring to *ἔτη δέκα*, because the richest men must have remained leaders of their symmories indefinitely, not for ten years only. Instead it is the object of *εἰσφέρων*, 'contributing as much as Phormion ...', one-fifth of the symmory's *eisphora*, giving much the same sense as 27.7 *ὅσονπερ Τιμόθεος ... εἰσέφερον*. I therefore place a comma after *εἰσφέρων* rather than (as other editors) before it.

Φορμίωνι: there are two possible identifications of this Phormion:

A. Phormion son of Ktesiphon, of the deme Peiraieus, is mentioned several times in connection with naval and mercantile affairs, as a trierarch, as the donor of a set of oars, and as a witness to a mercantile contract (*IG* 2^2 1622.472–3, 1623.246–8, 1629.647, D. 35.13–14). His son Arkhippos also performed liturgies (*IG* 2^2 1623.103–4, 1623.300–1, 2318.336).

B. Phormion, successively slave, banker, and merchant, is prominent in several orations in the Demosthenic corpus (36, 45, 46, 52, 53). He was a slave of Pasion the banker, but in due course was freed, took over the management of the bank in Pasion's old age and during the minority of Pasion's younger son Pasikles, married Pasion's widow Arkhippe, and became an Athenian citizen. Much of the information about him comes from speeches concerning his disputes with Pasion's elder son Apollodoros, but there is also a fragment of a speech *Against Pasikles* (Hyp. fr. 134; the title indicates that the reference here is to the same family) accusing Phormion of trying to get off the list of the 300 richest men after D.'s naval reform in 340. For detailed discussion of his career see R. Bogaert *Banques et banquiers dans les cités grecques* (Leiden 1968) 74–8, Davies *Families* 431–2, 435–7, S. Isager and M. H. Hansen *Aspects of Athenian Society in the Fourth Century B.C.* (Odense 1975) 176–91.

Whether *A* and *B* were identical is discussed by Davies *Families* 436–7. Two considerations favour identity: (*a*) the wife of *B* was named Arkhippe and the son of *A* was named Arkhippos; (*b*) D. in **157** speaks as if there were one rich Phormion, not two. The only objection which Davies regards as serious is that *A* had a patronymic, *Κτησιφῶντος*, whereas *B* was an ex-slave. But this objection seems to me to have no weight, since *B*'s father, though presumably a slave, may, for all we know, have been named Ktesiphon, or have had a foreign name which could be translated into Greek as Ktesiphon. I therefore accept that all these references are to the same man.

Λυσιθείδῃ: Lysitheides, of the deme Kikynna, was awarded a gold crown for his services to Athens, including financial contributions (Isok. 15.94). No doubt he performed many liturgies, though the only one known is the trierarchy in 355/4; this led to an accusation of misappropriating money due to the state, but he was eventually exonerated (D. 24.11–13). He owned land in the mining area (*Hesperia* 19 (1950) 219 no. 6 line 10, 260 no. 19 lines 6–7, and his name is restored in other inscriptions in this series), which may have been the source of his wealth. He had been a student of Isokrates, and is mentioned as a friend of Pasion and as a respected arbitrator in the oration *Against Kallippos* (Isok. 15.93–5, D. 52.14–16, 52.30–1). I take the reference to 'Lysitheides nephew of Thrasyboulos' in Plu. *Eth.* 575f as meaning that his mother was the sister of Thrasyboulos, the democratic leader in 403. (In that passage the point is that Lysitheides had a friendly attitude to Thebes because Thrasyboulos gained support there in 403. Davies *Families* 239 misses this point, and denies the identification with our Lysitheides on grounds of 'the difference in political standpoint'; but really nothing is known of the political standpoint of Lysitheides.)

Καλλαίσχρῳ: a fairly common name. Holders of it in Athens in the fourth century include one (or possibly two) for whom Lysias wrote one or two speeches (Plu. *Eth.* 833a, Pol. 10.105), one against whom Deinarkhos wrote a speech (Harp. *Κηφισόδωρος, κύκλοι*), one of the deme Phegous (*IG* 2^2 1926.22), one of Thorai (*IG* 2^2 1927.173), and one of Aphidna who was a treasurer of the goddesses at Eleusis in 328/7 (*IG* 2^2 1672.250). But the one probably meant here was a metic from Siphnos who served as a trierarch (*IG* 2^2 1609.27); cf. Davies *Families* 591.

ὑπὸ γὰρ τῶν ἐπιτρόπων ἀπεστερήμην: for D.'s claim that his guardians deprived him of his property see **78** with notes.

δοκιμασθέντα: examined at the age of eighteen to check that he was entitled to be registered as an Athenian citizen.

ἐγὼ μὲν ...: for the shape of this sentence, combining *μέν* and *δέ* with *οὕτως* and a concluding *πῶς*, cf. 14.17–18 *καὶ τὰ μὲν σώμαθ' οὕτω συντετάχθαι φημὶ δεῖν* ... *τὰς δὲ τριήρεις πῶς;*

οὐδὲν: best taken as adverbial, 'not at all'. Yet the alternative 'deprived of no part of ...' is possible, and would not necessarily require the genitive; cf. 47.32 *ἀποστερῆσαι τὴν πόλιν τὰ σκεύη*.

158. πλὴν εἰ ...: the items which follow are of course not liturgies performed for Athens, but luxuries which Meidias has bought for himself. All are things visible in public places, which D. expects some at least of the jurors to have seen.

μυστήρια may refer to the lesser mysteries at Agrai as well as to the Eleusinian Mysteries.

τοῦ λευκοῦ ζεύγους: a carriage drawn by a pair of white horses. (Not a carriage painted white, because white paint was not particularly expensive or luxurious.) Sikyon was well known for its horses; cf. A. Griffin *Sikyon* (Oxford 1982) 30.

τρεῖς ἀκολούθους ἢ τέτταρας: an Athenian going to the Agora would often take a slave with him, to carry his purchases or to run errands; but to take a posse of three or four, D. implies, is overbearing.

σοβεῖ: used elsewhere of shooing away birds, this verb is best interpreted here as meaning that Meidias, by means of his attendant slaves, makes people get out of his way. Cf. Plu. *Sol.* 27.3 *σοβοῦντας ἐν ὄχλῳ προπομπῶν καὶ δορυφόρων*.

κυμβία: **133** n.

ῥυτὰ: a kind of cup out of which the liquor is poured into the drinker's mouth through a small hole. Originally it was a horn pierced at the tip, but by the classical period it was normally made of metal in the shape of a horn; later *ῥυτά* were made in other shapes. They are discussed in Ath. 496f–497e.

φιάλας: cups made of silver or gold. Various shapes of *φιάλη* are discussed in Ath. 500f–502b. '*φιάλαι* and such things are a sign of wealth' (22.75); that is why Meidias talks about them in an ostentatious manner.

ὀνομάζων: either in casual conversation, or in negotiating to buy such goods from traders in the Agora. E. J. Robson *CR* 23 (1909) 258 suggests the interpretation 'have booked to him', 'saying "put that down to my account"'; but I find no evidence that shopping in the Agora was done on account, nor that *ὀνομάζειν* can mean 'give one's own name'.

159. τοὺς τυχόντας: 'ordinary people'; cf. LSJ *τυγχάνω* A.I.2.b. Not 'those who happen to come in his way' (Goodwin), which would be *τοὺς ἐντυγχάνοντας* (cf. **88**).

ἀφικνούμενα: 'touching', 'affecting', a sense not recorded in LSJ. So also in 26.3 *τὰ δὲ τῶν ἀρχόντων καὶ πολιτευομένων* [sc. *ἁμαρτήματα*] *εἰς ἅπαντας ἀφικνεῖται*.

φιλοτιμίαν: a word highly characteristic of Greek pride and competitiveness, but hard to define or translate precisely. Many modern scholars refer to it briefly, but a full-scale study of the concept is still awaited. Meanwhile the best discussions are those of Dover *Morality* 229–34 and D. Whitehead *Cl. et Med.* 34 (1983) 55–74; the former is stronger on the literary, the latter on the epigraphical evidence.

Literally 'love of honour', the word refers not only to a state of mind but also to activity undertaken for the purpose of gaining honour; and honour (*τιμή*) means praise, admiration, deference, and sometimes material rewards, given by other people in acknowledgement of such activity successfully undertaken. But different opinions are possible about the kind of activity which should be honoured, and in this matter it is clear that a major change occurred in Athens in the fourth century. For the writers of the late fifth and early fourth century, *φιλοτιμία* is mainly selfish; it involves acquiring advantages (material or political) for oneself, so that other people think one superior. Such activity is generally antisocial, and consequently *φιλοτιμία* is regarded as harmful to the community; e.g. Th. 2.65.7 writes of the bad effect of conduct *κατὰ τὰς ἰδίας φιλοτιμίας καὶ ἴδια κέρδη*, and in Ar. *Thesm.* 383 a speaker is anxious not to be thought to be motivated by *φιλοτιμία*. This sense can be found sometimes even in D. (e.g. 8.71), but by the second half of the fourth century *φιλοτιμία* is usually not selfish but patriotic; it means performing service to the community, to which the community is expected to respond by conferring *τιμή* in gratitude. Such service may consist simply of donating a large amount of money to the state or of performing liturgies (a kind of service which was largely but not entirely financial), or it may be military or political or of some other kind.

This change is documented by Whitehead, who adduces a large number of honorific decrees in which men are praised and rewarded for the *φιλοτιμία* which they have shown towards the people of Athens or other public bodies; e.g. *IG* 2^2 360.13–16 *δεδόχθαι τῶι δήμωι ἐπαινέσαι Ἡρακλείδην Χαρικλείδου Σαλαμίνιον καὶ στεφανῶσαι χρυσῶι στεφάνωι εὐνοίας ἕνεκα καὶ φιλοτιμίας τῆς πρὸς τὸν δῆμον τὸν Ἀθηναίων*. These show that *φιλοτιμία* was a standard term of approval for public service from about 350 onwards. How did the change from the earlier usage occur? Whitehead maintains that 'the Athenians of the fourth century were doing something quite deliberate when they sought to "democratise" the concept of *philotimia* by defining it in terms of their collective profit'. I think it would be hard to show that the population as a whole had so sophisticated an intention; more likely it was deliberate only on the part of some enlightened political leaders. The importance of **159** is that it reveals D. as one of these, or at least as supporting the change. When he says that building a grand house for oneself is not *φιλοτιμία* but performing liturgies is, he is rejecting the old sense of the word and advocating acceptance of the new one.

καλά: the mss. are divided between *καλά*, *λαμπρά*, and *πολλά*. The latter two may be errors induced by the occurrence of *λαμπρῶς* and *πολλὰς* earlier in the sentence, whereas *καλά* cannot be explained in that way and so is more likely to be correct. Conjectures offered to explain the existence of variant readings: *πολλὰ καὶ καλά* (Spalding), *ποικίλα* (Buttmann).

160. τριήρη ἐπέδωκεν: between these two words *ἣν* intrudes in all the older mss. except A. Correctors have tried to make sense of this by altering the two previous words, to produce *ἀλλὰ μὴν διὰ τριήρη ἣν ἐπέδωκεν*, but this uncompleted sentence cannot be right.

D. is now moving on to the naval and military affairs of 349/8. *ἐπιδίδωμι* is the regular verb for free donations to state resources, as distinct from compulsory contributions; cf. LSJ *ἐπιδίδωμι* I.2.b. Did Meidias pay for a ship to be built, or did he donate an already existing ship which he possessed or bought for the purpose? The latter, it appears, for the ship was ready to sail almost immediately (**163**).

ἐγὼ ὑμῖν τριήρη ἐπέδωκα: repetition to mock Meidias' repeated boasting. This will be most effective if D. delivers it in a comic imitation of Meidias' own voice.

οὐδενὸς γὰρ πράγματος οὐδ' ἔργου: genitive of price; cf. **146** *οὐδενὸς τούτων*.

δειλίας καὶ ἀνανδρίας: why D. attributes the donation of a trireme to cowardice will be explained in **162–4**.

ἄνωθεν δέ, βραχὺς ...: for the phraseology cf. **77**. FYP here add *κἂν ἄνωθεν ἄρχεσθαι δοκῇ*: this probably results from a marginal note giving the wording of **77** for comparison. To explain the donation of a trireme in 349 it is not really necessary to go back to earlier years, but D. does so in order to point out that Meidias had not previously donated a trireme but D. himself had.

161. εἰς Εὔβοιαν: in 357. When the Thebans invaded Euboia, the Athenians, spurred on by a rousing speech by Timotheos in the Ekklesia, sent a naval and military force and within thirty days compelled the Thebans to leave Euboia under a truce (**174**, 8.74–5, Ais. 3.85, Diod. 16.7.2). On these events see Cawkwell *Cl. et Med.* 23 (1962) 34–40, Brunt *CQ* 19 (1969) 247–8. Athens then made an alliance with the cities of Euboia, and the inscription recording the alliance is dated 357/6 (*IG* 2^2 124 = Tod 153). The campaign therefore was either in that year or at the end of 358/7, but it remains uncertain which. Two possible lines of argument point to different years but are both inconclusive:

A. Ruschenbusch *ZPE* 67 (1987) 158–9 argues thus: in the emergency arising from the Theban invasion of Euboia every possible trierarch must have been compelled to serve unless exempt; D. claims credit for serving voluntarily, and so must have been exempt from compulsory service that year; exemption lasted for two years after a previous liturgy; D. was a trierarch in 360/59; therefore his voluntary service must have been in 358/7, not in 357/6 when his exemption would have expired; therefore the expedition to Euboia was in 358/7. The weakness of this argument is that it is not certain that exemption lasted for two years at this period; Isai. 7.38 merely implies that there had been a time when two years' exemption could be claimed, but in 355, at least, the period of exemption was only one year (20.8).

B. At the time of this expedition to Euboia Meidias was tamias of the *Paralos* (**174**); the incident involving some men from Kyzikos occurred while he held that office (**173**), and Σ **173** (586 Dilts) says it was during the Social War, which did not begin until 357/6; therefore the expedition to Euboia was also in 357/6. The weakness of this argument is that Meidias could, for all we know, have been tamias of the *Paralos* in more than one year.

ἐπιδόσεις παρ' ὑμῖν πρῶται: one fifth-century instance of a man donating ships to the Athenian navy is known (X. *Hell.* 2.3.40); but we should take D. as meaning that in 357, for the first time, a general appeal was made to rich men to donate ships in order to deal with the emergency in Euboia. In a later speech D. refers to this same occasion as the first on which voluntary trierarchs were appointed (18.99). This must have been part of the same appeal: if a man donated a ship, he was expected to serve as its trierarch.

τούτων: sc. *τῶν ἐπιδιδόντων.*

συντριήραρχος means only that Philinos shared the duties of trierarch with D., not that he shared the cost of donating the ship. The decree of Demokhares, praising D.'s services to Athens many years later, says that on this occasion he donated 8 tal. and a trireme (Plu. *Eth.* 850f).

Φιλῖνος: Philinos, of the deme Lakiadai, is mentioned as a trierarch in two inscriptions. He shared the trierarchy of the *Eutykhia* with Pheidippos, probably in 358/7 (*IG* 2^2 1611.363, 1612.282), and he shared the trierarchy of the *Heos* with D. (*IG* 2^2 1612.301). The date of the latter is not known (*IG* 2^2 1612 refers to past trierarchies over a period of several years); the *Heos* could have been the trireme which D. donated in 357, or he and Philinos may have shared a trierarchy on another occasion too. Nothing else is known about Philinos.

εἰς Ὄλυνθον: in the autumn of 349; see p. 4. From D.'s words it is clear that neither he nor Meidias contributed a trireme on this occasion.

ἐξετάζεσθαι: 65 n. *ἐχθρὸν ἐξεταζόμενον.*

τρίται: early in 348, not long before the Dionysia; for the sequence of events see pp. 6–7. Instead of *τοίνυν* (S) editors generally adopt *νῦν* (A). But *τοίνυν* has an inferential sense which suits the context: there were two previous rounds of donations, so that the recent one, to which Meidias contributed, was merely the third. The corruption is easily explained as haplography after *-ται.*

ἐν τῇ βουλῇ γιγνομένων ἐπιδόσεων must mean that the Boule resolved to appeal for donations and some men said immediately that they would make them.

παρὼν: presumably as a spectator; there is no evidence that Meidias was a member of the Boule in 349/8. Cf. **116**.

162. ἐν Ταμύναις: a town in the territory of Eretria. It used to be thought that it was on the coast, near modern Aliveri, but it now seems clear that it was inland, near modern Avlonari; cf. E. Kirsten *RE* 22.1 col. 342–4, Knoepfler *BCH* 105 (1981) 293. The reason why the Athenian force under Phokion had gone there was presumably that the opponents of Ploutarkhos were in that area.

ἐξηγγέλλετο: the imperfect implies that several reports arrived, perhaps over some days. The conjecture *ἐξήγγελτο* (Cobet *Nov. lect.* 582) is unnecessary.

προεβούλευσεν: for the avoidance of the crasis *πρου-* in this word cf. *IG* 2² 336a.7. The meaning is that the Boule resolved that the proposal to send out all the remaining cavalry be put before the Ekklesia at its next meeting. This must have been a later meeting of the Boule than the one at the end of **161**, but no meeting of the Ekklesia had been held meanwhile, so that the appeal for donations of ships and the proposal to send out the cavalry both came to the same meeting of the Ekklesia.

πρὶν καὶ προέδρους καθίζεσθαι: before the meeting of the Ekklesia formally began, Meidias spoke to the citizens already assembled for it; by making his offer before the proposal to send out the cavalry came forward, he hoped to avoid the imputation of cowardice which D. now makes. For the proedroi, **9** n. For the form *καθίζεσθαι*, **35** n.

163. ὡς οὐκ ἐδόκει ...: this may mean that news of Phokion's successful battle at Tamynai (described in Plu. *Phok.* 13) came in while the meeting of the Ekklesia was in progress.

ἀνεπεπτώκει: 'had fallen back', a military metaphor.

οὐκ ἀνέβαινεν: the imperfect, instead of the aorist, is here more emphatic, implying 'he did not begin to embark', 'he made no attempt to embark'. Naval service was less dangerous than military service, but still involved some risk and trouble. When he expected to be called up for military service, Meidias (according to D.) argued that his naval duties, as trierarch of the ship which he donated, must take priority; but when there was no likelihood of military service, he evaded naval service by paying a deputy to take charge of the ship. For the use of a deputy, **80** n. *τὴν τριηραρχίαν ἦσαν μεμισθωκότες*.

τὸν μέτοικον ... τὸν Αἰγύπτιον means that Pamphilos was of Egyptian origin but was licensed to reside permanently in Athens. Pamphilos was a common name, and nothing else is known for certain of this holder of it; but it is tempting to identify him with a shipbuilder named Pamphilos, who was active in Athens at this time (*IG* 2² 1612.156–85) and could have been the builder of Meidias' ship. No other instance is known of a metic taking charge of an Athenian trireme, but D.'s words do not suggest that this was illegal or unprecedented; cf. D. Whitehead *The Ideology of the Athenian Metic* (Cambridge 1977) 81.

164. Φωκίων: the distinguished general. For his life see Plutarch's *Phokion*, which includes an account of the campaign in Euboia (12–13). It is worth noting here that some years later it was Meidias' son, also named Meidias, who proposed the award of a gift to Phokion for his services (Plu. *Eth.* 850b).

τοὺς ἐξ 'Αργούρας ἱππέας: the cavalry which had returned to Athens from Argoura. The sentence does not mean 'Phokion ordered the cavalry to come from Argoura', which would be *μετεπέμπετο τοὺς ἱππέας ἐξ Ἀργούρας*: cf. Knoepfler *BCH* 105 (1981) 291. For the location of Argoura, **132** n.

ἐπὶ τὴν διαδοχὴν: to take over from the hoplites who had won the battle of Tamynai, who no doubt were considered to have earned a rest.

ἱππαρχεῖν ἠξίωσε: Meidias was a hipparch in an earlier year, not in 349/8 (**133** n. *ἐκληροῦ*).

παρ' ὑμῖν: 'in Athens', where the duties of a hipparch were ceremonial, such as riding in processions (**171**, **174**).

οὐ συνεξῆλθεν: in fact, as D.'s own account has already shown (**132–3**), Meidias did accompany the cavalry when they were first posted to Euboia. It was only when they were posted there for the second time that he went to sea instead.

165. Νικήρατος: greatgrandson of the famous Nikias. He must have been born in 390/89 or 389/8, because he is known to have served as a public arbitrator, and thus to have been in his sixtieth year, in 330/29 (*IG* 2^2 2409.21–2; cf. Lewis *BSA* 50 (1955) 27–9). He had interests in the mining area (*Hesperia* 19 (1950) 242 no. 15 line 46, 261 no. 19 line 24, etc.), and probably that was the basis of his wealth, as in the case of his greatgrandfather. He is mentioned in several naval inscriptions as a trierarch or a contributor (*IG* 2^2 1616.8, 1627.201, 1629.494, etc.). He also served as a treasurer of the army funds (*IG* 2^2 1443.13), as a hieropoios sent to Delphi (*SIG* 296.7), and as an *ἐπιμελητής* of the festival of the Amphiereia (*SIG* 298.27–8). Cf. Lewis *BSA* 50 (1955) 30–1, Davies *Families* 406–7.

ἀγαπητός: 'only son'. The point is that Nikeratos had no brothers or sons to continue his family if he were killed in action; the repetition of the article emphasizes the number of reasons for avoiding risks. We may infer that *ἀγαπητός*, more literally 'cherished', had come to mean an only son or only child, without any need to add *μόνος*. Other passages in which it may be so interpreted, even though *μόνος* is not added, are Hom. *Il.* 6.401, *Od.* 4.817, Ar. *Thesm.* 761, Arist. *EE* 1233b 2.

Εὐκτήμων ὁ τοῦ Αἰσίωνος: evidently another rich man who donated a trireme, but nothing else is known of him for certain. He is clearly not the Euktemon mentioned in **103** and **139**. There is no particular reason to identify him with the Euktemon who, with Diodoros, prosecuted Androtion and Timokrates in the 350s (D. 22 and 24). There was a trierarch named Aision in the 350s (*IG* 2² 1612.293); and there was an orator or rhetorician named Aision, with whom D. studied rhetoric, and who paid him the doubtful compliment of saying that D.'s speeches, when *read*, were better than those of earlier orators (Arist. *Rhet.* 1411a 25, Plu. *Dem.* 11.4, Souda δ 454). It is plausibly suggested by Davies *Families* 5 that one or both of these may have been the father of Euktemon.

Εὐθύδημος ὁ τοῦ Στρατοκλέους: this father and son performed trierarchies together in the 350s (*IG* 2² 1612.137–8, 1612.271–2). The fact that Stratokles is not said by D. here to have joined his son in 348 does not necessarily mean that he was now dead, as maintained by Davies *Families* 495 (though he may have been so); D.'s point here is about active service in Euboia, and Stratokles was probably too old for that now. The name of Euthydemos is restored as the khoregos of a boys' chorus in 342/1 (*IG* 2² 2318.294, 3041), but no more is known of him. His son, also named Stratokles, later became a prominent politician and was one of the accusers of D. in 324/3 (Dein. 1.1).

ταύτην τὴν στρατείαν means that Nikeratos, Euktemon, and Euthydemos each served in the campaign in Euboia, whether in the cavalry or in the infantry or in some other capacity, presumably paying deputies (in the same way as Meidias paid Pamphilos) to command the triremes which they had donated. Bekker's emendation *ταύτῃ*, 'in this way', is plausible but not necessary.

ἐπίδοσιν: each gave a trireme without expecting anything in return (such as exemption from military service). But to say 'they provided the donation as a gift' would be a tautologous expression, and it is probably right to delete *ἐπίδοσιν*. The article with *μέν* can be used as a pronoun (referring to *τριήρη*), and *ἐπίδοσιν* may have been added by someone who thought that obscure.

δωρεᾶς: this form, given by all mss. here and in **170** and **172**, is well attested in the fourth century, and there is no need to change it to *δωρει-*. Cf. Threatte *Grammar* 1.311.

πλέουσαν: afloat and equipped, ready for action.

λειτουργεῖν: metaphorical; the trierarchy was a *λειτουργία* in the literal sense, but these men served their country in another way too.

166. **ὁ ἵππαρχος**: D. sarcastically uses the title which Meidias was proud of having held.

τὴν ἐκ τῶν νόμων τάξιν: military service was a legal requirement, and D. suggests that Meidias ought to have been prosecuted for not performing it. This seems to show that trierarchs were not legally exempt from military service (a point of law on which we have no other evidence). But it is quite likely that in practice the strategoi refrained from calling up men who were serving as trierarchs, so that Meidias may have been justified in assuming that he would be excused service with the cavalry while he was commanding his ship. D. is right, however, to argue that a trierarchy undertaken for that ulterior motive does not deserve the special gratitude or credit which Meidias evidently claimed.

τελωνίαν: a normal method for the state to impose a tax was to sell the right to collect it to the highest bidder at an auction; the purchaser hoped to make a profit by collecting from the payers of the tax a larger total amount than the lump sum which he paid to the treasury. To enable him to collect the tax without interruption he was given exemption from military service for the year (59.27). D. suggests that Meidias has found that, for avoiding military service, paying for a trireme is as effective as buying a tax-collection.

πεντηκοστὴν: **133** n. *πεντηκοστολόγοι*.

λιποτάξιον: **103** n. The mss. here give *λιποταξίαν*, but that form is not found elsewhere, and it is rightly emended by Cobet *Nov. lect.* 79.

φιλοτιμίαν: **159** n.

Μειδίας: the name, unnecessary for the grammar or sense, is put in for sarcastic effect, as if to proclaim the brilliant inventor.

ἱππικῆς: the normal noun for the activity of being a *ἱππεύς* (not necessarily the skill of horsemanship; LSJ *ἱππικός* II.2 associate the word too closely with *τέχνη*). The genitive is used because D. treats *πεντηκοστή* as a word for 'exemption'. Some editors adopt *ἱππικήν* (S), but between other words in the accusative corruption of -*ῆς* to -*ήν* is likelier than the reverse.

πεντηκοστήν: metaphorical. Donating a trireme is not a *πεντηκοστή* in the literal sense, but it has turned out to be an equivalent means of procuring exemption from military service.

167. **καὶ γὰρ αὖ τοῦτο**: there is also a second respect in which Meidias' donation of a trireme was like a *πεντηκοστή*: he used it as a way of making money (*χρηματισμός*).

τῶν ἄλλων ἁπάντων only means all who donated triremes on the third occasion mentioned in **161**, not on the other two.

παραπεμπόντων: providing a naval escort for the ships carrying the soldiers home. This was probably done as an honour to the victors of Tamynai, rather than from fear of any attack during the short voyage.

ὑμᾶς: here, as often, this pronoun simply means 'the Athenians'. Of course D. does not know whether all, or any, of the jurors for the trial of Meidias were among the soldiers returning from the battle of Tamynai; indeed no individual can have been both in Athens (**132** *ὑμῖν*) and in Euboia (**167** *ὑμᾶς*) at the same time. But *ὑμεῖς* in an Athenian speech does not refer to particular individuals, but means the Athenian people or any representative part thereof, such as the Ekklesia, a jury, or an army.

Στύρων: Styra is a town and port in south Euboia.

χάρακας ...: since his trireme was after all not required for warfare, Meidias put it to profitable use by carrying cargo on it.

ξύλα: either for pit-props or for fuel.

καταπτύστῳ: **137** n.

168. Another spurious document, composed no doubt by the same forger as the other testimonies in this oration. This time it is primarily the list of names which gives him away. He seems not to have known many Athenian demotics, for he tends to repeat the same ones; *Σουνιεύς* was used in **121**, *Παιανιεύς* in **107** and **121**, *Σφήττιος* in **82** and **121**.

Πάμφιλος: the forger has thought that he should include the metic mentioned in **163**. But he has muddled the story; Pamphilos was not one of the trierarchs alongside Meidias, but Meidias' own deputy. A man named Pamphilos of Akherdous is mentioned in Ais. 1.110, and it has therefore been suggested by Boeckh *Urkunden über das Seewesen des attischen Staates* (1840) 247 that his name should be restored here, by deletion of *Νικήρατος*. That suggestion might be acceptable if there were other reasons for regarding the document as genuine.

Νικήρατος Ἀχερδούσιος, Εὐκτήμων Σφήττιος: the forger has decided to include two of the trierarchs mentioned in **165**, but has guessed at least one of their demotics wrongly. In fact Nikeratos was *Κυδαντίδης* (*IG* 2^2 1629.494–5, 2409.21–2, etc.); and if Euktemon was the son of Aision the trierarch, he was *Κόπρειος* (*IG* 2^2 1612.293).

τοῦ στόλου πλεόντων: in Attic, singular nouns referring to a number of persons do sometimes take plural verbs or epithets, especially if other words intervene, but this instance, if genuine, would be exceptionally flagrant.

169. ἅπερ ... τοιαῦτ': 'of the same kind as the ones which ...'. For this correlation cf. Pl. *Rep.* 349d *τοιοῦτος ἄρα ἐστὶν ἑκάτερος αὐτῶν οἷσπερ ἔοικεν*. But Reiske's conjecture *οἷάπερ* may be right.

καταλαζονεύσεται: 'he will be a thorough braggart' in his speech for the defence; cf. 36.41 *ἀλαζονεύσεται καὶ τριηραρχίας ἐρεῖ καὶ* *χορηγίας*. Here *κατ-* is merely intensive; it does not have the sense 'against' as in the other fourth-century instances of this compound (Isok. 15.5, 15.31).

κατὰ: 'in the manner of'; cf. LSJ *κατά* B.IV.3.

στήσαντες τρόπαια: a metonymy for winning victories on land. *καλὰ* means that the victories were glorious, not that the monuments were well designed.

170. οὐδ' ἂν δοίητε: 'nor would you give', combining the senses of probability and rightness.

ὅπου ἂν: *ὁπόταν* (A) would give equally good sense, but *ὅπου* is supported by Σ (578 Dilts), whose note *καὶ ἐν ἑταίροις καὶ ἐν θεάτροις* looks like an attempt to identify places rather than times. (The attempt is misguided, however, since D.'s words mean all places whatsoever.)

'Αρμοδίῳ καὶ 'Αριστογείτονι: the conspirators who in 514 attempted to assassinate the tyrant Hippias, though they succeeded only in killing his brother Hipparkhos. They themselves were put to death (Th. 6.57.4, *AP* 18.4), but after the tyranny ended they were honoured posthumously: statues of them were erected in the Agora, and their descendants (the one nearest surviving relative of each) had the perpetual privileges of free dinners at the Prytaneion, front seats and other honours on religious occasions, and exemption from some taxes (19.280, 20.70, 20.159–60, Isai. 5.47, Dein. 1.101, Paus. 1.8.5, *IG* 1^3 131.5–7). The *στήλη* to which D. refers must be the one bearing the official inscription of those privileges. He blurs the distinction between the tyrannicides themselves and their posterity; strictly *αὐτοῖς* has to refer to the latter, but the former must be the subject of *ἔπαυσαν*.

171–4. *Offices which Meidias has held, and his poor showing in them.*

171. χάριν: D. here regards the offices which Meidias has held as rewards which he did not deserve. Meidias no doubt would have preferred them to be regarded as services which he performed for Athens. Then, as now, opinions could differ about whether the honour of holding a particular office outweighed the burden of performing its functions, but D. skilfully makes it appear that Meidias has received more than he has given.

καταπτύστῳ: **137** n.

τῆς Παράλου ταμίαν: in 358/7 or 357/6 (**161** n. *εἰς Εὔβοιαν*). Meidias' activities while holding the office are described in **173–4**. The *Paralos*, like the *Salaminia*, was a sacred ship (cf. **174** *τὴν ἱερὰν τριήρη*), used by the Athenians not only for conveying their representatives to religious celebrations outside Attika, but also for military errands, especially those on which particular speed was needed (e.g. Th. 8.74.1, X. *Hell.* 2.1.29). Its crew, the *Πάραλοι*, were a clearly defined section of the Athenian citizenry (Th. 8.73.5, Ar. *Lys.* 58). Two inscriptions show that they had their own religious cult in honour of a hero named *Πάραλος* at a shrine named *Παράλιον* which was probably located in Peiraieus (*IG* 2^2 1254, and an inscription published by A. Dain in *RÉG* 44 (1931) 296–301; cf. D. 49.25–6, Harp. *Πάραλος*). It seems likely that membership was hereditary and the hero Paralos was the supposed common ancestor, so that the Paraloi were a *γένος* with the unusual special function of manning a sacred ship; cf. Jordan *Navy* 173–6. The inscriptions show that they had officials called *ἐπιμεληταί* and *ἱεροποιοί*, but also that the tamias was their chief official, because his name is used for giving the date, like the arkhon's name in Athens generally. **171** and *AP* 61.7 show that he was elected not by the Paraloi themselves but by the whole Athenian people.

For the functions of the tamias the main evidence is **174**. Meidias was given 12 tal. of public money and told to sail to Euboia; yet he was overtaken by one of the privately financed triremes, which showed that he had prepared the *Paralos* inadequately. This means that, whereas the cost of maintaining other triremes was paid by their trierarchs (and later by the contributors through the symmories), money for maintaining the *Paralos* was provided by the state, and the tamias had the duty of spending it so as to keep the ship in tiptop condition. Whether the sum of 12 tal. was intended to cover the crew's pay also is not stated; very likely it was. In any case it was a large sum; and the financial responsibility of the tamias to the state may explain why he was appointed by the whole people, not merely elected by the Paraloi from among themselves. (Jordan *Navy* 182 is surely wrong to assume that a member of the Paraloi was always chosen. There is no evidence that Meidias, or any other known tamias, was himself one of the Paraloi. Ar. *Frogs* 1071–2, a jocular reference to antagonism between the Paraloi and their commanders, points rather the other way.) But the main problem is whether his responsibility was financial only, or whether he was also the commander of the ship, like the trierarch of an ordinary ship. Jordan *Navy* 181–3 maintains that the tamias was not the commander; apart from **174** he adduces three arguments, but none of them is strong. (*a*) Besides references to the tamias, there are references to the trierarch of the *Paralos* (Isai. 5.6, 5.42, *IG* 2^2 2966). However, they leave open the possibility that the tamias and the trierarch were the same man; the tamias may have been called the trierarch while actually at sea in command of the ship. (*b*) Jordan finds three references in the naval inventories to a tamias of the *Paralos* who was a trierarch of another ship (*IG* 2^2 1623.225, 1628.8, 1628.79). But the interpretation of these inscriptions is difficult, and in none of the three cases is the identification clear; what is recorded may be, perhaps, transfers of gear between the *Paralos* and another ship. (*c*) The tamias of the *Paralos* was elected, while there are no other instances of election of trierarchs. But this argument is of no value, since it is anyway clear that the sacred ships were treated differently from other ships. (*d*) So we are left with **174**, in which D. says that Meidias was ordered to sail and to escort the soldiers to Euboia. Such an order could appropriately be given only to the commander of the ship. Jordan is wrong when he says that D. here blames Meidias only for the poor condition of the *Paralos*; he blames him not only for the slowness of the voyage but also for lateness in setting sail. (*οὐκ ἐβοήθησεν, ἀλλ' ... ἧκεν. καὶ τότε ἡττᾶτο* ... means 'He did not go to join the troops when he was ordered to, but much later. And even when he did set sail, he was overtaken by another ship.') So, although the picture might be changed by the discovery of new evidence, the balance of probability at present seems to indicate that the tamias of the *Paralos* was also its commander.

πάλιν: 'also', passing on to the next item in the list of offices to which Meidias was elected. It does not mean that he had been a hipparch before.

ἵππαρχον: the two hipparchs, elected annually, were the commanders of the whole Athenian cavalry, ranking above the ten phylarchs (4.26, *AP* 61.4). It is not known in which year Meidias held the office.

διὰ τῆς ἀγορᾶς: the point is that riding through the Agora was much easier than riding into battle, but Meidias was incompetent even at that. He did not possess a horse of his own (**174**) and probably had little experience of riding.

μυστηρίων ἐπιμελητὴν: administrator of the Eleusinian Mysteries. These officials are not known to have existed before the mid-fourth century, and the most important evidence about them is an inscribed law of this period, published by K. Clinton in *Hesperia* 49 (1980) 258–88. There were four epimeletai of the Mysteries elected annually, one each from the Eumolpidai and the Kerykes (the two *γένη* who also provided the priests of the Eleusinian cult) and two from all Athenian citizens over thirty (cf. *AP* 57.1). They had to assist the basileus in organizing the festival and punish disorderly persons by a fine up to a certain amount (not legible in the inscription); if they considered that an offender deserved a more severe penalty, they could take him into court for trial, but if they failed to impose appropriate fines they were liable to be fined themselves. It is not clear whether this law establishes the epimeletai for the first time or merely redefines the functions of epimeletai who already exist (cf. Clinton op. cit. 272). Thus we cannot say whether Meidias was one of the first men to hold the office; nor is it known in which year he held it. Later inscriptions show that the epimeletai of the Mysteries were responsible for performing sacrifices (*IG* 2^2 661, 683, 807, 847) and that they also helped to organize the Lenaia (*IG* 2^2 1496.74–5; cf. *AP* 57.1), but it is not known that they performed those functions in Meidias' time. On the apparent reduction in the number of the epimeletai from four to two in the third century see Rhodes *Comm. on AP* 636–7.

ἱεροποιόν: 115 n. D. does not say what ritual Meidias arranged or performed. There is no need to take *μυστηρίων* with *ἱεροποιόν*, and thus no reason to assume that Meidias was an Eleusinian hieropoios, as Rhodes *Boule* 128 n. 13 does. Hieropoioi of the Mysteries are not known to have existed in the fourth century; cf. K. Clinton *Hesperia* 49 (1980) 281–2.

βοώνην: boonai were officials appointed to assist hieropoioi by purchasing, with public funds, cattle for a sacrifice. They are mentioned in inscriptions in connection with the Panathenaia, the Dionysia, and other festivals (*IG* 2^2 334.18, 1496.81, etc.).

172. ἐπανορθοῦσθαι: the character of Meidias, which was bad by nature, has been raised to a proper standard of *ἀρετή* by the honours which the Athenians have given him. The passage is a striking illustration of the Greek belief that *ἀρετή* consists, at least in part, of the respect which one receives from other people.

173. Κυζικηνῶν: according to Σ (586 Dilts), during the Social War (which began in 357/6) the Athenians passed a decree authorizing plunder of enemy merchandise at sea; Meidias made a seizure from some merchants of Kyzikos, who thereupon went to Athens and pointed out (presumably when prosecuting him) that Athens was not at war with Kyzikos, but Meidias argued against them so successfully that he was able to keep the money or goods which he had seized. The scholiast does not name his source, but the story does not look like his own invention and it should probably be accepted.

πλεῖον: the usual Attic forms are *πλέον* and *πλεῖν*, but *πλεῖον* is guaranteed by the metre in Ar. *Ekkl.* 1132 and *PSI* 1281.7, and so need not be emended here.

τὰ σύμβολα: the normal term for a treaty between states regulating legal procedures in cases which involved citizens of both. There is no other evidence for this treaty between Athens and Kyzikos. In 362 Prokonnesos, a city allied to Athens, had been attacked by Kyzikos (50.5), and Meidias may possibly have maintained that consequently the treaty was no longer valid. Cf. P. Gauthier *Symbola* (Nancy 1972) 169–70.

νόμους: regulations or routines for the cavalry, allegedly made by

Meidias on his own authority as hipparch. Nothing else is known of this incident.

τεθεικέναι: possibly to be emended to *τεθηκέναι*. Mss. of Attic authors give *τεθεικ-*, but fourth-century inscriptions *τεθηκ-* (*IG* 2² 1534.76, 2490.7). Yet from the analogous verb *ἵημι* we have *ἀφεῖκε* (*IG* 2² 1631.365).

174. εἰς Εὔβοιαν: **161** n.

παραπέμπειν τοὺς στρατιώτας: the *Paralos* was to sail alongside the troop-carrying ships from Peiraieus to Euboia, perhaps to be ready to carry dispatches from the generals back to Athens. The Athenian force set out five days after the Thebans had landed in Euboia (Ais. 3.85). Meidias, it appears, could not get the *Paralos* ready to sail as soon as that.

ἤδη τῶν σπονδῶν γεγονυιῶν: the campaign was completed within thirty days (Ais. 3.85).

Διοκλῆς: Diokles, of the deme Alopeke, was one of the strategoi in 357/6 (*IG* 2² 124.23). In other years he served as a trierarch and as a naval contributor (*IG* 2² 1612.279, 1615.135, 1615.156, 1616.25). The only other information about him is an anecdote about how, as a general on some campaign, he got his disorderly soldiers to march properly by making them think that the enemy was near (Polyainos 5.29).

ἡττᾶτο πλέων: though the sacred ships were supposed to be faster than the others, on this voyage the *Paralos* was overtaken by one of the ordinary triremes. That ship must have been even later than the *Paralos* in setting out.

ἰδιωτικῶν: maintained by contributions from individuals; **155** n.

τί οἴεσθε: evidently a colloquial idiom for 'extraordinary', like the commoner *πῶς οἴει* or *πῶς δοκεῖς* for 'extraordinarily' (Ar. *Clouds* 881, *Frogs* 54, etc.).

ἐτόλμησεν: a sarcastic comment on Meidias' meanness: he did not do anything as rash as buying a horse.

τὰς πομπὰς ἡγεῖτο: *ἡγέομαι* meaning 'lead' regularly takes the dative of the person led, but the accusative of the journey or route; cf. X. *Kyr.* 3.2.28 *οἵτινες αὐτῷ τήν τε ὁδὸν ἡγοῖντο ἂν καὶ συμπράττοιεν.* The reading *ἐποιεῖτο* (in the margin of S) may be a conjecture made in a later period when this was not understood.

Φιλομήλου: Philomelos, of the deme Paiania, belonged to a well-to-do family, prominent in Athens for several generations; most of the evidence is assembled by Lewis *BSA* 50 (1955) 19–20, Davies *Families* 548–50. His father, Philippides, appears in Plato alongside such men as Kallias and the sons of Perikles in attendance on the visiting sophist Protagoras (Pl. *Prot.* 315a), and Philomelos himself was a pupil of Isokrates (Isok. 15.93). He was generally considered to have more merit than wealth (Lys. 19.15 *ὃν οἱ πολλοὶ βελτίονα ἡγοῦνται εἶναι ἢ πλουσιώτερον*); he nevertheless performed trierarchies and other liturgies (listed by Davies) and was awarded a gold crown for his services to Athens (Isok. 15.94). Since he is mentioned incidentally in two speeches of the years 393–87 (Isok. 17.9, Lys. 19.15), his age was probably seventy by the time of *Meidias*, but he did not die until 336 (*IG* 2^2 1628.373–6, 1629.892–5). His son, Philippides, appears in **208** and **215**.

καὶ ταῦτα: the addition of *ἐξιόντων* in SYP is an unexplained puzzle.

175–83. *Instances of men punished severely for offences less serious than those of Meidias.*

Here D. adduces previous cases of men prosecuted by *probole* for offences concerning festivals; for the legal procedure of *probole* see pp. 13–16. In **182** he adds some cases brought by other procedures for other offences. These are not wholly irrelevant, because the *a fortiori* argument rests on the penalty rather than the procedure: if men whose offences were less serious than Meidias' were punished severely, all the more should Meidias be.

175. ὅσων ... ἀδικεῖν: these words may be taken both with *καταχειροτονήσαντος* and with *κατεγνώκατε*. Both verbs regularly take a genitive (with *κατα-*) and an infinitive (of *oratio obliqua*), giving the sense 'decide against them that they offended'; cf. 39.38 *αὐτὸς αὑτοῦ κατέγνω δικαίως εἶναι Βοιωτός*.

τὴν ἑορτὴν: two different festivals will be mentioned, the Mysteries (**175**) and the Dionysia (**178**), but D. uses the singular here because *περὶ τὴν ἑορτὴν ἀδικεῖν* is a quotation of the wording used in the charge on each occasion.

Euandros and Menippos are not otherwise known, but the fact that one was from Thespiai (in Boiotia) and the other from Karia (in Asia) shows that *probole* was not restricted to Athenian citizens.

Καρός τινος ἀνθρώπου: a slightly contemptuous expression; so also **176** *ἄνθρωπος*. It is implied that one has heard of Euandros, but not of Menippos. However, Goodwin's suggestion that the word *Κάρ* here does not mean 'Karian' but 'an insignificant fellow' should be rejected because, as he himself concedes, this sense is not found outside the proverbial expression *ἐν Καρὶ ὁ κίνδυνος*, which means 'try it on the dog' (Pl. *La.* 187b with schol., *Euthd.* 285c, E. *Kyk.* 654, Kratinos 16 Kock = 18 Kassel and Austin, Philemon 18).

ὁ αὐτὸς νόμος: not the same enactment (as the rest of the sentence shows), but a later enactment making the same provision for *probole* at the Mysteries as already existed for the Dionysia. D. speaks here as if there were only one law about *probole* at the Dionysia, though he has actually quoted two (**8**, **10**). Probably (though there is no other evidence) the provisions for *probole* at the Dionysia were made in two stages, some years apart, and then later, because they seemed to work well, similar provisions for *probole* were made in a single enactment about the Mysteries.

ὕστερος: laws tend to be esteemed for age, and so the point here is that, if the more recent law is strictly enforced, all the more should the older law be.

176. τοῦτ' ἀκούσατε: a good example of a superfluous phrase inserted for the rhetorical purpose of prolonging the suspense in which the listeners await the next sentence.

ὅτι ... ἐπελάβετο: previous editors have put a full stop at *ἐπελάβετο*. But it is awkward to take this clause with the preceding *τί ποιήσαντος*, and easier to take it with the following *διὰ ταῦτα*: cf. **206** *ὅτι προσκέκρουκεν ἐμοί, διὰ ταῦτα τοῦτον ἐξαιτήσεται.*

δίκην ἐμπορικὴν: 'a mercantile case' concerning the import or export of goods, especially grain. Legal proceedings for such cases in Athens were drastically reorganized by 'the mercantile laws' (35.3 *τοὺς ἐμπορικοὺς νόμους*) in order to speed them up. Previously mercantile cases involving Athenian citizens were dealt with by the nautodikai (Lys. 17.5), while foreigners had to apply to the polemarch or the xenodikai (cf. MacDowell *Law* 221–4, 229–30). Henceforth merchants, whether Athenian or not, could apply to the thesmothetai for trial. The most important feature of the new procedure was that cases now became 'monthly' during a certain part of the year, so that merchants did not need to spend a long time in Athens awaiting trial: 33.23 *αἱ δὲ λήξεις τοῖς ἐμπόροις τῶν δικῶν ἔμμηνοί εἰσιν ἀπὸ τοῦ Βοηδρομιῶνος μεχρὶ τοῦ Μουνιχιῶνος, ἵνα παραχρῆμα τῶν δικαίων τυχόντες ἀνάγωνται.* That probably means that applications to bring cases were accepted every month, except during four months of the summer; but that interpretation is disputed. On this and other details of the procedure see E. E. Cohen *Ancient Athenian Maritime Courts* (Princeton 1973), MacDowell *Law* 231–4, Hansen *Symposion 1979* (1981) 167–75.

Was the case of Euandros and Menippos tried by the old or the new procedure? The old procedure was still in use in 355 or soon after, when X. *Poroi* 3.3 criticizes it, and the earliest mention of the new procedure is in 342 (D. 7.12). Hansen *Symposion 1979* (1981) 172 n. 15 assumes that **176** *δίκην ἐμπορικὴν* must be a reference to the new procedure; but that is not so, because Lys. 17.5 shows that mercantile cases were already distinctive in the early fourth century. In fact we can say confidently that this case was tried by the old procedure, for one feature of the new procedure was that a defendant condemned to make a payment was imprisoned until he paid up: 33.1 *τοῖς ἀδικοῦσιν δεσμὸν ἔταξεν τοὐπιτίμιον, ἕως ἂν ἐκτείσωσιν ὅ τι ἂν αὐτῶν καταγνωσθῇ.* If Menippos had been imprisoned, Euandros would not have been trying to catch him at the Mysteries. But this hardly helps us to date the reform of mercantile cases, because D. does not say how long ago the incident occurred; calling this the most recent instance of conviction in a *probole* (**175** *τῆς τελευταίας γεγονυίας*) may imply a date within the last two or three years, but does not necessarily do so.

λαβεῖν: 'to find'; cf. LSJ *λαμβάνω* A.I.3.

τοῖς μυστηρίοις ἐπιδημοῦντος: the dative is both temporal and local: 'when he was in Attika at the Mysteries'. The great Eleusinian Mysteries were held at Athens and Eleusis in the month Boedromion, and the little Mysteries (*τὰ μικρὰ μυστήρια*) were held at Agra (on the south-east side of Athens) in the month Anthesterion; D.'s words could refer to either. There is nothing surprising in the attendance of a Thespian and a Karian at these ceremonies, since initiation was open to all Greek-speaking people, though not to barbarians (Isok. 4.157); cf. G. E. Mylonas *Eleusis and the Eleusinian Mysteries* (Princeton 1961) 247–8.

ἐπελάβετο: this verb can be used of seizure either of a person (cf. **60**) or of goods (cf. **133**). The law (if the law about the Mysteries had the same wording as the law in **10**) forbade seizure of money or goods at the festival, but here the juxtaposition of *ἐπιδημοῦντος* with *ἐπελάβετο* encourages the hearer to take the genitive with the verb (rather than as genitive absolute) with the meaning 'collared Menippos' rather than 'made a seizure of property'.

κατεχειροτονήσατε ... ἠβούλεσθε: the two verbs refer respectively to the Ekklesia and to a jury. At the trial by jury, after Euandros was convicted, the jury will have fixed the penalty in the usual way by voting to choose between alternatives proposed by the prosecutor and by the defendant. According to D. the jury would have voted for death, if Menippos had proposed it, but he was 'persuaded' (*πεισθέντος* hints at private threats or bribery without alleging them specifically) by Euandros and his friends to propose a less severe penalty instead. (There is no ground for Goodwin's suggestion that, in a *probole* case, the court could impose a penalty not proposed by either the prosecutor or the defendant, and in fact the present passage refutes it: D.'s point is precisely that the jury could not impose the death penalty because the prosecutor did not propose it.) How then can D. know that the jury would have voted for death? Strictly he cannot; but Athenian juries were not silent, and booing and cheering may have made clear the jury's hostility to Euandros and support for Menippos.

ἀφεῖναι: *ἀφίημι* is the standard verb for giving up a claim to which one is entitled. Here Euandros was ordered to forfeit his entitlement to receive 2 tal. from Menippos, awarded to him in the mercantile case.

τὰς βλάβας: 'damages', the amount of the expenses, or loss of trading profits, which Menippos had incurred by having to wait in Athens for the trial.

ἐπὶ τῇ καταχειροτονίᾳ: the preposition is both temporal and causal: after the vote in the Ekklesia, and because of its result, Menippos waited for a trial to be held. (Not 'waiting for the vote', because the vote was held immediately after the festival; cf. the law in **8**.)

ἅνθρωπος: 'the fellow', Menippos; cf. **175** *Καρός τινος ἀνθρώπου.*

177. ἰδίου: Menippos' prosecution was not a *δίκη ἰδία* (a *probole* was a *δίκη δημοσία*) but the dispute which led to it arose from a private debt. A contrast is implied with the case of D. and Meidias, which was of public concern because D. was a khoregos (**31–5**). But the contrast is false: in both cases alike the dispute was originally personal but became a matter of public concern when it was pursued at a festival.

αὐτοῦ: 'by itself', 'only'.

τοὺς νόμους, τὸν ὅρκον: for the asyndeton, **81** n. *τῇ δίκῃ, τοῖς νόμοις*. For the jurors' oath, **4** n. *ὀμωμοκὼς*.

ἀεὶ with the participle, as in **37**, **131**, **194**, **223**: 'you who are jurors on any particular occasion'. In this sense it is usually placed between the article and the participle; for the position after the participle cf. Isok. 4.52 *τοῖς ἀδικουμένοις ἀεὶ τῶν Ἑλλήνων*.

παρακαταθήκην: money or valuable property entrusted to another person to keep it safe. The same word is used for the same image of justice and law entrusted to jurors in 25.11, Ais. 1.187.

178. παρεδρεύοντος ἄρχοντι: the arkhon, the basileus, and the polemarch each had two *πάρεδροι* of their own choice (*AP* 56.1). It may have been common to choose a relative; one inscription (*Hesperia* 40 (1971) 257 no. 5) records a basileus named Exekestides son of Nikokrates and his two paredroi, one of whom is Nikokrates son of Exekestides, presumably either the son or the father of the basileus. The functions of a paredros are nowhere defined, but they must have been of some importance, since he was subject to *dokimasia* before taking up the office and to *euthyna* on demitting it (*AP* 56.1). Cf. S. Dow in *In Memoriam O. J. Brendel* (ed. L. Bonfante and H. von Heintze, 1976) 80–4, Rhodes *Comm. on AP* 621–2. In the present case evidently the paredros was assisting the arkhon to perform his duty of organizing the Dionysia.

Kharikleides was arkhon in 363/2, which must therefore be the date of the incident mentioned here. Nothing else is known about him to explain why D. gives him the complimentary epithet *βέλτιστος*.

179. εἰ ... εἰ: the double condition is clumsy, but gives a convincing air of unstudied indignation.

ἄνθρωπε: addressed to the paredros during the hearing of the *probole* in the Ekklesia. This vocative (in place of the man's name) has a tone of patronizing remonstrance; so also in 19.94, 32.15, 34.15. The English vocatives 'man' and 'mate' can have a similar tone.

τοῖς κηρύγμασιν: presumably announcements that certain rows of seats were reserved for certain persons; the man was taking a good seat to which he was not entitled.

τοῖς ὑπηρέταις: attendants or beadles acting under the direction of the arkhon and his paredroi; compare the *ὑπηρέτης* of a general who conveys an order in 50.31. Elsewhere we hear of *ῥαβδοῦχοι*, 'staff-bearers', who disciplined the competitors in a dramatic or athletic contest (Ar. *Peace* 734, Th. 5.50.4, Pl. *Prot.* 338a), but it is not clear that they were the same men as the arkhon's *ὑπηρέται*.

οὐδ' οὕτω πείθομαι: asyndeton (*οὐδ'* is adverbial, not connective): **101** n. *ἅπασι προσήκει*.

ἐπιβολὴν: there is no other evidence that a paredros could impose a fine on his own authority, and probably the procedure strictly would have been that he would get the arkhon to impose a fine. The arkhon, like other magistrates, had authority to impose a fine up to a specified limit on offenders within his sphere of authority; cf. Harrison *Law* 2.4–7, MacDowell *Law* 235–7, Rhodes *Comm. on AP* 634–5.

πρὸ τοῦ μὴ ...: *πρό* means 'in preference to', and hence 'to prevent'. *μή* is added with the infinitive as often after expressions of prevention, though I have not found another instance after *πρό*.

ἕκαστον may be either masculine ('every man being assaulted as to his person') or neuter ('every person being assaulted').

οὗτος: the paredros. As Σ (606 Dilts) points out, the fact that the condemnation was never confirmed in a trial by jury makes this precedent weaker than those adduced in **175–7** and **180**. No doubt that is why D. has sandwiched it between the stronger precedents.

180. Κτησικλέα: a fairly common name; there is no reason to identify this one with any other known holder of it.

διὰ τί δή: for the question introduced at this point in the story cf. **176** *τί οὖν ποιήσαντος ...;* For the use of *διὰ τί;* cf. **24**, **186**. The purpose is to increase the listeners' sense of expectation. The words are omitted by S, but their intrusion, if they are not genuine, is hard to explain. On the other hand, the addition of *ἀπεκτείνατε τοῦτον* (A) is easily explained as a gloss on *διὰ τί δή;*

σκῦτος ἔχων: Ktesikles was riding in the procession.

ἐδόκει ... τύπτειν: like verbs of prosecution (**1** n. *προυβαλόμην*) *δοκέω* in the sense 'am found guilty' takes the present infinitive; cf. **182** *δόξαντι παράνομα γράφειν*.

ὕβρει καὶ οὐκ οἴνῳ: **74** n.

ἐπὶ τῆς πομπῆς: 'in the circumstances of the procession'; **2** n. *ἐπὶ τῶν ἄλλων*.

τοῖς ἐλευθέροις: a generalizing plural; Ktesikles struck only one man.

181. ἁπάντων ... τούτων: there is an unobtrusive zeugma, since this genitive (of comparison) has to be taken as masculine with *ὧν ὁ μὲν* ..., but as neuter with *πολλῷ δεινότερ'* ...

ὁ μὲν ὧν εἷλεν ἀποστάς: Euandros; cf. **176** *τὴν δίκην ... ἀφεῖναι ... ἣν ᾑρήκει*.

ὁ δὲ: Ktesikles (**180**). The father of Kharikleides (**178–9**) is omitted here because no penalty was imposed on him before his death.

πομπεύων ... δίκην ᾑρηκὼς ... παρεδρεύων: the excuses respectively of Ktesikles, Euandros, and the father of Kharikleides. The reason why riding in a procession is a partial (though inadequate) excuse for hitting someone is not that the rider has a whip, as Σ (619 Dilts) suggests, but that a procession is privileged traffic and bystanders ought to keep out of its way.

182. ἀλλὰ instead of *δέ*, responding to *μέν*: cf. 8.52 *τὰ μὲν ἄλλ' ἐάσω· ἀλλ' ἐπειδάν* ..., and Denniston *Particles* 5–6.

Πύρρον ... τὸν Ἐτεοβουτάδην: the Eteoboutadai (earlier called the Boutadai) were an aristocratic *γένος*, of which the chief distinction was that the priestess of Athena Polias was always appointed from among its members (Ais. 2.147); cf. J. Toepffer *Attische Genealogie* (Berlin 1889) 113–33. Pyrrhos cannot be identified with certainty, though Davies *Families* 353 n. 2 suggests identifying him with Pyrrhos of Lamptrai, who was a claimant to an estate about 350 (Isai. 4.9).

ἐνδειχθέντα δικάζειν: for the present infinitive after a verb of prosecution, **1** n. *προυβαλόμην*. Any citizen who was recorded as owing money to the state was regarded as disfranchised (*ἄτιμος*) from the date when the payment was due until he paid up, and if during that period he exercised any of the functions of a citizen, such as sitting on a jury (*δικάζειν*), he was liable to prosecution by *endeixis* (58.48–9, And. 1.73); cf. Harrison *Law* 2.172–5.

τινες ὑμῶν: some members of the jury which tried Pyrrhos. Contrast **180** *ὑμεῖς ... ἀπεκτείνατε*. Here D. prefers a form of words which implies that the jury was not unanimous, and he goes on to make an excuse for Pyrrhos (*δι' ἔνδειαν*). Possibly he was a friend of Pyrrhos and supported him at the trial, so that he was now reluctant to admit that he had been totally defeated.

τοῦτο τὸ λῆμμα: the juror's fee of 3 obols a day. If Pyrrhos continued the offence indefinitely, this would give him a regular, though small, income.

διὰ governs *πράγματα*, while *πολλῷ* goes with *ἐλάττω*. The interval between the preposition and its noun is unusually great, but is tolerable because the word immediately following the preposition is part of the phrase qualifying the noun.

τούτων: the actions of Meidias.

δ': omitted by U and deleted by Weil, on the ground that the punishment of Smikros and Skiton is not a further item but an instance of what has just been mentioned. But it can be retained if these two offenders are regarded as additional to *πολλοὺς ἑτέρους*.

Σμίκρῳ ... Σκίτωνι: not known.

δέκα ταλάντων: genitive of price, as in **102**, **151**, **152**. The amount of the fine is greater than most Athenians would have been able to pay, and the prosecutor's intention was undoubtedly to disfranchise Smikros and Skiton by making them debtors to the state. That is what happened to the father of the speaker of D. 58 (*Theokrines*), who was fined 10 tal. in a *γραφὴ παρανόμων* (58.31 *τῷ μὲν πατρὶ δέκα ταλάντων ἐτίμησαν*) and was consequently disfranchised (58.66 *ἀπεστερήμεθα ταύτης τῆς πόλεως*).

παράνομα γράφειν: for the tense, **180** n. *ἐδόκει ... τύπτειν*. For the offence, **5** n. *παρανόμων*. The plural does not necessarily mean more than one decree.

παιδία: for the use of children to arouse pity for a defendant, **99** n.

183. μὴ negatives the combination of *ἐὰν μὲν ... φαίνεσθε* and *ἐὰν δὲ ... διάκεισθε*.

εἴπῃ ... λέγῃ: these are normal verbs for proposing a decree in the Ekklesia, and D. here prefers them to the equally normal *γράφειν* (used in the previous sentence) to enable him to contrast speech with action. As Σ (624 Dilts) points out, it is not really true that speech is always less serious than action; making an illegal proposal in the Ekklesia might well be considered worse than some of the more trivial items of Meidias' misconduct.

μὴ τοίνυν introduces a conclusion, and it is surprising to find the phrase twice in **183**. Weil suggests, perhaps rightly, that we have here two alternative conclusions between which D. intended to choose when making the final version of the speech.

ὡς ἄρα introduces a reported statement which is not, or may not be, true; cf. Denniston *Particles* 38–9.

τῶν μετρίων ... καὶ δημοτικῶν: 'ordinary people', like **209** *ὑμῶν τῶν πολλῶν καὶ δημοτικῶν*. Elsewhere in the speech *μέτριος* has a purely moral sense, 'well behaved' (e.g. **128**, **185**), but here it is primarily social, 'middle class'; cf. 18.10, where D. refers to his ancestors when claiming to be as good as any of the *μέτριοι*.

184–5. *Meidias' previous life has not earned lenience.*

Most of this passage is a revised version of **101**. See the notes there on the details, and see pp. 26–7 on the bearing of these passages on the problem of the composition of the oration.

184. βραχέα: there is a good deal more of the speech yet, but it is a common rhetorical device to pretend that one is going to say very little and the end is not far off, in order to encourage the listeners to go on attending: *τὸ γὰρ τέλος πάντες βούλονται καθορᾶν* (Arist. *Rhet.* 1409a 31–2). Besides **77** and **160**, cf. 20.154 *ἐγὼ δ' ἔτι μικρὰ πρὸς ὑμᾶς εἰπὼν καταβήσομαι*, 23.215. Such expressions do not necessarily mark any formal division, and Erbse *Hermes* 84 (1956) 139 is wrong to insist that **184** begins a formal epilogue.

καταβήσομαι: although the exact form of Athenian courts is not known, it is clear that speakers (including witnesses, in the period when they gave evidence orally) mounted some kind of raised platform or stand to deliver their speeches. Thus *ἀναβαίνω* is used for 'begin a speech' and *καταβαίνω* for 'end a speech'; cf. **205**, 19.32, 19.289–90, 20.154, 23.215, 58.70, Ar. *Wasps* 963 (a witness), 979 *κατάβα* (shouted by the jury), etc. The speaker's stand must be distinguished from the (presumably lower) platforms, one for the prosecutor and one for the defendant, on which the litigants and their supporters sat when not speaking (48.31, Ais. 3.207).

μερὶς: 'contribution' to the defendant's case; so also in **70**.

185. ἀγῶνα: primarily, though not exclusively, a lawsuit.

οὑτοσί τις: 'here is someone'. As in **101**, this phrase gives a hypothetical example. Strictly, it does not mean Meidias, who is not brought into the picture until the last sentence of the paragraph; but of course D. hopes that his listeners will think of Meidias during this sentence.

πτωχούς: 'paupers'. The word signifies a greater degree of poverty than *πένης* (**83** n.) and is often contemptuous in tone. One to whom it is wrongly applied may react indignantly; cf. Ar. *Akh.* 593–5 *Λα. ταυτὶ λέγεις σὺ τὸν στρατηγὸν πτωχὸς ὤν; Δι. ἐγὼ γάρ εἰμι πτωχός; Λα. ἀλλὰ τίς γὰρ εἶ; Δι. ὅστις; πολίτης χρηστός* ... Rich men like Meidias cause annoyance by applying the word to anyone less rich than themselves; so also in **198**, **211**.

καθάρματα: in its literal sense *κάθαρμα* belongs to the language of religious purification, and refers to what is thrown away or expelled in the process of cleansing from pollution. It may be the refuse of a sacrifice or other ceremony of purification, as in A. *Kho.* 98, *IG* I^3

257.10, Paus. 8.41.2, and probably Hipp. *Sacred Disease* 6.362 Littré = 2.148 Jones (*καθαρμῶν* codd.: *καθαρμάτων* conj. Jones); cf. R. Parker *Miasma* (Oxford 1983) 229–30. Or it may be a 'scapegoat', a criminal or worthless person expelled from the city to purify it; however, *κάθαρμα* in this sense (equivalent to *φαρμακός*) occurs only in scholia to Ar. *Knights* 1136, *Frogs* 733, *Wealth* 454, and it is not certainly a classical usage. But in D. and his contemporaries the word is just a general term of abuse, not to be taken literally; so it is here and in **198**, 18.128, 19.198, Ais. 3.211, Dein. 1.16, Men. *Sam.* 481. Cf. Wankel *Kranz* 683–4.

οὐδὲν: 'of no value'; cf. LSJ *οὐδείς* II.2. A similar list in **198** ends *οὐδ' ἄνθρωποι*, but it is not necessary to assimilate this passage to that by emending *οὐδὲν* to *οὐδ' ἀνθρώπους*, as proposed by Markland (in Taylor's edition) and Herwerden *Mn.* II 3 (1875) 136–7.

ἐπίῃ: 'it occurs'; cf. LSJ *ἔπειμι* (B) I.2.b. So also **192** *ἐπῄει*. A and other mss. add the redundant word *ὀρθῶς*: cf. pp. 50–1 on extra words in A.

186–8. *Pleas from Meidias' children should be ignored.*

For the use of children to arouse pity, **99** n.

186. **ταπεινοὺς**: this adjective, like the English 'humble', may mean either that a man's circumstances are poor or that his demeanour is self-effacing; the two things are assumed generally to go together. Here D. means that today Meidias, instead of boasting of his wealth as he usually does (cf. **153**), will assert that at the present moment (cf. **187** *τὸν παρόντα καιρόν*) he is in financial or other difficulties, and will humbly plead that he should not suffer a penalty in addition.

ἐλεεινότατον: in fifth-century Attic poets the contracted form *ἐλειν-* is sometimes required by the metre (e.g. S. *Phil.* 870, Ar. *Frogs* 1063), but in the fourth century the uncontracted form seems to prevail. It is always used in the mss. of D. and is given by the papyri in Men. *Dys.* 297, *Sam.* 371.

μηδαμῶς δυνηθεὶς: D.'s point is the sound one that a person who does wrong because he knows no better deserves less severity than one who does wrong in full knowledge of the seriousness of his offence. For other instances in Greek literature of the distinction between nature (*φύσις*) and wish (*βούλησις*) as causes of wrongdoing, see Dover *Morality* 150. It is closely related to the distinction between unintentional and intentional wrongdoing, expounded in **42–5**.

τι: adverbial, *ἀνεῖναι* being intransitive here: 'to relax from one's anger to some extent'. Cf. Ar. *Frogs* 700 *τῆς ὀργῆς ἀνέντες*. Two of the later mss. (RLp) accent differently, *ἀντὶ τῆς ὀργῆς*, giving the sense 'to give way instead of being angry'. That could be correct, since *ἄν* is not necessary with an expression of obligation or rightness.

διακρούσηται: not 'practise evasions' (as LSJ say under *διακρούω* II) but 'evade punishment', as in **128**, **201**.

187. ἐξεπίτηδες: **56** n.

πλάττεται: Meidias' tale about his misfortunes is a lie; cf. **196** *ἐξαπατᾷς*, which refers to the same story.

ἐμοὶ παιδία οὐκ ἔστιν: D. subsequently had two sons and one daughter (Ais. 3.77, Plu. *Eth.* 847c). For the asyndeton, **101** n. *ἅπασι προσήκει*.

188. ὅταν ... ἀξιοῖ (subjunctive): 'whenever' is not frequentative here; Meidias will probably make the appeal only once, but it is not yet known at what point in his speech he will make it.

τούτοις ... δοῦναι τὴν ψῆφον: 'to give your votes to the children' means to acquit Meidias because his children want you to do so. The singular *τὴν ψῆφον* is normal, because each juror has one vote.

τοὺς νόμους ἔχοντα: contrasted with *ἔχων τὰ παιδία*. The laws are almost personified here; the notion that they require defence, like injured persons, is developed in **224–5**.

πλησίον ... παρεστάναι: 'stand alongside Meidias'. While Meidias is speaking, D. will not be on the platform and so will not (or at least not prominently) be visible to the jury. In order to avoid forgetting him and the strength of his case, the jurors are to imagine him and the laws standing on the platform alongside Meidias and his children.

καὶ τὸν ὅρκον ὃν ὀμωμόκατε: rightly deleted by Dobree *Adversaria* 1.465. The whole of **188** is about the importance of the laws, not about the importance of the jurors' oath, which is adduced (at *καὶ γὰρ ὀμωμόκατε* ...) merely as one of the items of evidence which prove the importance of the laws. The intrusion of this phrase may have been encouraged not only by *καὶ γὰρ ὀμωμόκατε* ... but also by **177** *τοὺς νόμους, τὸν ὅρκον*.

τούτοις: for the dative with *ψηφίζομαι* cf. 24.35.

ὀμωμόκατε: for the jurors' oath, **4** n.

πείσεσθαι: restored by Herwerden *Mn.* II 1 (1873) 312, Cobet *Misc. crit.* 515. All mss. have *πείθεσθαι*. With *ὄμνυμι* the future infinitive is normal for an oath concerning the future. There are a few instances of the present or aorist, all usually emended: Lyk. *Leo.* 126 *ἀποκτείνειν συνώμοσαν*, ibid. 127 *διομωμόκατε δ' ἐν τῷ ψηφίσματι τῷ Δημοφάντου κτείνειν τὸν τὴν πατρίδα προδιδόντα*, X. *Hell.* 7.4.11 *ὀμόσαντες ἐπὶ τοῖς αὐτοῖς τούτοις εἰρήνην ποιήσασθαι*, *An.* 2.3.27 *δεήσει ὀμόσαι ἦ μὴν πορεύεσθαι ὡς διὰ φιλίας*. A rather different instance, where the infinitive may be felt as partly jussive, is 23.170 *ἀναγκάζει τὸν Κερσοβλέπτην ὀμόσαι πρός θ' ὑμᾶς καὶ τοὺς βασιλέας εἶναι μὲν τὴν* *ἀρχὴν κοινὴν τῆς Θρᾴκης εἰς τρεῖς διῃρημένην, πάντας δ' ὑμῖν ἀποδοῦναι τὴν χώραν.* It is probably right to leave 23.170 unaltered but to emend the other instances.

τῶν ἴσων: **67** n.

189–92. *It is irrelevant to say that Demosthenes is an orator and has prepared his speech.*

Originally *ῥήτωρ* was a neutral word for 'speaker' or 'politician', but by the time of D. it was often used opprobriously, implying that a man used rhetoric in the Ekklesia to obtain power or profit for himself; cf. Dover *Morality* 25–6. D. himself criticizes *ῥήτορες* elsewhere, e.g. in *Timokrates*: 'But orators in Athens, men of the jury, in the first place legislate to their own advantage nearly every month; next, they imprison private individuals whenever they are in office, but they do not think the same sanction should apply to themselves; next, they repeal the long-approved laws of Solon which our ancestors made, and they think you ought to obey their own laws, which they make to the detriment of the city' (24.142). But the sort of allegation which D. now expects Meidias to make against him, and which he is therefore trying to counter in advance, is financial rather than political, as the end of **189** shows: he is afraid of being accused of using his oratorical skill to become rich at the city's expense. Having denied this, he goes on in **190** to dissociate himself from *ῥήτορες* altogether and blacken Meidias instead with that brush.

189. ἐμὲ: 'about me'. For the accusative cf. X. *Hell.* 3.5.12 *Κορινθίους δὲ καὶ Ἀρκάδας καὶ Ἀχαιοὺς τί φῶμεν;*

καὶ τοῦτ' has the same sense here as the more usual *καὶ ταῦτα* (e.g. **199**), 'adding a circumstance heightening the force of what has been said' (LSJ *οὗτος* C.VIII.2).

ἄχρι τοῦ μηδὲν ὑμῖν ἐνοχλεῖν: 'only as long as he causes you no annoyance'. Usually *ἄχρι* means 'as far as the point of'; I have found no exact parallel to this instance. For D.'s claim that his political style is considerate, cf. 19.206 *οὐδὲν γὰρ πώποτ' οὔτ' ἠνώχλησα οὔτε μὴ βουλομένους ὑμᾶς βεβίασμαι.*

ἐξ ὑμῶν: it is not essential to adopt *ἀφ' ὑμῶν*, as most editors do. (The passage is quoted by Photios and the Souda (s.v. *ῥήτωρ*) in the form *εἰ δέ, οἵους ἐγὼ καὶ ὑμεῖς δὲ ὁρᾶτε, ἀπαιδεύτους καὶ ἀφ' ὑμῶν πεπλουτηκότας, οὐκ ἂν εἴην οὗτος ἐγώ*, but that is so inaccurate that it cannot be used as reliable evidence for the details of D.'s text.) For the use of *ἐκ* cf. Lys. 32.25 *ῥᾳδίως δὲ ἐκ τῶν ἀλλοτρίων αὐτὸς πλουτήσει*, X. *Poroi* 4.14 *πολλοὺς πλουτιζομένους ἐξ αὐτῆς* (sc. *τῆς πόλεως*).

ἅπαντ' ἀνήλωκα: the liturgies performed by D. have been recounted in **154–7**.

ἦν: **26** n.
ἔδει: *ἄν* is omitted, as usually with expressions of obligation.

190. τῶν λεγόντων: those who frequently speak in the Ekklesia.
συναγωνίζεται: speaking in support of D. in this trial. The verb has this sense also in 29.15; in other legal contexts it refers to joint or collusive litigation (43.9–10, 48.43) or to a favourable reception given to a litigant by a jury (23.4). The fact that none of the *ῥήτορες* is supporting D. is not a further supposed charge by Meidias (as King and Goodwin take it), but a further part of D.'s rebuttal of the charge that he is one of the *ῥήτορες*.
ἔγνων with infinitive: 'resolved'. Cf. *Pr.* 28.1 *ἐγὼ δ' οὐδεπώποτ' ἔγνων ἕνεκα τοῦ παραχρῆμ' ἀρέσαι λέγειν τι πρὸς ὑμᾶς, ὅ τι ἂν μὴ καὶ μετὰ ταῦτα συνοίσειν ἡγῶμαι.*
πάντας: vague and exaggerated. Probably D. did not know which politicians other than Euboulos (**205–7**) would support Meidias.
συνεξεταζομένους: **65** n. *ἐχθρὸν ἐξεταζόμενον.*
αὐτὸν: S has *αὑτὸν* and this is accepted by Goodwin. But *ἀξιόω* does not normally take a reflexive accusative with the infinitive, and it is better to keep *αὐτὸν* in its intensive sense.

191. τάχα: **141** n.
ἐσκεμμένα: passive, but the other instances of *ἐσκεμμένος* and *ἐσκέφθαι* in **191–2** are all middle. At the end of **192** (as usual in Attic) *σκοπεῖν* is used in preference to *σκέπτεσθαι* for the present tense.
μεμελετηκέναι: practising delivery of the speech aloud; cf. 19.255 *λογάρια δύστηνα μελετήσας καὶ φωνασκήσας.*
ἄθλιος: 'foolish', as in **66**.
ἦν: **26** n.
ἤμελλον: **20** n. *ἠδυνήθησαν.*
γεγραφέναι: Meidias has 'written' D.'s speech in the sense that his activities have provided plenty of material for it. Cf. S. *El.* 624–5 *σύ τοι λέγεις νιν, οὐκ ἐγώ· σὺ γὰρ ποεῖς τοὔργον, τὰ δ' ἔργα τοὺς λόγους εὑρίσκεται.*

192. μεριμνήσας: a mainly poetic verb, not found in Attic prose writers except Xenophon (though it became common in later Greek prose). Its use with an infinitive is not paralleled, but is analogous to the use of *μελετάω* with an infinitive.
ἐπῄει: **185** n. *ἐπίῃ.*
διημάρτανε τοῦ πράγματος: 'got the matter wrong'; cf. 51.2 *δοκοῦσί μοι παντὸς διημαρτηκέναι τοῦ πράγματος.*

193–204. *Meidias' hostility to ordinary citizens.*

A sure route to unpopularity in democratic Athens was to criticize the people in the Ekklesia. By saying that Meidias proposes to do this

D. hopes to involve him in such unpopularity. Really Meidias' point is that many citizens were absent from the Ekklesia on the day of the *probole* hearing there, so that the meeting was not truly representative of the people; but that is a more complex argument which will be harder to put across than D.'s crude one. D. then proceeds to generalize, asserting that Meidias regularly expresses views opposite to most people's, never takes pleasure in Athenian successes (**202**), and displays ill will to 'the many' (**204**). All this is intended to make the jurors think of Meidias as their opponent.

193. ὅτ' ἦν ἡ προβολή: on the day of the *probole* hearing after the Dionysia in 348.

ἐξιέναι: on the expedition to Euboia.

τὰ φρούρια: primarily the forts guarding the frontier between Attika and Boiotia. In 348 there may have been apprehension that the Thebans would attack in order to divert Athenian forces from Euboia. The forts may have been manned mainly by ephebes (cf. *AP* 42.4); but, if so, this passage does not imply that ephebes were excluded from the Ekklesia (as suggested by M. H. Hansen *Demography and Democracy* (Herning 1986) 93 n. 39) but rather that they could attend it.

ἐξεκλησίασαν: there is doubt about the correct augment in the imperfect and aorist of this verb. In the other two instances in D. the mss. give ἠκκλ- (18.265, 19.60), but in earlier authors they give ἐκκλ- or ἐξεκλ- or ἐξεκκλ- (Th. 8.93.1, X. *Hell.* 5.3.16, Lys. 13.73, 13.76; but in Lys. 12.73 the present tense ἐκκλησιάζετε should be retained). If the verb is regarded as a compound the augment should produce ἐξεκλ-, not ἐξεκκλ-. So it seems preferable to adopt ἐξεκλ-here, even though its support from mss. is weak (in S the second ε is written over an erasure).

χορευταὶ: exempted from military service to perform at the Dionysia (**15** n.).

ξένοι: aliens were not entitled to attend the Ekklesia. If Meidias really said that they did, he was alleging that the law had been infringed. (Goodwin is wrong in saying 'these must have been naturalized foreigners'. A man who had been naturalized ceased to be a ξένος, and was liable to military service in the same way as other citizens.)

194. ἀεὶ: 'at each moment', as in **37**, **131**, **177**, **223**.

ᾗ: 'wherefore' (LSJ ᾗ II.2); not found elsewhere in D. in this sense.

τὰ νῦν, οἶμαι, δάκρυα: Meidias' attempts to make himself seem pitiable today (cf. **186** δακρύων) are inconsistent with his bullying pretensions at the hearing of the *probole* in 348. The phrase is a good instance of the use of parenthetic οἶμαι to give prominence to the word which precedes it.

195. κεφαλή: **117** n.

τούσδε: object of *ἀξιώσεις*, referring to the jury.

σπουδάζειν: **2** n. *ἐσπούδασεν*.

ὑπερηφανίας ... πάντων ἀνθρώπων: **83** n. A and other mss. insert *καὶ ὑπεροψίας*, perhaps to give an easier construction for the objective genitive *πάντων ἀνθρώπων*: cf. pp. 50–1 on extra words in A. In fact *ὑπερηφανία* can take a genitive; cf. Pl. *Rep.* 391c *ὑπερηφανίαν θεῶν τε καὶ ἀνθρώπων*. Taken thus, *πάντων ἀνθρώπων* can be retained; Blass deleted it, perhaps because he assumed that it should go with *φανερώτατος* and then found it redundant after *μόνος τῶν ὄντων ἀνθρώπων*.

μεστὸς A: *πλήρης* S. Either could be correct, but elsewhere D. uses *μεστός*, not *πλήρης*, when saying that a person is full of some quality, e.g. 19.206 *πλείστης ἀναιδείας καὶ ὀλιγωρίας μεστόν*, 19.218, 22.31, *Ep.* 3.34.

ἔσει: for the 2nd. sing. pres. or fut. ind. mid. or pass. usually S gives -*ει* and the other mss. -*ῃ*, though in **207** all give *βούλει* and *πολιτεύῃ*. I have put -*ει* in the text throughout, but we cannot know whether D. wrote *EI* or *HI*.

τὴν σὴν θρασύτητα ...: it is not entirely clear why some nouns in this list have the article and possessive adjective while others do not. Possibly the text needs emendation; Weil omits *τὸ σὸν*. Otherwise the intention may be to produce three phrases referring to (*a*) Meidias' bullying talk, (*b*) his physical posturing, (*c*) the effects of his wealth.

φωνὴν ... θεωροῦντας: a zeugma, strictly illogical.

τοὺς σοὺς ἀκολούθους: cf. **158**.

196. μεντἂν: 'you *would* have ...!'; cf. Denniston *Particles* 402, 'expressing lively surprise or indignation'.

ἐξαπατᾷς: sarcastically substituted for some verb like *ἀτυχεῖς*, because D. maintains that Meidias' story about his present misfortunes is false; cf. **187** *πλάττεται*, **204** *ἐξαπατῶν καὶ φενακίζων*.

οὐκ ἔστιν: asyndeton, because this is the answer to the question asked in **195**. The intervening sentence takes the form of a comment on the question before the answer is given.

197. κατηγόρει: imperfect tense, referring to the occasion in 348, already mentioned in **132**. On the sequence of events see p. 6.

πάλιν without a connective particle: cf. **64**.

μείνας: 'after remaining in Athens', not joining the cavalry at Olynthos.

τοὺς ἐξεληλυθότας: those who went on the expeditions to Euboia and Olynthos. D. assumes that the present jury comprises some men who went overseas in 348 and others who remained in Athens at that time; each of these groups either has heard or (D. predicts) will shortly hear Meidias criticizing the other. *τοῦ δήμου* is used instead of *τῶν μεινάντων* in order to make Meidias' conduct sound more offensive.

ἐάν τε μένητε ...: the subjunctive (where one might expect *εἴτε ἐμείνατε* ..., referring to the occasion in 348 only) generalizes; D. implies that Meidias regularly criticizes both those who go on campaigns and those who stay at home.

συνάρχοντες: presumably colleagues in the various offices mentioned in **171**, but D. has nowhere given any specific information about their dislike of Meidias.

φίλοι: it sounds paradoxical to say that his friends cannot bear him; the next sentence explains.

198. **ἐμοί**: with *ἔνδηλοι*.

νὴ τὸν Δία καὶ τὸν Ἀπόλλω καὶ τὴν Ἀθηνᾶν: an oath with two or three accusatives is used by D. to introduce a statement cautiously or apologetically; cf. 9.65 *ὅ νὴ τὸν Δία καὶ τὸν Ἀπόλλω δέδοικ' ἐγὼ μὴ πάθηθ' ὑμεῖς*, 18.129 *ἀλλὰ νὴ τὸν Δία καὶ τοὺς θεοὺς ὀκνῶ μὴ* ... Here he appears to be worried that he may be giving offence to one or more of Meidias' supporters (Euboulos perhaps) by saying that they are insincere.

εἰρήσεται γάρ ...: an iambic trimeter, perhaps a quotation from a tragedy, but not found elsewhere.

ἡδέως cannot mean that they find conversation with Meidias enjoyable; that would be inconsistent with the context. It must mean that they cultivate his friendship in order to obtain his support in politics or other activities, but do not really like him, and have therefore been hoping to see him lose his case against D.

πλουτεῖ μόνος ...: Meidias' boasts. *μόνος* means not that Meidias says he is the only rich man, but that he is the only man who says he is rich. For the boast about wealth cf. **153** *ἡμεῖς οἱ πλούσιοί ἐσμεν*.

καθάρματα καὶ πτωχοί: **185** n.

οὐδ' ἄνθρωποι: cf. **101** *οὔθ' ὅλως ἄνθρωπον*.

199. **ἐπὶ ταύτης τῆς ὑπερηφανίας**: **2** n. *ἐπὶ τῶν ἄλλων*.

νῦν could be taken with *ὄντα*, but is better taken with the ensuing conditional clause, because it must be so taken in **201**, where the same point is made again.

τὴν καταχειροτονίαν: the vote against Meidias in the Ekklesia in 348.

θεωρήσετε in protasis, with *ἂν εἰδείητε* in apodosis: 'you can know this ... if you will consider it ...'. Cf. 1.26 *τῶν ἀτοπωτάτων μεντἂν εἴη, εἰ ... ταῦτα δυνηθεὶς μὴ πράξει*, and Goodwin *Moods and Tenses* §505. Emendation to *θεωρήσαιτε* (Bekker), though easy (Byzantine confusion of *αι* and *ε*), is not necessary.

καταχειροτονηθὲν: impersonal passive, accusative absolute. This is the reading of both S (before correction) and A. The alternative -*θέντος* gives a genitive absolute which is not compatible with *ὅστις*. Corruption of the accusative to the genitive was easy, because of the presence of *αὐτοῦ* and because the accusative absolute of this verb was unfamiliar.

ἀσεβεῖν instead of *ἀδικεῖν*: see p. 18.

κατέδυ: trying to hide himself from shame. For this sense of the verb cf. 24.182 *παρακάθηται καὶ οὐ καταδύεται τοῖς πεπραγμένοις*, X. *Kyr.* 6.1.35 *δακρύειν ὑπὸ λύπης, καταδύεσθαι δ' ὑπὸ τῆς αἰσχύνης*.

τόν γε δὴ μέχρι τῆς κρίσεως χρόνον is strictly inconsistent with *εἰ καὶ μηδεὶς ἄλλος ἐπῆν ἀγὼν ἔτι*, but D. is thinking of Meidias, for whom another trial did impend.

πάντα: sc. *χρόνον*, 'for ever'.

200. Striking use is made of asyndeton here, to give an impression that Meidias' activities occur in rapid and precipitate succession. Plutarch in his *Platonic Questions* quotes this passage (and **72**) to show how speech without connective particles *πολλάκις ἐμπαθεστέραν καὶ κινητικωτέραν ἔχει δύναμιν* (*Eth.* 1010e–1011a).

χειροτονεῖταί τις: equivalent to 'whenever an election is held', but expressed paratactically for stylistic effect. D. is probably thinking here of the minor festival offices mentioned at the end of **171**. (*τις* just means 'some official', such as a *ἱεροποιός* or a *βοώνης*. Σ (664 Dilts) is wrong to take it as some other candidate whom Meidias then ousts; that interpretation does not fit the perfect *προβέβληται*).

προβέβληται: passive, not middle; **15** n. *προβαλλόμενος*. The perfect tense is used to suggest a *fait accompli*: when the names of candidates are read out in the Ekklesia, it is found that Meidias is among them, or perhaps is even the sole nominee, so that he is inevitably appointed. His name is given here with his demotic because that is the form in which it is read out in the list of nominations. (Anagyrous was a coastal deme of the Erekhtheis tribe, located at Vari, halfway between Peiraieus and Sounion; cf. C. W. J. Eliot *Coastal Demes of Attika* (Toronto 1962) 35–46.) The sentence implies, of course, that Meidias himself took the initiative in getting someone to nominate him.

Πλουτάρχου: cf. p. 5. *προξενεῖ* means that Ploutarkhos had appointed Meidias to represent his interests at Athens.

τἀπόρρητ' οἶδεν must be an allusion to some recent occasion when Meidias claimed to possess confidential information, but it is not otherwise known. The context suggests, but does not prove, that he made this claim in the course of a speech in favour of Ploutarkhos.

οὐ χωρεῖ: 'cannot contain him'. Cf. 9.27 *οὔθ' ἡ Ἑλλὰς οὔθ' ἡ βάρ-*

βαρος τὴν πλεονεξίαν χωρεῖ τἀνθρώπου (referring to Philip), Ais. 3.164 *τὴν δὲ σὴν ἀηδίαν ἡ πόλις οὐκ ἐχώρει* (referring to D. himself).

201. ἀπολωλέναι δεκάκις: 118 n. At the end of this question A adds *ἐγὼ μὲν ἡγοῦμαι*, 'I think so'. See pp. 50–1 on extra words in A.

202. D. suggests that Meidias deserves punishment for the political speeches which he makes. To us this passage may well seem unconvincing, not merely because we do not think people should be punished for expressing unpopular opinions, but also because the opinions which Meidias is said in **203** to have expressed (that military setbacks are due to the Athenians' own failure to go out on campaigns and to contribute money to pay for them) are much the same as those which D. himself expressed in his *Philippics* and *Olynthiacs*.

ἐν οἷς καιροῖς: equivalent to *τῶν καιρῶν ἐν οἷς δημηγορεῖ*. Meidias chooses inappropriate occasions for his speeches.

ἐξητάσθη: 65 n. *ἐχθρὸν ἐξεταζόμενον*. The criticism made of Meidias here, that he is silent at the Athenians' successes and loquacious at their failures, is also made of Aiskhines in the speech *On the Crown*, where D. finds more vivid language to express it: *πράττεταί τι τῶν ὑμῖν δοκούντων συμφέρειν· ἄφωνος Αἰσχίνης. ἀντέκρουσέ τι καὶ γέγον' οἷον οὐκ ἔδει· πάρεστιν Αἰσχίνης. ὥσπερ τὰ ῥήγματα καὶ τὰ σπάσματα, ὅταν τι κακὸν τὸ σῶμα λάβῃ, τότε κινεῖται* (18.198).

203. μηδεὶς, instead of *οὐδείς*, because the clause is generic.

οὐ γὰρ ἐξέρχεσθε . . .: compare D.'s own admonitions in the *Olynthiacs*: 2.24 *νυνὶ δ' ὀκνεῖτ' ἐξιέναι καὶ μέλλετ' εἰσφέρειν*, 1.6, 2.13, etc.

εἰσφέρειν: 153 n. *οἱ προεισφέροντες*.

νεμεῖσθε: the distribution of public money as *θεωρικά*, doles for attending performances in the theatre. D. too had repeatedly suggested that this money should be used for military purposes (1.19–20, 3.11, etc.). Apollodoros actually got a decree passed about it in 349/8, but it was overturned by a *γραφὴ παρανόμων* (59.4–5). Cf. M. H. Hansen *GRBS* 17 (1976) 235–46.

The mss. here all give the future indicative, but Σ **2** (12 Dilts) quotes this sentence with the wording *ὑμᾶς δὲ νεμεῖσθαι*, and this reading receives some support from the use of infinitives in the parody of this passage at the end of **204**. However, the absence of *μέν* shows that the antithesis here is less sharp than in **204**, and it may be better to keep the indicative. If so, the question about the citizens' own behaviour is a question about their will and intention, not a question of fact subordinated to *οἴεσθε*: 'If you are going to share out money among yourselves, do you think that I shall continue paying it?' A sentence in the *Third Philippic*, where the mss. are divided between *ἀποδράσεσθαι* and *ἀποδράσεσθε*, is similar: *εἰ δ' οἴεσθε Χαλκιδέας τὴν*

Ἑλλάδα σώσειν ἢ Μεγαρέας, ὑμεῖς δ' ἀποδράσεσθε τὰ πράγματα, οὐκ ὀρθῶς οἴεσθε (9.74).

ἐμβήσεσθε: to embark as sailors.

204. περιέρχεται: when this verb is used with a participle, its point is not so much to describe circumambulation as to refer to daily social intercourse and going about one's normal business; cf. 4.48, 18.323, 19.288, 54.36.

οὕτω: in the same manner as Meidias speaks to you. The remarks which D. now puts into the mouth of the ordinary citizens are a parody of those which he attributes to Meidias in **203**.

ἐξαπατῶν: the hard-luck story which Meidias will tell when producing his children in court (**186–8**, cf. **196** *ἐξαπατᾷς*).

ὑποβάλλειν: interrupt (*ὑπο-*) in order to oppose (*-βάλλειν*). Cf. 49.63 *ἐὰν ταύτῃ τῇ ἀπολογίᾳ καταχρῆται, ὅτι οὐκ ἐπεδήμει ..., ὑποβάλλετε αὐτῷ ὅτι "ἔλαβες μὲν ἐπιδημῶν ..."*, Ais. 3.16, 3.23, etc.

κακὸς κακῶς ἀπολεῖ: a cliché, found also in 7.45, 18.267, 32.6, and in other authors; cf. R. Renehan *Studies in Greek Texts* (Göttingen 1976) 114–16.

ἀνέξεσθαί σου: *σου* is omitted in A, but it should be retained, because without it the inappropriate sense 'put up with your death' would be implied from the previous sentence. For the genitive with *ἀνέχομαι* cf. 19.16 *τῶν τὰ τρόπαια καὶ τὰς ναυμαχίας λεγόντων ἀνέχεσθαι*, Pl. *Prot.* 323a *εἰκότως ἅπαντος ἀνδρὸς ἀνέχονται*. The origin of the usage is probably that *τὰ ἔργα* or *τοὺς λόγους* is understood. (Passages in which *ἀνέχομαι* is accompanied by a genitive absolute are different.)

σέ: **74** n. *ἐμαυτὸν*.

205–18. *A number of prominent men will speak in support of Meidias, but they are self-interested and the jury must not be deflected from voting against him.*

This passage, especially the remarks about Euboulos (**205–7**), provides the main evidence in favour of the view that the dispute between D. and Meidias was political, not just personal, but it does not really prove that that was so; see pp. 11–13. The main emphasis is rather social and moral: D. is trying to arouse the jury's hostility to Meidias by portraying him as belonging to a group of men who are selfish and contemptuous of ordinary people.

205. μὰ τοὺς θεούς: such oaths usually come near the beginning of a sentence or clause. K. J. Dover *CQ* 35 (1985) 332 includes this one in a list of instances which come relatively late, and comments (ibid. 341) on 'the forceful, man-to-man tone' of the passage. While agreeing that D. may have used such a tone here, I suggest that the main influence on the placing of this oath was a wish to emphasize *τούτῳ χαρίζεσθαι* in contrast with *ἐπηρεάζειν ἐμοὶ* ...

ἐπηρεάζειν: **14** n.

οὗτος must refer to Euboulos, as the sequel shows, though his name does not appear in the text until **206**. Several explanations are possible; I favour *b*, but there can be no certainty about it.

(*a*) D. may have pointed at Euboulos, sitting in the court awaiting his turn to speak. This is the least likely explanation, because probably not all the jury would have been able to see who was the object of the gesture.

(*b*) *οὗτος* may have displaced *Εὔβουλος* from the text. In fact *Εὔβουλος* is written above *οὗτος* in S and in the margin of P, but that is probably a gloss rather than a variant reading.

(*c*) Some words, including the name of Euboulos, may have been lost from the text earlier in the sentence. Cobet *Misc. crit.* 517–18 suggests that there is a lacuna before *οὐχ οὕτω*, where we should supply something like 'Euboulos will also speak for Meidias'; then *βουλόμενοι* is to be emended to *βουλόμενος* and *οὗτος* is to be deleted.

Euboulos was evidently the leading politician in Athens at this time, though detailed evidence of his activities and policies is scanty. The best attempt to reconstruct them is Cawkwell's article in *JHS* 83 (1963) 47–67. The whole passage **205–7** may be compared with a passage near the end of the speech *On the False Embassy* (19.288–97), where D., expecting Euboulos to speak in support of the defendant Aiskhines, criticizes him even more strongly than here for arrogance and lack of principle.

βιάζεται: 'asserts vehemently'. So also in 23.100, 59.28; cf. LSJ *βιάζω* II.4.

οὐδ' ἀφιέντα ἀφίησιν: 'does not release me (from his hostility) even though I release him'. The use of polyptoton emphasizes the reciprocity which is normal in a hostile relationship, but in this case it takes only one to make a quarrel. Cf. 19.118 *οὐδ' ἀφιέντων ἀφίεται.*

τοῖς ἀλλοτρίοις ἀγῶσιν: probably just a generalizing reference to the present case. It is not known that Euboulos had spoken against D. in any other trial. Σ (687 Dilts) says that it was Euboulos who prosecuted Aristarkhos for the murder of Nikodemos and included D. in his accusation; but that is probably just a conjecture by the scholiast, who has evidently looked back through the speech to find a trial in which D. was implicated, and has found that one in **104** (cf. p. 330).

ἀναβήσεται: **184** n. *καταβήσομαι.*

τῆς κοινῆς τῶν νόμων ἐπικουρίας: 'the legal protection which is available to everyone'. These, of course, are not the words which

Euboulos would use to describe the conviction of Meidias, but D.'s interpretation of it.

ἡμῶν ἑκάστῳ: for the citizens of Athens it is a bad thing if any politician becomes too powerful. The remark is in line with the traditional Athenian fear of tyranny. Cf. the stronger comment on Euboulos in the *Embassy* speech: *οὐ γὰρ ἔστιν, οὐκ ἔσθ' ὅ τι τῶν πάντων μᾶλλον εὐλαβεῖσθαι δεῖ ἢ τὸ μείζω τινὰ τῶν πολλῶν ἐᾶν γίγνεσθαι* (19.296).

206. Cf. the passage in the *Embassy* speech, where Euboulos is going to speak for Aiskhines although he did not speak for the defence in an earlier trial: *τί γὰρ δή ποτ', Εὔβουλε, Ἡγησίλεῳ μὲν κρινομένῳ … καὶ Θρασυβούλῳ πρώην … οὐδ' ὑπακοῦσαι καλούμενος ἤθελες …; εἶθ' ὑπὲρ μὲν συγγενῶν καὶ ἀναγκαίων ἀνθρώπων οὐκ ἀναβαίνεις, ὑπὲρ Αἰσχίνου δ' ἀναβήσει;* (19.290).

ἐν τῷ θεάτρῳ: the meeting of the Ekklesia held in the theatre of Dionysos; cf. **8–9**.

οὐκ ἀνέστη: 'he sat tight', ostentatiously not coming forward to speak in support of Meidias.

μηδὲν ἠδικηκότος may be either an objective genitive with *τὴν προβολὴν* (cf. **10** *προβολαὶ αὐτοῦ*) or a genitive absolute with *τούτου* understood.

τόν γε φίλον δήπου: *γε* gives a causal sense, 'since he is Meidias' friend'; but *δήπου*, 'surely', adds a sarcastic touch, implying doubt whether the friendship is genuine.

καταγνοὺς ἀδικεῖν: for the construction, **175** n. *ὅσων … ἀδικεῖν*. Here *καταγνοὺς* refers not to a formal verdict but to Euboulos' private opinion.

διὰ ταῦτ' … διὰ ταῦτα: emphasizing that it is the motives, rather than the words, of Euboulos which should be scrutinized.

ὑπήκουσεν: this compound of *ἀκούω* normally means not just 'hear' but to speak or act in response, e.g. 8.75 *ὁ μὲν εἶπεν …, ὑμεῖς δ' ἀπερρᾳθυμήσατε καὶ μηδὲν ὑπηκούσατε*.

χαρίσασθαι: to acquit Meidias as a favour to Euboulos.

207. τηλικοῦτος: 'so powerful', as in **142**. The same point is made in the *Embassy* speech: *μή μοι σῳζέσθω μηδ' ἀπολλύσθω μηδείς, ἂν ὁ δεῖνα βούληται* (19.296). In both passages the third-person imperative gives the effect of a solemn warning.

ὑβρίσθαι, not *ὑβρίζεσθαι*: Euboulos did not cause D. to suffer the act of insolence, but if he causes Meidias to be acquitted that will leave the effect of it still in existence.

ὡς is relative to *κακῶς*, and *ἐμὲ βούλει ποιεῖν* is understood with it.

ἀνθ' ὅτου implies that harm is normally done in retaliation for harm, but D. claims to have done no harm to Euboulos.

πολιτεύει: not just 'you are a citizen', but 'you are active in politics'; cf. 18.18, 39.3, etc. For the form, **195** n. *ἔσει*.

κρίνων: **25** n. *κρίνειν*. The phrase implies that Euboulos prosecutes rather often. In the *Embassy* speech D. goes further by giving two examples (Moirokles and Kephisophon) of men whom Euboulos prosecuted for comparatively minor offences (19.293).

208. Φιλιππίδην: son of Philomelos of Paiania (**174** n.). He had a long and distinguished life, and in 293/2 was decreed special honours for his services to Athens: a gold crown, meals at the Prytaneion, a front seat at festivals, a bronze statue of himself in the Agora, and two inscriptions of the decree cataloguing his merits, one of which was erected beside the statue in the Agora and the other on the Akropolis. The latter copy has been found, cut into two parts; they were reunited and fully published by W. B. Dinsmoor *The Archons of Athens in the Hellenistic Age* (Cambridge Mass. 1931) 3–15. The decree praises Philippides for his liturgies and for his service as strategos, basileus, agonothetes, ambassador, and in other capacities. Other evidence about his liturgies is listed by Davies *Families* 549–50; the earliest known was performed in the 350s, when he served as a syntrierarch with D. himself as his partner (*IG* 2^2 1613.191–2). Elsewhere he is mentioned as a witness for the prosecution of Theokrines (D. 58.33) and as proposer of a decree honouring Poseidippos for services in connection with an embassy to Kassandros in 299/8 (*IG* 2^2 641).

The question arises whether Philippides was simply a rich man who deployed his money so as to do service to Athens and thus obtain honour for himself, or was also a political leader. **208–9** has generally been taken to mean that he was a politician of oligarchic tendency in the 340s, and P. Treves *RE* 19.2 (1938) 2201–4 presents his career as a gradual movement from oligarchic to democratic views. By itself **208–9** is rather weak evidence for this, but it is greatly strengthened if we accept the suggestion, made by Lewis *BSA* 50 (1955) 20, that the speech of Hypereides *Against Philippides* is directed against him. The Philippides attacked in that speech is on trial in a *γραφὴ παρανόμων* in the 330s. He has been convicted in *γραφαὶ παρανόμων* twice before, and is clearly an active politician and orator; Hypereides accuses him of thinking he can get himself acquitted by his usual jigs and jokes in court: *κορδακίζων καὶ γελωτοποιῶν, ὅπερ ποιεῖν εἴωθας ἐπὶ τῶν δικαστηρίων* (7). He belongs to a group who, Hypereides says, are always looking out for anti-democratic opportunities: *καιροφυλακοῦντες τὴν πόλιν, εἴ ποτε δοθήσεται ἐξουσία λέγειν τι ἢ πράττειν κατὰ τοῦ δήμου* (8). They were friendly to the Spartans in the days of the Spartans' power (1 *τότε φίλοι ὄντες Λακεδαιμονίων*, viz. before Leuktra in 371), and Davies *Families* 550 questions whether Philippides son of Philomelos could have been old enough for that; but I agree with G. Bartolini *Iperide* (Padua 1977) 77 that the plural expression may refer to his family rather than to himself personally. I am therefore inclined to accept the identification, and with it the view that the Philippides supporting Meidias was an active politician of anti-democratic tendency.

Μνηαρχίδην: named as a trierarch in two of the surviving naval lists (*IG* 2² 1609.80, 1612.364). He could be identical with a Mnesarkhides who served one year as arkhon's paredros (D. 58.32), and with Mnesarkhides son of Mnesarkhos of the deme Halai Araphenides, whose gravestone has been found (*IG* 2² 5501; cf. Davies *Families* 392–3). Nothing else is known of him.

Διότιμον: Diotimos son of Diopeithes of Euonymon (a city deme of the Erekhtheis tribe) belonged to a rich family which had been distinguished for several generations; its members are discussed by Davies *Families* 161–5, though there is some uncertainty about the precise relationships. His wealth probably came mainly from the silver mines, the records of which mention him several times (*Hesperia* 19 (1950) 274 no. 26, 26 (1957) 3–4 no. S2 lines 15–16 and 24, *IG* 2² 1582.65–6). The naval records mention him as a trierarch, as a strategos commanding ships, and as one of the sureties for a loan of some ships to Khalkis (*IG* 2² 1623.194–5, 1623.257, 1628.397, 1629.539–40, 1629.622, etc.). After the battle of Khaironeia he made a voluntary donation of shields, and was rewarded for it by the honour of a crown (D. 18.114, cf. *IG* 2² 1496.22–5). In 335/4 as a strategos he led an expedition against some pirates; Lykourgos proposed both the decree authorizing the expedition and a decree in 334/3 honouring Diotimos, no doubt for carrying it out successfully (*IG* 2² 1623.276–85, Plu. *Eth.* 844a; cf. E. Schweigert *Hesperia* 9 (1940) 340–1). He was regarded as a leading opponent of the Macedonians (Arr. *Anab.* 1.10.4, Plu. *Eth.* 844f), and after his death D. spoke of him as one of the *δημοτικοί* (*Ep.* 3.31). Thus it seems that Diotimos, like Philippides, changed to a more democratic standpoint over the years, though in this case there is no evidence other than **208–9** that his views ever were anti-democratic; cf. Longo *Eterie* 90 n. 2. He may have been a poet too: an epitaph for a piper named Lesbon is attributed to him, though doubtfully (*Anth. Pal.* 7.420; cf. A. S. F. Gow and D. L. Page *The Greek Anthology: Hellenistic Epigrams* 2.270–1).

τοῦτον (A) is preferable to *αὐτὸν* following *λιπαρήσειν παρ' ὑμῶν* (SFYP), because normally *λιπαρέω* is either intransitive (as in **206**) or takes the accusative of the person entreated.

ἀξιοῦντας δοθῆναι τὴν χάριν: they will say that the trierarchies and other services they have done for Athens give them a right to ask for a favour in return.

καὶ γὰρ ἂν μαινοίμην: a compliment to the three men just named, inserted by D. in the hope of preventing them from taking offence at his other remarks. Similar phrases are used in 35.40, 52.11, *Pr.* 45.4.

209. ὃ μὴ γένοιτ', οὐδ' ἔσται: a wish, and then a prediction. Cf. Pl. *Laws* 918d *ὃ μή ποτε γένοιτο, οὐδ' ἔσται.*

οὗτοι κύριοι τῆς πολιτείας: D. is asking the jury to imagine Athens being taken over by an oligarchy once again, as in 411 and 404, though he declares that there is no chance of its happening in fact.

μή: some imperative such as *λέγε* is understood. D. hastens to dispel any implication that any of the jurors might ever be guilty of *hybris.*

πεπληρωμένον ἐκ τούτων: a jury drawn entirely from the class of rich citizens.

ἐλέου: S alone has *λόγου*: **90** n. *μήτε λόγου ... τυχεῖν.*

ταχύ γ' ἄν: ironic, 'they *would* be in a hurry to ...!' So also in 25.95, 58.15. *οὐ γάρ;* also marks irony; cf. Denniston *Particles* 86.

τὸν δὲ βάσκανον, τὸν δὲ ὄλεθρον: strictly *βάσκανος* is a malignant person (a sorcerer or a slanderer) and *ὄλεθρος* is one who destroys, but both words are used colloquially as terms of abuse with no very exact meaning. In D. there are other abusive instances of *βάσκανος* in 18.119, 18.132, 18.139, 25.80, 25.83, and of *ὄλεθρος* in 9.31, 18.127, 23.202. The accusative of exclamation (without an infinitive) is not usual in Greek, but it is easy to understand some imperative such as *ὅρα*. For *δέ* 'in passionate or lively exclamations, where no connexion appears to be required' see Denniston *Particles* 172. (There seems to be no good reason for Denniston to put this instance under a separate heading on p. 175.)

τοῦτον δὲ ὑβρίζειν, ἀναπνεῖν δέ: accusative and infinitive of exclamation, 'to think that he ...!' Cf. 25.91 *τοῦτον δὲ ταῦτα ποιεῖν, καὶ ταῦτ' ὀφείλοντα τῷ δημοσίῳ.* 'Breathe' is just a vivid expression for 'live', implying that anyone guilty of *hybris* deserves death.

ἀγαπᾶν: 'be satisfied', without expecting exemption from other penalties; cf. LSJ *ἀγαπάω* III.1.

210. τούτοις: for the dative with *ἔχω* and an adverb in the sense 'have a certain attitude to', cf. 20.135 *τοῖς ἄλλοις χαλεπῶς τις ἔχων ὁρᾶται*, 20.142.

θαυμάζετε: here equivalent to *αἰδεῖσθε*, 'treat with respect and consideration'. Cf. 19.338 *ἐγὼ Φίλιππον μὲν οὐκ ἐθαύμασα, τοὺς δ' αἰχμαλώτους ἐθαύμασα.*

ἀγαθά: property owned by rich individuals, in contrast to legal protection, which is common property in a metaphorical sense.

211. ἐλεεινὸν: 186 n. ἐλεεινότατον.

ἡμῶν S: ὑμῶν A. Either could be right, but ἡμῶν is preferable after ἡμῖν in the previous sentence.

πτωχοὺς: 185 n.

περιόντ' (nom. pl. neut.): 'excessive'. This makes an appropriate contrast with ἴσα and should therefore be preferred to the alternative reading περιιόντα (acc. sing. masc.), 'as he goes about his daily business'.

περιαιρεθῇ: subjunctive, because still subordinate to ἄν. Meidias is the subject, and the antecedent of ἃ is understood in the accusative; cf. **138** περιαιρεθεὶς οὗτος τὰ ὄντα.

εὐορκεῖτε: referring to the jurors' oath (**4** n. ὀμωμοκὼς).

212. εἰς τὰ μάλιστ': 'to the highest degree', not a normal Demosthenic expression. Elsewhere he uses τὰ μάλιστα without εἰς (e.g. **62**), but the phrase with εἰς occurs in other authors; cf. LSJ μάλα III.2.

καὶ καλῶς ποιοῦσι: this phrase is used to express approval, 'very properly', 'and quite right too', but usually only as a preliminary to an objection or critical suggestion which follows. Cf. 18.314 τῶν πρότερον γεγενημένων ἀγαθῶν ἀνδρῶν μέμνησαι, καὶ καλῶς ποιεῖς· οὐ μέντοι δίκαιόν ἐστιν ..., 20.149, 25.97.

προοῖντ': 3rd. plur. aor. opt. mid. of προίημι (contracted from προέοιντο: hence the accentuation is properly προοῖντ', though the mss. give πρόοιντ': cf. H. W. Chandler *Greek Accentuation* (second edition, Oxford 1881) §807). προίημι (but not other compounds of ἵημι) seems always to have -οι- rather than -ει- in the aor. opt. mid.; there are other instances in 5.15, 6.8, 18.254, Th. 1.120.2, X. *An.* 1.9.10, Pl. *Gorg.* 520c, Isok. 17.6.

ἂν with a negatived optative gives a sense of refusal: 'they *would* not (if they were asked)'.

213. This section introduces afresh the topic of Meidias' rich supporters, leading on to a mention of Philippides and Mnesarkhides among others (**215**), as if they had not been mentioned before. It seems that **208–12** and **213–18** are alternative versions of this part of the speech; cf. p. 27.

τινες: 'persons of importance'; cf. LSJ τις A.II.5.a. The nominative is used because the unexpressed subject of δοκεῖν εἶναι is the same as the subject of the main verb.

προῆσθε: 2nd. pl. aor. subj. mid. of προίημι: cf. **212** προοῖντ'.

ἐφ' ὑμᾶς καταπεφευγότος: a slightly melodramatic way of saying that D. has brought the present case for trial by jury.

ἐφ' ἧς νῦν ἐστε: D. is telling the jurors that they have already made up their minds in his favour. This is rather like his device of telling them that they already know the facts (**1** n. *οὐδένα ... ἀγνοεῖν*). For the sense of *ἐπί*, **2** n. *ἐπὶ τῶν ἄλλων*.

214. ὕβριν μὴ γεγενῆσθαι: the infinitive needs a subject. The meaning cannot be 'that the events had not happened', because everyone knew that Meidias had punched D.; the question was whether that act was an offence. This sentence, imagining the effect of a verdict in favour of Meidias in the Ekklesia, finds its converse in **215–16**, recounting the actual verdict against him, where the relevant words are *κεχειροτόνηται μὲν ὕβρις τὸ πρᾶγμα εἶναι*. We should therefore assume that *ὕβριν* has been lost from the text here, as suggested by Reiske and by Herwerden *Mn.* II 3 (1875) 137.

This seems to be the only passage in which *παραμυθέομαι* takes acc. and inf., giving the sense 'console by saying that ...'. The negative is *μή*, as with some other verbs which refer to speech accompanied by some sort of emotion or determination (*ἐλπίζω, ὄμνυμι, μαρτυρέω*, etc.).

τις: not 'any friend of his' (Goodwin), since Meidias' friends would have needed no consolation if he had been acquitted. The meaning is that D. himself, and other right-minded people disappointed at the acquittal, would be consoled by the thought that the verdict of the Ekklesia showed that the offence was, after all, not as bad as it had seemed.

215. νῦν δὲ: 'but as it is in fact', since the Ekklesia did vote against Meidias.

εἰ παρ' αὐτὰ μὲν ...: as often, *δεινὸν* (or *δεινότατον*) *εἰ* is followed by *μέν* and *δέ*: what is dreadful is not either consequence individually but the combination. Here *δέ* is not reached until the middle of **216**; the principal verb with *μέν* is *ἐφαίνεσθε*, which controls the participles *ἔχοντες* and *λέγοντες*, and the principal verb with *δέ* is *ἀποψηφιεῖσθε*. This gives a reasonable sentence. Weil, followed by other editors, noting that *μέν* and *δέ* are omitted by most of the mss. (though F does have them), deletes *ἐφαίνεσθε* also, making the participles dependent on *ἀποψηφιεῖσθε*: but that is a needless emendation and produces a more clumsily constructed sentence. Editors have criticized the structure, e.g. 'this cumbrous sentence, which the orator could never have intended to leave as it now stands' (Goodwin), but unjustly. The accumulation of vivid incidents, including the humorous picture of D. losing his cloak, effectively gives the impression that there is an overwhelming number of reasons to expect Meidias to be condemned.

παρ' αὐτὰ ... τἀδικήματα: **26** n.

ὥστε governs both *ἐβοᾶτε* and *ἀνεκράγετε*, which is then followed by another *ὥστε*. These incidents occurred during the meeting of the

Ekklesia which considered the *probole* in 348; D. and Meidias had already spoken, but the vote had not yet been taken, and Meidias' friends were each in turn speaking in his support.

Νεοπτολέμου: identified as Neoptolemos son of Antikles of the deme Melite. In the *Crown* speech D. says that he had been a superintendent of many works, and had been honoured for his donations (18.114); this suggests that his public services were civil rather than military. One which is known is the gilding of the altar of Apollo in the Agora, for which Lykourgos proposed a decree that he should be honoured with a crown and a statue (Plu. *Eth.* 843f). Other evidence of his donations is collected by Davies *Families* 399–400. As a rich man he must have performed liturgies, but none is specifically attested. He had one minor religious appointment: he went to Delphi as a *ἱεροποιός* (*SIG*³ 296; cf. Lewis *BSA* 50 (1955) 34–5).

Μνησαρχίδου καὶ Φιλιππίδου: 208 n.

τινος, if correct, means 'anyone you care to name', and thus 'all'. This seems possible, and may be preferred as *lectio difficilior* to *τῶν ἄλλων* (A) and *τινων* (F). Sauppe conjectures *τίνος οὐ*, 'everyone'.

ἐβοᾶτε μὴ ἀφεῖναι: the citizens in the Ekklesia (or some of them) were afraid that D. might be persuaded to withdraw his accusation against Meidias, and shouted out to him not to do so.

Βλεπαίου: Blepaios the banker is mentioned also in 40.52, where he is said to have made a loan to the speaker of that oration (Mantitheos). In Alexis 227 (Kock) there is a reference to Blepaios as a rich man. An inscription names Blepaios son of Sokles as contractor for some work at the temple at Eleusis in the 330s (*IG* 2² 1675.32); that may well be the same man, since Sokles is also known to have been the name of a banker (D. 36.29). Nothing else is known of him, but our passage shows that he was a well known figure: when the crowd in the Ekklesia saw him go up to D. and speak privately to him (presumably while the speeches of Meidias' supporters were in progress), they recognized him as a financier and therefore assumed that he was negotiating a financial arrangement with D. to induce him to drop the accusation.

τοῦτ' ἐκεῖνο: 'that's it!', 'the same old story!', a colloquial expression of recognition; cf. Ar. *Akh.* 820, *Peace* 289, Pl. *Euthd.* 296b, etc. Here it is parenthetic; *ὡς* goes with the genitive absolute to give the sense 'on the assumption that I was going to accept money'.

216. **ὥστε** governs the infinitive clauses *με ... προέσθαι καὶ ... γενέσθαι*.

θοἰμάτιον προέσθαι: the *ἱμάτιον* was draped round the body and over one or both shoulders without any kind of fastening, so that a sudden or careless movement might easily cause it to fall off; cf. A. G.

Geddes *CQ* 37 (1987) 312–13. Here Blepaios catches hold of a corner of D.'s cloak to attract his attention (*ἕλκοντά με*), the onlookers shout, D. jumps away, and his cloak falls to the ground, leaving him in an embarrassing state of undress.

μικροῦ: 'almost' (LSJ *μικρός* III.2); so also in **110**. D. was still wearing his tunic, but his expression 'almost nude' implies that it was not respectable to appear in the Ekklesia without a cloak. Passages in Aristophanes imply that it was standard practice to put on a cloak for attendance at the Ekklesia, e.g. *Wasps* 33, *Ekkl.* 352–3.

χιτωνίσκῳ: originally a diminutive form, but by the fourth century *χιτών* had fallen out of use and *χιτωνίσκος* was the normal word for 'tunic', with no diminutive sense; cf. 25.56, Ais. 1.131, Pl. *Hipp. Min.* 368c, *Laws* 954a.

ἕλκοντα: taking hold of a man's hand or arm or (in this case) cloak, to prevent him from going away before one has spoken to him. Cf. Ar. *Wasps* 793 *Φι. κᾆθ' εἷλκον αὐτόν. Βδ. ὁ δὲ τί πρὸς ταῦτ' εἶφ';* Most mss. have δ' before this word, and Weil conjectures *φεύγοντα, ἐκεῖνον δ' ⟨ἕπεσθαι⟩ ἕλκοντά με*.

ἀπαντῶντες: subordinate to *λέγοντες*, referring to casual meetings in the street. The point here is the same as in **2** *πολλοί μοι προσιόντες*.

λέγοντες: participle with *ἐφαίνεσθε*. Since speech is not primarily a matter of appearances, the combination is not particularly appropriate, and Gebauer *De hypotacticis* 192 proposes the emendation *τοιαῦτ' ἐλέγετε*. If emendation is really needed, I should prefer to postulate the loss of an indicative earlier in the sentence and write *μετὰ ταῦτα ⟨παρεκελεύεσθ'⟩ ἀπαντῶντες* (haplography, from *απα* to *απα*: for the verb cf. **2**) or even *μετὰ ταῦτα ⟨ἅπαντες παρεκελεύοντ'⟩ ἀπαντῶντες* (an easier haplography). But probably no change is necessary.

κεχειροτόνηται: by the Ekklesia. There is no other evidence that the verdict of the Ekklesia stated explicitly that the offence was *hybris*, and D. may be over-interpreting it.

ἐν ἱερῷ: in the theatre of Dionysos (**8–9**); cf. **227** *εἰς ἱερὸν καθεζόμενος*.

καθεζόμενοι: for the form, **56** n. *καθίζεσθαι*.

οὐ προύδωκα: cf. **222** *μὴ προδῶτε*. D. regards the vote of the Ekklesia as having the effect of a compact between himself and the Athenians, by which he would prosecute Meidias in court and they, through a jury, would condemn him; either party, by failing to carry out this agreement, would be betraying the other.

τηνικαῦτ' ἀποψηφιεῖσθε ὑμεῖς: the climax of the long sentence, expressing the upshot which would be inconsistent with all that has gone before. The future indicative is not compatible in strict grammar with *ἄν* ... *συμβαίη*, but the sentence is long enough for that to have been forgotten. Or perhaps D. simply breaks off the conditional construction and ends with an indignant question; A does give a question mark after *ὑμεῖς*.

217. αἴσχιστα: it will be shaming for D. if his prosecution fails.
ὑμῖν: 'from you'; cf. LSJ *ἄξιος* II.2.b.
κρίνων: **25** n. *κρίνειν*.
τῶν Ἑλλήνων: **74** n. *ξένων*.
ἤκουσεν: asyndeton; **101** n. *ἅπασι προσήκει*.

218. ἄλλου τινὸς: a totally inexplicit expression, allowing each listener to imagine any kind of bad influence on the jury; but political prejudice, bribery, and intimidation are the ones most likely to come to mind.

ἡττῆσθαι: it is not clear where Lambinus found this reading. It may be only a conjecture, but it should be accepted, with *δόξετε* understood from the earlier part of the sentence. Sykutris retains the perfect indicative *ἥττησθε* (SYP), which gives a kind of vivid anacolouthon: 'but if you acquit him—you've been defeated!' But that is unsatisfactory, because D. does not wish (and would be foolish) to suggest that the jurors may actually succumb to some bad influence; he is talking about what other people will think of them, and therefore the understood *δόξετε* is essential.

ἐκ πολιτικῆς αἰτίας: elsewhere (e.g. 18.311) *πολιτικός* and *κοινός* are virtually synonyms, in the sense 'public'; but here there is a contrast between *τἀδικήματα ... κοινά*, 'the offences concern everyone', and *πολιτικῆς αἰτίας*, 'a political accusation', meaning a charge arising from political rivalry, of little concern to ordinary people. The sentence interweaves two antitheses: *ἐκ πολιτικῆς αἰτίας* is contrasted with *ἐξ ὕβρεως* (because *hybris* is an offence which, according to D., concerns everyone), and *ὥσπερ Ἀριστοφῶν ... ἔλυσε τὴν προβολήν* (an offence which was rectified) is contrasted with *ἐκ τοῦ μηδὲν ... ἀναλῦσαι δύνασθαι* (Meidias' offence cannot be undone). The two negatived items are linked by *οὐδ'*, and the two positive items should therefore be linked by *καί*, even though most mss. omit it; the scribe of F adds it in abbreviated form at the end of a line, possibly by his own conjecture.

'Αριστοφῶν: presumably the well known Aristophon of the deme Azenia, who in his heyday was the leading politician in Athens (cf. 19.297); on his political career see D. Whitehead *CP* 81 (1986) 313–19. By now he was elderly, but still politically active. The incident mentioned here is undated, but since D. assumes that the jurors remember it, it may have been fairly recent. His words imply that Aristophon, holding a financial or religious office, was required to supply some crowns for a festival and failed to do so, and he was therefore accused by *probole* of committing an offence concerning the festival, but the charge was dropped when he delivered the crowns. Probably this means that the vote went against him in the Ekklesia, but he delivered the crowns before the trial by jury was held. Σ (716 Dilts) gives an account which adds a little more: *οὗτος φορολόγος ὢν κατέσχε παρ' αὐτῷ τὰς δεκάτας τῆς θεοῦ, ἀφ' ὧν ἔδει στεφάνους ποιῆσαι καὶ ἀναθεῖναι τῇ θεᾷ Ἀθηναίων. κατηγορηθεὶς δὲ ὑπὸ Εὐβούλου φθάσας τὴν εἴσοδον ἀνέθετο τοὺς στεφάνους καὶ πέπαυται ἡ προβολή.* However, much of this could be mere inference from what D. says, and the points which are new arouse some suspicion: Athens had no official called *φορολόγος*, which is a Hellenistic word; *τῇ θεᾷ Ἀθηναίων* is not how the Athenians themselves refer to Athena; a mere religious offering would not call for prosecution by *probole*, which was concerned with the major festivals. Thus it seems possible that the scholiast just made up his account without having any other evidence; and his statement that the prosecutor was Euboulos, though quite plausible (for opposition between Aristophon and Euboulos cf. 19.291), may be only a guess.

τούτου γενομένου: i.e. now that Meidias is on trial. But the phrase seems otiose as it stands, and there may be some fault in the text.

αὖθις: 'at some time in the future' (cf. LSJ *αὖθις* II.3), when he offends again.

219–25. *All ordinary people will be at risk if the law is not upheld.*

D. tries to get the jurors to identify themselves with his case, so that they will feel that a vote for him is a vote for themselves against a common enemy. He drives the point home in a masterly manner by getting them to imagine themselves walking home from the court at the end of the trial—just what some of the more bored among them are likely to be thinking about spontaneously by now. (Compare 59.110–11, where the jurors are invited to think what they will say to their wives when they get home.) Their own personal safety is protected by the rule of law, but the rule of law will break down unless it is maintained by juries. This leads to a powerful statement of the interdependence of the jurors and the laws, which must prevail over any excuses that Meidias may offer.

219. τῇ διανοίᾳ: physically Meidias struck only D., but his intention was that he should be able to strike others too.

ἧττον ἐμοῦ δύνασθαι: this vague expression can refer to wealth, political influence, oratorical ability, or a combination of them, but D. tactfully avoids attributing any of these things to himself specifically. The reasoning attributed to Meidias is *a fortiori*: if he can get away with hitting D., he will also be able to get away with hitting all who are less capable than D.

220. ἀδικησόμενον: passive; **30** n.

ἐλθεῖν: sc. τὰ τοιαῦτα. The acc. and inf. with περιμένω (instead of ἕως ἂν ἔλθῃ) is rare, but cf. Pl. *Rep.* 375c οὐ περιμενοῦσιν ἄλλους σφᾶς διολέσαι.

μισεῖ: asyndeton; **101** n. ἅπασι προσήκει.

221. ὁρᾶτε δέ: δέ has been lost in SFYP, but should be retained; cf. 18.322 ὁρᾶτε δέ, 10.18 ὁρᾶτε γάρ. The sentence which follows is one of D.'s most effective periods, but it consists largely of co-ordinate phrases with no great amount of subordination. These phrases, linked by μέν and δέ, οὐδέ, οὔτε, or ἤ, give the impression of an exhaustive list of possibilities; the point conveyed is that absolutely everyone is free from absolutely every interference—except D.

μεταστρεφόμενος: not 'glancing behind him' (Vince), but going in a different direction to avoid meeting someone. Cf. Ar. *Lys.* 125 τί μοι μεταστρέφεσθε; ποῖ βαδίζετε;

εἰ: the three participles which precede all imply uncertainty, and so the rest of the sentence takes the form of indirect questions.

οὐδ' εἰ ἰσχυρὸς: Reiske's emendation. Clearly ἰσχυρὸς ἢ ἀσθενής is a separate question from μέγας ἢ μικρός; strength and size are not alternatives. Butcher prefers asyndeton, μέγας ἢ μικρός, ἰσχυρὸς ἢ ἀσθενής, but it is better to retain οὐδέ because the phrases before and after have it.

τῇ πολιτείᾳ: the laws and their enforcement by the courts.

ὑβριεῖν SFYP: ὑβρίσειν A. The future of ὑβρίζω is rare and its form is uncertain, but Ar. *Ekkl.* 666 ὑβριεῖται is supported by the metre and Ar. *Thesm.* 719 ἐνυβρίσεις is not.

222. ἐφ' ἧς ἀδείας: **2** n. ἐπὶ τῶν ἄλλων. This emendation should be accepted, because ἄδεια denotes the circumstances in which the citizens go about their business. Corruption to the accusative was easy because a preposition with the accusative is normal with verbs of motion. The accusative is defended by Benseler *De hiatu* 113, who gives ἐφ' the sense 'in accordance with'; but this sense (LSJ ἐπί C.III.4) seems not to be Attic.

περιεῖναι SFYP: ζῆν AF. The two readings are clearly alternatives, and the latter, being the simpler word, is more likely to be a gloss on the former than *vice versa*.

με: sc. ταῦτα παθόντα.

ἐὰν δέ: **138** n. εἰ δ' ἄρα.

223. ἐάν τε διακοσίους ...: by the time of D. a jury always had an odd number of members to avoid a tie in the voting; so the figures here are round ones, and the actual number of jurors would be 201, 1001, etc. For most kinds of case the number was fixed by law; but when a particular trial was ordered by decree of the Ekklesia, the number was fixed *ad hoc*. 201 is the smallest number ever mentioned; this was the size of the jury for a private case under the jurisdiction of the forty tribe judges, if the amount at issue was not more than 1000 dr. (*AP* 53.3). A jury of 1001 members is specified in a decree quoted in D. 24.27 (cf. 24.9); even larger juries are mentioned in And. 1.17, Lys. 13.35, Dein. 1.52, 1.107, *AP* 68.1. Cf. MacDowell *Law* 39–40, Rhodes *Comm. on AP* 728–9.

τοὺς δικάζοντας: a gloss which has got into the text. The nominatives *συντεταγμένοι μόνοι* and *νεώτατοι* show that the subject of the infinitives is not expressed in the accusative, but is 'you', understood from *εὕροιτ'*.

τοῖς νόμοις: dative of means or instrument. The alternative readings *τῇ τῶν νόμων ἰσχύι* (S) and *τῷ τοὺς νόμους ἰσχύειν* (a variant recorded by Lambinus) should be rejected, because this sentence is about the power of the jurors; the next sentence passes on to the power of the laws. *τοῖς νόμοις* here is confirmed by the last words of **224**.

224. ὑμῶν: citizens in general, rather than jurors only; but as usual D. does not distinguish the two categories sharply.

γράμματα ... ἐστίν: the subject understood should be *οἱ νόμοι*, but the verb has been attracted into the sing. by the neuter complement, and then *δύναιτ'* follows suit.

225. τὰ τῶν νόμων ἀδικήματα κοινά: the wording is similar to that at the end of **218**, but the context gives it a wider implication. **218** means that Meidias' offences concern everyone because they were committed against a public official at a public festival, but **225** implies that all transgressions of laws (even those which might be dealt with by private cases) concern everyone because they undermine the laws' authority.

ἐφ' ὅτου: 'in any case in which'; **2** n. *ἐπὶ τῶν ἄλλων*.

μήτε λειτουργίας ...: kinds of defence or excuse which Meidias will offer. Liturgies: **151–9**. Attempts to arouse pity: **186–8**. Powerful men: **205–16**. Rhetorical skill: **196**.

μήτ' ἄλλο μηδὲν: omitted in S by haplography after *μήτε ... μηδεμίαν*. It should be retained to justify the neuter singular *δι' ὅτου*.

226–7. *You have already shown your disapproval of Meidias on other occasions. Now is the time to convict him.*

226. εἰσιόντα εἰς τὸ θέατρον: presumably on a subsequent day of the festival, after the day on which he hit D.

ἐκλώζετε: not in any ms. of D., but an entry in Harpokration's lexicon reads: *ἐκλώζετε· Δημοσθένης κατὰ Μειδίου. κλωσμὸν ἔλεγον τὸν* *γιγνόμενον ἐν τοῖς στόμασι ψόφον, ᾧ πρὸς τὰς ἐκβολὰς ἐχρῶντο τῶν ἀκροαμάτων ὧν οὐχ ἡδέως ἤκουον.* **226** must be the place where Harpokration found the word. Scribes to whom it was unfamiliar have either substituted a word which they knew, *ἐκεκράγετε*, or simply omitted it. Its restoration to the text was first suggested by Lambinus in a note on p. 795 of his edition of D. The verb is used in conjunction with *συρίττειν* in Alkiphron 3.71 (= 3.35.3 Schepers, Benner and Fobes), where a novice actor asks his friend to assemble an applauding claque, so as to drown any disapproving noise from the city lads: *ἵνα ... μὴ λάβῃ χώραν τὰ ἀστικὰ μειράκια κλώζειν ἢ συρίττειν.* Just what kind of sound *κλώζειν* was is not clear from either Alkiphron or Harpokration; and its inclusion by Pol. 5.89 in a list of vocabulary for bird calls, as being appropriate for the sound of a jackdaw, is not much help. We can say only that it was a sound conventionally used to show disapproval, equivalent in intention to a modern hoot or boo but not necessarily the same phonetically.

ἐλεγχθῆναι τὸ πρᾶγμα: in the present speech.

προυκαλεῖσθ' ἐπὶ τιμωρίαν: cf. **2** *παρεκελεύοντο ἐπεξελθεῖν*, **216** "*ὅπως ἐπέξει* ...".

227. προσεξήτασται refers to what D. has said in this speech about Meidias' life, not to any other investigation.

ἐστὶν ἐν: 'depends on'. This expression sometimes has a noun as subject, sometimes an infinitive, but not normally both. (E. *Alk.* 278 is perhaps the only clear exception; Pl. *Prot.* 313a is ambivalent.) Here *πάντ'* may be taken as accusative.

μιᾷ ψήφῳ: really there would be two votes, one to declare Meidias guilty and another to assess the penalty. But what D. means is that the result of this one trial ought to take account of all Meidias' misdeeds.

ἥλωκεν: **105 n.** *ἡλωκέναι.*

References to due punishment, and to setting an example to others, are naturally common at the close of prosecution speeches in public cases, and it is not surprising that similar phraseology is sometimes used. Cf. 19.343 *τιμωρησαμένους παράδειγμα ποιῆσαι πᾶσι, καὶ τοῖς πολίταις καὶ τοῖς ἄλλοις Ἕλλησιν*, 24.218 *πάντων οὖν ἕνεκα τῶν εἰρημένων ἄξιον ὀργισθῆναι καὶ κολάσαι καὶ παράδειγμα ποιῆσαι τοῖς ἄλλοις.*

APPENDIX

THE HYPOTHESES

THE hypotheses, or introductions, which are prefixed to the orations of Demosthenes in many manuscripts, are not written by Demosthenes but by scholars of later centuries, in some cases perhaps ones who also wrote some of the scholia. They might most logically be printed with the scholia; but since it is customary to print them with the orations, the hypotheses to *Meidias* are given here. They do not give us any factual information about Demosthenes beyond what may be discovered from the surviving orations, but they have some interest as a reflection of the manner in which Demosthenes was studied in the Roman period.

The first hypothesis belongs to the collection of hypotheses written by Libanios in the middle of the fourth century A.D. (edited by R. Foerster in volume 8 of the Teubner edition of Libanios). This collection seems to have been at first transmitted as a whole, separately from the orations. Of the older manuscripts, S and A do not contain Libanios' hypotheses; F and Y have them grouped together at the beginning of each manuscript; P is the first manuscript in which each hypothesis of Libanios is prefixed to the oration to which it refers, and this arrangement is followed in many of the later manuscripts. Libanios' brief account of *Meidias* does little credit to so distinguished a scholar. Besides a minor error in his preliminary description of the Dionysia (which did not include 'choruses of pipers'), he has fundamentally misunderstood the dispute between Demosthenes and Meidias: he wrongly thinks that Meidias admitted committing an act of *hybris*, and that the only point in dispute was whether it was also an act of *asebeia*. He is right, however, in saying that the question at issue (*στάσις*) is one of definition (for the sense of *ὁρικός* cf. Hermog. *Περὶ στάσεων* 2); this probably was already a commonplace of critical study of *Meidias* (cf. Hermog. *Περὶ εὑρέσεως* 3.2).

The second hypothesis is in far fewer manuscripts: only T and Bc (which evidently take it, like the oration, from the same exemplar), Pr, and later manuscripts derived from those. There are considerable differences between the text in Pr and that in TBc. At some points one is hardly more than a paraphrase of the other; where either might be correct, I give the version of

TBc in the text and that of Pr in the apparatus criticus. There is no clue to the identity of the author. He gives first an irrelevant account of the Panathenaia (1) and an inaccurate account of the Dionysia (2). His account of the dispute between Demosthenes and Meidias (3–5) has been compiled carelessly from the oration: he misinterprets **13** *τρίτον ἔτος τουτί* ('two years ago') as meaning that Pandionis had no khoregos for three years; he says that Meidias tore Demosthenes' festival dress in the theatre, though really this happened in the goldsmith's house (**16**); and he says that Meidias tried to steal the gold crowns, whereas Demosthenes says that he tried to destroy them (**16**). He then proceeds to rhetorical analysis. Like Libanios, he says that the issue is one of definition, but he interprets this more accurately as a question whether Meidias' offence is or is not an offence against the people as a whole (*δημόσιον ἀδίκημα*). This discussion, which may be based on the lost work of Caecilius (cf. hyp. v), is enlivened by the example of a tomb-robber (7) in a manner which suggests the possibility that the piece was originally a lecture.

The two brief passages which I have numbered as the third and fourth hypotheses have been treated by previous editors as part of the second, but that is surely wrong. They are not in the same manuscripts as the second. Hyp. iii is in Pr, like hyp. ii, but it is not in T and Bc; it is, however, in Y and X and some other manuscripts derived from them. It is a note making much the same point as hyp. ii 8 about the issue of either a more or a less comprehensive definition of the offence (*ὅρος κατὰ σύλληψιν*), but in different words and with a different example, that of a general accused of rape; it can hardly have been intended to form part of the same essay as hyp. ii 8. Hyp. iv is an isolated sentence which I have found only in LhNbFl, three manuscripts copied from the same exemplar, and in Mm.

The fifth hypothesis is the oldest, but has not hitherto been printed in editions of Demosthenes. It is found only in the *AP* papyrus, to which I refer as Π12, and I take it from Kenyon's edition (see p. 39). Its date cannot be later than the end of the first century A.D., when the papyrus was written. Its chief point of interest is the reference to Caecilius, a rhetorician of the first century B.C. whose work is known only from fragments. He, it appears, maintained that the oration is concerned with two questions: whether the offence is one against the people, and whether the acts committed are serious. The author of hyp. v insists, against Caecilius, that there is a third question: whether the acts are *hybris*. Certainly much of the oration is about *hybris*, just as some of it is about *asebeia* (not mentioned in hyp. v). But if Caecilius meant that *hybris* was not part of the formal charge, he was correct; see p. 16. Whether the rest of hyp. v is also based on Caecilius, as suggested by Blass *Jb. Cl. Ph.* 145 (1892) 33, is uncertain; there is no other evidence that he wrote hypotheses to Demosthenes, but he may well have done so.

I

Ἑορτὴν ἦγον οἱ Ἀθηναῖοι Διονύσῳ, ἣν ἐκάλουν ἀπὸ τοῦ θεοῦ
Διονύσια· ἐν δὲ ταύτῃ τραγικοὶ καὶ κωμικοὶ καὶ αὐλητῶν χοροὶ
διηγωνίζοντο. καθίστασαν δὲ τοὺς χοροὺς αἱ φυλαὶ δέκα τυγχάνου-
σαι· χορηγὸς δὲ ἦν ἑκάστης φυλῆς ὁ τὰ ἀναλώματα παρέχων τὰ
περὶ τὸν χορόν. ὁ τοίνυν Δημοσθένης τῆς ἑαυτοῦ φυλῆς, τῆς Παν- 5
2 διονίδος, ἐθελοντὴς ὑπέστη χορηγός. ἐχθρῷ δὲ κεχρημένος τῷ
Μειδίᾳ, τῶν πλουσίων ἑνί, φησὶ μὲν καὶ ἄλλα παρὰ τὴν χορηγίαν
ὑπ' αὐτοῦ πεπονθέναι κακῶς, τὸ δὲ τελευταῖον ἐπὶ τῆς ὀρχήστρας
κονδύλους ἔλαβεν ἐναντίον πάντων τῶν θεατῶν. ἐπὶ τούτῳ
κατηγόρησεν ἐν τῷ δήμῳ τοῦ Μειδίου ὡς ἠσεβηκότος εἰς τὴν 10
ἑορτὴν καὶ τὸν Διόνυσον· ἐκαλεῖτο δὲ ἡ τοιαύτη κατηγορία προβολή. ὁ μὲν οὖν δῆμος κατέγνωκε τοῦ Μειδίου τὴν ἀσέβειαν, ἀγωνί-
ζονται δὲ νῦν ἐν δικαστηρίῳ περὶ τῆς τοῦ δήμου καταχειροτονίας·
ἔδει γὰρ καταγνόντος τοῦ δήμου δικαστήριον κρῖναι δεύτερον. ἔστιν
οὖν ὁ ἀγὼν περὶ ὑποτιμήσεως· οὐ γὰρ περὶ τοῦ μηδὲν ἀδικεῖν ὁ 15
Μειδίας ἀγωνίζεται, ἀλλὰ περὶ τοῦ τιμήματος, πότερον ὕβρεως ἢ
3 ἀσεβείας ὀφείλει δίκην. ὁρικὸς οὖν ὁ λόγος τῇ στάσει, τοῦ μὲν
Μειδίου λέγοντος ὕβριν εἶναι τὸ πραχθέν, ἐπειδὴ τετύπτηκεν ἄνδρα
ἐλεύθερον, τοῦ δὲ Δημοσθένους ἀσέβειαν, ἐπειδὴ χορηγὸς ὁ
τετυπτημένος καὶ ἐν Διονυσίοις καὶ ἐν τῷ θεάτρῳ. διὰ γὰρ τούτων 20
καὶ ἠσεβηκέναι τὸν Μειδίαν φησίν· ὡς εἶναι διπλοῦν ὅρον κατὰ
σύλληψιν. ⟨ἔστι δὲ κατὰ σύλληψιν⟩ ὅταν, μὴ ἐκβάλλοντες τὸ ὑπὸ
τῶν ἀντιδίκων εἰσαγόμενον ὄνομα, καὶ ἕτερον αὐτῷ προστιθῶμεν,
ὥσπερ ἐνταῦθ' ὁ Δημοσθένης, τοῦ Μειδίου λέγοντος ὑβρικέναι, οὐκ
ἐκβάλλει μὲν οὐδὲ τὴν ὕβριν, προστίθησι δὲ αὐτῇ καὶ τὴν ἀσέβειαν. 25

Hyp. i in FYP   1 ἦγον] ἐπετέλουν Lamb^{vl} ἣν YP: ἐν ᾗ F   6 χορη-
γεῖν F   7 μὲν γὰρ καὶ F   παρὰ Lamb^{vl}: ὑπὸ FYP: ἐπὶ Vf^{pc}Ald   8 ὑπ'
Fh^{sl}: om. FYP   αὐτοῦ Lamb   9 λαβεῖν Cd^{pc} Lamb^{vl} τούτῳ Cd^{pc}VbX:
τοῦτο FYP   11 διονύσιον F   12 -κεν F   τὴν om. LpVf   13 ἐν
τῷ Y   14 δικαστηρίου F   16 τοῦ Ald: om. FYP   18 ἔτυψεν Ald
19 χορηγὸς ἦν TBc   20 τυφθεὶς Ald   διονύσοις P   τῷ om. P   τοῦτο
VsOdWd   22 ἔστι δὲ κατὰ σύλληψιν add. Sauppe   ἀπὸ P^{pc}
23 προστίθεμεν F   24 -θα F   25 -σιν F

II

Διάφοροι παρ' Ἀθηναίοις ἤγοντο ἑορταί, ἐν αἷς ἦν τὰ Παν-
αθήναια, ἅπερ ἦν διπλᾶ, μικρά τε καὶ μεγάλα· καὶ τὰ μὲν μεγάλα
κατὰ πενταετηρίδα ἐτελεῖτο, κατὰ τριετηρίδα δὲ τὰ μικρά. ἐν τοῖς
μεγάλοις δὲ γυμνάσιά τινα ἐγίνοντο, καὶ προεβάλλετο ἀφ' ἑκάστης
5 *φυλῆς εἷς γυμνασίαρχος, λαμβάνων χρήματα εἰς τὸ γυμνάζειν τοὺς*
μέλλοντας ἐπιτελέσαι τὴν ἑορτήν, καὶ διδόναι τὰς τούτων δαπάνας
τοῖς τῆς ἑαυτοῦ φυλῆς. ἤγετο δὲ παρ' αὐτοῖς καὶ τὰ Διονύσια, καὶ 2
ταῦτα διπλᾶ, μικρά τε καὶ μεγάλα. καὶ τὰ μὲν μικρὰ ἤγετο κατ'
ἔτος, τὰ δὲ μεγάλα διὰ τριετηρίδος ἐν τοῖς ληνοῖς, ἐν οἷς προε-
10 *βάλλετο χορηγὸς ἀφ' ἑκάστης φυλῆς πρὸς τὸ τρέφειν χοροὺς παί-*
δων τε καὶ ἀνδρῶν· ἐλάμβανε δὲ χρήματα εἰς τροφὴν τῶν τοῦ χοροῦ.
ἐπιστάσης δὲ τῆς ἑορτῆς ἠγωνίζοντο πρὸς ἀλλήλους οἱ χορηγοὶ καὶ
ἤριζον ὕμνους εἰς τὸν Διόνυσον ᾄδοντες, καὶ τῷ νικῶντι τρίπους τὸ
ἆθλον ἦν, ἐπειδὴ τὸν αὐτὸν Ἥλιον καὶ Ἀπόλλωνα καὶ Διόνυσον
15 *ᾤοντο. παυομένης δὲ τῆς ἑορτῆς ἐν τῷ πρώτῳ μηνὶ προεβάλλοντο*
οἱ χορηγοὶ τῆς μελλούσης ἑορτῆς.

Ἐν τοίνυν τῷ παρόντι καιρῷ προεβλήθησαν οἱ χορηγοὶ ἑκάστης 3
φυλῆς, ἐσπάνιζε δὲ ἡ Πανδιονίς, ἡ τοῦ Δημοσθένους φυλή,
χορηγοῦ, καὶ ἠμέλησε τὸ πρῶτον ἔτος, τὸ δεύτερον, τὸ τρίτον. ἔθος
20 *δὲ ἦν πρὸ μηνὸς τῆς ἑορτῆς τὸν ἄρχοντα συνάγειν τοὺς χορηγοὺς*
ἑκάστης φυλῆς εἰς τὸ λαχεῖν περὶ τῶν αὐλητῶν. καὶ ἐλθόντων τῶν
χορηγῶν ἑκάστης φυλῆς πλὴν τῆς Πανδιονίδος, ηὐτελίζετο ὑπὸ
πάντων. καὶ ἰδὼν ὁ Δημοσθένης τὴν ἑαυτοῦ φυλὴν ἀτιμαζομένην,
ἐθελοντὴς ὁ ῥήτωρ αὐτοχειροτόνητον ἤτοι αὐτεπάγγελτον ἑαυτὸν
25 *χορηγὸν ὑπὲρ τῆς φυλῆς προεβάλετο, καὶ ἐπῃνεῖτο παρὰ πάντων*
διὰ τοῦτο. καὶ δὴ λαχόντος τοῦ Δημοσθένους περὶ τῶν αὐλητῶν 4
συνέπραξεν ἡ τύχη τῇ προθυμίᾳ, καὶ ἔλαχεν αὐτῷ ὁ κάλλιστος τῶν
αὐλητῶν ὁ Τηλεφάνης. καὶ δὴ ὁ Δημοσθένης θέλων πλέον τῶν

Hyp. ii in TBcPr 1 -ρὰ Pr *ἀθηναίους* TBc 2 *ἦν* Pr: *ἦσαν* TBc 2 *μεγάλα* (post *μὲν*) om. Pr 3 *πεντεετηρίδα* Bc *ἐπετελοῦντο* TBc 3–4 *τὰ δὲ μικρὰ κατὰ τριετίδα· ἐν μὲν οὖν τοῖς μεγάλοις γυμνάσιά* Pr 4 *προυβάλετο* Pr 5–6 *τοὺς ... ἐπιτελέσαι*] *τοὺς ἐπιτελέσοντας* Pr 6 *διδόναι τὰς*] *δώσοντας* Pr 7 *ἑαυτοῦ*] *αὐτοῦ* TBc *αὐτοῖς* Bc: *αὐτῆς* T: *αὐτῶν* Pr 8–9 *καὶ τὰ μὲν ... μεγάλα* om. Pr 9 *τετραετηρίδος* Bc 9–10 *προυβάλετο* Pr 10 *ἀπὸ* T 11 *τροφὴν ... χοροῦ*] *τοῦτο* Pr 15 *προυβάλλοντο* Pr 19 *δεύτερον καὶ* Bc 20 *δὲ* om. T *πρὸ ἑνὸς μηνὸς* Bc 22 *πλὴν τοῦ τῆς* Pr *ὑπὸ* Dindorf: *ἀπὸ* TBc 22–3 *ηὐτελίζετο ... καὶ* om. Pr 23–5 *ἀτιμαζομένην παρὰ πάντων τῷ μὴ κεκτῆσθαι χορηγόν, αὐτεπάγγελτον καὶ αὐτοχειροτόνητον αὐτὸν χορηγὸν τῇ φυλῇ καθίστησι, κἀντεῦθεν ἐπῃνεῖτο* Pr 25 *προεβάλλετο* T 26 *διὰ τοῦτο* om. Pr *δὴ* om. Pr *λαχόντος*] *κληρουμένου* Lamb^{vl} *τοῦ Δημοσθένους*] *αὐτοῦ* Pr 27 *αὐτῷ* (post *συνέπραξεν*) Pr: *αὐτοῦ* Lamb^{vl} 27–8 *ἔλαχεν αἱρεῖσθαι αὐτῷ τὸν κάλλιστον τῶν αὐλητῶν τὸν Τηλεφάνη* Lamb^{vl} 28–428.1 *καὶ δὴ ... χορόν*] *θέλων οὖν κοσμῆσαι τὸν αὑτοῦ χορὸν πλέον τῶν ἄλλων* Pr

ἄλλων κοσμῆσαι τὸν ἑαυτοῦ χορόν, ἐποίησεν αὐτοὺς φορέσαι χρυσοῦς στεφάνους. Μειδίας δέ, τῶν πολιτευομένων τις, σφόδρα πλούσιος καὶ πολλὰ δυνάμενος, ἐχθρὸς τῷ Δημοσθένει γεγονὼς διὰ τὰς αἰτίας ἃς ἐρεῖ μετὰ μικρὸν ἐν τῷ λόγῳ, πολλάκις καὶ ἄλλα παρηνώχλει καὶ ἐπηρέαζε, καὶ δὴ καί, ὡς ὁ Δημοσθένης λέγει, ὅτι ὀμνυόντων τῶν κριτῶν τῷ καλῶς ᾄσαντι δοῦναι τὴν νίκην, νύττων αὐτοὺς ὁ Μειδίας ἔλεγε "πλὴν Δημοσθένους"· ὅθεν ὁ Δημοσθένης 5 ἐβόα ἐλέγχων αὐτόν. καὶ τελευτῶν εἰς τοιαύτην ἦλθε μανίαν ὁ Μειδίας, ὥστε ἐν τῷ θεάτρῳ κόνδυλον αὐτῷ παρασχεῖν καὶ τὴν ἱερὰν περιρρῆξαι ἐσθῆτα. καὶ ἰδὼν ὁ δῆμος ἐπεσύριττεν· ὅστις συρισμὸς παρὰ τοῖς παλαιοῖς ἐπὶ κακοῦ ἐλαμβάνετο. ἀπελθὼν δὲ ὁ Δημοσθένης ἐσκέψατο τὸν παρόντα λόγον, κατηγορῶν αὐτοῦ δημοσίων ἀδικημάτων· ἐν ᾧ καὶ διαβάλλει τὸν Μειδίαν ὡς κλέψαντα ἀπὸ τῶν χρυσῶν στεφάνων παρὰ τοῦ χρυσοχόου.

6 Ἄγει τοίνυν αὐτὸν ἐπὶ τὴν κρίσιν ὁ ῥήτωρ, καταφορᾷ πλείστῃ καὶ τόνῳ σφοδρῷ προσχρησάμενος· ἡ γὰρ τοῦ Μειδίου προπέτεια καὶ ἡ τῶν πραγμάτων ποιότης τῇ καταδρομῇ συμμαχεῖ. ἡ δὲ στάσις ὁρική, ζητούντων ἡμῶν τί ἴδιον ὄνομα τῷ ἐγκλήματι· ὁ μὲν γὰρ Μειδίας ἰδιωτικόν, ὁ δὲ ῥήτωρ δημόσιον εἶναι κατασκευάζει. ὅρος γάρ ἐστιν, οὗ τὸ μὲν πέπρακται, τὸ δὲ λείπει πρὸς αὐτοτέλειαν τοῦ 7 ὀνόματος τοῦ ἐπιτεθειμένου τῷ πράγματι, ὡς ἐπὶ τοῦ κενοτάφιον ὀρύξαντος καὶ κρινομένου τυμβωρυχίας. ἐνταῦθα γὰρ πέπρακται μὲν τὸ ὀρύξαι, λείπει δὲ τὸ τάφον ὀρύξαι. λέγει γὰρ ὁ φεύγων "ὤρυξα μέν, οὐ τάφον δέ· οὐ γὰρ εὗρον νεκρόν". ὁ δὲ διώκων ἀντιφέρει ὅτι "τὸ δ' ὀρύξαι κενοτάφιον τυμβωρυχίαν λέγω· οὐ γὰρ αὐτὸς ᾔδεις ὅτι κενοτάφιόν ἐστιν, ἀλλ' ὡς τάφον ὀρύττων, ἐπεὶ κενοτάφιον εὕρηται, ἀξιοῖς μὴ δοῦναι δίκην". οὕτω κἀνταῦθα πέπρακται μὲν τὸ τύψαι τὸν Δημοσθένην, λείπει δὲ τὸ καλέσαι τὸν αὐτοχειροτόνητον χορηγόν. ὁ γὰρ Δημοσθένης λέγει ὅτι "χορηγὸν ἔτυψας", ὁ δὲ Μειδίας ὅτι "χορηγὸν ἁπλῶς οὐκ ἔτυψα, ἀλλὰ Δημοσθένην ἰδιώτην ὄντα· τὸ δὲ τύψαι ἰδιώτην οὐκ ἔστι δημόσιον ἀδίκημα".

8 Διπλοῦς δὲ ὁ ὅρος εἴδους τοῦ κατὰ σύλληψιν. κατὰ σύλληψιν δέ

1 αὐτοὺς] αὐτὸν Pr 2 τις ὢν Pr 6 δίκην Pr 9 αὐτοῦ T 10–11 ὅστις συρισμὸς] ὃ Pr 14 ἀπὸ om. Pr στεφάνων ὁντινοῦν Pr παρὰ Pr: ἀπὸ TBc 15 καταφορᾷ] ὁρμῇ Lamb[vl] 16 τόνῳ Bernhardy: τόπῳ TBcPr 16–17 προπέτεια ... ποιότης Bekker: ποιότης ... προπέτεια TBcPr 18 τί om. Pr 20 αὐτοτέλειαν] τὸ τέλειον Pr 21 ἐπιτεθησομένου Pr τοῦ τὸ κεν- Pr 22 τυμβωρύχου Bc 23 τάφον ὀρύξαι, οὐ κενοτάφιον TBc 24 ὤρυξα ... δέ] οὐ τάφον ὤρυξα TBc 25 ἀντιφέρει Pr: λέγει TBc δ' ὀρύξαι G. H. Schaefer: διορύξαι TBcPr κενοτάφιον Pr: τάφον TBc λέγω] καλῶ Pr 25–7 οὐ ... δίκην om. TBc 27 καὶ ἐν- TBc 28 δημοσθένη Pr 30–2 χορηγὸν οὐκ ἔτυψα· αὐτοχειροτόνητος γὰρ ἦσθα· ἀλλ' ἰδιώτην· τοῦτο δὲ οὐκ ἔστι δημόσιον ἀδίκημα Pr 33 ὁ om. Pr

ἐστιν, ὅταν ὁ διώκων τὴν ἑαυτοῦ δικαιολογίαν καὶ τὴν τοῦ φεύγοντος εἰς ἓν συναγάγῃ· ἔνθα γὰρ οὐ τὸ μὲν ἐκβάλλει τις, τὸ δὲ δέχεται, ἀλλ' ἀμφότερα συγκροτεῖ καὶ συλλαμβάνει, τούτῳ ὑπάγομεν τῷ εἴδει. φαίνεται τοίνυν ἐν πολλοῖς μέρεσιν ὁ Δημοσθένης 5 τοῦτο ποιῶν, καὶ φάσκων ἅμα τῷ Δημοσθένει καὶ τὴν πόλιν ὑβρίζεσθαι. κεφάλαια δὲ τὰ τῇ στάσει προσήκοντα. τὸ δὲ προοίμιον καταφορικόν, περιβολὴν ἔχον πολλὴν καὶ τῶν περιστατικῶν αὔξησιν· τὸ γὰρ "{οὐ} πρὸς ἅπαντας καὶ οὐ πρὸς ἐμὲ μόνον" καὶ τὸ "ἀεί" τὴν μελέτην τῆς ἀτοπίας καὶ οὐ πρὸς ἅπαξ ἐκ τύχης 10 ἡμαρτηκότα δείκνυσι.

Κεφάλαια δὲ τοῦ λόγου εἰσὶ ταῦτα· ὅρος, ἀνθορισμός, γνώμη νομοθέτου, συλλογισμός, πηλικότης, πρός τι, καὶ μία τῶν ἀντιθετικῶν, μεθ' ἣν ἐμπίπτει τὸ μεταληπτικὸν καὶ ἀντιληπτικόν. ἐνταῦθα διὰ τεσσάρων ὅρων ὁ Δημοσθένης πλέκει τὴν κατηγορίαν, 15 δεικνύων ὅτι δημοσίᾳ Μειδίας ἠδίκησεν. ἔστι δὲ ὁ πρῶτος ὅρος οὗτος, ὅτι οἱ ἐν ἑορτῇ ἀδικοῦντες δημόσιον ἀδίκημα ποιοῦσι. δεύτερος ὅρος, καὶ μάλιστα οἱ χορηγὸν ἀδικοῦντες. τρίτος ὅρος, ὅτι πᾶσα ὕβρις δημόσιόν ἐστιν ἀδίκημα. παραλογίζεται δὲ ἐνταῦθα ἐκ τῆς ὁμωνυμίας τῆς ὕβρεως· λέγεται γὰρ ὕβρις ἡ δι' αἰσχρουργίας 20 γινομένη, λέγεται ὕβρις καὶ ἡ διὰ λόγων, λέγεται πάλιν ὕβρις καὶ ἡ διὰ πληγῶν. δημόσιον δὲ ἀδίκημα ἡγοῦντο τὴν αἰσχρουργίαν· τῇ οὖν ὁμωνυμίᾳ παρελογίσατο. τέταρτος ὅρος, ὅτι ὁ πάντας ἀεὶ ὑβρίζων δημοσίᾳ ἀδικεῖ· εἰ γὰρ τὸ δημόσιον ἐκ πάντων συνίσταται, ἄρα δημόσιον τὸ ἀδίκημα. τίθησι δὲ σπερματικῶς ἐν τῷ προοιμίῳ τοὺς 25 τέτταρας ὅρους. καὶ ἐκ τούτων εἰσὶν ἐν τοῖς ἀγῶσι τρεῖς, τὸν δὲ τέταρτον ὅρον τίθησιν ἐν τῇ παρεκβάσει, καὶ δικαίως· λέγων γὰρ ὅτι ὁ πάντας ὑβρίζων δημοσίᾳ ἀδικεῖ, παρεξέρχεται λέγων τὸν πρότερον αὐτοῦ βίον. ἔχει δὲ ὁ παρὼν λόγος δύο προοίμια. καὶ εἴληπται τὸ πρῶτον προοίμιον ἐκ διαβολῆς τοῦ ἐναντίου, καὶ ἐκ 30 συστάσεως τοῦ οἰκείου προσώπου, καὶ ἐκ προσοχῆς. ἔστι δὲ ἡ πρότασις διμερής, καὶ τὸ μὲν πρῶτον μέρος ἐστὶν ἀκατάσκευον, τὸ δὲ δεύτερον καὶ αὐτὸ διμερές, καὶ κατασκευάζει τούτων ἑκάτερα. εἶτα ἐπιφέρει τὸ συμπέρασμα, ἐν ᾧ ἐστιν ἡ προσοχή.

(margin: 9, 10, 11)

1 διώκων Pr: κατήγορος TBc αὐτοῦ Pr 3 ἄμφω Pr τοῦθ' Pr 4–5 φαίνεται οὖν ὁ ῥήτωρ ἐν πολλοῖς τοῦτο ποιῶν μέρεσι Pr 5 τῷ Δημοσθένει] αὐτῷ Pr: αὐτῷ Aq 6–7 τὰ τῇ τάξει προσήκοντα ἔστι τῷ λόγῳ. τὰ δὲ προοίμια καταφορικὰ Pr 7 ὑπερβολὴν Ald ἔχον MacDowell: ἔχοντα TBcPr πολλὴν ἔχοντα τὴν περιβολὴν Bc 8 οὐ del. Iurinus 11 εἰσὶ om. Pr 12 καὶ om. Pr 13 ἐμπίπτει om. Pr 14 ἐνταῦθα τοίνυν Bc ὁ ῥήτωρ ἐμπλέκει Pr 15 ἔστι BcPr: ἔτι T: ἔστη Ald 17 χορηγοὶ Pr ὅρος (post τρίτος) om. Pr 19 αἰσχρουργίας Aq^pc Ald: αἰσχρολογίας TBcPr 20 γινομένη, λέγεται ὕβρις om. Pr λέγεται πάλιν ὕβρις om. Pr πάλιν om. Bc 24 τάδ- Pr 25 τέσσαρας Pr καὶ om. Pr 26 ὅρον om. Pr καὶ om. Pr 27 ὅτι πάντας ὑβρίζει Bc 28 ὁ λόγος οὗτος Pr 28–9 καὶ εἴληπται om. Pr 29 πρότερον Bc προοίμιον om. Pr διαβολῆς εἰλημμένον Pr 31 ἐστὶν om. Pr

III

Ὅρος κατὰ σύλληψιν. λέγεται δὲ οὕτως, ὅταν τοῦ φεύγοντος ἀντονομάζοντος ὁ διώκων καὶ τούτῳ κἀκείνῳ ὑπεύθυνον αὐτὸν εἶναι λέγῃ τῷ ὀνόματι, ὥστε διπλοῦς ἐστιν, ἐπεὶ δύο περιέχει ἐγκλήματα. παράδειγμα ὁ στρατηγὸς ὁ βιασάμενος τὴν παρατεθεῖσαν κόρην ὑπὸ τοῦ πρεσβευτοῦ, καὶ δημοσίων ἀδικημάτων κρινόμενος, καὶ ἀπο- 5 κρινόμενος μὴ δημοσίᾳ ἠδικηκέναι, ἀλλὰ βιάσασθαι, ὁ δὲ πρεσβευτὴς ἀμφοτέροις αὐτὸν φάσκων ὑπεύθυνον εἶναι.

IV

Τὸ προοίμιον ἀπὸ τοῦ ἀντιδίκου, ὁ δὲ λόγος δι' ἑνὸς εἴδους προάγεται, τοῦ δικανικοῦ· τούτου γὰρ καὶ τὸ τέλος ⟨τὸ⟩ δίκαιον καὶ ἡ κατασκευὴ διὰ τοῦ δικαίου. 10

V

Μειδίας εἰς τὰ μάλιστα ἐχθρὸς ἦν τῷ Δημοσθένει, καὶ διὰ πολλῶν μὲν καὶ ἄλλων ἐνεδείξατο εἰς αὐτὸν τὴν ἔχθραν, καί ποτε χορηγὸν ὄντα αὐτὸν τῆς Πανδιονίδος φυλῆς ἐν μέσῃ τῇ ὀρχήστρᾳ κονδύλοις ἔλαβεν. ὁ δὲ ἐγράψατο αὐτὸν δημοσίων ἀδικημάτων, συμπεριλαβὼν τοῖς δημοσίοις ἀδικήμασι τὴν ἑαυτοῦ ὕβριν· ἐπεὶ ἐξῆν ἐκείνῳ λέγειν 15 2 ὅτι "ὑβρίσθης· λαβὲ τῆς ὕβρεως τὸ πρόστιμον". ἔχει δ' ἡ ὑπόθεσις κατὰ μὲν Καικίλιον δύο κεφάλαια, εἰ δημόσιόν ἐστιν ἀδίκημα, καὶ εἰ μεγάλα τὰ πεπραγμένα ἐστίν. προσθετέον δὲ κἀκεῖνο, εἰ ὕβρις ἐστὶν ἡ γενομένη· ὅπερ ἀθετεῖ Καικίλιος, κακῶς· ἔσται γὰρ ἐναντίως αὐτῷ γεγραμμένον τὸ προοίμιον καὶ ἡ τοῦ χρυσοχόου μαρτυρία. ὅτι δὲ 20 δῆλός ἐστι συμπεριλαβὼν τοῖς δημοσίοις ἀδικήμασι τὴν ἑαυτοῦ ὕβριν ἐξ ἐκείνου φανερόν, ὅταν λέγῃ "ἐπειδὰν ἐπιδείξω Μειδίαν τοῦτον μὴ μόνον εἰς ἐμὲ ἀλλὰ καὶ εἰς ὑμᾶς καὶ εἰς τοὺς ἄλλους ἅπαντας ὑβρικ- 3 ότα" καὶ τὰ ἑξῆς. αἱ δ' ὑποθέσεις ὅταν μὴ ἔχωσιν ζητήματα μηδ' ἀμφισβητήσεις λελυμέναι εἰσί, καὶ τόπον τῷ ῥήτορι οὐ καταλείπουσι· 25 οἷον περὶ φόνου τις ἐγκαλεῖται καὶ λέγει "ἀπέκτεινα μὲν τὸν δεῖνα, δικαίως δέ", τότε ὁμολογήσαντος αὐτοῦ τὸν φόνον ζητεῖται πότερα δικαίως ἢ ἀδίκως ἀπέκτεινε· ὅταν δὲ λέγῃ ὁ ἐγκαλούμενος ὅτι ἀπέκτεινε καὶ ἀδίκως ἀπέκτεινε, τότε λέλυται ἡ ὑπόθεσις. οὕτως [καὶ περὶ ταύ]της τῆς ὕβρεως ῥηθήσεται. 30

Hyp. iii in YXPr 3 *λέγει* XPr *διπλῶς* Ald *ἐπὶ* X 4 *παρακαταθεῖσαν* Lamb[vl]

Hyp. iv in MmLhNb 9 *τοῦ* Buttmann: *ἤτοι* Mm: *ἤ τι* LhNb *τὸ* add. Bekker

Hyp. v in Π12 13 *κονδυλοις αυτον* Π12[ac] 20 *μαρτυρια δημοσιων αδικηματων ουκ οφειλε* Π12[ac] 28 *ἀπέκτεινε* Kenyon: *απεκτεινα* Π12 28–9 *απεκτεινα* (post *ὅτι*) Π12[ac] 29–30 *καὶ περὶ ταύ-* Kenyon

INDEXES

I. ENGLISH

II. GREEK